Basic Microbiology

WITH APPLICATIONS

Thomas D. Brock

E.B. FRED PROFESSOR OF NATURAL SCIENCES
UNIVERSITY OF WISCONSIN, MADISON

Katherine M. Brock

RESEARCH ASSOCIATE IN BACTERIOLOGY
UNIVERSITY OF WISCONSIN, MADISON

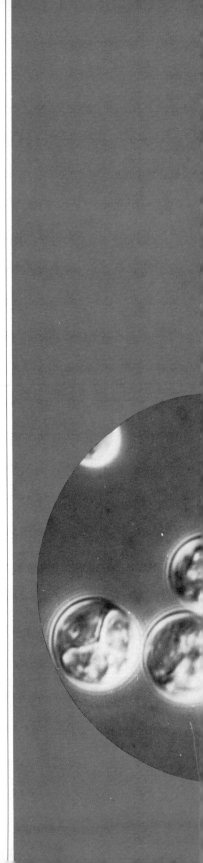

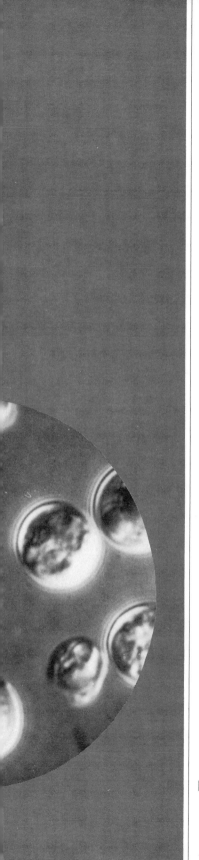

Basic Microbiology

WITH APPLICATIONS

Second edition

Prentice-Hall, Inc., Englewood Cliffs, New Jersey

For Emily and Brian

Editorial and production supervisor: Joyce Fumia Perkins
Interior and cover designer: Natasha Sylvester
Cover photograph: Marine diatom by Manfred Kage from Peter Arnold

10 9 8 7 6 5 4 3 2 1

Library of Congress Cataloging in Publication Data
Brock, Thomas D
Basic microbiology with applications.

Includes bibliographies and index.
1.   Microbiology.   I.   Brock, Katherine M., joint
author.   II.   Title.
QR41.2.B76 1978   575   77-13641
ISBN 0-13-065284-9

BIOLOGICAL SCIENCE SERIES
William D. McElroy and Carl P. Swanson, editors

Prentice-Hall International, Inc., London
Prentice-Hall of Australia Pty. Limited, Sydney
Prentice-Hall of Canada, Ltd., Toronto
Prentice-Hall of India Private Limited, New Delhi
Prentice-Hall of Japan, Inc., Tokyo
Prentice-Hall of Southeast Asia Pte. Ltd., Singapore
Whitehall Books Limited, Wellington, New Zealand

PREFACE

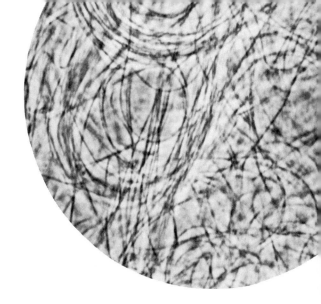

Basic Microbiology with applications, second edition, is intended to serve as an introduction to microbiology for students interested in science, as well as for liberal arts students who are not going to major in biology or microbiology. The book is oriented in a very practical way and deals primarily with those aspects of microbiology that directly affect human affairs. Such fields as nursing, environmental protection, agriculture, food technology, and public health are given special attention.

We have been pleased that students and instructors alike found the first edition of this book to be readable as well as informative. A primary objective in writing the second edition was to bring the same level of accessibility to the discussion of several new, user-suggested topics, including chemistry, molecular biology, and genetics. Some exposure to chemistry is inevitable in the field of microbiology, and so we have added a chapter on chemistry sufficient for comprehending this text. Previous college chemistry is neither assumed nor required.

Chapter 5, "Chemical and Physical Background," also serves to prepare students for the discussions of microbial metabolism and biosynthesis and microbial genetics in Chapters 6 and 7. This important new triad of chapters comprises Part 2, "Molecular Biology." We have attempted to present a readily understandable introduction to this aspect of microbiology for students at this level.

The history of microbiology is introduced in Chapter 1 not only because it is a fascinating topic but also as a means to present to readers ideas and themes that are amplified throughout the book. Thus, although this chapter is rather long, it need not be read as a unit; relevant sections may be read to accompany various topics in later chapters.

Material on host-parasite relationships, immunology, and epidemiology has been considerably expanded. Emphasis is given in this edition to infectious disease and disease-causing organisms. We have chosen to organize this material around the organisms themselves rather than around organ systems affected or portals of entry. We elect this approach because it seems to us more logical from a practical microbiological point of view. In diagnostic microbiology, the main emphasis is on identification of the causal agent. The pathologist may be concerned with organ systems and the epidemiologist with portals of entry but the microbiologist is concerned with microorganisms.

Suggested readings lists are descriptive and direct readers to noteworthy publications, both basic and specialized, that further explore the subject of each chapter. The glossary, classification appendix, and index are useful for readers during the course of study and even, we hope, beyond.

Another important concern in the choice of material and manner of presentation has been to keep the size of the book within reasonable bounds. We have had clearly in mind in this regard the fact that this text will be used for courses of varying lengths and emphases, and we have tried to select material with the broadest interest and appeal. We have included a great many photographs, in particular, of real-world situations.

This book has a distinctly ecological flavor. Our own research interests lie in this direction, so that it has been easy for us to weave environmental thinking into the fabric of the text; but we have been further encouraged to do so by the obvious need for a textbook with more ecological emphasis. We hope that this book will stimulate students to turn to careers in the increasingly crucial fields of environmental studies and public health.

Thomas D. Brock
Katherine M. Brock

CONTENTS

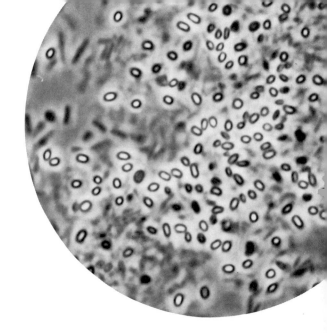

PART

GENERAL MICROBIOLOGY 32

PART

MOLECULAR BIOLOGY 130

PART

INFECTIOUS DISEASES 214

**Chapter 8
Infection and immunity** 216

PART

ENVIRONMENTAL MICROBIOLOGY 388

Introduction: the roots of microbiology

Chapter

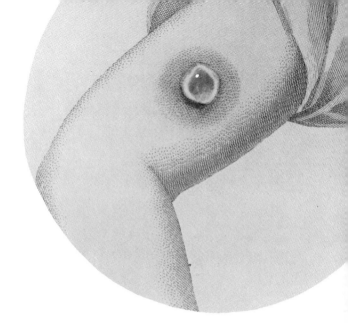

In health and in disease, the activities of microorganisms greatly affect human life. Whether in country or city, tropics, midlatitudes, or the arctic, human beings are continually influenced by microbes. The science that deals with the study of microorganisms is called *microbiology* and is a branch of biology parallel to *botany*, the study of plants, and *zoology*, the study of animals. However, the procedures and practices by which microorganisms are studied are quite different from those used to study plants and animals, and it is for this reason that microbiology has developed as a science independent of botany and zoology. The goal of the microbiologist is to understand the beneficial and harmful activities of microorganisms and through this understanding to devise ways that benefits may be increased and damages curtailed. Microbiologists have been successful in achieving this goal, and microbiology has played a major role in the advancement of human health and welfare.

Microbiology may be the most applied of the biological sciences. At the same time, it is one of the most basic of the biological sciences, because microorganisms have provided the most suitable experimental materials for studies on the nature of life itself, studies now classified under the heading of *molecular biology*. Molecular biology has developed into an independent science, but microbiology as an applied science has remained intact and is no less important now than it was before the rise of molecular biology. One apprecia-

1

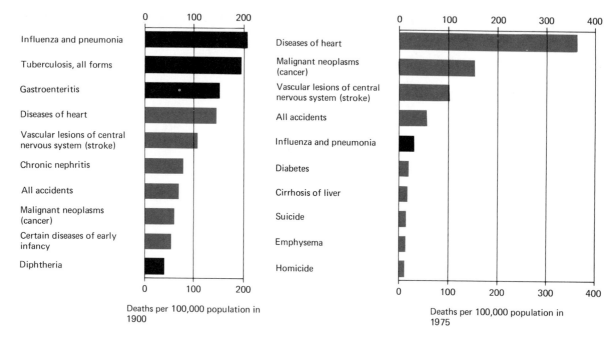

	Deaths per 100,000 population in 1900
Influenza and pneumonia	
Tuberculosis, all forms	
Gastroenteritis	
Diseases of heart	
Vascular lesions of central nervous system (stroke)	
Chronic nephritis	
All accidents	
Malignant neoplasms (cancer)	
Certain diseases of early infancy	
Diphtheria	

	Deaths per 100,000 population in 1975
Diseases of heart	
Malignant neoplasms (cancer)	
Vascular lesions of central nervous system (stroke)	
All accidents	
Influenza and pneumonia	
Diabetes	
Cirrhosis of liver	
Suicide	
Emphysema	
Homicide	

FIGURE 1.1
Death rates for the 10 leading causes of death: 1900 and 1975. Infectious diseases were the leading causes of death in 1900, whereas today they are much less important. From *U.S. Public Health Service Publ. No. 600* (revised 1967) and *Statistical Abstract of the United States, 1975.*

tion of the importance of microbiology for human health is shown by the statistics in Figure 1.1, which compares death rates in the United States in 1900 and 1975. In 1900, the major causes of death were all infectious diseases; currently, infectious diseases are of only minor importance. Control of infectious disease has come as a result of our vast scientific understanding of disease processes. Microbiology had its beginnings in these studies of disease.

In this chapter, we introduce the subject of microbiology through the presentation of a series of brief historical essays. Most of these sections concern studies that were done, primarily in the nineteenth century, to understand and control infectious disease. However, the striving to control infectious disease was not the only impetus for the development of microbiology. We also find that some interesting and important advances were made through studies on food and agricultural problems. And one of the most significant advances, the discovery of microorganisms themselves, occurred as a result of basic research done without any preconceived practical goal, but merely because of an interest in using microscopes to see the very small.

Although the existence of creatures too small to be seen with the eye had long been suspected, their discovery was linked to the invention of the microscope. Robert Hooke, using elegantly ornate micro-

1.1
The discovery of microorganisms

FIGURE 1.2
The early microscopic observation of a microorganism by Robert Hooke. *a* Robert Hooke's microscope, as illustrated in his great book, *Micrographia*, published in 1665. This is a compound microscope having two lenses, one near the eye and the other near the object. *b* Hooke's drawing of a blue mold growing on the surface of leather; the round structures contain spores of the mold.

scopes (Figure 1.2*a*), described the fruiting structures of molds in 1664 (Figure 1.2*b*), but the first person to see microorganisms in any detail was the Dutch amateur microscope builder Antoni van Leeuwenhoek, who used simple microscopes of his own construction (Figure 1.3). Leeuwenhoek's microscopes were extremely crude by today's standards, but by careful manipulation and focusing he was able to see organisms as small as bacteria. He reported his observations in a series of letters to the Royal Society of London, which published them in English translation. Drawings of some of Leeuwenhoek's "wee animalcules" are shown in Figure 1.4. His observations were confirmed by other workers, but progress in understanding the nature of these tiny organisms came slowly. Only in the nineteenth century did improved microscopes become available and widely distributed. At all stages of its history, the science of microbiology has taken the greatest steps forward when better microscopes have been developed, for these enable scientists to penetrate ever deeper into the mysteries of the cell.

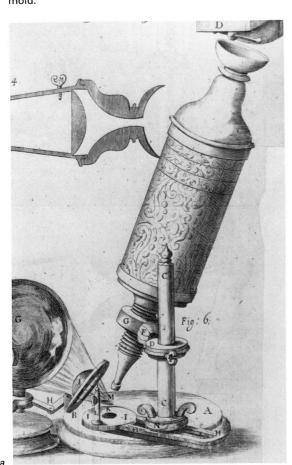

a

b

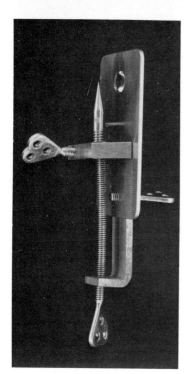

FIGURE 1.3
A replica of the microscope Leeuwenhoek used. The object to be viewed was placed on the pointed tip at the end of the screw and was moved back and forth by turning the screw. This is a simple microscope, composed of only a single lens. Although a simple microscope theoretically provides less resolution than a compound microscope such as Hooke's (Figure 1.2a), actually Leeuwenhoek was a superb lensmaker, and his microscopes resolved better than Hooke's.

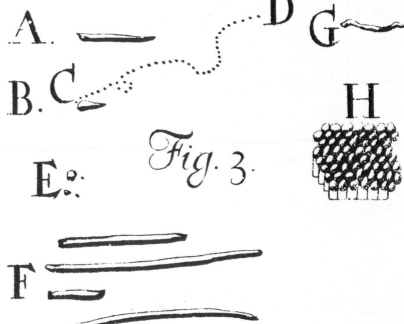

FIGURE 1.4
Leeuwenhoek's drawings of bacteria, published in 1684. Even from these crude drawings we can recognize several kinds of common bacteria. Those lettered A, C, F, and G are rod-shaped; E, spherical or coccus-shaped; H, coccus-shaped bacteria in packets. From A. van Leeuwenhoek, *Phil. Trans. Roy. Soc.*, London, **14**, 568, 1684.

Microbiology did not develop as a science until the latter part of the nineteenth century. This long delay occurred because, in addition to microscopy, certain other basic techniques for the study of microorganisms needed to be devised. In the nineteenth century, investigation of two perplexing questions led to the development of these techniques and laid the foundation of microbiological science: (1) Does spontaneous generation occur? (2) What is the nature of contagious disease? These two questions were studied simultaneously, and sometimes the same people worked on both. By the end of the century, both questions were answered, and microbiology was firmly established as a distinct and growing field of science.

1.2 Spontaneous generation

The basic idea of spontaneous generation can easily be understood. If food is allowed to stand for some time, it putrefies, and when the putrefied material is examined microscopically, it is found to be teeming with bacteria. Where do these bacteria come from, since they are not seen in fresh food? Some people said they developed from seeds or germs that had entered the food from the air, whereas others said that they arose spontaneously.

Spontaneous generation would mean that life could arise from something nonliving, and many people could not imagine something so complex as a living cell arising spontaneously from dead materials. The most powerful opponent of spontaneous generation was the French chemist Louis Pasteur, whose work on this problem was the most exacting and convincing. Pasteur first showed that there were structures present in air that closely resembled the microorganisms seen in putrefying materials. He did this by passing air through guncotton filters, the fibers of which stop solid particles. After the guncotton was dissolved in a mixture of alcohol and ether, the particles that it trapped fell to the bottom of the liquid and were examined on a microscope slide. Pasteur found that in ordinary air there exists constantly a variety of solid structures ranging in size from 0.01 millimeter (mm) to more than 1.0 mm. Many of these structures resembled the spores of common molds, the cysts of protozoa, and various other microbial cells. As many as 20 to 30 of them were found in 15 liters of ordinary air, and they could not be distinguished from the organisms found in much larger numbers in putrefying materials. Pasteur concluded, therefore, that the organisms found in putrefying materials originated from the organized bodies present in the air, which are constantly being deposited on all objects. If this conclusion was correct, it meant that food treated to destroy all the living organisms contaminating it would not putrefy. In fact, Nicholas Appert had already devised a method for food preservation based on heat treatment (see Section 1.5) but did not understand the principle upon which his method worked.

Pasteur used heat to eliminate contaminants, since many workers had shown that if a nutrient infusion was sealed in a glass flask and heated to boiling, it never putrefied. The proponents of spontaneous generation had criticized such experiments by declaring that fresh air was necessary for spontaneous generation and that the air inside the sealed flask was affected in some way by heating so that it would no longer support spontaneous generation. Pasteur skirted this objection simply and brilliantly by constructing a swan-necked flask, open to the air, now called the *Pasteur flask* (Figure 1.5). In such a flask, putrefying materials can be heated to boiling; after the flask is cooled, air can reenter but the bends in the neck prevent particulate matter, bacteria, or other microorganisms from getting in. Material sterilized in such a flask did not putrefy, and no microorganisms ever appeared so long as the neck of the flask remained intact. If the neck was broken, however, putrefaction occurred, and the liquid soon teemed with living organisms. This simple experiment effectively settled the controversy of spontaneous generation.

Killing all bacteria or germs is a process we now call *sterilization*, and the procedures that Pasteur and others used were eventually carried over into microbiological research. Disproving the theory of spontaneous generation thus led to the development of effective sterilization procedures, without which microbiology as a science could not have developed.

It was later shown that flasks and other vessels could be protected from contamination by cotton stoppers, which still permit the exchange of air. The principles of aseptic technique, developed so effectively by Pasteur, are the first procedures learned by the novice microbiologist. Food science also owes a debt to Pasteur since his principles are applied in the canning and preservation of many foods (see Section 1.5).

FIGURE 1.5
One of Pasteur's swan-necked flasks. *Courtesy of the Musée Pasteur, Paris.*

1.3
The germ theory of disease

Proof that microorganisms could cause disease provided a great stimulus for the development of the science of microbiology. As early as the sixteenth century, it was thought that something could be transmitted from a diseased person to a well person that induced disease. Many diseases spread through populations and were called *contagious*; the unknown thing that did the spreading was called the *contagion*. After the discovery of microorganisms, it was more or less widely held that these organisms might be responsible for contagious diseases, but proof was lacking. In 1846, Miles Joseph Berkeley provided the first clear demonstration that microorganisms caused diseases by showing that a mold was responsible for the Irish potato blight (Figure 1.6). Discoveries by Ignaz Semmelweis and Joseph Lister provided some evidence for the importance of microorganisms in causing human diseases, but it was not until the work of Robert Koch, a German physician, that the germ theory of disease was placed on a firm footing (Figure 1.7).

KOCH'S EARLY WORK

In his early work, published in 1877, Koch studied anthrax, a disease of cattle, which sometimes also occurs in man. Anthrax is caused by a bacterium now called *Bacillus anthracis*, and the blood of an animal

INTRODUCTION: THE ROOTS OF MICROBIOLOGY

FIGURE 1.6
Berkeley's drawing of a disease-causing microorganism, done in 1846. The disease was Irish potato blight, which was responsible for the great famine in Ireland, leading to the massive Irish immigration to North America. From M. J. Berkeley, *J. Roy. Hort. Soc., London*, **1**, 9, 1846.

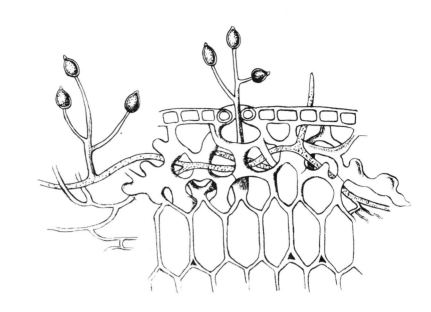

FIGURE 1.7
Robert Koch's study, showing his microscope, staining supplies, and photomicrographic camera (at left, horizontal position). From The Bettmann Archive.

infected with anthrax teems with cells of this large bacterium. Koch established by careful microscopy that the bacteria were always present in the blood of an animal that had the disease. However, he knew that the mere association of the bacterium with the disease did not prove that it caused the disease; it might instead be a result of the disease. Therefore, Koch demonstrated that it was possible to take a small amount of blood from a diseased animal and inject it into another animal, which in turn became diseased and died. He could then take blood from this second animal, inject it into another, and again obtain the characteristic disease symptoms. By repeating this process as often as 20 times, successively transferring small amounts of blood containing bacteria from one animal to another, he proved that some living agent did indeed cause anthrax. The twentieth animal died just as rapidly as the first, and in each case Koch could demonstrate by microscopy that the blood of the dying animal contained large numbers of the bacterium.

Koch carried this experiment further. He found that the bacteria could also be cultivated in nutrient fluids outside an animal's body, and even after many transfers in culture, the bacteria could still cause the disease when reinoculated into an animal. Bacteria from a diseased animal and bacteria in culture induced the same disease symptoms upon injection. The experiment thus proved that the bacterium *Bacillus anthracis* was the cause of the disease anthrax. This was the first proof that an infectious disease could be caused by a specific bacterium. Koch's work was important in elucidating the nature of anthrax, and even more important, it provided the experimental basis for the isolation and culture of a number of other infectious agents.

KOCH'S PURE CULTURE METHODS

In order to identify a microorganism successfully as the cause of a disease, one must be sure that it alone is present in culture; that is, the culture must be *pure*. With material as small as microorganisms, ascertaining purity is not easy, because even a very tiny sample of blood or animal fluid may contain several kinds of organisms that may all grow together in culture. Koch realized the importance of pure cultures and developed several ingenious methods of obtaining them, of which the most useful is that involving the isolation of single colonies. Koch observed that when a solid nutrient surface, such as a potato slice, was exposed to air and then incubated, bacterial colonies developed, each having a characteristic shape and color. He inferred that each colony had arisen from a single bacterial cell that fell on the surface, found suitable nutrients, and began to multiply. Because the solid surface prevented the bacteria from moving around, all the offspring of the initial cell had remained together, and when a large enough number were present, the mass of cells became visible to the naked eye. He assumed that colonies with different shapes and colors were derived from different kinds of organisms. When the cells of a single colony were spread out on a fresh surface, many colonies developed, each with the same shape and color as the original.

Koch realized that this discovery provided a simple way of obtaining pure cultures, since streaking mixed cultures on solid nutrient surfaces spread the various organisms so far apart that the colonies they produced did not mingle. Many organisms could not grow on potato slices; so he devised *semisolid media*, in which gelatin was added to a nutrient fluid, such as blood serum, in order to solidify it. When the gelatin-containing fluid was warmed, it liquefied and could be poured out on glass plates; after cooling, the solidified medium could be inoculated. Later, *agar* (a material derived from seaweed) was found to be a better solidifying agent than gelatin, and this substance is widely used today to obtain colonies.

KOCH AND TUBERCULOSIS

Probably the most important single discovery of a bacterial disease agent was Koch's discovery of the causal agent of tuberculosis, *Mycobacterium tuberculosis*. Tuberculosis has been one of the great scourges of mankind, and was still the second greatest cause of death at the beginning of the twentieth century (see Figure 1.1). Also called "consumption" or "phthisis," the disease has been known for centuries. Tuberculosis can involve many parts of the body, including the bones and the skin, but the organ most commonly affected is the lung. In advanced cases there is considerable degeneration of the lungs, difficulty in breathing, spitting of blood, and emaciation; death is often cruelly slow in coming as the patient gradually wastes away. Two of the most famous heroines of grand opera, Violetta in *La Traviata* and Mimi in *La Bohème*, die of tuberculosis, and their lingering deaths provide the essential pathos for the last acts of these operas.

In contrast to some other infectious diseases, the communicability of tuberculosis was by no means clear. This was because although the bacterium is a necessary condition for the development of the disease, it alone is not sufficient, since the general health of the person, heredity, and social factors strongly influence whether an infection leads to disease. Many people infected with the bacterium never develop the illness. But once it had been clearly established that *some* infectious diseases could be caused by bacteria, it was natural to turn to tuberculosis and look for its causal agent. Before Koch, the most important work on tuberculosis had been done by a French surgeon, Jean-Antoine Villemin, who showed that material from tubercular lung tissue could be used to initiate an infection in rabbits or guinea pigs and could be transmitted from one infected animal to another without any decrease in virulence. His work, published in 1868 before Koch began his studies on anthrax, showed clearly that tuberculosis was infectious but did not show what the infectious agent was. This Koch did, in a brilliant series of studies conducted between 1880 and 1882. He succeeded in culturing the organism on an artificial medium and maintaining the virulence of his cultures, which retained their infectivity for experimental animals. He also developed a specific staining procedure that permitted microscopic examination of tissues for the presence of the bacteria; such examination is an important diagnostic tool.

The magnitude of Koch's achievement is hard to appreciate without knowing some of the difficulties inherent in working with *M. tuberculosis*. First, it is a slow-growing bacterium; whereas other pathogens usually produce visible colonies on culture media after a day or two of incubation, *M. tuberculosis* requires 10 days to 2 weeks, and even then the colonies are small and hard to see. Other workers were probably unsuccessful in culturing the organism because they were impatient, discarding their cultures too early. The second peculiarity of the organism is the difficulty with which it can be stained by dyes for microscopy. When examining infected tissues, some staining procedure is essential because the bacteria are virtually impossible to see among the large amount of deteriorating tissue cells and other debris. Most pathogens can be stained easily with dyes, but *M. tuberculosis* is refractory to staining by normal methods. Koch discovered through a lucky accident (as Pasteur said in other circumstances, "Chance favors the prepared mind") that the bacteria could be stained if the dye solution was made alkaline with potassium hydroxide; staining was also faster if the material was heated during the staining process. After this treatment, both the bacteria and the tissues were heavily stained. The rest of the procedure required a decolorization of the stained tissues under conditions that did not decolorize the bacteria.

Once Koch had developed his staining procedure, he used it to examine a variety of pathological materials for the presence of the bacteria. In Koch's words:

> On the basis of my extensive observations, I consider it as proven that in all tuberculous conditions of man and animals there exists a characteristic bacterium which I have designated as the tubercle bacillus, which has specific properties which allow it to be distinguished from all other microorganisms. From this correlation it does not necessarily follow that these phenomena are causally related. In order to prove that tuberculosis is brought about through the bacilli, the organisms must be isolated from the body and cultured so long in pure culture that they are freed from any diseased production of the animal organism which may still be adhering to the bacilli. After this

the isolated bacilli must bring about the transfer of the disease to other animals, and cause the same disease picture which can be brought about through the inoculation of healthy animals with naturally developing tubercle materials.[1]

Koch met these requirements with his staining and culture methods and was able to prove beyond all doubt that tuberculosis was indeed an infectious disease caused by a specific bacterium. More importantly, the experimental requirements that Koch set for himself provided a framework for the study of *any* infectious disease. We now call these experimental requirements *Koch's postulates*, and they are outlined in more detail in the next section.

KOCH'S POSTULATES

As Koch's work on bacterial diseases and the discovery of the main pathogenic organisms progressed, he formalized his procedures by delineating a series of experimental guideposts. These postulates are as follows:

1 The organism must always be found in animals suffering from the disease and must not be present in healthy individuals.
2 The organism must be cultivated in pure culture away from the animal body.
3 Such a culture, when inoculated into susceptible animals, must initiate the characteristic disease symptoms.
4 The organism must be reisolated from these experimental animals and cultured again in the laboratory, after which it must still be the same as the original organism.

Koch's postulates not only supplied a means of demonstrating that specific organisms cause specific diseases but also provided a tremendous impetus for the development of the science of microbiology by stressing laboratory culture.

The development by Koch of the proper procedures and criteria for the isolation and study of pathogens opened the way for subsequent discoveries in many other laboratories around the world. In the 20 years following the formulation of Koch's postulates, the causal agents of a wide variety of contagious diseases were isolated. The major pathogenic microorganisms and their discoverers are given in Table 1.1. These discoveries led to the development of successful treatments for the prevention and cure of many contagious diseases and contributed to the development of modern medical practice. The impact of Koch's work has been worldwide.

---

[1] T. D. Brock, *Milestones in microbiology*, reprint, American Society of Microbiology, Washington, 1975, p. 111.

TABLE 1.1

The discoverers of the main
pathogenic microorganisms

Year	Disease	Organism	Discoverer
1877	Anthrax	*Bacillus anthracis*	Koch
1878	Suppuration	*Staphylococcus*	Koch
1879	Gonorrhea	*Neisseria gonorrhoeae*	Neisser
1880	Typhoid fever	*Salmonella typhi*	Eberth
1881	Suppuration	*Streptococcus*	Ogston
1882	Tuberculosis	*Mycobacterium tuberculosis*	Koch
1883	Cholera	*Vibrio cholerae*	Koch
1883	Diphtheria	*Corynebacterium diphtheriae*	Klebs
1884	Tetanus	*Clostridium tetani*	Nicolaier
1885	Gastroenteritis	*Escherichia coli*	Escherich
1886	Pneumonia	*Streptococcus pneumoniae*	Fraenkel
1887	Meningitis	*Neisseria meningitidis*	Weichselbaum
1888	Food poisoning	*Salmonella enteritidis*	Gaertner
1892	Gas gangrene	*Clostridium perfringens*	Welch
1894	Plague	*Yersinia pestis*	Kitasato; Yersin
1896	Botulism	*Clostridium botulinum*	van Ermengem
1898	Dysentery	*Shigella dysenteriae*	Shiga
1900	Paratyphoid	*Salmonella paratyphi*	Schottmüller
1903	Syphilis	*Treponema pallidum*	Schaudinn and Hoffmann
1906	Whooping cough	*Bordetella pertussis*	Bordet and Gengou

1.4
Cholera, typhoid fever, and water purification

Cholera is no longer a problem in the Western world. It wasn't always so. In the days before adequate water purification (see Chapter 14), cholera was widespread in Europe and North America, flourishing especially in large cities where houses were supplied with water from central sources. In fact, it was the discovery that cholera was a disease transmitted by polluted water that led to the first development of water-purification systems. Later developments in the proper treatment of water led to the control of another serious bacterial disease of humans, typhoid fever.

The disease cholera is so much associated with Asia that it is frequently called *Asiatic* cholera. The disease has been present in India for centuries and is still commonplace on that subcontinent. However, in the nineteenth century the disease spread out of Asia, and vast areas of Europe were ravaged by violent epidemics. (An epidemic that affects wide geographical areas at one time is usually called a *pandemic*.) In the pandemic of 1832 and 1833, 4,000 deaths occurred in London alone, and 7,000 in Paris. This pandemic was truly worldwide, as the disease spread from Europe to the eastern United States and Canada, brought by Irish immigrants. In the later pandemic of 1846 to 1862, the disease was also brought to the United States, this time through New Orleans, by sailors; it spread up the Mississippi Valley to the midwest. Since the beginning of the twentieth century, the Western world has been relatively free of cholera, although the disease has remained constantly present in large areas of India and Southeast Asia. (A disease that is constantly present in a population is called *endemic*.) The decrease in incidence of cholera was accomplished by the development of adequate water purification. The disease remains endemic in areas where water treatment is less dependable. The need for water purification was made obvious through the studies of two men, John Snow working in the 1850s and Robert Koch in the 1880s and 1890s.

SNOW AND CHOLERA

The importance of drinking water as a vehicle for the spread of cholera was first shown by Snow, a British physician, even without any knowledge of the bacterial causation of the disease. Snow's study is one of the great classics of epidemiology (*epidemiology* is the study of disease in populations, rather than in individuals; the latter is the province of medicine) and serves as a model for how a careful study can lead to clear and meaningful conclusions. Snow studied several cholera epidemics in London and deduced that the causal agent was transmitted in two ways, by person-to-person contact and by polluted drinking water. He showed most clearly the importance of polluted drinking water from his study of a great outbreak in London in 1853 and 1854.

In those days in London, the water supplies to different parts of the city were from different sources and were transmitted in different ways. In a large area south of the Thames River (in fact, just across the river from Westminster Abbey and the Houses of Parliament), the water was supplied to homes by two competing private water companies, the Southwark and Vauxhall Company and the Lambeth Company. It was the water of the former company that was the major vehicle for the transmission of cholera. When Snow began to suspect the water supply of the Southwark and Vauxhall Company, he made a careful survey of the residence of every death in the district and determined which company supplied the water to that residence. In some parts of the area served by the two companies, each had a monopoly; but in a fairly large area, the two companies competed directly, each having run independent water pipes along the various streets. Homeowners had the option of connecting with either supply, and the distribution of houses between the two companies was random. The results of Snow's survey were so clear-cut as to be completely convincing, even to those sceptical about the importance of polluted water in the transmission of cholera: In the first 7 weeks of the epidemic, there were 315 deaths per 10,000 houses supplied by the Southwark and Vauxhall Company and only 37 deaths per 10,000 houses supplied by the Lambeth Company. In the rest of London, there were 59 deaths per 10,000 houses, showing that those supplied by the Lambeth Company had fewer deaths than the general population. In the districts where each company had exclusive rights, it could, of course, be argued that it was not the water but some other factor (soil, air, general layout of houses, etc.) that might have been responsible for the differences in disease incidence; but in the districts where the two companies competed, all other factors were the same, yet the incidence was high for those supplied with Southwark and Vauxhall water and low for those supplied with Lambeth water. Snow attempted to relate the differences in disease incidence to the sources of the water used by the two companies. Because he knew that the excrement and evacuations from cholera patients were highly infectious, he considered that sewage contamination of the water supply might exist. In those days, there was no sewage treatment, and raw sewage was dumped directly into the Thames River. The Southwark and Vauxhall Company obtained its water supply from the Thames right in the heart of London, where the opportunity for sewage contamination was great, whereas the Lambeth Company obtained its water from a point on the river considerably above the city, and hence the water was relatively free of pollution. Snow was almost certain that it was this difference in source that accounted for the difference in disease incidence. In his words:

> As there is no difference whatever, either in the houses or the people receiving the supply of the two Water Companies, or in any of the physical conditions with which they are surrounded, it is obvious that no experiment could have been devised which would more thoroughly test the effect of water supply on the progress of cholera than this.... The experiment, too, was on the grandest scale. No fewer than three hundred thousand people of both sexes, of every age and occupation, and of every rank and station, from gentlefolk down to the very poor, were divided into two groups without their choice, and, in most cases, without their knowledge; one group being supplied with water containing the sewage of London, and, amongst it, whatever might have come from the cholera patients, the other group having water quite free from such impurity.[2]

When Snow's work was made known, people generally were convinced that drinking water could be a primary agency for the spread of cholera (Figure 1.8). However, Snow did not know what the causal agent actually was and made no strong case for a bacterial involvement. It was only Koch's work, much later, that clarified this issue.

Another study by Snow, less far-reaching but more dramatic, dealt with the spread of cholera via a single pump in Broad Street, in the heart of London. In this part of London, houses were not served with water from a central source, and residents had to go to

---

[2] J. Snow, *Snow on cholera*, reprint, The Commonwealth Fund, New York, 1936, p. 75.

FIGURE 1.8

A satirical cartoon showing the source of water used by the Southwark and Vauxhall Company to supply that part of the city of London that had a high incidence of cholera. *Courtesy of the British Museum.*

hand-operated pumps on the street in order to obtain water. Although there were a number of pumps providing water in the area, the Broad Street pump apparently was unusually contaminated with sewage via underground seepage (the water developed a marked sulfide smell after being in a bottle for a few hours—a good sign of sewage pollution). Snow showed, by making a survey of the sources of drinking water for those who died of cholera, that most of them in this district had been using the Broad Street pump. As a result of his study, the authorities removed the handle from the pump, thus preventing further use of the contaminated water.

KOCH AND CHOLERA

Although Snow's work represents outstanding epidemiological research and provided a basis for control of water-borne infectious disease, it was ahead of its time and had little impact on the overall problem. Not until it was proved that bacteria cause infectious disease could a rationale for water treatment be developed based on the killing or removal of disease-causing agents. And it required the specific discovery by Koch of the causal agent of cholera, and his careful study of a serious epidemic in Hamburg, Germany, in 1892 and 1893, to provide the impetus for the institution of modern water-purification systems. We have already discussed Koch's early contributions to the understanding of the microbial nature of infectious disease and his development of the principles known as Koch's postulates. It was late in his career that Koch became involved in research on cholera, following a journey to Egypt and India to study the disease where it was most severe. When the Hamburg epidemic broke out, Koch made the striking observation that the city of Altona, immediately adjacent to Hamburg, had a much lower incidence of cholera than Hamburg itself. Both cities obtained their water from the Elbe River, but Hamburg had the better source, obtaining its water from a point on the river above the city, whereas Altona obtained its water from below the city. Because its source was inferior, the city of Altona had to treat its water by running it through sand filters, whereas the city of Hamburg used its water unfiltered. Koch showed clearly that cholera was associated with only the Hamburg water supply, and even on streets which formed the border of the two cities, the Hamburg side was attacked by cholera while the Altona side remained free of it. In Koch's words:

> Here then we have a kind of experiment, which performed itself on more than a hundred thousand human beings, but which, despite its vast dimensions, fulfilled all the conditions one requires of an exact and absolutely conclusive laboratory-experiment.... The group supplied with unfiltered Elbe water suffers severely from cholera, that supplied with filtered water very slightly.... For the bacteriologist nothing is easier than to give an explanation of the restriction of the cholera to the sphere of the Hamburg water-supply. He need only point out that cholera-bacteria got into the Hamburg water from the outlets of the Hamburg sewers.... Altona received water which was originally much worse than that of Hamburg, but which was wholly or almost wholly freed of cholera-bacteria by careful filtration.[3]

TYPHOID FEVER AND WATER PURIFICATION

By the late nineteenth century, most of the severe epidemics of cholera in Europe and North America had come under control. Typhoid fever was another matter. This disease still raged out of control, causing large numbers of deaths, especially among urban populations. One of the main routes for transmission of typhoid fever is the water route, and a strong correlation existed between purity of water supply and typhoid fever incidence. In the United States, the state of Massachusetts was a pioneer in promoting research on public health problems, and a research station was established at the city of Lawrence, on the Merrimack River, to carry out research on methods for water filtration. In 1885, after this station had been established, an epidemic of typhoid fever struck the city of Lawrence, hitting especially hard at those households obtaining their water from the Merrimack River. A filtration plant for the city was built, with the aim of lowering the incidence of typhoid fever. The results were dramatic: deaths due to typhoid fever dropped 79 percent in the 5-year period after the filter was installed as com-

---

[3] R. Koch, *Professor Koch on the bacteriological diagnosis of cholera, water-filtration and cholera, and the cholera in Germany during the winter of 1892–93*, translated by G. Duncan, David Douglas Publisher, Edinburgh, 1894, p. 26–27.

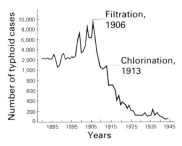

FIGURE 1.9
Incidence of typhoid fever in the city of Philadelphia, illustrating the dramatic decrease when filtration and chlorination were instituted. From E. D. Kilbourne and W. G. Smillie, *Human ecology and public health*, 4th ed., The Macmillan Company, New York, 1969.

pared to the pre-filter 5-year period. At about the same time, Louisville, Kentucky, using the muddy and polluted Ohio River as its source, set up experiments on water filtration, and methods were developed that were widely applicable in the midwest. Within a few years, cities throughout the United States had set up filtration plants, and there was a marked reduction in the incidence of typhoid fever. By the turn of the twentieth century, over 25 large U.S. cities filtered their water supplies. The dramatic fall in the incidence of typhoid fever in the city of Philadelphia (one of the pioneers in filtration) is shown in Figure 1.9.

The other major advance in the control of water-borne diseases was the introduction of chlorination. Chlorine was first used to deodorize sewage, and it was not used for disinfecting water until 1908. It became widely used by 1912, and its effect on the incidence of typhoid fever is also shown in Figure 1.9. Chlorination is essential for proper water purification because filtration is not 100 percent effective; the bacteria that escape removal by filtration are killed by chlorination. Details of water purification processes and other aspects of the microbiology of water are discussed in Chapter 14. Public-health measures such as water purification have played a major role in the reduction of disease incidence in populations. As shown in Figure 1.1, gastroenteritis (inflammation of the lining of the stomach and intestines), mostly due to water-borne organisms, was the third most important cause of death in the United States in 1900, whereas today, gastroenteritis is only of minor importance and does not even appear on the list of the top 10 causes of death. Snow's heritage is truly a great one for mankind.

1.5
Canning and pasteurization: the legacy of Nicolas Appert

In the days before electricity and refrigeration, the long-term preservation of food was a difficult matter. This was especially important to sailors at sea or to soldiers in battle, who were often long distances from their food supplies. In 1795, the French Directoire offered a prize for a method of preserving food that would permit its transport to soldiers and sailors without the usual spoilage. This prize offer stimulated Nicolas Appert (Figure 1.10), a French chef, confectioner, and distiller, to initiate studies on the problem.

APPERT'S METHOD

After more than 10 years of experimentation, Appert published his epochal work, *The book for all households, or the art of preserving animal and vegetable substances for many years.* His method, the first practical use of heat sterilization, was derived from completely empirical observations and was developed long before the role of microorganisms in food spoilage was known. It is not clear from Appert's writings just how he concluded that heat should be used, but his method was entirely effective in sterilizing and keeping sterile most types of food. In his own words:

> I became convinced by my experiments and perseverance, 1st, that heat has in itself the property not only of altering constituents of vegetable and animal products, but also, if not to destroy, at least to arrest for many years the natural tendency of these products to decompose; 2nd, that the application of heat in a proper manner to these products, after having deprived them as rigorously as possible of contact with air, allows perfect preservation, with all their natural qualities. . . .
>
> My process consists principally of:
>
> *1st.* To put products to be preserved in jars or bottles;
>
> *2nd.* To cork these different vessels with the greatest care because success depends chiefly on the closing;
>
> *3rd.* To submit these substances thus enclosed to the action of boiling water in a water-bath for more or less time according to their nature and in the manner that I shall indicate for each kind of food;
>
> *4th.* To remove the bottles from the water-bath at the time prescribed.[4]

FIGURE 1.10
Nicholas Appert, the father of canning. *Courtesy of Librarie Larousse.*

Appert then proceeds to describe in detail how he drives his corks firmly into the bottles, how he wires the corks on to prevent their popping out during the heating process, and how he immerses the bottles in water for the heating process. He had his own bottles made of the best glass, and he selected his corks with great care. He then determined proper heating times for a wide variety of meats, vegetables, and fruits. He was aware that fruits required considerably less heating time than vegetables. (We know now that this is because of their acidity; see Section 15.4.) He was also aware of the far-reaching importance of his method: "From this method a new branch of industry will arise involving French agricultural products, permitting export and import of the commodities with which nature has favored the different countries."[5]

As proof of the effectiveness of his technique, Appert provided samples of his preserved products for the French government. Agreeing that his method was quite effective, the French government awarded him a prize of 12,000 francs. More importantly, Appert established a factory for the preservation of food products, the first commercial cannery in the world, the House of Appert, at Massy, France. Although Appert died in 1841, the firm remained in business at Massy until 1933.

Appert's methods for preserving foods were introduced into the United States in 1821 by William Underwood in Boston and were followed until about 1840. Then, Peter Durand, an English-

---

[4] Translated from N. Appert, *Le livre de tous les ménages, ou l'art de conserver, pendant plusieurs années, toutes les substances animales et végétales*, Patris et Cie, Paris, 1810, p. 7.

[5] Ibid., p. 110.

man, conceived and patented the idea of using tin cans to replace glass jars. This was a major advance, as cans are unbreakable and readily sealed, thus eliminating the cumbersome corking procedure specified by Appert. Cans were first used in the United States by Charles Mitchell, of Maine, for packing corn and lobsters. The use of pressure sterilizers was introduced about 1870. These permitted processing the canned goods at temperatures above boiling, and the shorter heating times that could therefore be used resulted in less deterioration of the product. During the cooling of the cans after the heating process, a vacuum develops in the can; in the early days of microbiology, many thought the vacuum was responsible for preserving the quality of the product. However, once an understanding of the importance of microorganisms in disease was obtained, it was quickly realized that the crucial event occurring during the canning process was the killing of contaminating microorganisms by heat, a process called *sterilization*. There has always been a conflict between the desirability of short heating times to avoid deterioration of the product and the necessity of long heating times to ensure complete killing of all organisms present. If a can is heated a very long time, it will certainly be sterile, but the food will be unpalatable. The heating time must be as short as possible, while still ensuring complete killing of all organisms. Heating time is determined not only by the load of microorganisms present in the product, but also by the viscosity, acidity, sweetness, or saltiness of the product, and by the size of the can. We discuss the details of the canning process as it is currently used in Section 15.4.

PASTEURIZATION

Appert also seemed to be aware that his method could be used for the preservation of those kinds of wine that have a short shelf life, although he did not provide specific instructions: "This method will facilitate the exportation of the wines from many vineyards. In effect, those wines which are scarcely able to hold up for one year without changing, will be able to be sent to foreign lands and be preserved many years."[6] It remained for Pasteur, many years later, to perform the detailed studies necessary before the preservation of wine by heat could become a routine industrial procedure.

In the 1860s, as today, the French wine industry was an important part of the French economy. But spoilage of wine was a serious problem, and many wines could not be stored or shipped without marked deterioration in quality. Pasteur began his studies of wine because of his interest in alcohol fermentation; it was only natural to examine microscopically wines that had undergone spoilage. He saw organisms that looked exactly like the bacteria he had shown to be responsible for lactic and acetic acid fermentations, and

[6] Ibid., p. 110.

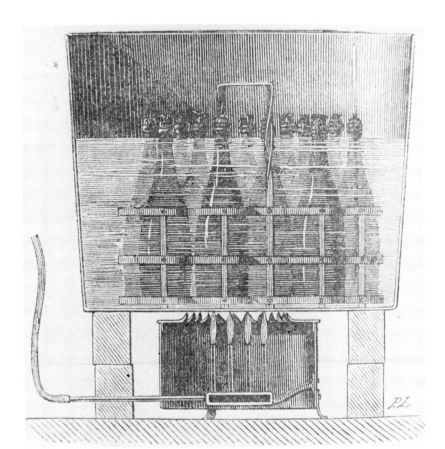

FIGURE 1.11
An experimental water bath used by Louis Pasteur during his experiments on the pasteurization of wine. The wine bottles are submerged with just their necks protruding from the water. The water is heated by a gas jet, and the temperature is measured by the thermometer immersed in one of the bottles. From L. Pasteur, *Études sur le vin: ses maladies, causes qui les provoquent*, Victor Masson et Fils, Paris, 1866.

it was logical to conclude that the spoilage of wine, which led to the development of acidity, was caused by these organisms. After considering various ways to preserve the wine so as to prevent such spoilage, Pasteur decided to use heat to destroy the spoilage organisms (Figure 1.11). The use of heat was not original with him, and Pasteur recounted in detail the history of the subject when he published the results of his research. Even in Roman times, heat was probably used to help in the preservation of wine, and Pasteur quotes this remarkable passage from Pliny's *Natural history*: "In Campania [present-day Naples region] they let their best wines lie abroad in vessels, even in the open aire, to take the Sunne, the Moone, raine, and wind, and all weathers that come: and this is thought to bee best for them." Pasteur reported that when he devised his method, he had forgotten about Appert's earlier work and rediscovered it after his own work was complete. A number of other people also had made attempts to preserve wine by heat before Pasteur's studies. Pasteur's great contribution was a far more systematic study of the problem and a careful determination of the minimum temperature that would preserve the wine without any significant alteration in its quality. This temperature he found to be 55 to 60° Celsius (C) for a few moments (exact time unspecified). Pasteur concluded that:

> ... the use of heat is applicable either to wine in barrels or in bottles and it is sufficient to heat the wine to 55–60°C in order for it to be completely protected from all problems of spoilage. This treatment has no effect on the delicate qualities of the wine, and works equally well for common table wine as for wine of the highest quality; it affects neither the color, the flavor, the bouquet, nor the most delicate nuances which have developed during the ripening process...[7]

As a result of Pasteur's work, the French wine industry was placed on a firm footing; its wines could be prepared, heated, and then stored for long periods or transported to distant places. After these studies on "diseases" of wine, it is not surprising that Pasteur moved on to studies of diseases of animals, initiating the work that would occupy him for the rest of his life.

Pasteur's name today is a household word because we call the heating process that he developed *pasteurization*. He used his process only for wine; it remained for others who followed to apply the pasteurization process to milk and milk products.

Milk pasteurization was first proposed in 1886 by V. H. and F. Soxhlet, German chemists, who conceived of the idea of preserving milk for infant feeding by brief heat treatment. They invented a special infant-feeding bottle and divided a day's supply of milk into separate bottles, which were then pasteurized and stored. The milk was not sterile; it would still spoil if kept for a long enough time. However, the pasteurization process did reduce the incidence of infant sickness. Pasteurization of milk was introduced into the United States in 1889 by Dr. Abraham Jacobi and was adopted by the pediatrician Henry Koplik for preparing milk for infants. A New York philanthropist, Nathan Straus, also became interested in the process and established milk depots where pasteurized milk was made available for infants and babies. A dramatic drop in infectious disease was found in babies who drank the pasteurized milk; on the basis of this outcome, Straus initiated an extensive publicity campaign throughout the United States and Europe to promote the routine pasteurization of milk. By the early 1900s, about 25 percent of the milk in New York was pasteurized, and the process had been introduced into Boston and Chicago, among other places.

Harry Russell, at the University of Wisconsin, carried out systematic studies on the time and temperature requirements of pasteurization in the early 1900s. The most heat-resistant pathogenic organism present in milk was known to be *Mycobacterium tuberculosis*, the causal agent of tuberculosis; Russell determined the minimum

---

[7] Translated from L. Pasteur, *Études sur le vin: ses maladies, causes qui les provoquent*, (1866), in *Oeuvres de Pasteur*, **3**, Masson & Cie., Paris, 1924, p. 261.

time and lowest temperature at which this organism was killed. He advised against the use of higher temperatures or longer heating times because they led to deterioration in the quality of the milk. Although the engineering aspects of milk pasteurization have changed considerably since their beginnings, the time-temperature relationships for milk pasteurization have not changed significantly. We discuss the current practices of milk pasteurization in Section 15.5.

Vaccination is a treatment that renders an individual resistant or immune to a particular infectious disease. The development of procedures for immunization against infectious diseases has been one of the most significant practical outcomes of our understanding of the microbial nature of infectious disease. Louis Pasteur is closely identified with the early development of immunization methods; he developed and promoted methods for immunizing against fowl cholera, anthrax, and rabies. But he was not the first to use immunization, nor did he coin the word vaccination. The procedure had been known for many years before Pasteur's work and was used to control smallpox before the microbial causation of infectious disease was even suspected.

Smallpox, known since antiquity, is an acute disease characterized by fever, malaise, and a rash. Smallpox is caused by a virus that enters through the respiratory tract and multiplies in the lymph and other tissues. The virus particles then circulate through the blood, become localized in the skin and produce characteristic skin lesions that are rich in virus particles. The lesions form crusts that fall off after 2 to 4 weeks, leaving permanent scarring. Because they contain virus, the skin lesions are infectious, and material taken from the lesions can infect susceptible individuals. Smallpox is a disease only of humans, and severe cases are fatal; a closely related disease, cowpox, occurs in cows and horses but is only mildly infectious for humans.

The process of purposely inoculating people with infectious smallpox virus to produce a localized skin lesion was practiced in the Far East for centuries. Such an inoculation produced a relatively mild case of the disease in most cases, followed by recovery and lifelong immunity. Occasionally, a person inoculated in this way would succumb to a serious infection and die, but in general, the process of inoculation was far safer than chancing the disease itself, which frequently resulted in death. Inoculation for smallpox was introduced into England from Turkey in 1721 to 1722 by Lady Mary Montagu, the wife of the British ambassador. Although the process was first greatly resisted, it eventually became established.

1.6
Vaccination and the prevention of infectious disease

FIGURE 1.12
Jenner's drawings of pustules formed after inoculation with cow pox viruses. From E. Jenner, "An inquiry into the causes and effects of the *variolae vaccinae*," published privately by the author, 1798.

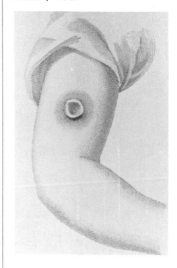

A British physician, Edward Jenner, made the acute observation that inoculation with the cowpox virus rather than the smallpox virus initiates a safe and mild infection (see Figure 1.12) that confers lifelong immunity to smallpox. Although Jenner's observations were originally disputed, inoculation with the safer cowpox eventually replaced inoculation with smallpox. The procedure came to be called *vaccination*, although it was not Jenner who first coined the word. He had referred to smallpox by its Latin name, *Variolae vaccinae* ("small pocks from the cow"), and from this the French derived the word vaccination about 1800. The material inoculated was called the *vaccine*. When Pasteur began his work on immunization, he adopted the word vaccination (as he said, in honor of Jenner) to use with other diseases as well as smallpox, and so it has come to us today: vaccination is a specific immunization procedure, using any sort of immunizing agent.

PASTEUR'S WORK ON VACCINATION

Almost 20 years after Pasteur began his career, he started to investigate infectious disease. His early work on spontaneous generation, which we discussed earlier in this chapter, was extremely important to the fundamental understanding of infectious diseases. Many scientists realized this, most importantly Joseph Lister in England, who was led to the development of antiseptic surgery after reading Pasteur's work on spontaneous generation. Pasteur was encouraged by many in the French scientific community to undertake studies on diseases, but he resisted for a long time because he was not trained in medicine. Finally, after much persuasion by his old teacher, Jean Baptiste Dumas, Pasteur agreed in 1865 to undertake a study of the diseases of the silkworm to help save the French silk industry from destruction. This study occupied him for 3 years and met with considerable success. He learned how to determine microscopically whether silkworms were infected and could prescribe methods for the examination of silkworm stocks that permitted the elimination of diseased animals. When he began his work on silkworms, he was rather diffident, but his success gave him encouragement to study other animal diseases. As he said in the preface to his book on silkworm diseases:

> I was hardly prepared for the studies which form the subject of this book, when in 1865 the ministry of Agriculture charged me with a study of the diseases which were decimating the silkworm. At that time I had not even seen a silkworm, so that I hesitated to accept this difficult mission. In addition to my uncertainty of achieving success, I regreted having to abandon, for a rather long period, the studies of fermentation which were dear to me and which excited me so much because they were meeting with great success. It was just at this moment that the results of my research on organized ferments, animal

and plant, were opening to me a vast future. As an application of these studies, I had come to understand the true mechanisms of the formation of vinegar, and had discovered that diseases of wine were caused by the presence of microscopic fungi. My studies had thrown a new illumination on the question of spontaneous generation. It was clear to me that *the role of the infinitely small was infinitely large*, either as the cause of diverse diseases, notably contagious diseases, or in their contribution to decomposition and to the return to the atmosphere of all the elements which had come from it.

One day, around the beginning of October 1868, I met Monsieur Dumas upon leaving a session of the Academy of Sciences: "Ah," I said, "I made a really great sacrifice for you in 1865." During the session, there had been a great deal of heated discussion about several questions related to fermentation and contagion, and this had revived all my regrets about abandoning my studies of fermentation. It was M. Dumas who was responsible for my beginning the studies which are published here. How could I accede to his confident request, in spite of my inexperience, in spite of the attraction of the studies I had underway? Indeed, I found it difficult to resist the invitation of an illustrious colleague and a revered teacher.[8]

Pasteur did not go directly from his studies on silkworm diseases to those of higher animals. He wished to do so, and wrote to the French government requesting a larger laboratory where he could have the special facilities he deemed necessary for working on highly contagious diseases, but soon after this, the Franco-Prussian War broke out, and the French were defeated disastrously. Pasteur, an ardent French patriot, began studies on spoilage of beer, with the goal of putting the French beer industry on a firm scientific footing (as he had already done with the French wine industry). He hoped that the French beer industry could surpass the German industry, so that France could regain some of its self-esteem. Thus, it was not until 1877, 12 years after he began his silkworm studies and 20 years after his first paper on fermentations, that he published his first work on a disease of higher animals, anthrax. Earlier the same year, Koch had published his pivotal work on anthrax, which had not only established clearly the bacterial cause of the disease (see Section 1.3) but had also launched Koch on the career that would carry him to the pinnacle of success. Throughout the next 15 years, while Koch was developing his methods for the culture and study of pathogenic organisms, leading to the isolation of a vast number of disease-causing agents, Pasteur was completing his own career in a most dramatic and practical way by developing important vaccination methods for anthrax, fowl cholera, and (most importantly) rabies.

After some work on anthrax, Pasteur turned to fowl cholera, a bacterial disease that was at that time affecting 10 percent of the

---

[8] Translated from L. Pasteur, *Études sur la maladie des vers a soie*, Gauthier–Villars, Paris, 1870, pp. ix–x.

chickens in France. He was able to isolate the causal agent (now known as *Pasteurella multocida*), but most importantly, he discovered a method of achieving immunity to the disease by inoculation with a culture whose virulence had been decreased. Being aware of Jenner's work on smallpox and cowpox, he interpreted the development of cultures with reduced virulence in the following way:

> Without wishing to make a definite assertion on the relationship between the small pox and the cow pox viruses, it seems ... that in fowl cholera there exists a state of the virus relative to the most virulent virus, which acts in the same way as cow pox virus does in relation to small pox virus. Cow pox virus brings about a benign illness, cow pox, which immunizes against a very serious illness, small pox. In the same way, the fowl cholera virus can occur in a state of virulence that is sufficiently attenuated so that it induces the disease but does not bring about death, and in such a way that, after recovery, the animal can undergo an inoculation with the most virulent virus.[9]

Pasteur used the word *attenuation* to refer to a laboratory-induced process by which the virulence of a pathogen was reduced. Actually, attenuation is a complicated process, the ramifications of which Pasteur never realized, and which was not explained until much later when an understanding of microbial genetics had been developed. Pasteur's main contribution was the realization that once an attenuated culture was available, it could be used to vaccinate healthy animals against the disease. He showed this readily with fowl cholera, although vaccination of chickens was probably not an especially practical procedure. However, having derived a general principle, Pasteur was quick to apply it elsewhere; he returned to anthrax and developed an attenuated culture that could be used as a vaccine. He then applied this attenuated culture in a dramatic public experiment in 1881 at Pouilly-le-Fort. On 5 May 1881, 24 sheep, 1 horse, and 6 cows were vaccinated. On 17 May, the animals were revaccinated, and then on 31 May, all the vaccinated animals plus 29 unvaccinated animals (24 sheep, 1 horse, 4 cows) were inoculated with a virulent culture. Pasteur invited a large number of government officials, scientists, veterinarians, and journalists to view the results. The visitors arrived on 2 June, 2 days after the inoculation with virulent organisms. The vaccinated animals appeared completely healthy, whereas 21 of the unvaccinated sheep and the horse had died. Two other unvaccinated sheep died before the eyes of the spectators, and the last died by the end of the day. The experiment was thus virtually a complete success and demonstrated to the world the importance of this procedure. As Pasteur said: "The discovery of the anthrax vaccine constitutes a significant advance on the Jennerian procedure, since attenuation of

[9] Translated from L. Pasteur, "De l'atténuation du virus du choléra des poules," *Comptes rendus de l'Académie des Sciences*, **91**, 1880, 673–680.

small pox has never been obtained experimentally, as I have done with anthrax."[10]

Pasteur's final work, for which he is perhaps most famous, was concerned with the disease rabies and vaccination against it. Rabies is a disease primarily of wild animals, but it can affect domestic animals and man. It is an acute disease, ordinarily fatal, in which the virus multiplies in the central nervous system and becomes excreted in the saliva, from which it can be transmitted to another animal by bite. In the early stages of the disease, the animal is often irritable and vicious (mad dogs typically have rabies); such vicious animals bite readily and thus transmit the virus. Humans become infected most commonly by the bite of a rabid dog although the disease is endemic in many wild animals, as well. The incubation period is fairly long, about 4 to 6 weeks, but once the infection is firmly established in a person, the disease is almost always fatal.

Pasteur discovered in 1881 that the infective agent of rabies could be recovered from the brain of an animal that had died of the disease and could be propagated by intracerebral inoculation in rabbits. After many passages from rabbit to rabbit, the virus became modified and then initiated an infection showing a short, fixed incubation period. This modified virus was called "fixed" virus, to distinguish it from the natural, or "street," virus. The fixed virus was not virulent for dogs and was thus used as the basis for developing a vaccination procedure. Rabbit spinal cord infected with the fixed virus was dried at room temperature (Figure 1.13) for various periods of time, during which the virulence of the fixed virus fell; by varying the drying time, a series of preparations could be developed ranging from no infectivity to maximum infectivity. Using this graded series, injections were made daily over a 2-week period, beginning with the nonvirulent and progressing to the most virulent. Dogs so treated were immunized to the natural virus and would not succumb even if inoculated directly.

Because the disease in humans has a long incubation period, Pasteur theorized that even a human who had already been infected, for instance by the bite of a rabid dog, could be protected if immunization were begun as soon as possible after the bite. Pasteur had a chance to test his theory in 1885 when Joseph Meister, a 9-year-old boy who had been severely bitten by a rabid dog, was brought to him. Because of the extensive lesions, it was almost certain that the boy would succumb if untreated. Three days after the accident, Pasteur initiated his series of vaccinations, beginning with the nonvirulent material and progressing to the most virulent,

[10] Translated from L. Pasteur, "Compte rendu sommaire des experiences faites a Pouilly-le-Fort, pres Melun, sur la vaccination charbonneuse," *Comptes rendus de l'Académie des Sciences*, **92**, 1881, 1378–1383.

THE RABIES VACCINE

FIGURE 1.13
An original preparation of Pasteur's that illustrates the manner by which he dried spinal cord from rabbits to attenuate the virulence of the rabies virus. *Courtesy of the Musée Pasteur, Paris.*

for a total of 12 injections. On the last day of the series, the boy was inoculated with the most virulent virus, a preparation that would have induced rabies in dogs within 8 or 9 days. Joseph Meister remained completely healthy, and this experiment was widely publicized.

Soon Pasteur was swamped with those seeking treatment against rabies, and within the next year he treated over 700 people from many countries. Mortality from rabies dropped by about 94 percent. To promote the method further, the Pasteur Institute was set up in Paris in 1886, funded by an international subscription, and branches were established in other cities in France and Europe. The institute prepared and tested the rabies vaccine and shipped material where needed. The Pasteur Institute did much more than simply perform vaccination for rabies; workers there initiated research on other diseases and on other problems in microbiology, and the institute exists today as one of the primary research facilities in France. Probably the most significant contribution of Pasteur's research on rabies was to further publicize the value of vaccination. Soon, discoveries of other vaccination procedures were made, most importantly for diphtheria, a serious childhood disease. Because of the success of immunization programs against diphtheria, that disease is all but eradicated today.

The discipline of immunology was founded to discover the mechanisms by which immunization against specific diseases is brought about. Today, we have a vast understanding of immunological processes, and many practical aspects of medicine and public health have resulted from studies in this field. We discuss the current status of immunology and vaccination practices in Section 8.15.

1.7 Antibiotics and chemotherapy

Although immunization and public health sanitation procedures had brought many infectious diseases under control during the period 1890 to 1930 (see Sections 1.4 and 1.6), many infectious diseases remained that were not amenable to control by these methods. In 1929, a paper was published that was eventually to have a greater impact on the medical control of infectious disease than anything that had gone before. Alexander Fleming, a Scottish physican engaged in research at St. Mary's Hospital in London, begins his paper announcing the discovery of penicillin with the following words:

> While working with staphylococcus variants a number of culture-plates were set aside on the laboratory bench and examined from time to time. In the examinations these plates were necessarily exposed to the air and they became contaminated with various micro-organisms. It was noticed that around a large colony of a contaminating mould

[fungus] the staphylococcus colonies became transparent and were obviously undergoing lysis. Subcultures of this mould were made and experiments conducted with a view to ascertaining something of the properties of the bacteriolytic substance which had evidently been formed in the mould culture and which had diffused into the surrounding medium.[11]

So begins the quiet announcement of the discovery of the first antibiotic agent.

Fleming characterized the product that the mold produced but did not show that penicillin could be used in the therapy of infectious disease, nor did he develop a process for large-scale production of the antibiotic.

Meanwhile, the search for other drugs to combat bacterial disease continued, and in 1935, a group of chemists at the Bayer works in Germany, headed by Gerhard Domagk, announced the discovery of the first sulfa drug, a chemical agent that could be used to control bacterial infections. The sulfa drugs were discovered by the systematic testing in infected animals of a wide variety of synthesized chemical compounds. The early sulfa drugs were effective in controlling only streptococcal infections. They were ineffective against two much more common and more serious infections, those caused by staphylococcus and pneumococcus. The demonstrated effect of Fleming's penicillin against these bacteria encouraged a group of British scientists at Oxford University, headed by Howard Florey, to investigate penicillin further. They began their work in 1939, just as World War II began, and were motivated partly by the knowledge that a chemical agent effective against staphylococcal and pneumococcal infections could play a significant role on the battle field. Two problems presented themselves to Florey and his colleagues: to produce large amounts of penicillin and to test this agent against bacterial infections in humans. Production was accomplished by culturing the fungus in large bottles, extracting the agent with solvents, and purifying it by chemical fractionation. The first purified penicillin was tested in humans, with dramatic success. With the effectiveness of penicillin demonstrated and the war in Europe becoming more intense, Florey came to the United States in 1941 carrying cultures of the penicillin-producing fungus with him and induced the U.S. government to create a large-scale research program for penicillin production. Progress was rapid and dramatic: high-yielding cultures of the penicillin-producing fungus were discovered; bottle production was replaced by production in large aerated vats; new culture media were discovered that greatly increased yields; and new chemical purification methods were

[11] A. Fleming, "On the antibacterial action of cultures of a penicillium, with special reference to their use in the isolation of *B. influenzae*," *Brit. J. Exptl. Pathol.*, **10**, 1929, 226–236.

developed. The research was carried out jointly by the pharmaceutical industry, the U.S. Department of Agriculture at its laboratory in Peoria, Illinois, and several universities, as well as by the English pharmaceutical industry. This cooperation, with all information exchanged freely, greatly aided the success of the mission. By the end of World War II, penicillin was available in large amounts, not only for military medicine but also for civilians. As soon as the war was over, pharmaceutical companies entered into commercial production on a competitive basis. Yields increased further, the price of penicillin decreased, and it became available wherever needed. Other antibiotics were soon discovered, such as streptomycin, chloramphenicol, and tetracycline, which were effective against certain bacteria not affected by penicillin.

The importance of the discovery of antibiotics for modern medicine is by now quite obvious. Infant and child mortality have been greatly reduced, and many diseases that at one time had high fatality rates are now no more than medical curiosities. As shown in Figure 1.1, pneumonia, once the major cause of death in the United States, is now only of minor importance, and tuberculosis, now adequately controlled by streptomycin and other chemotherapeutic agents, no longer stands among the top 10 causes of death. We discuss the current use of antibiotics in medicine in Section 8.16 and discuss the laboratory study of antibiotics, including the testing of cultures for antibiotic sensitivity, in Section 4.6.

Antibiotics have had another important impact on microbiology by encouraging the growth of industrial microbiology (see Chapter 17). At the time Florey began his work on penicillin, the only large-scale microbial processes used were in the brewing and distilling industries for the production of alcoholic beverages. It is not a simple task to scale up a microbial process from the laboratory to large tanks (often as large as 30,000 gallons). New techniques of sterilization and culture are necessary. Once these techniques were worked out for penicillin, it was relatively easy to apply them to other microbial processes. Today, we have a large number of microbial processes carried out economically by industry, not only for the production of antibiotics but also for the production of other substances (see Chapter 17), and the potential for the discovery of even more processes is great.

1.8
Root-nodule bacteria and the rise of agricultural microbiology

Some of the most important species of crop plants used in agriculture are legumes: peas, beans, soybeans, alfalfa, clover, vetch. It has been known for centuries that these legumes have peculiar structures, called *root nodules*, on their roots (Figure 1.14). Although for many years they were considered to be insect galls or some other pathological manifestation, the beneficial importance of root nodules for the nitrogen nutrition of leguminous plants was demonstrated in 1886 to 1888 by the German agricultural scientist Hermann Hellriegel. He showed that legumes grow readily in soils that are completely devoid of nitrogen fertilizer while other crop plants, such as oats, do not. He concluded that the legumes were obtaining the nitrogen they needed for growth from the atmosphere. Since the atmosphere contains 76 percent nitrogen gas (N_2) but only traces of combined nitrogen (in the form of nitrate, NO_3^-, and ammonia, NH_3), it was reasonable to conclude that the legumes were using N_2 as a source of nitrogen for growth. This use of N_2 as a source of nitrogen is called *nitrogen fixation*. Nitrogen fixation is of great importance in agriculture, since nitrogen is the element in which plants are most commonly deficient. In agricultural practice, nitrogen fertilizers are almost always required to produce excellent growth of all plants other than legumes. Since N_2 is the most widely available nitrogen source on earth, the ability of legumes to use that as their nitrogen source and make combined nitrogen compounds is of prime importance to soil fertility.

Microscopic examination of nodules revealed the presence of a dense population of bacterialike structures. The probable importance of bacteria in the formation of nodules was shown by B. Frank in 1879, when he showed that legume seeds allowed to grow in sterilized soil do not become nodulated. The British scientist H. M. Ward then showed in 1887 that crushed nodules may be used to inoculate leguminous seeds and induce nodulation even in sterilized soil. The next year, the great Dutch microbiologist Martinus Beijerinck developed methods for the isolation and culture of the root-nodule bacteria. His pure cultures could be used to inoculate legumes and to induce the formation of root nodules, so that it was possible to fulfill Koch's postulates (Section 1.3) with regard to the role of the bacteria in nodule formation. Although Beijerinck made many attempts, he could not demonstrate that his cultures were able to use atmospheric nitrogen when growing alone, in pure culture, and he concluded that the ability to use N_2 developed only after the bacteria had become associated with the plant in the root nodule. This conclusion has been verified by extensive research by many workers; with rare exceptions, it is known that only when the plant and bacterium are associated does nitrogen fixation occur. Further, each kind of legume plant has a specific bacterium that will cause nodulation.

Symbiosis is the word used to describe a mutually beneficial relationship between two organisms, and the legume root nodule is one of the best examples of symbiosis. The plant benefits because it receives useful nitrogen for growth, and the bacterium benefits because it is provided with a protected environment and a rich source of food. The bacteria take in N_2 from the air and convert it into organic nitrogen compounds; the compounds then move out of the nodule into the root of the plant and spread from there

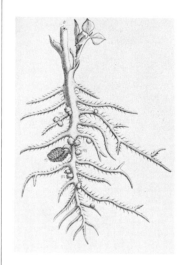

FIGURE 1.14
Root nodules of the broad bean, *Vicia faba*, as shown by Malpighi in 1679. From E. B. Fred, I. A. Baldwin, and E. McCoy, "Root nodule bacteria and leguminous plants," *University of Wisconsin Studies in Science*, no. 5, 1932.

throughout the plant. Beijerinck and others showed that root-nodule bacteria can live free in the soil, although when doing so, they must compete with all other soil bacteria. In the specialized structure of the root nodule, however, root-nodule bacteria have a habitat where they can obtain food and grow without competition.

The importance for agriculture of understanding the nature and interaction of root nodules and root-nodule bacteria was considerable. When planting a legume in a field, a farmer could not always be sure that the proper root-nodule bacterium was present in the soil. If it were absent, the plants would not become nodulated, and they could thus suffer from nitrogen deficiency. Today we know that it is a simple matter to inoculate the seed with the appropriate bacterium before planting; as soon as the seeds germinate, the bacteria inoculated onto the seeds infect the roots and start the nodulation process. Thus inoculation, a simple and inexpensive process, has proved extremely useful in agricultural practice.

An understanding of root-nodule bacteria was one of the first major developments in agricultural microbiology, and it raised considerable interest in other possible roles of microorganisms in soil fertility and plant growth. In the twentieth century, agricultural colleges have carried out extensive research on these questions; some of the areas in which microbiology has benefited agriculture are discussed in Chapter 16.

Suggested readings

Brock, T. D., *Milestones in microbiology*, reprint, American Society for Microbiology, Washington, 1975. The key papers of Pasteur, Koch, and others are translated, edited, and annotated for the beginning student.

Bulloch, W., *The history of bacteriology*, Oxford University Press, London, 1938. This book, still in print, is the standard history of bacteriology.

Dixon, B., *Magnificent microbes*, Atheneum Publishers, New York, 1976. An extremely well-written book outlining the beneficial aspects of microbes and showing the various ways in which they are vital for our welfare. The book also presents a usefully simplified account of microbial physiology.

Dobell, C., ed. and trans., *Antony van Leeuwenhoek and his "little animals,"* reprint, Dover Publications, Inc., New York, 1960. An excellent introduction, originally published in 1932, to the life and times of Leeuwenhoek.

Dubos, R. J., *Louis Pasteur: free lance of science*, Little, Brown and Co., Boston, 1950. A beautifully written elementary account of Pasteur's life and work.

Dubos, R. J., and J. Dubos, *The white plague: tuberculosis, man, and society*, Little, Brown and Co., Boston, 1952. An excellent history of tuberculosis and its impact on society, with a good elementary treatment of the scientific aspects.

Snow on cholera, being a reprint of two papers by John Snow, M.D., The Commonwealth Fund, New York, 1936. This reprint of Snow's work, complete with maps, should be available in many libraries. It is an excellent "detective story," making completely fascinating reading. It is a classic study in medical epidemiology.

Part 1 introduces the student to the general biology of micro-organisms, providing essential background for an understanding of subsequent parts. It also contains considerable practical material on the laboratory study of microorganisms.

Chapter 2 begins with a discussion of the microscope and how it is used to study microorganisms. This chapter gives a brief description of how cells are constructed, with the differences between pro-caryotic and eucaryotic cells being emphasized. Procaryotic *cells (bacteria and blue-green algae) have simple cell structures and lack a true nucleus;* eucaryotic *cells (fungi, protozoa, and most algae) have a nucleus, and their internal cell structures are more complex. Chapter 2 concludes with a brief overview of the kinds of microorganisms that are known, giving some idea of the diversity in the microbial world.*

Chapter 3 deals with the care and feeding of microbes. Only when microbes can be obtained in pure laboratory cultures can many aspects of their structures and functions be studied. The proof that a

GENERAL MICROBIOLOGY

specific microbe causes a specific disease, following Koch's postulates, requires that pure cultures be obtained. In Chapter 3, we describe the isolation and maintenance of pure cultures, the nutrition of microorganisms, and the procedures and methods used in proper aseptic technique. The principles of aseptic technique outlined in Chapter 3 will also be applicable to practical situations in infectious-disease control, as discussed in Chapters 4 and 9.

Chapter 4 deals with microbial growth and its control. Because microbes generally have harmful or beneficial effects only after they have undergone growth, an understanding of how microbial growth is measured and expressed is essential in interpreting the probability of microbial effects on man. Also, by understanding how growth occurs, we can begin to find ways to control it. Thus, Chapter 4 describes the details of the control of microbial growth by heat, chemicals, and other agents. A major portion of Chapter 4 deals with antibiotics: what they are, how they act on microorganisms, and how they are studied and assayed.

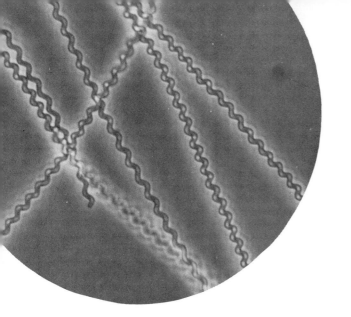

What are microorganisms?

Microorganisms are so small that at first glance one might think that they were unimportant. That is not the case, however; we may say that they are small but powerful. Just how small are they? How can we see them? What are they composed of? What do they do? To answer some of these questions, let us begin our study by looking at the structure and function of microorganisms.

Each individual microorganism is an independent unit that we call a *cell.* Animals and plants are composed of many cells, but microorganisms are single cells. Each microbial cell is a self-sufficient unit, able to carry out all life functions independently of other cells. In contrast, single plant and animal cells are not self-sufficient and can survive and reproduce in nature only when associated with other cells in the plant or animal body. Although most microorganisms exist as single cells, some do occur as aggregates of individual and independent cells.

All cells are constructed in similar ways. On the outside is a barrier separating the inside of the cell from its surroundings. This barrier may be simple and composed of a single layer, or it may be complex and composed of several layers. The most important layer, found in all cells, is the *cell membrane.* It is through the membrane that all food materials and other substances of vital importance to the cell pass in, and it is through the membrane that waste materials and other cell products pass out. If the membrane is damaged the interior contents of the cell leak out, and the cell dies. As we shall see, some drugs and other chemical agents damage the cell membrane and, in this way, act to destroy cells.

The membrane is a very thin, highly flexible layer and is generally rather weak. By itself, it usually cannot hold the cell together, and an additional layer of a stronger nature is necessary for many organisms. Most often, a rigid layer, called the *cell wall*, is present outside the membrane and makes the cell strong; all plant cells have rigid walls. Animal cells, however, do not have walls; they may have some other structure or may have no layer at all external to the membrane. Among microorganisms, most bacteria, all algae, and all fungi have cell walls, whereas protozoa do not.

The inside of a cell, the *protoplasm*, consists of a complicated series of substances and structures embedded in a salt solution. These materials and structures carry out the functions of the cell. The word *metabolism* is often used to refer to the collection of vital processes that cells carry out. The reactions and pathways of microbial metabolism will be presented in Chapter 6.

Upon careful study of the structure of microbial cells, two basic types that differ greatly are recognized: procaryotic and eucaryotic cells. *Procaryotes* are simpler and are thought to be the more primitive. *Eucaryotes* are more complicated, and it is generally hypothesized that eucaryotes probably arose from procaryotes during evolution. Bacteria and blue-green algae are procaryotes, while all other algae, fungi, and protozoa are eucaryotes, as are the cells of all higher animals and plants. Later on in this chapter, we shall present some of the basic differences between procaryotes and eucaryotes and list the kinds of microorganisms that are of each type (Figure 2.1).

2.1
Seeing microorganisms

The cells of microorganisms are so small that it is virtually impossible to see a single microbial cell without a *microscope*. The microscope is the most important tool of the microbiologist and is used so frequently that every microbiology laboratory must have one. The microscope permits us to see individual microbial cells and hence to be sure that we are really dealing with microorganisms. Further, since various microorganisms differ in size, shape, structure, and other properties discernible under the microscope, we can use the microscope as the first means of telling one microorganism from another.

THE LIGHT MICROSCOPE

The earliest microscopes were simply designed by present-day standards and provided only a very crude idea of what was observed. We discussed the early use of the microscope in Chapter 1. During the

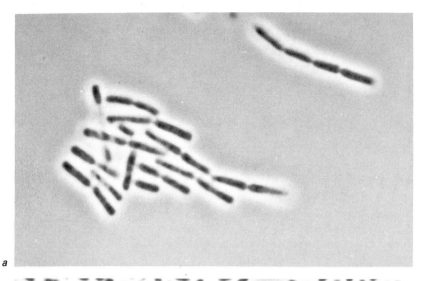

a

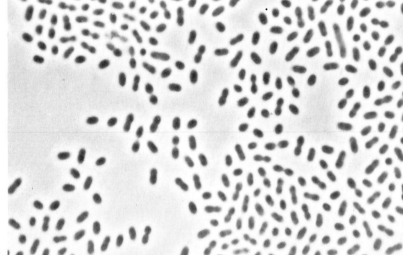

b

c

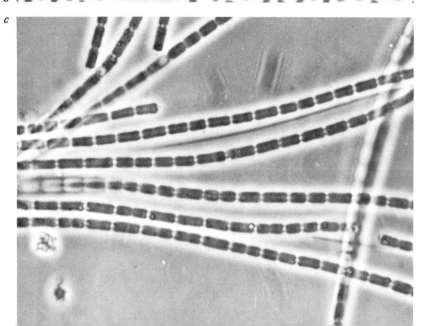

FIGURE 2.1
Typical representatives of the microbial world. *a* Large rod-shaped bacterium, *Bacillus cereus*, magnification 2,200×, phase contrast. *b* Short rod-shaped bacterium, *Acetobacter aceti*, magnification 2,200×, phase contrast. *c* Filamentous blue-green algae, *Anabaena*, magnification 850×, phase contrast. *d* Yeast cells dividing by budding, magnification 3,200×, phase contrast. *e* Fungus, magnification 500×, phase contrast. *f* Stalked diatom, a freshwater alga, magnification 600×, phase contrast. *g* Protozoan, *Paramecium multinucleatum*, magnification 220×, phase contrast.

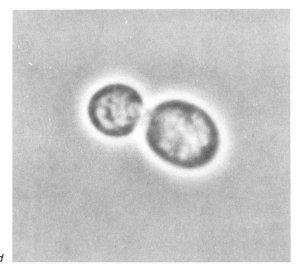

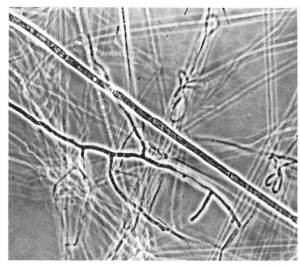

d

e

f

g

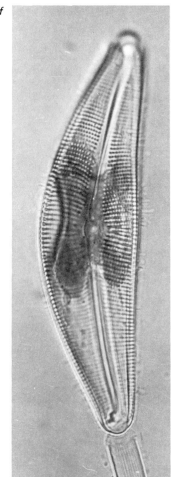

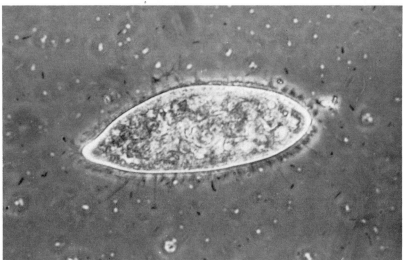

intervening years, microscopes have become increasingly sophisticated, and the image seen has improved accordingly. All of the early microscopes, as well as most of those used in laboratories today, are called *light microscopes* because the specimen is illuminated with visible light.

The two key attributes of a light microscope are magnification and resolution. *Magnification* refers to the ability of the microscope to enlarge objects, and *resolution* refers to the ability of the microscope to resolve into distinct images the various parts of the object being viewed. In crude or poorly designed microscopes, magnification may occur without resolution, resulting in rather fuzzy images. The microscope used today is always of a type called a *compound microscope* because it has two lenses. The *objective* lens is placed close to the object to be viewed, and the *ocular* lens, or *eyepiece*, is placed close to the eye (Figure 2.2). The compound microscope is able to achieve considerably greater powers of resolution than is a microscope composed of only a single lens, such as a hand lens or magnifying glass. The total magnification of a compound microscope is calculated by multiplying the magnification of the objective by the magnification of the eyepiece. Thus, with an objective magnifying 40 times (40×) and an eyepiece magnifying 10 times (10×), the total magnification is 400 times (400×).

Most microscopes used in microbiology have eyepieces that magnify about 10× and have at least three objectives magnifying 10×, 40×, and 100×, giving final magnifications of 100×, 400×, and 1,000×. The 100× magnification is used for scanning the specimen quickly to find objects of interest. The 400× magnification is used to study large microorganisms such as algae, protozoa, and fungi, and the 1,000× magnification is used for the study of bacteria. The 100× objective (giving 1,000× final magnification) is always an *oil-immersion* objective. This means that it can only be used with a drop of special immersion oil between it and the object, to eliminate light scattering.

In order to see properly with a light microscope, the light must be carefully adjusted and focused on the specimen. This is especially important with oil-immersion objectives. To achieve satisfactory light, a *condenser lens system* is essential, and it must be properly adjusted. It cannot be emphasized too strongly that proper adjustment of the light is crucial to good microscopy, especially at higher magnifications.

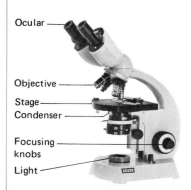

FIGURE 2.2
A modern light microscope.
Courtesy of Carl Zeiss, Inc., New York.

SPECIMEN PREPARATION

The best and simplest procedure for observing microorganisms is to make a *wet mount*. This involves placing a drop of water containing the specimen on a glass slide and covering it with a thin square of glass called a *cover slip*. If the specimen is to be observed for a long time, the cover slip should be ringed with mineral oil or melted paraffin to prevent drying. If an oil-immersion lens is to be used, it is preferable to locate the organism with the 40× objective first, then place a drop of oil on the cover slip, and turn to the oil-immersion objective.

Bacterial specimens can be dried before examination, although this usually leads to some distortion. The drop of water containing the specimen is spread thinly over the slide, and the water is allowed to evaporate. The dried slide is passed several times quickly through a flame to fix the cells to the surface, and the slide is then rinsed in water to remove salts. Usually, dried material is stained before examination, as described below. If an oil-immersion objective is to be used, a drop of oil is placed directly on the dried specimen, and the objective is immersed in the oil.

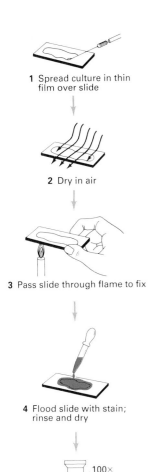

1 Spread culture in thin film over slide

2 Dry in air

3 Pass slide through flame to fix

4 Flood slide with stain; rinse and dry

100×

Slide   Oil

5 Place drop of oil on slide; examine with 100× objective

FIGURE 2.3
Staining cells for microscopic observation.

The size of most bacterial cells is such that they are difficult to see well with the light microscope. The main difficulty is the lack of contrast between the cell and the surrounding medium, so that the cell is virtually invisible. The simplest way of increasing contrast is through the use of *dyes*, which color the cells and hence make them darker than the surrounding medium. A variety of dyes are used in microbiology; the most useful are *methylene blue, crystal violet*, and *safranin* (a red dye). It is often enough to do a *simple stain*, a staining procedure in which only one dye is used. A drop of the dye may be added to living cells in a wet mount, and methylene blue is frequently used in this manner. However, for most studies, the dyes are used with dried preparations (Figure 2.3). The slide containing fixed and washed organisms is flooded for a minute or two with a dilute solution of a dye, then rinsed several times in water, and blotted dry. It is almost essential to observe dried stained preparations under oil immersion; a drop of oil is added directly onto the surface of the dry slide to avoid light scattering, and the oil-immersion objective is swung into position. If the slide is to be observed with lower-power objectives, a drop of water and a cover slip must be added first.

Another effective way of giving contrast to an organism is by *negative staining*. In this procedure, the organism itself is not stained, but its surroundings are, so that the organism appears light on a dark background. Common negative stains are the black dye *nigrosin* and plain *india ink*.

A large number of staining procedures have been developed with specific applications either in visualizing particular groups of organisms or in determining the presence of organisms in certain kinds of habitat. One of the most widespread and useful staining procedures is the *Gram stain*, named for the Danish bacteriologist Hans Christian Gram, who developed it in 1884. The Gram stain is called a *differential* staining procedure because it can be used to divide bacteria into two groups, called *Gram positive* and *Gram negative*. After Gram staining (Figure 2.4), the Gram-positive bacteria appear purple, and the Gram-negative bacteria appear red. The Gram stain is one of the most useful staining procedures in the bacteriological laboratory; it is almost essential in identifying an unknown bacterium to know first whether it is Gram positive or Gram negative.

The *phase-contrast microscope* makes it possible to see small cells easily even without staining. Cells differ in refractive index from their surrounding medium, and because of this, when they are observed in a phase-contrast microscope, they have greater contrast than when observed with a normal light microscope. The phase condenser forms a hollow cone of light from the light source, and this light is then shifted slightly in phase by the phase objective lens. Light passing through an object (such as a bacterium) that has a refractive index different from that of the medium is retarded, while the rest of the light passes through unretarded. Interference between retarded and unretarded light then produces the image of the specimen, which usually appears as dark cells against a light background. Because phase-contrast microscopy permits visualization of living unstained organisms, it is increasingly used as a research tool and is also becoming more widely used in routine bacteriological work. Although phase-contrast microscopes cost more than conventional light microscopes, the greater clarity and visibility they provide often make the difference in cost worthwhile.

A *dark-field microscope* is an ordinary light microscope with the condenser system modified so that light is directed at the specimen from the sides. Direct light does not reach the objective, and the only light seen is that scattered from the cells. The cells thus look light on a dark background. The principle involved is the same as that in seeing dust particles in a shaft of sunlight. Dark-field microscopy makes possible the observation in the living state of particles and cells so tiny that they are invisible in a conventional light microscope. In medical bacteriology, dark-field microscopy has been used most often to study the tiny cells of the causal agent of syphilis, *Treponema pallidum*. These tiny spiral-shaped organisms move rapidly and can readily be seen in the dark field microscope.

The *fluorescence microscope* is used to see specimens that fluoresce. Fluorescence is a property of certain chemicals and dyes. When they are subjected to light of one color, they give off light of another color. Some microorganisms contain naturally fluorescing substances (for example, the *green* pigment chlorophyll fluoresces *red*); others can be treated with fluorescent dyes. In the fluorescence microscope, the colored light source that shines on the specimen does not reach the eyepiece because a filter is placed between specimen and eyepiece that absorbs this light but permits the fluorescing light to pass. The fluorescing organism thus looks colored on a black background. The fluorescence microscope is widely used in medical bacteriology.

Light microscopy is unsuitable for seeing viruses or for studying the detailed structure of cells, and the *electron microscope* must be used (Figure 2.5). With this microscope, electron beams are used instead of light rays, and magnets are used as lenses. With the electron microscope it is even possible to see individual molecules. However,

1 Flood heat-fixed smear with crystal violet (1 minute)

All cells purple

2 Add iodine solution (3 minutes)

All cells still purple

3 Decolorize with alcohol (about 30 seconds)

Gram-positive cells are purple; Gram-negative cells are colorless

4 Counterstain with safranin (1 to 2 minutes)

Gram-positive cells are purple; Gram-negative cells are red

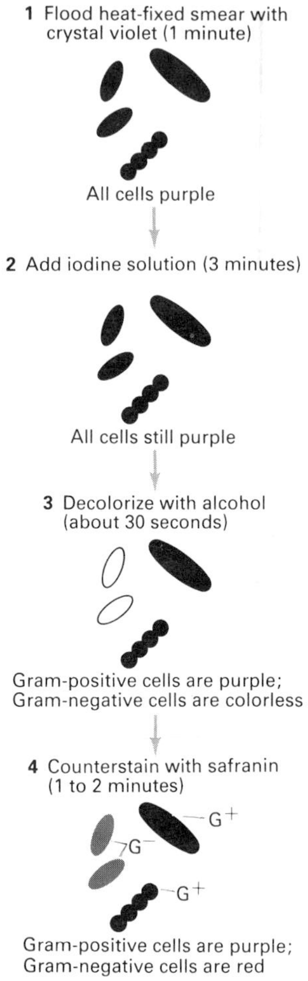

FIGURE 2.4
Steps in the Gram stain procedure.

THE ELECTRON MICROSCOPE

WHAT ARE MICROORGANISMS?

FIGURE 2.5
A modern electron microscope.
Courtesy of Carl Zeiss, Inc., New York.

the specimen must be examined dry and in a very high vacuum, conditions that greatly alter its properties. Instead of a glass slide, the specimen is mounted on a thin film of plastic. The electron beam is able to penetrate this thin plastic film but is scattered by the specimen. The image is seen on a small screen resembling a television screen, and for a permanent record photographs must be taken. Electron microscopes are not only much more expensive and complicated to use than are light microscopes but also must be carefully aligned and maintained if they are to yield good results.

Only very thin objects can be examined with the electron microscope, and if whole cells are observed only the outline of the specimen may be seen. Thus for much work, thin sections of the organism must be examined. Such sections are prepared by embedding the cells in plastic and making very thin slices with a glass or diamond knife. Then from a study of these sections, the observer must attempt to deduce what the whole organism is like. The preparation of thin sections is a complicated technical matter, but electron microscopy has greatly increased the resolution with which cellular structures can be seen (Figure 2.6).

For routine work often only the shape of the virus or organism need be seen in the electron microscope. For this purpose, the organism or virus is *shadowed*. A thin deposit of metal is cast upon

a

b

FIGURE 2.6
The yeast cell as seen with microscopes of increasing resolution.
a Leeuwenhoek's drawing of yeast, dating from 1694. From A. van Leeuwenhoek, *Ondervindingen en beschouwingen der onsigtbar geschapene waarheden*, 2nd ed., Henrik van Kroonevelt, Delft, 1694. *b* Pasteur's drawings of yeast, made in 1860. From L. Pasteur, *Ann. Chim. Phys.* **58**, 323, 1860. *c* Drawing of the idea of a yeast cell in 1910. From H. Wager and A. Peniston, *Ann. Botan.* **24**, 45, 1910. *d* A thin section of a yeast cell as seen with a modern electron microscope. From S. F. Conti and T. D. Brock, *J. Bacteriol.* **90**, 524, 1965.

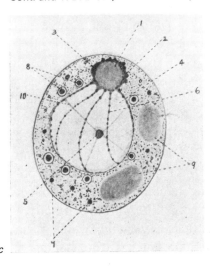

c

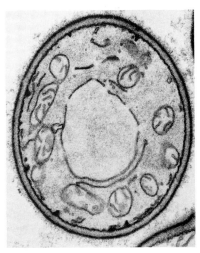

d

the mounted specimen from one side; a shadow is created on the side of the specimen away from the metal beam, and the size and shape of this shadow indicate the size and shape of the specimen (refer to Figure 2.9). Metals used for shadowing are gold, palladium, and platinum, since these heavy metals give very good contrast in the electron microscope.

Another useful and relatively simple procedure for electron microscopy is *negative staining*. The specimen is mixed with a solution of phosphotungstic acid or uranyl acetate, substances that do not penetrate into the cells or viruses but scatter electrons. The specimen then looks light on a dark background. Good detail of surface structures of cells and viruses can often be seen with negative staining.

Another type of electron microscope, designed especially for looking at surfaces of cells, is the *scanning electron microscope*. The electron beam is directed down on the surface, and the electrons scattered from the cell are caught and made to form an image of the surface. Two scanning electron micrographs are shown in Figure 2.11*b* and *c*.

The size of an organism is one of the first things to establish for identification purposes. This can usually be determined with a light microscope. A slide is prepared and the organism measured. Because of their small size, however, ordinary units of measurement cannot be used. The basic unit of size for microscopy is the micrometer (abbreviated μm), a unit that is one-millionth the length of a meter. Bacteria vary in size from cells as small as 0.1 μm in width to those more than 2 μm in width; length can vary from 0.1 μm to greater than 5 μm (Figure 2.7). Algae, fungi, and protozoa are usually much larger than bacteria. Some protozoa are as large as 1 to 2 mm (1,000 to 2,000 μm), and most fungi and algae have cells from 10 to 50 μm in diameter.

How is the size of a microscopic object measured? A small glass disc (called an *ocular micrometer*), which has a ruler engraved upon it, is placed in the eyepiece of the microscope, and the cell to be measured is lined up next to the ruler. The ocular micrometer must have been previously calibrated for the particular objective and eyepiece used, and on the basis of these calibrations, the size of the cell can be determined. The size of a single type of microbial cell does not always remain constant but varies within limits. Measurement of the size of a number of cells should be made, and the range of sizes observed should be reported.

2.2
Cell size

Careful study with light and electron microscopes has revealed the detailed structure of microbial cells. Starting at the outside and

2.3
Procaryotic cell structure

WHAT ARE MICROORGANISMS?

Oscillatoria (blue–green alga)
5 × 40 μm

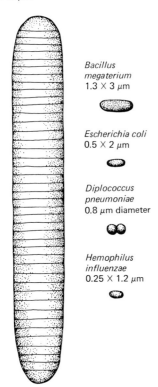

Bacillus megaterium
1.3 × 3 μm

Escherichia coli
0.5 × 2 μm

Diplococcus pneumoniae
0.8 μm diameter

Hemophilus influenzae
0.25 × 1.2 μm

FIGURE 2.7
Drawings to scale of procaryotic cells of various sizes.

FIGURE 2.8
Structure of a typical procaryotic cell. *a* Internal cell structure. *b* Electron micrograph of a thin section of a *Bacillus* cell, magnification 35,100×; *courtesy of William A. Samsonoff.*

proceeding inward, the following structures are seen (Figure 2.8*a* and *b*) in a typical procaryotic cell: (1) cell wall; (2) cell membrane; (3) internal membranes; (4) ribosomes; (5) nuclear region. What are their functions?

The *cell wall* confers rigidity on the cell and determines the cell shape. The cell wall also acts like armor, protecting the cell from many external influences. If the cell wall is removed, the cell usually bursts, a process called *lysis.*

The *cell membrane* is a very thin structure through which all of the food materials pass in and through which waste products and other products of metabolism pass out. Connected to the cell membrane and extending into the interior of the cell are usually many other thin membrane structures, called *intracellular membranes,* serving as surfaces onto which other substances of the cell attach and upon which many important cell functions take place. These intracellular membranes serve as organizers for many cell functions, which otherwise would take place randomly within the cell.

Some procaryotes contain the green pigment *chlorophyll.* Such organisms, which use light as a source of energy, are called *photosynthetic* and include the blue-green algae and the photosynthetic bacteria. Chlorophyll is present in a special membrane called the *photosynthetic membrane.*

The *ribosomes* are small particles composed of protein and ribonucleic acid (RNA). They are usually attached to intracellular membranes, and a single cell may have as many as 10,000 ribosomes. Synthesis of cell proteins takes place upon these structures, and different types of ribosomes are involved in the synthesis of different kinds of cell proteins.

The *nuclear region* of the procaryotic cell is rather ill-defined, in contrast to that of the eucaryotic cell to be discussed in the next section. Procaryotic cells do not possess a true nucleus with many chromosomes as do eucaryotes. Instead, the functions of the nucleus are carried out by a single long strand of deoxyribonucleic acid (DNA), and this nucleic acid is present more or less in a free state within the cell. The procaryote is thus said to have a nuclear region (the place where the

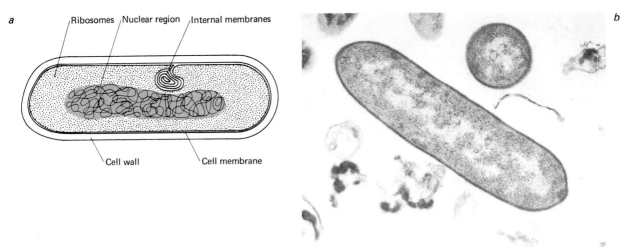

a Ribosomes / Nuclear region / Internal membranes

Cell wall   Cell membrane

b

DNA is present) rather than a nucleus. DNA is the key substance bearing information that determines production of proteins and other cellular substances and structures; it transmits this information to new cells during cell reproduction.

Various other structures are found in some, but not all, procaryotes, including flagella, spores, heterocysts, capsules, and sheaths. These are described in reference to the organisms in which they are found (see Section 2.6).

2.4 Eucaryotic cell structure

Eucaryotic cells are larger and more complex in structure than procaryotic cells. The greatest distinction is that eucaryotic cells contain within them distinct cellular structures called *organelles* within which important cellular functions occur (Figure 2.9; see also Figure 2.6*d*). These structures are lacking in procaryotes, where similar functions, if they occur, are not restricted to special organelles.

The most characteristic eucaryotic organelle is the *nucleus*, a special membrane-surrounded structure within which the cell's genetic material, DNA, is located. The DNA in the nucleus is organized into *chromosomes*, which are invisible except at the time of cell division. When cell division occurs, nuclear division also occurs, and the chromosomes become visible and undergo division also. The process of nuclear division in eucaryotes is called *mitosis* and is a complex but highly organized process.

Another organelle found in most eucaryotes is the *mitochon-*

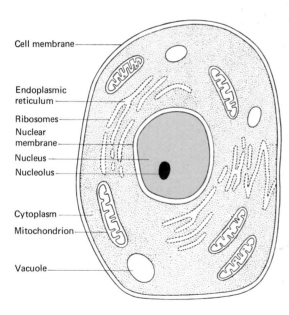

Cell membrane
Endoplasmic reticulum
Ribosomes
Nuclear membrane
Nucleus
Nucleolus
Cytoplasm
Mitochondrion
Vacuole

FIGURE 2.9
Eucaryotic cell structure. Compare this diagram with the electron micrograph of a yeast cell shown in Figure 2.6*d*.

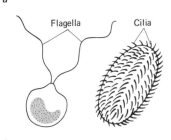

a

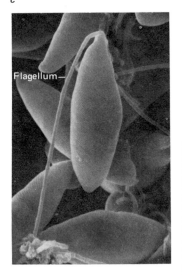

c

Flagellum—

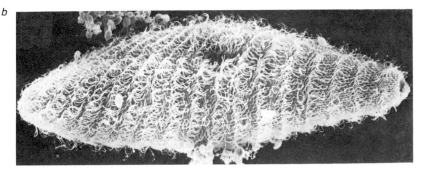

b

FIGURE 2.10

a A comparison of cilia and eucaryotic flagella. *b* A scanning micrograph of the ciliate *Paramecium*, magnification 650×. From S. L. Tamm, *J. Cell Biol.* **55**, 250–255, 1972. *c* A scanning electron micrograph of the flagellate *Euglena*, magnification 1,800×. From R. G. Kessel and C. Y. Shih, *Scanning electron microscopy in biology*, Springer-Verlag, New York, Inc., New York, 1974.

drion (plural, *mitochondria*). Mitochondria are the organelles within which the energy-generating functions of the cell occur. Respiration occurs in the mitochondria, and the energy produced is then made available to other parts of the cell.

In algae, an additional organelle is found: the *chloroplast*. The chloroplast is green, being the site of chlorophyll, and it is within this organelle that the light-gathering functions involved in photosynthesis occur.

The *cell membrane* of eucaryotes differs chemically but not functionally from that of procaryotes. The *cell wall* of eucaryotes is usually much thicker than that of procaryotes and also has a different chemical structure. Protozoa do not have cell walls, but algae and fungi do.

Many eucaryotic cells are motile, and two types of organelles of motility are recognized: flagella and cilia (Figure 2.10). *Flagella* are long filamentous structures, attached to one end of the cell, that move in a whiplike manner. It should be emphasized that the flagella of eucaryotes are quite different in structure from those of procaryotes, which we shall discuss in Section 2.6, even though the same name is used for both. The flagellum of a eucaryote is composed of two central fibers surrounded by nine outer fibers, with each of the latter composed of two subfibrils. Eucaryotic flagella are large enough to be seen with the light microscope, and their movements can be easily followed. *Cilia* (singular, *cilium*) are similar to eucaryotic flagella in structure but are shorter and more numerous. A single cell may have over 10,000 cilia. These organelles operate like oars to propel the cell through water. They are found in the microorganisms only among the protozoa.

To further emphasize the vast structural differences between

	Procaryotes	Eucaryotes
Nuclear body	No nuclear membrane; no mitosis	True nucleus, nuclear membrane; mitosis
DNA arrangement	Single molecule, not in chromosomes	In several or many chromosomes
Respiratory system	Part of plasma membrane; mitochondria absent	In mitochondria
Photosynthetic apparatus	In internal membranes; chloroplasts absent	In chloroplasts
Cytoplasmic movement	Cytoplasmic streaming rare or absent	Cytoplasmic streaming often occurs
Cell wall	Relatively thin	Thick; chemically different
Flagella	Submicroscopic size	Microscopic size; each flagellum composed of fibrils in a distinct pattern, 9 + 2

TABLE 2.1
Comparison of the procaryotic and eucaryotic cell

procaryotes and eucaryotes, a comparison between the cells of these two kinds of organisms is given in Table 2.1. It should be emphasized, however, that although procaryotes and eucaryotes differ greatly in structure, in terms of chemical composition and metabolism they are quite similar. At present it is believed that procaryotic organisms evolved first and that eucaryotes arose from procaryotes.

2.5
Kinds of microorganisms

Not all microorganisms are alike. We have already seen that microorganisms can be divided into two large groups: procaryotes and eucaryotes. Within each of these two large groups, a vast number of different organisms exist. Among the procaryotes is the very large group of microorganisms known as bacteria and also a smaller, more specialized group, the blue-green algae. There are three major groups of eucaryotes, the algae, fungi, and protozoa, as well as higher forms of plants and animals. It is one of the main tasks of the microbiologist to be able to recognize and distinguish among different kinds of organisms. The importance of this is that different microorganisms do different things of interest to man, some beneficial, some harmful. By being able to recognize different microorganisms and knowing their characteristics, the microbiologist can control their activities for the benefit of humanity. We shall describe very briefly the properties of these various major groups of procaryotes and eucaryotes in the following pages.

WHAT ARE MICROORGANISMS?

Because of the diversity of microorganisms, it is useful to give different organisms names. Microbiologists use the *binomial system* of nomenclature first developed for plants and animals by the Swedish botanist Carolus Linnaeus in the eighteenth century. In the binomial system, each organism has two names. The first name is the *genus* name, and the second is the *species* name. Thus, the name of a common yeast is *Saccharomyces cerevisiae*, *Saccharomyces* being the genus and *cerevisiae* the species. There may be several species with the same genus name, for example, *Saccharomyces cerevisiae*, *S. pastorianus*, and *S. carlsbergensis*. We can refer to the genus *Saccharomyces* without designating a specific organism, and when doing so, we are referring to a whole group of related organisms. When we wish to refer to a particular species of *Saccharomyces*, we always use both names; thus *S. cerevisiae* is a single species of the genus *Saccharomyces*.

The names of microorganisms are usually derived from the Latin or Greek and hence are unfamiliar to most students and may be difficult to learn. This is unfortunate, but there is no way of avoiding learning at least some names of microorganisms. After all, we cannot discuss the cast of characters without calling them by name. Students who know some Latin or Greek may be able to work out the meaning of a name. *Saccharomyces cerevisiae*, the beer yeast, got its name in the following way: yeasts convert sugar to alcohol, and *saccharo* means "sugar"; a yeast is a fungus, and *-myces* derives from the Greek meaning "fungus"; *cerevisiae* derives from the Latin word meaning "beer."

How do we tell one microorganism from another? We begin by studying its cellular structure and arrangement, using the microscope. In many cases, microscopic examination alone may permit us to classify the organism. All of the major groups—bacteria, blue-green algae, eucaryotic algae, fungi, and protozoa—can be distinguished using a microscope. In most cases, this distinction can be made immediately, without any special effort. It is only when we wish to determine the genus and species of the organisms that more refined observations are needed. In the rest of this chapter, we will describe briefly the important characteristics of the major groups of microorganisms and indicate how they can be told one from another with a light microscope.

2.6 Procaryotic microorganisms
BACTERIA

Bacteria are the smallest microorganisms and are the most difficult to see well under the microscope. In unstained preparations, bacteria can be seen only with difficulty in the conventional light microscope. With the phase-contrast microscope such observation is easier, but careful attention to adjustment of light and phase optics must still be made. If a phase-contrast microscope is not available,

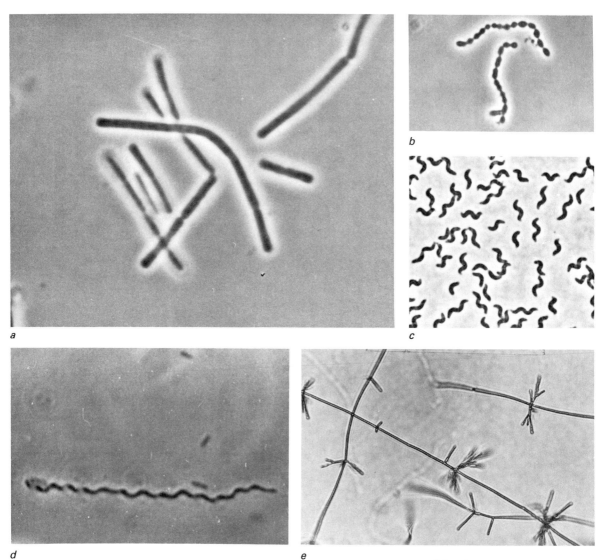

simple staining with methylene blue or negative staining with nigrosin or india ink will greatly improve the visibility of bacteria.

Cell shape is a very useful characteristic in identifying bacteria (Figure 2.11). Several different shapes are recognized: spherical (generally called *cocci*; singular, *coccus*); cylindrical, or *rod*-shaped; helical, or *spiral*. Some bacteria form rods that are very long, many times their width, and these structures are called *filaments*.

In many microorganisms, the cells remain together in groups or clusters, and the arrangements in these groups are often characteristic of different organisms. For instance, cocci or rods may occur in long chains. Some cocci form thin sheets of cells, while others occur in three-dimensional cubes or irregular grapelike clusters.

FIGURE 2.11

Bacteria of various shapes.
a Rod; *Bacillus*, magnification 2,500×, phase contrast.
b Coccus; *Streptococcus*, magnification 1,600×, phase contrast. *c* Spiral; *Rhodospirillum*, magnification 1,600×, phase contrast; *courtesy of Norbert Pfennig*. *d* Helix; *Spirochaeta*, magnification 2,000×, electron micrograph. *e* Filament; *Streptomyces*, magnification 420×, phase contrast; *courtesy of Peter Hirsch*.

The dimensions of bacteria are generally small. As discussed earlier, the diameter of the cell may vary from 0.5 to 1.0 μm, and only a few genera have cell diameters larger than 1.0 μm. A bacterial cell with diameter greater than 2.0 μm would be considered a giant. The hardest bacteria to see are the spirochetes, a group of actively motile but very thin organisms. Fortunately, they can be seen in the dark-field microscope. The lengths of bacterial cells vary greatly. Cocci are about as long as they are wide. Rod-shaped bacteria can vary in length from 1 or 2 μm to as much as 10 μm. Filaments, of course, may be very long. Certain groups of bacteria are characteristically filamentous, whereas other species, in which the organisms are normally rod-shaped, may form filaments under abnormal conditions.

Although size, shape, and cell arrangement are relatively constant for a single organism, variations do occur, often influenced by environment. Some organisms are more variable than others. For instance, in the genus *Arthrobacter*, cells of a given species are sometimes rods and at other times cocci. Cells with a variable morphology are called *pleomorphic*, meaning "many-shaped." A great many organisms show some pleomorphism with age, so it is wise to use cultures of known, and preferably young, age when studying morphology.

With small bacteria, it is virtually impossible to observe any special structures within the cells, although in many cases shiny granules can be seen. These granules usually consist of a substance called poly-β-hydroxybutyric acid, abbreviated simply as PHB, which is a storage material that the bacteria make when they are well fed and which they then use in times of starvation. Another substance seen in certain bacteria is *sulfur*, which is visible in the light microscope as dark bodies and in the phase-contrast microscope as bright shiny bodies with a faint reddish hue. Another type of granule formed by some bacteria is *volutin*, also called the metachromatic granule. Volutin granules are composed of phosphate and are formed generally when cells grow in environments rich in phosphate.

There are several kinds of specialized structures found only in certain bacteria. One such structure is the *flagellum* (plural, *flagella*), an organ of motility. Flagella are long, thin filaments free at one end and attached to the cell at the other end. By their movement they propel the cell through the water. They are so thin that they can never be seen directly with the light microscope, although they can easily be seen with the electron microscope or with the light microscope after special staining (Figures 2.12 and 2.13). Two types of flagellar arrangement exist: polar and peritrichous. Polarly flagellated organisms have a single flagellum or a group of flagella at one or both poles of the cell, whereas peritrichously flagellated organisms have flagella attached at many places around the surface of the

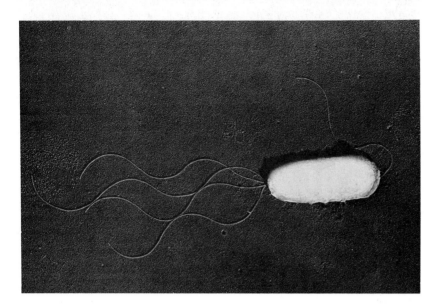

FIGURE 2.12
Electron micrograph of a
bacterial cell, shadowed to show
the flagella, magnification
15,000×. *Courtesy of Arthur
Kelman.*

FIGURE 2.13
Bacterial flagella as revealed by
a special flagella stain. *a, d, e, f*
Polar flagella. *b, c* Peritrichous
flagella. From E. Leifson, *Atlas of
bacterial flagellation.* Academic
Press, Inc., New York, 1960.

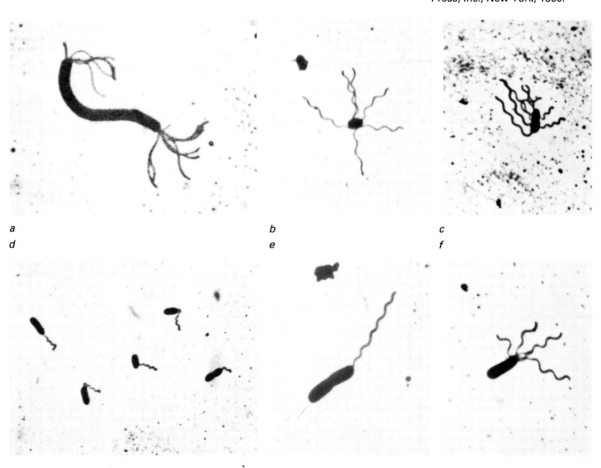

a

b

c

d

e

f

WHAT ARE MICROORGANISMS?

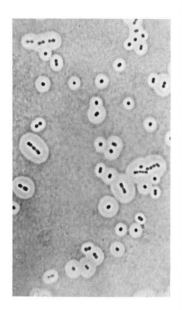

FIGURE 2.14
The bacterial capsule as revealed by a negative stain. *Courtesy of Elliot Juni.*

cell. The motions that polarly and peritrichously flagellated organisms make are different. Polarly flagellated organisms generally move rapidly with a spinning or tumbling motion, whereas peritrichously flagellated organisms move more slowly, usually in a straight line and perhaps with the cell rotating along the long axis while moving.

When watching motile bacteria under the microscope, it is important to distinguish motility, which is a living process, from *Brownian motion*, which is exhibited even by nonliving matter. Brownian motion is a random agitation caused by vibrating water molecules striking the particle, which thus usually moves haphazardly, not making any headway in the water. A motile bacterium, on the other hand, moves in a directed manner.

Since motility is often used in classifying bacteria, it is important to be aware of the proper way to study this process in the laboratory. Microscopic examination is done using a wet mount, preferably with a phase-contrast microscope, since stained preparations cannot be used. Only freshly grown cultures should be examined, since in old bacterial cultures, many of the cells may have lost their flagella or may be only weakly motile. The medium and conditions of culture are important, as organisms often do not make flagella under conditions in which they show otherwise normal growth.

Spirochetes are helical-shaped bacteria that move by wiggling in the manner of a snake. They do not have flagella but have instead a bundle of fibers wrapped around the cell that contract and move the cell through the water. Other procaryotic organisms are motile but have no flagella. These organisms exhibit *gliding motility* and move only when in contact with a surface. Many blue-green algae glide, as do the myxobacteria and certain other bacteria. The manner by which gliding motility occurs is not completely understood.

Another structure formed by some bacteria and blue-green algae is the *capsule*. This is a gummy material secreted onto the surface of the cell, forming a compact layer. Capsules can be visualized in the light microscope by use of a negative staining procedure (Figure 2.14). Many disease-causing bacteria produce capsules.

Some bacteria produce special structures called *endospores*, which are very resistant to heat and cannot be killed easily even by boiling. Because of endospores, heat sterilization of foods and other perishable products is difficult, and a knowledge of their nature and properties is of considerable importance in applied microbiology. Endospores can be seen easily with a phase-contrast microscope; they appear as bright shining objects in contrast to the dark appearance of the bacterial cell (Figure 2.15). They can also be seen in the normal light microscope by use of special spore-staining procedures, but in modern practice, the phase-contrast microscope is considered

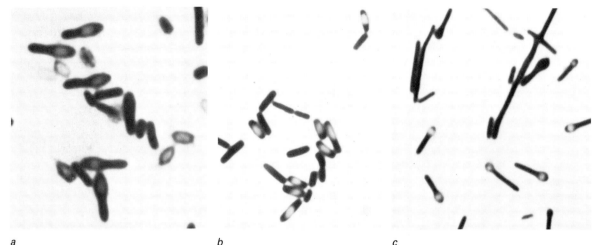

a   b   c

FIGURE 2.15
The bacterial endospore.
a Subterminal endospores.
b Endospores located centrally.
c Terminal endospores.
Courtesy of the Wellcome Research Laboratories, Beckenham, Kent, England.

more reliable. Also, preparation of a wet mount for phase-contrast examination is much easier than staining the cells.

Spore-forming bacteria are found most commonly in the soil, and virtually any sample of soil will have some bacterial spores present. Since soil or dust containing soil particles contaminates nearly all materials, it must be assumed that spores are present in any item to be sterilized.

Bacteria that are actively growing do not form spores, but when growth ceases due to starvation or some other cause, spore formation may be initiated. Spores are more resistant than normal cells to drying, radiation, and drugs, as well as to heat. Spores are able to remain alive but dormant and inactive for many years; however, they can convert back to normal cells in a matter of minutes, given proper conditions. This process, called *spore germination*, involves swelling of the spore, breakage of the spore coat, and outgrowth of a new cell from within the spore (Figure 2.16). As soon as germination begins, resistance to heat and to other harmful agents is lost.

Under the microscope, bacteria only very rarely appear colored. This is not because colored bacteria do not exist, since many colored forms are known. Rather it is because single bacteria are so small that the colors are usually not visible. Only when a large number of bacteria are collected together do they appear colored. The best way to tell whether a bacterium is colored is to observe the color of colonies formed on agar plates, as will be described in Section 3.7. A few bacteria are large enough and have such intense hues that their colors can be seen under the light microscope. This is true for some of the photosynthetic bacteria, which have chlorophyll and other colored pigments. When color is being observed under the microscope (as with such cells or with stained preparations), the phase-contrast microscope should not be used, for it imparts false

         WHAT ARE MICROORGANISMS?

colors to cells. Further, the light source used to examine cells for color must never have any colored filters between it and the specin.en.

IDENTIFICATION OF BACTERIA

For observation of bacteria, a microscope objective of at least 40× is required, and even better is a 100× oil-immersion objective. Eyepieces of 8 to 12× are normally used. The usual procedure with a phase-contrast microscope is to examine wet mounts at 40× in order to obtain some idea of the frequency and general diversity of bacteria present and then examine single cells under oil immersion. If cell size is to be measured, using an ocular micrometer, an oil-immersion lens should always be used to obtain the best accuracy. Since the sizes of cells vary, measurements should be made on at least 15 to 20 cells and an average taken. Size measurements should never be made on stained or dried preparations, because these treatments may alter the size of the cell.

If endospores are present, they can usually be recognized by phase-contrast microscopy, as described earlier in this chapter, or alternatively, a spore stain can be done and light microscopy used. Motility is observed using a wet mount, preferably with phase contrast. If a capsule is suspected, a negative stain should be performed.

Primarily because of the rather limited range of morphological forms possible and also because of the small size and the difficulty of

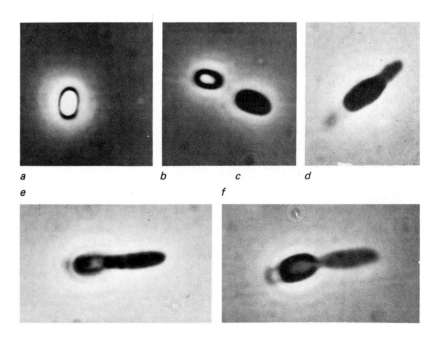

a   b   c   d

e   f

FIGURE 2.16
Stages of endospore germination. From J. F. M. Hoeniger and E. L. Headley, *J. Bacteriol.* **6**, 1835, 1968.

observing details of structure under either the light or phase-contrast microscope, bacteria can rarely be identified as to their species or even genus on the basis of microscopic observation alone. The usual procedure is to make observations of size, shape, and cell arrangement, look for motility and endospores, and perform a Gram stain. From these characteristics, a preliminary idea of the kind of organism being dealt with can be determined, but further work on the nutrition, metabolic products, and environmental requirements and tolerances of the organism must be carried out (as will be described in Chapter 3) to enable the microbiologist to affix a name to the isolate.

In contrast to the bacteria, the blue-green algae are relatively easy to see under the microscope, since they are the largest procaryotic microorganisms. Blue-green algae use light as their main source of energy and have two pigments, chlorophyll and phycocyanin, which capture light energy. Chlorophyll is green, and phycocyanin is blue; together they produce the blue-green color of these algae.

BLUE-GREEN ALGAE (CYANOBACTERIA)

The fact that the blue-green algae are called "algae" is a source of some confusion. They have very little in common with the true algae, which are eucaryotes, and in fact are more similar to the other procaryotes, the bacteria. It has been suggested that they might be more appropriately called *blue-green bacteria* or *cyanobacteria* (*cyano* means "blue-green"). We will continue to use the more common term *blue-green algae* but will remember the bacterial nature of these organisms.

The smallest blue-green algae are the size of large bacteria, 1.5 to 2 μm in diameter, and the cell diameter of large blue-green algal cells ranges up to 10 to 20 μm. Some blue-green algae are spherical or rod-shaped, but many are filamentous, and often the filaments are so large and long that they can be seen even with a lower-power microscope. Thus, when examining samples, it is advisable to begin with a 10× objective and then, after locating specimens of interest, to switch to a 40× objective. Rarely is a 100× oil-immersion objective necessary, nor is a phase-contrast microscope required. Most observations can be made with a light microscope, since the large size and pigmented nature of the organisms make them readily visible (Figure 2.17).

Classification of blue-green algae is based on size and shape of cells or filaments, arrangement of cells or filaments, presence or absence of branches in filamentous species, manner in which branches develop, and presence and structure of spores and hetero-cysts (discussed in the next paragraph). It is usually possible to classify blue-green algae as to genus by careful microscopic study, although difficulty is sometimes met with the smaller forms and with genera that have variable structure.

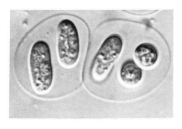
a

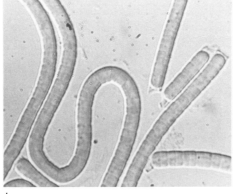
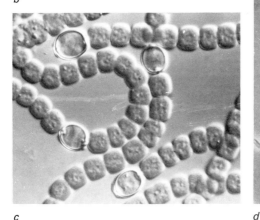
b

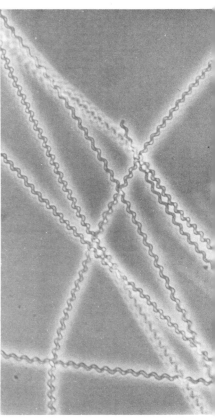

c

d

FIGURE 2.17
Blue-green algae. *a Gloeocapsa*, magnification 1,000×, Normarski interference contrast. *b Oscillatoria*, magnification 1,100×, Nomarski interference contrast. *c Anabaena*, magnification 1,000×, Nomarski interference contrast. *d Spirulina*, magnification 650×, phase contrast.

Two special structures are sometimes seen in filamentous blue-green algae: *spores* (also called *akinetes*) and *heterocysts*. Spores in blue-green algae are enlarged cells of dense appearance. They are not to be confused with bacterial endospores; the blue-green algal spores are formed in a different manner and are much less resistant to heat and other environmental factors than are bacterial endospores. Blue-green algal spores are resistant to drying and freezing and probably function in the dispersal of the species from one place to another. Heterocysts are rounded, seemingly more or less empty cells, which arise from cells of filaments. The shape of the heterocyst and whether it is found in the middle of the filament or at the end are characteristics used in classification. Heterocysts participate in the process of nitrogen fixation (the conversion of gaseous nitrogen to combined form) and thus enable blue-green algae to develop in nitrogen-deficient environments.

Some blue-green algae are motile and can be seen to glide slowly along surfaces. Motility in blue-green algae is never due to flagella but occurs by gliding. The organisms must be adherent to a surface (such as a glass slide or a piece of rock) in order to move.

The eucaryotic algae are quite different from the procaryotic, or blue-green, algae described in the previous section. Unless otherwise stated, throughout the rest of this book, whenever the word "alga" is used without modification, the reference is to a eucaryotic alga. Eucaryotic algae contain chlorophyll, a green pigment that serves as a light-gathering substance and makes it possible for them to use light as a source of energy. Because of chlorophyll, most algae are green, and they can be recognized first by their green color. However, a few kinds of common algae are not green but appear brown or red because in addition to chlorophyll other pigments are present that mask its green color. Many seaweeds (marine algae) are brown or red, although all contain chlorophyll as well.

The first problem to decide when one has an algal specimen is whether it is a blue-green alga or a eucaryotic alga. The color is of course the first clue, the former alga usually being blue-green and the latter grass-green. But because of the various hues that occur, structural study is necessary. Probably the best way of telling the two kinds of algae apart is by looking under the microscope for the

2.7
Eucaryotic microorganisms
ALGAE

FIGURE 2.18
Eucaryotic algae. *a Micrasterias*, magnification 250×, Nomarski interference contrast. *b Scenedesmus*, magnification 1,000×, Nomarski interference contrast. *c Volvox* colonies, magnification 100×, bright field. *d Ulothrix*, magnification 350×, Nomarski interference contrast.

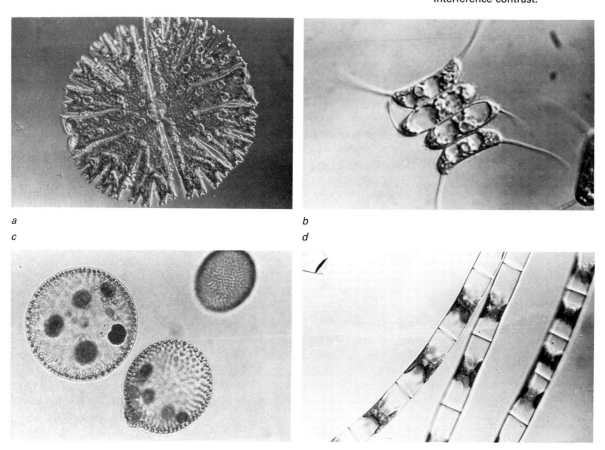

a   b
c   d

presence of chloroplasts, since eucaryotic algae have these structures and blue-green algae do not. In blue-green algae, the green or blue-green color appears to be spread more or less evenly throughout the cell, whereas in eucaryotic algae the green color is distinctly localized in certain organelles of the cell, the chloroplasts. The chloroplasts of eucaryotic algae are easily seen in Figure 2.18*b* and *d*; they should be compared with the photomicrographs of blue-green algae in Figure 2.17.

Eucaryotic algae can be either unicellular or filamentous (Figure 2.18). Among the unicellular forms are those in which the single cells occur in groups, called *colonies*; whereas the cells of other unicellular forms are always separate and individual. Among the filamentous forms, both unbranched filaments and more complicated branched filaments occur. In many cases, the filamentous algae form such large masses that they can be seen easily without a microscope. Small ponds or streams or even fish tanks frequently show accumulation of masses of green algae.

The identification of algae is based primarily on microscopic study. The main characteristics used in identification are (1) color and kind of pigments; (2) cellular organization, whether unicellular, colonial, filamentous, or of a more complex form; (3) motility, and if motile, the shape, number, and position of the flagella, or whether motile by gliding motility; (4) structure and appearance of the cell wall (some groups have complex walls composed of silica or calcium carbonate, and the markings on the walls are distinctive); (5) presence and type of sexual reproduction, and the kinds of structures involved in or formed during sexual reproduction; and (6) habitat, especially whether fresh-water or marine, since some kinds of algae are found only in marine habitats.

Detailed classification of algae is a job for highly trained specialists. The beginner can, however, carry out preliminary classification and at least obtain an idea of the major group to which the alga belongs.

FUNGI

Although the fungi are a large and rather diverse group, only three kinds of fungi are of major practical importance, and these are relatively easy to identify. These three are the molds, the yeasts, and the mushrooms. Fungi can be distinguished from algae because the fungi do not have chlorophyll and thus are not green. Fungi can be differentiated from bacteria by the fact that fungal cells are much larger, and vacuoles, nuclei, and other intracellular organelles can usually be seen inside, even with the ordinary light microscope.

The *molds* are filamentous fungi (Figure 2.19). They are commonly seen developing on stale bread, cheese, or fruit. An individual mold filament may have cross walls, or they may be absent. The filament grows mainly at the tip, by extension of the existing cell.

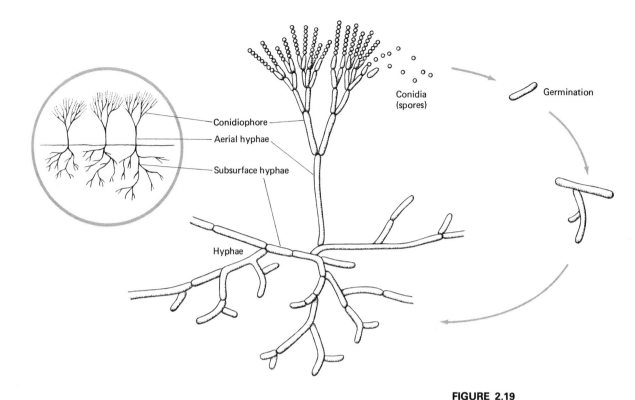

Conidia
(spores)

Germination

Conidiophore
Aerial hyphae
Subsurface hyphae

Hyphae

FIGURE 2.19
Filamentous fungus: structure and growth.

The single filaments, called *hyphae*, usually grow together across a surface to form rather compact tufts, collectively called a *mycelium*, which can be easily seen without a microscope. The mycelium arises because the individual hyphae form branches as they grow, and these branches intertwine and fuse with other branches, leading to the formation of the compact mat, or felt. From the mycelium, other branches form that reach up into the air above the surface, and on these aerial branches spores (also called *conidia*) are formed. They are round structures, often highly pigmented, that are resistant to drying, are very lightweight, and permit the fungus to be dispersed to new habitats. When the spores form, the white color of the mat changes, and the mat takes on the color of the spores, which may be black, blue-green, red, or brown. The presence of the spores gives the mat a rather dusty appearance. The spores just described are called *asexual spores*, since no sexual reproduction is involved in their formation. Because these spores are formed profusely and are spread so easily through the air, mold spores are common laboratory contaminants. They are also responsible for some airborne allergies in people.

Some molds also produce *sexual spores*, normally tiny but rather complicated structures, as a result of sexual reproduction. These sexual spores are usually resistant to drying, heat, freezing, and

WHAT ARE MICROORGANISMS?

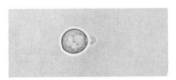

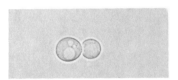

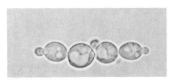

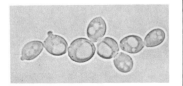

some chemical agents and play a role similar to that of the bacterial endospore, although they are not so resistant to heat as endospores. Either an asexual or sexual spore can germinate and develop into a new filament and mycelium.

The classification of molds is based on hypha and spore characteristics. If the fungus has hyphae with cross walls it is classified in a different group than if it has hyphae without cross walls. The morphology of the spore-bearing structures and the arrangement of the spores, both asexual and sexual, also provide a major basis of classification. Although not extremely complicated, the identification of molds is sufficiently difficult that it is usually done in specialized microbiology laboratories.

The *yeasts* are unicellular fungi. The cells are usually spherical, oval, or cylindrical, and cell division takes place by *budding* (Figure 2.20). In this process, a new cell forms as a small outgrowth of the old cell and gradually enlarges and then separates. Neither filaments nor a mycelium result, and the population of yeast cells remains a loose amorphous mass. Yeast cells are considerably larger than bacterial cells and can be told from bacteria by their size and by the obvious presence of internal cell structures. For the most part, yeasts spread from place to place as ordinary vegetative cells rather than as spores.

Classification of yeasts is based partly on the kinds of sexual spores formed and partly on the basis of nutrition and biochemistry. The classification of yeasts is even more specialized than the classification of molds. The common yeast, *Saccharomyces cerevisiae*, used in the making of bread, beer, whiskey, and other spirits, is one of the most important organisms affecting mankind; we shall discuss its industrial uses extensively in Sections 17.2 and 17.3. Most other yeasts are of little human importance, although a few cause human diseases.

The *mushrooms* are a group of filamentous fungi that form large complicated structures called *fruiting bodies* (the fruiting body is commonly called the "mushroom"). The fruiting body is formed through the growing-together and specific aggregation of a large number of individual hyphae. If the mushroom fruiting body is cut and examined under the microscope, the individual hyphae can be easily seen pressed tightly together.

During most of its existence, the mushroom fungus lives as a simple mycelium, usually buried within soil, in leaf litter, or in a decaying log. When environmental conditions are favorable, the fruiting body develops, beginning first as a small button-shaped structure underground and then expanding into the full-grown fruiting body that we see aboveground. Sexual spores called *basidiospores* are formed, borne on the underside of the fruiting body, either on flat plates called *gills* or within deep pores. The spores are often colored and impart a color to the underside of the fruiting body cap. Some mushrooms, called *puff balls*, produce their spores

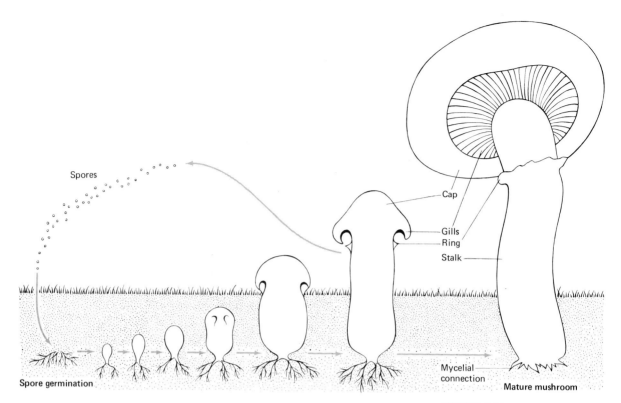

Spores

Cap

Gills

Ring

Stalk

Spore germination

Mycelial connection

Mature mushroom

FIGURE 2.21
Life cycle of a mushroom.

within spherical fruiting bodies instead of on gills or pores, the spores puffing out through cracks or holes that develop in the fruiting body as it dries. The spore is the agent of dispersal of mushrooms and is carried away by the wind. If it alights in a favorable place, the spore will germinate and initiate the growth of new hyphae, mycelium, and fruiting body (Figure 2.21).

Classification of mushrooms is based on the size, shape, and color of the fruiting body, the arrangement of the gills or pores, and the color of the spores. There is a wide variety of mushroom species, of which some are edible, while a few are poisonous. It is often difficult to distinguish poisonous from edible mushrooms, however, and one should, therefore, be very sure before eating mushrooms collected in the wild! The commercial mushroom is cultivated in vast beds containing a mixture of sterilized soil and horse manure. The mushrooms are harvested just as the mushroom buttons are expanding, since full-grown fruiting bodies are tougher and less tasty than immature fruiting bodies. Commercial production of mushrooms is discussed further in Section 17.10.

PROTOZOA

Protozoa are unicellular, colorless, generally motile organisms that lack a cell wall (Figure 2.22). They are distinguished from bacteria by their size, from algae by their lack of chlorophyll, and from yeasts

WHAT ARE MICROORGANISMS?

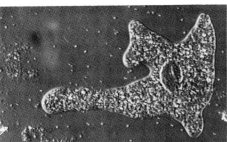 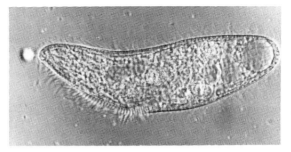 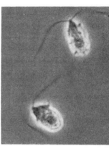

a   b   c

FIGURE 2.22

Protozoa of various types. *a Amoeba*, magnification 150×, Nomarski interference contrast. *b Blepharisma*, a ciliate, magnification 150×, Nomarski interference contrast. *c Dunaliella salina*, a flagellate, magnification 1,200×, phase contrast.

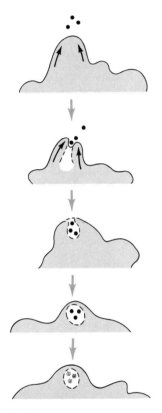

FIGURE 2.23

An amoeba obtains its food by engulfing small particles; digestion follows.

VIRUSES

and other fungi by their motility and absence of cell wall. Protozoa usually obtain food by eating other organisms or organic particles. They eat by surrounding the food particle with a portion of their flexible membrane and engulfing the particle (Figure 2.23) or by swallowing the particle through a special structure called the gullet.

The various groups of protozoa move in a number of different ways. The *amoebas* move by what is called *amoeboid motion*; the cytoplasm of the cell flows forward in a lobe of the cell, called a *pseudopodium* (false foot), and the rest of the cell flows toward this lobe. Amoeboid motion requires a solid substrate and is rather slow. A few amoebas are disease-causing agents, but most are harmless and live in soil or water. The *flagellates*, another group of protozoa, move by use of flagella, usually by a single long flagellum attached at one pole. The flagellates divide lengthwise instead of crosswise as do other protozoa. The *trypanosomes*, a subgroup of flagellates, cause African sleeping sickness and a number of other diseases of humans and animals. The *ciliates*, a third group of protozoa, move by the action of a large number of smaller appendages called *cilia*, which are attached all over the surface of the cell. The ciliates are the most complicated protozoa structurally, and in addition they have complicated sexual reproduction mechanisms as well as the more common vegetative cell division. A fourth group of protozoa, the *sporozoans*, comprises parasites of man or animals; the agent that causes malaria, *Plasmodium vivax*, is a member of this group. These nonmotile organisms do not eat food particles but absorb dissolved food materials directly through their membranes. Despite their name, sporozoans do not form true spores as do bacteria, algae, and fungi.

Viruses are not cells. Too small to be seen with the light microscope, viruses are particles that are inert by themselves, and they do not carry out any of the functions of cells. Only when a virus particle becomes associated with a cell does it begin to function. When becoming associated with a cell, the virus is said to infect it. Once inside the cell, a virus is able to reproduce itself, using the machinery of the cell for most essential functions. Cells that viruses can infect and in which they can reproduce are called *hosts*. The virus thus alternates between two states, the extracellular and the intracellular.

In the extracellular state, the virus particle, also called the *virion*, is composed of a molecule of nucleic acid, either ribonucleic acid (RNA) or deoxyribonucleic acid (DNA), surrounded by a coat composed of protein. When the virus particle infects a host cell, the nucleic acid separates from the protein coat, and the reproduction process within the cell begins (Figure 2.24). At the end of the reproduction cycle, molecules of nucleic acid and protein molecules join and reform new virus particles that become liberated from the dying cell. These virus particles can then infect other cells, and the process continues. When viruses do reproduce in cells, they usually damage or kill the cells, and in this way viruses are agents of disease. However, viruses do not always cause damage when they infect cells. Sometimes the virus nucleic acid becomes associated with the host nucleic acid, and a stable relationship occurs. Such viruses are called *latent*.

Viruses may vary widely in size, shape, chemical composition, hosts attacked, and kinds of cell damage induced. Viruses are known to infect animals, plants, bacteria, blue-green algae, and fungi. They cause a wide range of diseases in animals and plants, and because of this, an understanding of viruses is of great importance. Animal diseases caused by viruses include polio, measles, mumps, rabies, influenza, smallpox, and the common cold. A number of viruses cause tumors in animals, and viruses are probably responsible for some cancers in man. Plant diseases caused by viruses include tobacco mosaic, alfalfa mosaic, tomato bushy stunt, southern bean mosaic, and turnip yellow mosaic.

Although viruses are invisible in the light microscope, they can be easily observed in the electron microscope. In practice, viruses are recognized through their actions in causing disease. The study of viruses is much more complicated than is the study of microorganisms, and much more sophisticated equipment is required. More information about a number of viruses that are known to cause disease will be presented in Chapter 12.

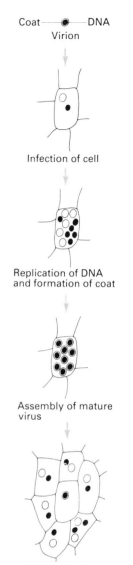

FIGURE 2.24
Virus replication in animal cells.

Summary

On the basis of the foregoing brief discussions, the student should, in most cases, be able to recognize under the microscope the major groups of microorganisms: bacteria, blue-green algae, eucaryotic algae, fungi, and protozoa. There will be times when a particular organism cannot be immediately recognized because it has some attributes of two groups. The only solution to this dilemma is to take the specimen to an expert. Eventu-

WHAT ARE MICROORGANISMS?

ally, with enough experience and after much looking at many microorganisms, the student will become the expert. In the meantime, patience and careful study are essential.

1 Trace the pathway of light through the photograph of the light microscope in Figure 2.2. What is the function of the condenser? of the objective?

2 Why are staining methods useful in the study of bacteria?

3 Why is the Gram stain of greater value than a simple stain? What is the color of a Gram-positive organism just after decolorization? after counterstaining? What is the color of a Gram-negative organism just after decolorization? after counterstaining?

4 Describe the different optical systems by which microbial cells are observed. What are the advantages and disadvantages of each?

5 Draw a diagram of a "typical" bacterial cell. Indicate the following procaryotic structures and describe the function of each briefly: cell wall, cell membrane, nuclear region, ribosomes, flagellum.

6 Describe and discuss the function of the following, which are found in some procaryotes under certain conditions: spore, flagellum, capsule, chlorophyll.

7 Compare and contrast procaryotic and eucaryotic organisms. How are they similar? how different?

8 Describe the cellular organelles unique to eucaryotic organisms, that is, not found in procaryotes. What is the function of each?

9 Cells range in size from such tiny cells as *Mycoplasma*, with a diameter of $0.125\,\mu$m, to an ostrich egg, with a diameter of $170,000\,\mu$m. A red blood cell has a diameter of $7\,\mu$m. Why is it that we give the name *cell* to things with such different sizes?

10 How would you differentiate the major groups of microorganisms: bacteria, blue-green algae, algae, fungi, protozoa, viruses? Include as many properties as possible to help distinguish each group.

11 The group designated *fungi* includes very different morphological types: molds, mushrooms, and yeasts. How do these differ? How are they similar?

12 Why are viruses not considered cells?

Suggested readings

Brock, T. D., *Biology of microorganisms*, 2nd ed., Prentice-Hall, Inc., Englewood Cliffs, N.J., 1974. Current textbook describing the structure and function of various groups of microorganisms.

Grell, K. G., *Protozoology*, Springer-Verlag, New York, 1973. The most complete advanced text on protozoology; outstanding illustrations.

Kudo, R. R., *Protozoology*, 5th ed., Charles C. Thomas, Springfield, Ill., 1971. Most widely available general textbook on protozoa.

Moore-Landecker, E., *Fundamentals of the fungi*, Prentice-Hall, Inc., Englewood Cliffs, N.J., 1972. Recent brief text on the fungi.

Prescott, G. W., *The algae: a review*, Houghton Mifflin Co., Boston, 1968. A useful brief review of the algae; emphasis on taxonomy and ecology.

Sieburth, J. M., *Microbial seascapes*, University Park Press, Baltimore, 1975. A pictorial version of marine microorganisms and their environments, done predominantly with the scanning electron microscope. Fascinating glimpses of the diversity of microbes in the sea.

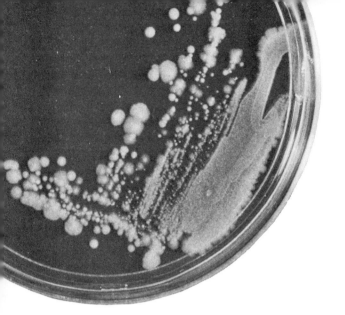

Chapter

3

Care and feeding of microbes

Our understanding of what things microorganisms do and how they do them depends upon our ability to study microorganisms in the laboratory. Laboratory study requires that we be able to culture microorganisms, providing them with the proper foods and conditions so that they can grow. It is also essential that we be able to isolate one microorganism from all others and maintain it in culture indefinitely in a pure state. Once we have isolated a pure culture, we can then proceed to a study of the characteristics of the organism and a determination of its capabilities. When a microorganism is cultured, it multiplies, and the cell number increases. This process is called *growth*.

The laboratory study of microorganisms is not unusually difficult to understand. In practice, however, great care must be taken to be sure that pure cultures remain pure and that appropriate nutrients and environmental conditions are provided. When disease-causing (*pathogenic*) microorganisms are to be studied, special precautions must be observed to prevent their spread and the resultant infection of people nearby. We will present here the basic principles and procedures involved in laboratory study of microorganisms.

Microorganisms are cultured in water to which appropriate nutrients have been added, usually in dissolved form. The aqueous solution containing such necessary nutrients is called a *culture*

3.1
Nutrition of microorganisms

medium (plural, *media*). The food materials present in the culture medium provide a microbial cell with those ingredients required to produce more cells like itself. We divide the nutrients of a culture medium into three major kinds: energy sources, cell structural components, and growth factors. There is a wide variety of food ingredients that provide these basic materials, but not all organisms require the same ones, nor can any one organism use all kinds. The skill of a microbiologist lies in selecting the proper components that will make up an adequate culture medium for a given organism.

ENERGY SOURCES

All organisms require energy to live and grow. We can divide energy sources into three groups: organic, inorganic, and light. *Organic energy sources* include sugars, starch, proteins, fats, and other organic materials. These are, in fact, the kinds of energy sources that we humans can use, and they are also energy sources for many microorganisms. An organic compound is a substance containing carbon (C) in an available form. To be "available," it must be in a form less oxidized than carbon dioxide (CO_2). CO_2 is the most completely oxidized form possible and can therefore never be used as an energy source. A compound such as glucose ($C_6H_{12}O_6$) has many hydrogen atoms and is thus more reduced than CO_2, which has none; glucose is one of the most common energy sources. The energy required for growth is released from the organic compounds by oxidation, or addition of oxygen (Figure 3.1). The manner in which microbes accomplish this will be discussed in Chapter 6. Organic energy sources are required by all fungi and protozoa and by the majority of bacteria. A *heterotroph* is an organism that uses some form of organic energy source. Thus fungi, protozoa, and most bacteria are heterotrophs, as are humans.

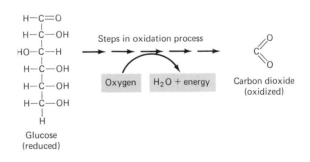

FIGURE 3.1
Oxidation of glucose, a common energy source.

Inorganic energy sources include ammonium (NH_4^+), nitrite (NO_2^-), ferrous iron (Fe^{2+}), hydrogen sulfide (H_2S), sulfur (S^0), thiosulfate ($S_2O_3^{2-}$), and hydrogen (H_2). Note that none of these compounds contains carbon, and thus they are all inorganic. The utilization (or *oxidation*) of an inorganic energy source also requires that oxygen (or its equivalent) be added to the compound, just as it is with organic energy sources. The end products of such oxidations are compounds containing more oxygen, such as nitrate (NO_3^-), sulfate ($SO_4^=$), and water (H_2O). Again, the manner in which microbes accomplish these reactions will be discussed in Chapter 6.

The organisms that use inorganic energy sources are all bacteria. These bacteria are usually given names based on the inorganic energy sources they use. Thus, we speak of the iron bacteria, the sulfur bacteria, the hydrogen bacteria, or the nitrifying bacteria (the latter because they produce nitrate, a process called *nitrification*). Collectively, organisms that use inorganic substances as energy sources are called *lithotrophs* (literally, "rock eaters").

Light is used as an energy source by photosynthetic organisms. The photosynthetic microorganisms include the eucaryotic and blue-green algae and the photosynthetic bacteria. All photosynthetic organisms contain at least one form of *chlorophyll*, the green pigment that is involved in the gathering of light. An organism that uses light as its energy source is called a *phototroph*.

Cells are constructed of a number of chemical elements. (A discussion of basic chemistry is presented in Chapter 5.) The major elements in cells are carbon, hydrogen, oxygen, nitrogen, sulfur, phosphorus, potassium, magnesium, calcium, iron, and sodium. In addition to these major elements, there are several elements present in very small amounts, which are called *trace elements*. The most important trace elements are cobalt, zinc, molybdenum, copper, and manganese. Another element required by only a few groups of microorganisms is silicon. In order to obtain microbial growth, all required elements must be provided in the culture medium in the proper form.

Carbon is provided either in organic form or as carbon dioxide. Heterotrophs, which use organic compounds as energy sources, also often use the same organic compounds as carbon sources. *Autotrophs* are organisms that use CO_2 as their sole carbon source, using either light or inorganic compounds as energy sources. *Photoautotrophs* use CO_2 as their sole carbon source and light for energy; algae, blue-green algae, and a few bacteria are photoautotrophs, as are all higher plants. *Chemoautotrophs* use CO_2 as their sole carbon source and have an inorganic energy source (they are thus lithotrophic); a few groups of bacteria are the only chemoautotrophs known.

Hydrogen and *oxygen* usually come from the water (H_2O) in

ELEMENTAL REQUIREMENTS

CARE AND FEEDING OF MICROBES

which the culture medium ingredients are dissolved. Most organisms can also obtain needed oxygen from the oxygen (O_2) of the air.

Nitrogen comes from either inorganic or organic compounds but is found in the cell almost exclusively in organic form. It occurs bonded to carbon, in organic nitrogen compounds, such as amino acids (the constituents of proteins), which are present in all cells. The principal inorganic compounds providing nitrogen are ammonium (NH_4^+) and nitrate (NO_3^-). If NO_3^- is used, it must first be converted to NH_4^+ or a derivative of NH_4^+ before entering organic form. Some bacteria and blue-green algae can obtain their nitrogen from the gaseous nitrogen (N_2) of the air, a process called *nitrogen fixation*. Nitrogen fixation, a rather special feat limited to only a few groups of procaryotes, is an important process agriculturally as it enriches the soil in the same way that addition of nitrogen-containing fertilizer does (see Section 16.4). Among organic nitrogen compounds, the most important biologically are amino acids. Proteins are built of amino acids, and when proteins are digested, the amino acids liberated serve as nitrogen sources for many microorganisms. Other organic nitrogen compounds include purines and pyrimidines, which are important constituents of nucleic acids. In general, heterotrophic microorganisms use organic nitrogen compounds as nitrogen sources, and autotrophic microorganisms use inorganic nitrogen compounds, but there are many exceptions to this statement. Many heterotrophic bacteria and fungi can use ammonium or nitrate as their nitrogen source along with an organic carbon source, such as glucose. On the other hand, protozoa almost always require organic nitrogen compounds, being unable to use ammonium or nitrate. Humans, too, require organic nitrogen and use proteins as the primary source. Autotrophic microorganisms, although mostly using ammonium or nitrate, can often also use an organic nitrogen source, such as an amino acid, if it is supplied to them. A single organic nitrogen compound can often carry out triple duty for a heterotroph, serving at once as energy source, carbon source, and nitrogen source.

Sulfur, like nitrogen, may also be supplied in either organic or inorganic forms. The principal inorganic sulfur source is sulfate (SO_4^{2-}), which must first be converted inside the cell to hydrogen sulfide (H_2S) before being incorporated into organic sulfur compounds in the cell. The principal organic sulfur compounds are the amino acids *cysteine* and *methionine*. In certain organisms, either of these two compounds can serve four roles: as energy source, carbon source, nitrogen source, and sulfur source. More normally, these requirements are filled by more than one compound.

Phosphorus comes from either organic or inorganic phosphates. However, before organic phosphates are used, they are usually converted to inorganic phosphate, which is then assimilated and converted back into organic form inside the cell. In nature, inorganic phosphates are found primarily as insoluble constituents of rocks and minerals or in animal bone in the form of calcium phosphate. In culture media, phosphate is usually provided in soluble inorganic form by the addition of compounds such as sodium or potassium phosphate.

The rest of the elements required by organisms are often collectively referred to as *minerals*. The major minerals, required in fairly large amounts, are potassium, sodium, magnesium, calcium, and iron. The trace elements are usually absolutely essential but are required in vanishingly small amounts. They often need not even be added to culture media because the minute quantities required are usually present in sufficient amounts as contaminants of the water or other ingredients used to prepare the culture medium.

Growth factors are specific organic compounds that are required in very small amounts and that the cell cannot make from other carbon sources. The best-known growth factors are *vitamins* and *amino acids*. It should be emphasized that many microorganisms do not require growth factors, but in the ones that do, failure to provide them will result in a complete lack of growth.

GROWTH FACTORS

The most commonly required vitamins in microorganisms are thiamine (vitamin B_1), biotin, pyridoxine (vitamin B_6), and cobalamin (vitamin B_{12}). Other vitamins sometimes required are folic acid, nicotinic acid (niacin), lipoic acid, pantothenic acid, riboflavin, and vitamin K. All these vitamins have been either synthesized chemically or purified from natural products and may be purchased in pure form from chemical companies, so that they can be added to culture media when necessary.

There are about 20 amino acids commonly found in proteins, and although many organisms can synthesize all of these from the nitrogen and carbon sources provided, not all can do so. Some may require one or more of these amino acids as growth factors. All of these amino acids are available in pure form and can be added individually to culture media. If it is desired to add all 20 at one time, this can most easily be done by adding what is called a *protein digest*, made by treating the protein with an enzyme capable of digesting it. The most common proteins from which digests for culture media are made are *casein*, the major protein of milk, and meat protein, from which the digest *peptone* is made.

Other growth factors required by certain microorganisms are the purines and pyrimidines, which are key building blocks of nucleic acid. These purines and pyrimidines can be provided as a nucleic acid digest but are most commonly provided as the individual compounds, which are available from chemical companies.

Certain disease-causing bacteria have an interesting growth factor requirement, growing only when blood or *heme*, the red pigment of blood, is present in the culture medium. There are many other growth factor requirements too specialized to discuss here.

A knowledge of growth factor requirements is of great importance for successful care and feeding of microorganisms in the laboratory. Without the proper growth factors, successful culture is impossible. Fortunately, we currently have a vast store of information on the growth factor requirements of different organisms, so that preparation of suitable culture media is almost always possible.

There are some organisms, however, which no one has been able to culture successfully. Many more probably occur whose existence we are still unaware of because we have never seen them in laboratory cultures. It is likely that these organisms fail to grow because the culture media used lack one or more required growth factors. Some may require growth factors that are unknown to us; their culture in the laboratory will need research just to determine their nutritional requirements.

3.2
Acidity, alkalinity, and neutrality

FIGURE 3.2
The pH scale.

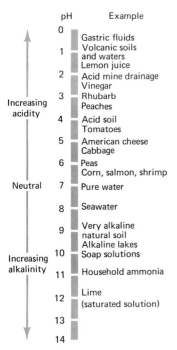

The proper acidity or alkalinity must be provided for each organism. Acidity or alkalinity of a solution is expressed by its *pH value* on a scale in which neutrality is pH 7 (Figure 3.2). Those pH values that are less than 7 are acidic, and those greater than 7 are alkaline. We shall discuss acids, bases, and pH in greater detail in Section 5.5. The pH values of solutions are usually measured with a pH meter, although pH-indicating dyes, which exhibit different colors at different pH values, can also be used. These indicator dyes are often added directly to culture media and can then indicate not only the initial pH of the medium but also any change in pH that results from growth or activity of the microorganism.

Each organism has a pH range within which growth is possible and usually also has a well-defined pH optimum. Organisms that have an acid optimum pH are called *acidophiles* (acid-loving), and those with alkaline optimum pH are called *alkalophiles*. It is an interesting point that a pH that may be highly detrimental to one organism may be the pH at which another organism thrives.

The pH of a culture medium is adjusted by adding an alkaline compound, such as sodium hydroxide, if the medium is too acid or an acidic compound, such as hydrochloric acid, if the medium is too alkaline. Because microorganisms usually induce changes in pH as they grow, it is often desirable to add to the culture medium a pH *buffer*, which acts to keep the pH relatively constant. Such pH buffers work only over a narrow pH range; hence, different buffers must be selected for different pH regions. For neutral pH ranges (pH 6 to 7.5), sodium phosphate and potassium phosphate make excellent buffers. For mildly acid pH ranges (pH 4 to 6), citric acid or glutamic acid may serve, and for very acid conditions (pH 1.5 to 3), sulfuric acid serves as a buffer. For alkaline pH ranges (pH 8 to 9), the organic compound tris(hydroxymethyl)aminomethane, abbreviated "tris," is often used, and for very alkaline conditions (pH 10 to 11), sodium silicate or sodium borate may be used. It is essential that the buffer be nontoxic to the organism.

We have now presented sufficient material on the principles of microbial nutrition so that we can discuss preparation of a culture medium.

3.3
Preparation of culture media

First, one begins with the formula for the culture medium, which usually consists of a recipe or list of ingredients and the concentrations needed. A measured amount of distilled or deionized water is placed in a flask or beaker, and the medium ingredients are added one at a time, each ingredient being dissolved before the next one is added. It is usually advisable to add and dissolve the ingredients in the same order as prescribed in the recipe. Changes in order of addition, or failure to dissolve one ingredient before adding the next, sometimes result in the formation of precipitates that may never redissolve. Also, to avoid metal precipitates, a compound called a *chelating agent* is often added, which combines with the metals and keeps them in solution. Examples of chelating agents are EDTA (ethylenediaminetetraacetic acid) and NTA (nitrilotriacetic acid).

In the usual order of addition of ingredients, the buffer compound is added first, followed by the major elements and then the trace elements. Vitamins and other growth factors are usually added last. The pH is then carefully adjusted and the medium sterilized. (Sterilization is described in Section 3.8.) In actual practice, many of the medium ingredients serve double or even triple duty. Potassium phosphate, for instance, serves not only as a buffer but also as a source of potassium and phosphorus. Ammonium sulfate may serve as a source of both nitrogen and sulfur. Glucose can be both a carbon and an energy source. In actual practice, carbon sources such as glucose are prepared and sterilized separately as concentrated solutions and added aseptically to sterilized media in order to avoid the carmelization that sometimes occurs when sugars are heated in solutions with salts such as phosphate.

It is common to distinguish between two types of culture media, synthetic and complex. In a synthetic medium, every essential nutrient is provided by a pure chemical of known composition, usually of synthetic origin and purchased from a chemical supply house. In a complex medium, certain of the ingredients used are plant, ani-

SYNTHETIC AND COMPLEX MEDIA

mal, or microbial extracts and consist of a complex and ill-defined mixture of ingredients that supplies all the essential nutrients for many microorganisms. Examples of such materials are malt extract (from barley malt), meat extract (from beef muscle), and yeast extract (from bakers' yeast). Other commonly used ingredients are tryptone and casitone, which are digests of the milk protein casein; peptone, a digest of beef muscle; and beef heart infusion, an extract of beef heart. These materials, called *extracts* or *digests*, usually contain all essential organic and inorganic nutrients for many microorganisms and are often used in preparing versatile culture media capable of supporting growth of a variety of microorganisms. These materials can be produced in the laboratory, but it is more convenient to purchase them fully prepared in dried form from commercial sources. Another advantage of using complex media is that the tedious weighing of many individual ingredients is avoided. The disadvantage, of course, is that the exact composition of the medium is not known. It should also be mentioned that there will be some variability in the medium from time to time as different batches of material are used. For instance, even the yeast extract provided by a single company will vary. One way of controlling this variability is to observe the *lot number* on the bottle, since a number of bottles containing the same lot number will have been prepared from the same batch.

Despite the tedium involved in their preparation, synthetic culture media are essential for many studies. One way to simplify the work involved in making up a synthetic culture medium is to make up stock solutions of mixtures of ingredients at 10 or 100 times the final concentration desired and then to add proper small volumes of these stock solutions to the water to achieve the desired final concentration. Stock solutions are especially useful for trace element and growth factor mixtures, since these ingredients are usually added in very tiny amounts that are difficult to weigh properly even if large batches of medium are to be prepared. Trace element stock solutions are usually prepared at acid pH because, in this way, precipitation is avoided. After the trace element solution is diluted to make the final medium, the pH can be adjusted without difficulty in the more dilute concentration. Growth factor solutions must be sterilized if they are to be stored for any length of time, since contaminating bacteria will otherwise grow in them. A convenient way of sterilizing a growth factor solution is to add a few drops of chloroform. When the final medium is heat-sterilized, the chloroform will be driven off and thus will not be part of the medium. It is also essential that growth factor solutions be stored in the refrigerator, since both light and heat will cause gradual decomposition of certain growth factors. A properly prepared growth factor solution can easily be stored for a year.

FIGURE 3.3
Technique for aseptic pouring of agar plates. The lip of the bottle is flamed before pouring, and the lid of the plate is raised just enough so that the agar can be added.

LIQUID AND SOLID MEDIA

Culture media can be prepared either in the liquid state or in the gel or solid state. Liquid culture media are dispensed in tubes, flasks, or bottles, which can then be inoculated and incubated. Microbial growth in liquid cultures is usually evenly dispersed throughout the liquid.

For many studies it is desirable to obtain microbial growth on or in a solid substrate. For this purpose, a liquid culture medium prepared as described above is converted into a solid form by adding to it a gelling agent. Gelling agents frequently used are agar, gelatin, and silica gel.

Agar, the most commonly used gelling agent, is manufactured from certain seaweeds. Agar is not a nutrient for most microorganisms and, hence, can be added to a culture medium without significantly modifying the quality of the medium. Another virtue is that agar is a solid at ordinary incubator temperatures but liquefies upon heating to boiling. Thus, a medium containing melted agar may be dispensed into tubes or plates; these are then cooled, causing the agar to solidify (Figures 3.3 and 3.4). *Agar slants* are prepared by allowing agar medium in tubes to solidify in a slanted position. The

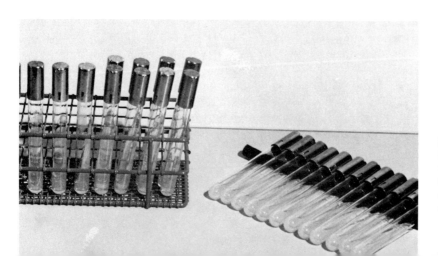

FIGURE 3.4
Preparation of agar slants. While the agar is still molten, the tubes are laid on the table in a slanted position and left until the agar is well solidified.

FIGURE 3.5
Use of a membrane filter as a
surface for microbial growth.
The filter with the adherent
microorganisms is placed in a
plate containing a culture
medium.

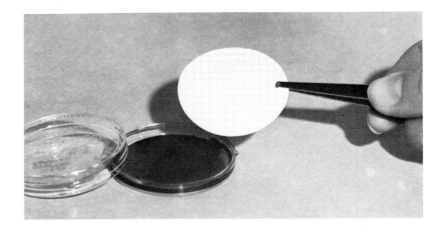

advantage of such slants is that a relatively large surface area is exposed, although the tube itself takes up very little space and can be handled easily. *Petri plates* are double glass dishes so named after their inventor, the German microbiologist R. J. Petri. Agar is poured into the lower half of the dish, which is then covered by the top half, and the agar is allowed to harden. Best results are obtained if the agar is cooled to just above the solidification point before pouring into plates, since if hot agar is poured, moisture condenses on the lid of the plate during cooling. The virtue of the petri plate is that a large surface is provided, and it is easy to reach any part of the surface with a needle or loop once the lid is removed. It should be noted that agar is not always an inert ingredient of culture media: a few bacteria can use it as an energy and carbon source.

Another way to provide a nutrient surface for microbial growth is to use a *membrane filter*. These filters are tough cellulose acetate discs, so manufactured that large numbers of tiny holes are present. The filter itself does not act as a nutrient but is laid on agar medium or on a cellulose pad saturated with liquid culture medium (Figure 3.5). The medium seeps through the holes, forming a thin layer on the surface of the filter, and the microorganism then grows in this thin film. Membrane filters remain solid at high temperatures (even up to the boiling point of water) and do not contribute any nutrients to the medium. Such filters are widely used in water-pollution studies, as will be described in Sections 4.2 and 14.3.

3.4
Environmental
requirements

In addition to the appropriate nutrient medium and the proper pH, several other environmental factors must be controlled in order to obtain growth of an organism. Of these, the most important are temperature and oxygen.

Organisms have a restricted range of temperatures within which they will grow, but the range differs greatly for different organisms. Within the temperature range from the *minimum* to the *maximum* is usually one temperature at which growth is best, the *optimum* temperature. This temperature is usually nearer the maximum than the minimum. Because of this, it is often safer to grow the organism at temperatures somewhat below the optimum, to ensure that any accidental increases in temperature do not lead to death of the organism.

Microorganisms are known that have their temperature optima as low as 5°C, and others are known with optima as high as 85°C. No one organism spans more than a small part of this temperature range, however. Organisms that cause disease in man have their optima near body temperature, 37°C. Organisms with low temperature optima are called *psychrophiles* (which means cold-loving), and organisms with high temperature optima are called *thermophiles* (heat-loving). Those with optima between room temperature (18 to 25°C) and body temperature (37°C) are called *mesophiles* (*meso* means middle).

To obtain the appropriate temperature for microbial growth, *incubators* or *water baths* are necessary. In an incubator, the culture vessels are placed in a temperature-regulated air chamber, whereas in a water bath they are immersed in water, which is heated or cooled. For low-temperature work, incubators or water baths must be refrigerated, and for work with mesophiles and thermophiles, heated incubators or water baths are used. It is important that the incubator be equipped with a *thermostat* so that the temperature is maintained relatively constant. Incubators suitable for most routine microbiological work may be purchased from various scientific and hospital supply companies. If funds are limited, an incubator can be constructed from simple components. Water baths can be purchased, or a simple water bath can be constructed by placing a thermostatically controlled immersion heater in a glass aquarium.

Most microorganisms require oxygen gas (O_2) for growth. These organisms are called *aerobes*, since the usual source of oxygen is the air, which contains about 20 percent O_2. Most eucaryotic microorganisms are aerobes, as are a large number of procaryotes. The energy-generating process requires O_2, as the energy source is *oxidized* by oxygen, with the release of energy. At the same time, the O_2 is converted to H_2O, a process called *reduction*. Oxidation and reduction are, thus, coupled processes in the generation of energy by living organisms. The process by which O_2 is used in energy generation is called *respiration* (see Chapter 6).

The simplest way to provide air for a culture is to use a closure for the culture vessel through which air can pass. As will be described in more detail later in this chapter, a closure must be so designed that foreign microorganisms that might enter the culture from the air cannot pass through. A cotton plug or metal closure allows free passage of air into the culture vessel; the oxygen diffuses into the medium. For many purposes, it is necessary to provide more extensive aeration. This is because O_2 is only poorly soluble in water, and as the O_2 is used up by the organisms during growth it is not replaced fast enough by diffusion from the air. Forced aeration of cultures is therefore frequently desirable and can be achieved either by vigorously shaking the flask or tube on a shaker or by bubbling sterilized air into the medium through a fine glass tube or porous glass disc. It is usually found that aerobes grow much better with forced aeration than when O_2 is provided by simple diffusion.

ANAEROBES

In some organisms, mainly bacteria, O_2 may not be needed in the energy-generating process. Organisms that do not need O_2 are called *anaerobes*. There are two kinds of anaerobes, obligate and facultative. *Obligate anaerobes* not only cannot use O_2 but are actually poisoned by it and must be cultured in the complete absence of oxygen. *Facultative anaerobes* can use O_2 when it is present but can use some other substance when O_2 is absent.

In anaerobes, another substance replaces O_2 in the energy-generating process. Inorganic compounds replacing O_2 are nitrate (NO_3^-), sulfate (SO_4^{2-}), and carbon dioxide (CO_2). All of these compounds contain oxygen, and it can be considered that the oxygen in these compounds replaces O_2. As with O_2, when these compounds are used in the energy-generating process they are reduced:

Oxidized		Reduced
Oxygen	$O_2 \rightarrow H_2O$	Water
Nitrate	$NO_3^- \rightarrow N_2$	Nitrogen
Sulfate	$SO_4^{2-} \rightarrow H_2S$	Hydrogen sulfide
Carbon dioxide	$CO_2 \rightarrow CH_4$	Methane

Note that in each of these cases, the reduced compound contains no oxygen or less oxygen than the oxidized compound. Those bacteria that use nitrate are called *denitrifying bacteria*, those that use sulfate are called *sulfate reducers*, and those that use CO_2 are called *methanogenic* (that is, methane-producing) *bacteria*. The denitrifying bacteria are facultative anaerobes, but the sulfate reducers and the methanogenic bacteria are obligate anaerobes. All these bacteria are of considerable importance in agriculture and in water pollution, as will be described in later chapters. When nitrate, sulfate, or carbon dioxide are used instead of O_2 in energy generation, the process is called *anaerobic respiration*.

In addition to the inorganic substances just discussed, some *organic compounds* can also serve as O_2 substitutes in the energy-generating process. In fact, the same organic compound can sometimes be used both as an energy source and as an O_2 substitute. A compound serving this dual role is said to be *fermented*, and the process by which it is acted upon is called *fermentation*, or *anaerobic fermentation*. A good example of a compound that is fermented by many organisms is the sugar glucose, $C_6H_{12}O_6$. When glucose is fermented, some of the molecules are oxidized to CO_2, and others are reduced to a compound that has less oxygen than glucose. The precise compound formed in this reduction depends on the organism, one common product being ethanol (ethyl alcohol).

When glucose is fermented to ethanol, the reaction is

$$\text{Glucose } (C_6H_{12}O_6) \rightarrow 2 \text{ ethanol } (C_2H_5OH) + 2CO_2$$

This fermentation is carried out by yeast and is a highly important process economically, as the alcohol product is a key constituent of beers, wines, whiskeys, and other spirits. The same fermentation occurs in bread making, but the important constituent is then the CO_2 which, as a gas, causes the rising of the bread dough (see Section 17.2).

Many organisms that ferment organic compounds can also use them aerobically. When yeast ferments glucose, the products are ethanol and CO_2, but when yeast uses glucose aerobically, only CO_2 is produced, the equation being

$$\text{Glucose } (C_6H_{12}O_6) + 6O_2 \rightarrow 6CO_2 + 6H_2O$$

Since in this reaction all of the glucose carbon is converted to CO_2, more energy is released. Yeast thus grows better aerobically on glucose than anaerobically.

One of the more difficult techniques in microbiology is the maintenance of anaerobic conditions: oxygen is ubiquitous in the air and air surrounds us, our culture vessels, our incubators, and other utensils. Since a number of important organisms are obligate anaerobes, it is necessary to have effective techniques for maintaining oxygen-free conditions. Obligate anaerobes vary in their sensitivity to oxygen, and a number of procedures are available for reducing the O_2 content of cultures—some simple and suitable mainly for less sensitive organisms, others more complex but necessary for the most fastidious obligate anaerobes.

Glass-stoppered bottles filled completely to the top with culture medium and provided with tightly fitting glass stoppers will provide anaerobic conditions for organisms not too sensitive to small amounts of oxygen. Instead of glass-stoppered bottles, screw-capped

MAINTENANCE OF
ANAEROBIC CONDITIONS

FIGURE 3.6
Special jar for incubating
cultures under anaerobic
conditions.

tubes or bottles filled to the top with medium can be used. Another way of excluding oxygen in tubes that cannot be filled to the top is to place a $\frac{1}{2}$-in. layer of sterile petroleum jelly on top of the liquid. In all of the methods, the oxygen dissolved in the medium should be driven off first by boiling it and inoculating and sealing as soon as it has cooled.

To completely remove all traces of O_2 for the culture of very fastidious anaerobes, it is possible to burn an O_2-consuming gas in a jar holding the tubes or plates. One of the most useful devices for this is the *anaerobic jar*, a heavy-walled jar with a gas-tight seal, within which tubes, plates, or other containers to be incubated are placed (Figure 3.6). The air in the jar is replaced with hydrogen gas (H_2) or illuminating gas, and the gas is burned by use of an electric heater or a chemical catalyst. The traces of O_2 left in the vessel or culture medium are consumed during the burning, thus leading to anaerobic conditions.

Another way of depleting a vessel of O_2 is by use of a solution of pyrogallic acid made alkaline by adding potassium hydroxide. Such a solution rapidly absorbs all traces of O_2 left in the system. A convenient procedure for eliminating O_2 in tubes by use of pyrogallic acid is to push a cotton plug down into the tube so that it is just above the culture medium, sprinkle powdered pyrogallic acid on top of the cotton, moisten the powder with 10 percent sodium hydroxide or potassium hydroxide, and quickly seal the tube with a gas-tight rubber stopper, inverting the tube to prevent mixing of the alkali with the agar. The oxygen is thus depleted within the tube, which can then be incubated as desired (Figure 3.7).

For the most fastidious anaerobes, it is necessary not only to carefully remove all traces of O_2 but also to carry out all manipula-

FIGURE 3.7
Simple methods for anaerobic culture of microbes. *a* The plate containing agar medium is streaked and sealed to the base; the dish is tipped to allow mixing of the chemicals, which results in the removal of O_2. *b* Slant is inoculated, cotton stopper is pushed down, pyrogallol crystals and NaOH are added, and then tube is closed *rapidly* and inverted immediately. Incubation is in inverted position. *c* Air-tight seal over liquid medium. *d* Broth contains a reductant, thioglycollate; only the top is aerobic.

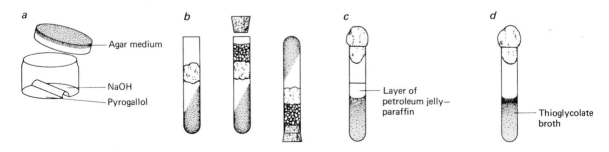

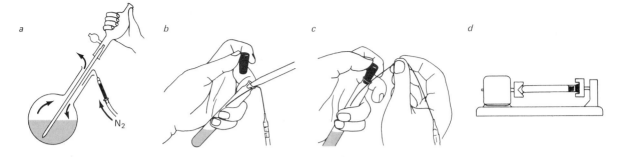

a   b   c   d

FIGURE 3.8

The Hungate technique for culturing fastidious anaerobes. *a* Medium is gassed with nitrogen (N_2) while medium is removed. *b* Tube is gassed first to remove all air, and medium is then added as shown and plug replaced immediately (*c*), leaving gassing needle in place until plug is to be pushed in. *d* In tubes receiving agar, roll tubes can be made, spinning tube as shown while agar solidifies. Inoculum is usually added to the liquid agar just before solidifying. *e* A typical roll tube, after incubation. Colonies can be picked as desired, using a gassing needle to maintain anaerobic conditions while the transfer is made.

e

tions of cultures in an anaerobic atmosphere, as the organisms may be killed by even brief exposure to O_2. To do this, a tiny jet of O_2-free hydrogen or nitrogen gas is directed into the culture vessel during manipulations, thus driving out any O_2 that might enter. This procedure, pioneered by the rumen microbiologist R. E. Hungate, is sometimes called the *Hungate technique* (Figure 3.8). By the use of the Hungate technique, a variety of anaerobic bacteria that had not been cultured before have been isolated from humans. Some of these anaerobes are pathogenic, and their study is currently of considerable interest.

The best seal for a culture vessel in order to prevent the entry of O_2 from the air is oxygen-impermeable rubber. It is important that the stopper be fitted very tightly into the vessel and that it not loosen during incubation.

In addition to maintenance of anaerobic conditions as described, it is usually advisable to add to the culture medium some kind of chemical *reducing agent*. Such a reducing agent helps to keep the O_2 content of the culture medium depleted. Suitable nontoxic reducing agents are sodium thioglycolate, cysteine, and ferrous sulfide. For anaerobic photosynthetic bacteria, hydrogen sulfide is sometimes used.

To be certain that anaerobic conditions are maintained, it is desirable to incorporate into the culture medium an oxygen-indicating dye such as methylene blue or resazurin. These dyes are colored when O_2 is present and colorless in the absence of O_2. The development of color in a culture medium containing one of these dyes will provide an instant warning that anaerobic conditions no longer exist.

3.5
Enrichment cultures

Organisms vary widely in their nutritional and environmental requirements. It is possible, by choosing the appropriate growth conditions, to select out of a mixture an organism of interest. The procedure of adjusting culture conditions to select a particular organism is called *enrichment culture* and is a procedure of considerable importance in microbiology.

The usual practice in enrichment culture procedures is to prepare a culture medium that will favor the growth of the organism of interest. The medium is then inoculated with a sample of material thought to contain the organism, such as soil, water, blood, or tissue. The appropriate conditions of aeration, temperature, and pH are provided and the inoculated medium then incubated. After an appropriate period of time, usually several days, the enrichment culture is examined visually and microscopically for evidence of microbial growth. Cultures in which growth has taken place are then used further for isolation of *pure cultures* (as will be described later in this chapter).

Some examples of enrichment cultures follow:

1 Photosynthetic organisms can be enriched by using nutrient salts media devoid of any organic or inorganic energy sources and incubating the cultures in the light. Inocula from rivers, lakes, the sea, or soil may be used.
2 Organisms using ammonium as energy source can be enriched by preparing medium with ammonium as sole energy source and incubating in the dark. Inocula can be from soil, sewage, or mud.
3 Organisms using specific organic compounds as energy sources can be enriched by preparing culture media in which the organic compound in question is the sole energy source. As an example, if we are interested in an organism that can utilize the herbicide 2,4-D (2,4-dichlorophenoxyacetic acid), we can prepare a culture medium in which 2,4-D is the sole carbon and energy source. Using appropriate enrichment cultures, microorganisms have been isolated that are capable of breaking down a tremendous variety of natural and man-made organic compounds.
4 If thermophilic organisms are desired, enrichment cultures can be prepared in which high temperatures are used for incubation.
5 If anaerobic organisms are desired, enrichment cultures devoid of O_2 can be prepared.

The examples are just a few of many that might have been given. The limitation on enrichment cultures is determined mainly by the imagination of the investigator.

3.6
Selective and differential media

Selective culture media are designed for the isolation of specific organisms. It has been found that some culture media can be prepared that allow the growth of single organisms or a group of

FIGURE 3.9
Differential culture media. A dye is added that changes color when acid is produced. An inverted vial is placed in each tube to indicate production of gas. Each of the three tubes contains a different sugar and all three are inoculated with the same organism: *a* no acid or gas; *b* acid, no gas; *c* both acid and gas formed.

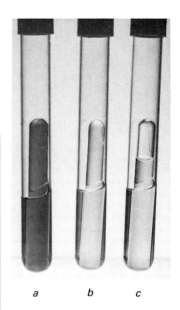

a   b   c

closely related organisms and prevent the growth of virtually all others. Such selective media are quite useful in showing the presence of the organism of interest. For instance, in water-pollution studies, the presence of the intestinal pathogenic bacteria *Salmonella* and *Shigella* is of great interest. Several selective media have been developed that, when inoculated with a water sample and incubated at the proper temperature, permit growth of only these two organisms. Selective media usually contain one or more substances inhibitory to the growth of organisms except those to be selected.

A *differential medium* is one that allows several or many organisms to grow but contains dyes or other ingredients upon which different organisms act in various ways to produce variations in color (Figure 3.9). Differential media are widely used in the identification of unknown microorganisms in medical, food, dairy, and other microbiological laboratories.

Various commercial suppliers make selective and differential culture media of a variety of types and usually provide directions for their use as well. Catalogs can be obtained from such firms listing these media and explaining the principles of their use.

In nature, microorganisms almost always live in mixtures. However, before most properties and characteristics of a particular microorganism can be determined, the organism must first be isolated in pure culture. The isolation and maintenance of pure cultures is one of the most important procedures in microbiology, and one with which all students must be familiar.

3.7
Isolation and maintenance of pure cultures

PURIFICATION

The easiest and most useful procedure for isolating pure cultures is *agar streaking*. This procedure, first devised by Robert Koch in the 1880s, involves preparation of petri plates containing a suitable culture medium solidified with agar. A sterile inoculating loop is placed in a mixed culture containing the organism of interest and then lightly streaked across the surface of the agar plate. The pattern of streaking is important, and there are several suitable variations. As the plate is streaked, organisms are gradually dislodged from the loop, and in the final parts of the streak single organisms that are well separated from each other will be deposited.

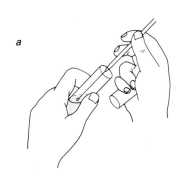

a

b

c

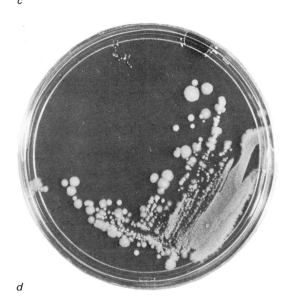

d

FIGURE 3.10
Method of making a streak plate to obtain pure cultures. *a* Loopful of inoculum is removed from tube. *b, c* Streak is made carefully to spread out organisms. *d* Appearance of streaked plate after incubation. In the portion of the plate streaked first, confluent growth and closely packed colonies occur; whereas in the parts streaked later, well-isolated colonies develop. It is from these isolated colonies that pure cultures usually can be obtained.

The streaked plate is then incubated so that the organisms will multiply and produce colonies (Figure 3.10). In the initial parts of the streak these colonies will be very close together, but in the final part well-isolated colonies should be obtained. It is assumed that a colony well isolated from all other colonies will have arisen from a single cell. One of the well-isolated colonies is then streaked on a fresh agar plate, which is incubated. If all of the colonies obtained are of similar size, shape, color, and texture, it is presumed that they are all alike and that a pure culture has been obtained. One of the colonies from the second streaking is then used to establish a stock culture of the organism. A culture that has been derived from a single cell is by definition a pure culture. (The maintenance of stock cultures will be described in a subsequent section.)

A variant of the above procedure is to prepare *pour plates*, in which a diluted inoculum is mixed with the melted agar in the plate before it cools and hardens. When the inoculated poured plates are incubated, isolated colonies should be obtained from which pure cultures can be prepared.

Another variation is the use of membrane filters as the solid support instead of agar. A dilution of the inoculum can be passed through the filter, and the filter is then placed on an appropriate culture medium for incubation. Isolated colonies developing on the filter can then be picked to prepare pure cultures. Other procedures are sometimes used in advanced research work for organisms that will not form colonies on solid surfaces, but for virtually all routine work agar streaking serves quite satisfactorily for isolating pure cultures.

It is important to verify that cultures isolated as pure are indeed so. A check of purity begins with careful microscopic examination to ensure that only one cell type is present. However, two organisms may look alike under the microscope but still be quite different. A second check is to ensure that all colonies obtained upon agar streaking are alike. As a final check, one can select several colonies from an agar streak and determine their nutritional and environmental requirements, which should be identical in pure cultures.

VERIFICATION OF PURITY

Once a pure culture is obtained, it must be kept pure. One of the most frequent ways in which erroneous results and conclusions are obtained in microbiology is by use of faulty or contaminated cultures. Cultures of organisms of interest that are maintained in the laboratory for study and reference are called *stock cultures*. The stock culture must be maintained so that it is free from contamination, retains viability, and remains true to type.

The most common way of maintaining stock cultures is on *agar slants*. The slants are inoculated and incubated until reasonable growth is obtained, and then they are usually placed in a refrigerator, care being taken that the culture tubes are well sealed so that they will not dry out. With hardy organisms such as spore-forming bacteria, Gram-positive cocci, yeasts, and fungi, agar slant cultures maintained in this way may retain viability for weeks or months. It is the usual practice to transfer cultures to fresh slants at intervals and to discard the old slants, although an old slant should never be discarded until it has been ascertained that good growth of the right organism has occurred on the fresh transfer.

With less hardy organisms, more frequent transfers may be necessary, perhaps weekly or even daily. The only way to determine how to maintain stock cultures is by experience. It is always advisable

STOCK CULTURES

in the early phases of a study to make transfers at frequent intervals, so that cultures of interest are not lost.

For some organisms, *agar stabs* are better for preserving viability than agar slants. The tube for an agar stab, called an *agar deep*, is allowed to stand vertically while the agar is solidifying rather than at an angle as for slants. Inoculation is done with a straight needle which, after being touched to the culture, is stabbed into the agar all the way to the bottom. During incubation, the culture grows in a thin zone along the line of the stab. The advantage of this procedure is that waste products such as acids that are produced by the organism spread out into the agar from the zone of growth and are diluted. Accumulation of waste products is one of the main reasons for the dying off of many stock cultures. Although less total growth is obtained in a stab than on a slant, enough growth occurs so that new cultures can be initiated when needed.

It is usually inadvisable to maintain stock cultures on agar plates; they become contaminated relatively easily, since so much surface is exposed to the air.

If agar cannot be used, stock cultures may be maintained in liquid medium. A disadvantage of liquid medium is that detection of contamination is less easy, since the contaminant mixes with the culture rather than forming discrete colonies as it would on agar. For this reason, frequent microscopic examination of liquid stock cultures is necessary to be sure that they are pure.

CULTURE PRESERVATION

For long-term storage of stock cultures, several procedures are available, including freezing, freeze-drying, and storage under oil. In these procedures, the activity of the culture is reduced to a very low level, and the organism hence remains viable for long periods of time. Not all organisms can be so treated, but in those that can, such procedures offer the best means of culture preservation.

With hardy organisms, *freezing* can be done directly in the culture medium, but for most organisms it is necessary to add a protective agent before freezing. Suitable protective agents are glycerol, sucrose, skim milk, and dimethylsulfoxide. For most microorganisms, viability is best preserved by freezing rapidly. To achieve this, the culture suspended in the protective agent is placed in small tubes, which are sealed and plunged rapidly into an acetone–dry ice bath (about $-70°C$). (Dry ice is frozen carbon dioxide.) Even better preservation is obtained if liquid nitrogen is used (about $-200°C$), although this is more expensive than dry ice. If neither dry ice nor liquid nitrogen is available, freezing can be done in an ordinary refrigerator freezer, a beaker of acetone being placed in the freezing compartment a few hours before use. Acetone acts as a heat transfer agent. It will remain liquid at temperatures below freezing and, because of the fact that heat transfer is far better in liquid than in air, will cool the cultures very rapidly. Conversely, some organisms survive freezing better if they are cooled very slowly rather than very rapidly. This is especially true of large organisms and those with fragile cell envelopes. Slow freezing can be done most simply by placing the tubes to be frozen in a freezing compartment without any heat transfer agent such as acetone. For very sensitive organisms, a programmed freezing device can be obtained in which the temperature of the compartment is slowly reduced to very low temperatures.

For storage of frozen cultures, an ordinary refrigerator freezer operating at about $-20°C$ can be used, but even better is a special freezer operating at about $-70°C$. Such an electrically operated freezer should be equipped with a battery-operated alarm system so that if the temperature rises because of power failure or some other cause, steps can be taken to transfer the frozen cultures to another chamber. If a steady supply of dry ice is available cheaply, as in many urban areas, it is possible instead to use a heavily insulated box containing dry ice in order to maintain the temperature at about $-70°C$. Arrangements must be made to have the dry ice replenished two or three times a week.

In reviving a frozen culture, it is preferable to thaw the tube rapidly by placing it in a beaker of water at 25 to 30°C. Then the tube can be opened and the contents transferred to a suitable culture medium.

Freeze-drying, also called *lyophilization*, involves removal of the water from the culture under high vacuum after the culture has been frozen at a low temperature, usually about $-70°C$. With the high vacuum used, the water evaporates from the culture in the gaseous state (sublimation) without ever melting. Thus the culture is maintained in the frozen state throughout the whole drying process. Once the culture is dried, it is sealed while still evacuated and then stored at room temperature. The advantage of freeze-drying is that the culture can be stored simply in the laboratory for a long period of time without the necessity of constant maintenance of low temperature. Not all organisms survive freeze-drying, but a surprisingly large number do. Most bacteria and fungi can be freeze-dried, as can a few algae. Protozoa in general do not survive freeze-drying, probably because they lack the protection of a cell wall.

Some protective agent is usually necessary for successful freeze-drying of a culture. The most useful protective agents are skim milk, blood serum, and sucrose.

Freeze-dried cultures should be rehydrated by adding a small amount of appropriate culture medium to the dried material and making a series of several tenfold dilutions in the same culture medium. All dilutions and the original suspension should be incubated. If no growth is obtained in any of the dilutions after an appropriate incubation time, transfers should be made from the original suspension into fresh medium and further incubations done. Freeze-drying virtually always results in killing of many of the cells in the population. However, when a large number of cells are freeze-dried, sufficient cells will survive to initiate new cultures even if 99 percent of them are killed. Death commonly occurs during the initial freezing and drying process, but once drying has been completed, death during long-term storage is usually minimal.

Storage under oil is suitable for fungus cultures. The agar slant culture is covered with a layer of sterile mineral oil. This prevents drying and reduces metabolism by preventing entry of oxygen. This procedure is inexpensive and simple and is suitable for many of the more hardy fungi. It is not often used for other microorganisms.

CULTURE COLLECTIONS

A culture collection either can be a specialized collection of a restricted group of organisms or it can be a generalized collection that attempts to maintain a wide range of organisms of interest. The purpose of a culture collection is to maintain cultures in a viable, contaminant-free state and to have them available for immediate shipment at nominal cost to research workers, teachers, students, or industry. The individual worker thus need not maintain cultures that he uses rarely.

The maintenance of a large and diverse culture collection is not simple, since different organisms must be stored in different ways. A high degree of skill and knowledge is necessary. The price charged for a stock culture may often seem high, but it should be noted that the purchaser is paying for the immediate availability of a contaminant-free culture (Figure 3.11).

It must be emphasized, however, that the investigator should not accept blindly the reliability and purity of a stock culture received from a culture collection. Immediately upon receipt, the culture should be carefully checked, and if it does not meet specifications, it should be returned. Since culture collections often maintain thousands of different cultures, it is only natural that errors may occur, leading to the distribution of faulty cultures. A number of countries have good collections from which cultures of bacteria, fungi, algae, protozoa, and viruses may be obtained. Catalogs are available from most of the culture collections.

Organisms are listed in culture collections by species name, but they are also often given numerical designations. Not all pure cultures of the same species are identical in all ways. Each isolate of the same species is usually designated as a *strain*, and one strain is distinguished from another by use of arbitrary numbers or letters. The culture number used by the culture collection is a unique number referring to a single strain of the species. Since strains of the same species may differ in important ways, it is necessary to specify

FIGURE 3.11
Storage of stock cultures in the frozen state in liquid nitrogen. *Courtesy of the American Type Culture Collection.*

the exact strain number when referring to work carried out with a specific stock culture.

As we have noted, pure cultures of microorganisms are essential for microbiological study. Only when one organism is cultured in the complete absence of other organisms is it possible to determine the unique properties of the organism under study. This is of maximum importance in studying organisms that cause disease. The availability of pure cultures of disease-causing organisms has made it possible to develop procedures for prevention and treatment of diseases. However, pure cultures do not exist in nature but must be created in the laboratory. This requires the use of precise methods for sterilization of culture media, culture vessels, and everything with which the culture comes in contact. We generally speak of this sterilization procedure as *aseptic technique*, and the student will be able to carry out a successful experiment in microbiology only after having mastered this technique.

STERILIZATION OF CULTURE MEDIA

The most common way of sterilizing culture media is by heat. However, since heat will also cause harmful changes in many of the ingredients of culture media, it is desirable to keep the heating time as short as possible. Although most microorganisms are quickly killed by temperatures near boiling, bacterial endospores are very resistant to such heat and may survive hours of boiling. Since such spores are ubiquitous, being common in soil, dust, and air, all heat sterilization procedures are so designed that the destruction of any contaminating spores is ensured. The best procedure is to heat under pressure, since at pressures above atmospheric the temperature can be increased above 100°C, thus decreasing the time necessary to make the medium sterile. Devices for heating under pressure are called *autoclaves* or *pressure cookers*. The usual procedure is to heat at a pressure of 15 lb/sq in. (1.1 kg/sq cm), which yields a temperature of 121°C, for 15 to 20 minutes.

Most microbiology laboratories have one or more autoclaves (Figures 3.12 and 3.13). The autoclave may be fed by steam from the building's steam line, or steam may be generated within the autoclave by boiling water. If the latter is done, the water may be heated by either gas or electricity. The steps in operation of an autoclave are as follows: (1) Place material to be sterilized in the chamber and shut the door, making sure the door is sealed tightly so that the steam does not subsequently leak out. (2) Make sure that the vent to the outside is open. (3) Turn on the steam, and allow the chamber to fill. The steam will drive the air out of the chamber to exit through the vent. (4) When steam is flowing freely out of the

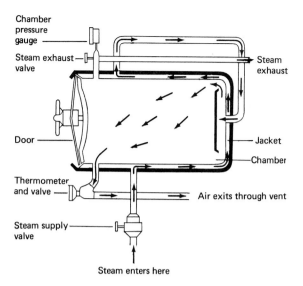

Chamber pressure gauge

Steam exhaust valve

Steam exhaust

Door

Jacket

Chamber

Thermometer and valve

Air exits through vent

Steam supply valve

Steam enters here

FIGURE 3.12
Schematic diagram of an autoclave.

FIGURE 3.13
An installation of autoclaves in a typical microbiology laboratory. The chamber of the autoclave and all the pipes are built into the wall so that the heat produced is carried away from the room.

vent, close it and allow the pressure in the chamber to build up. Many autoclaves have a built-in device for controlling the pressure at 15 lb/sq in. If not, adjust the steam supply so that the proper pressure is maintained within. (All autoclaves have safety valves that release if pressure rises much above 15 lb/sq in., but it is undesirable to use these valves to control pressure.) (5) Notice the temperature reading on the thermometer provided on the autoclave. Since it is not pressure but heat that sterilizes, it is important to be sure that the temperature in the chamber has reached 121°C. (6) Time the sterilization, allowing at least 15 to 20 minutes at 121°C. If large volumes of medium are being sterilized, longer sterilization times should be used, since the interiors of such containers will not reach sterilization temperature as rapidly as will the outside. (7) At the end of the sterilization time, turn off the steam supply, and allow the pressure in the autoclave to reduce *slowly* to atmospheric pressure. Do not try to speed up the process by opening the vent, because a rapid decrease in pressure will cause any liquids, such as culture media, to boil over. (8) As soon as the pressure has reached atmospheric, open the vent, then slowly open the door, being careful to avoid being burned by any remaining steam that may rush out of the chamber. (9) Remove the sterile items, preferably using tongs or a pair of asbestos gloves to avoid being burned.

Many modern autoclaves operate automatically and will go through a complete sterilization cycle once they are programmed. Such autoclaves allow for pressure buildup, time the period at sterilization temperature, and turn off and allow the pressure to reduce to atmospheric level. When the cycle is complete, a buzzer rings, and a light goes on.

In small laboratories, or if only a limited budget is available, a large pressure cooker, such as is used for home canning of foods, provides a very inexpensive and perfectly satisfactory autoclave (Figure 3.14). The pressure cooker works on exactly the same principle as described above. A small amount of water is placed in the bottom of the pressure cooker, the material to be sterilized is placed on a rack just above the water, and the lid is closed. The pressure cooker is heated on a gas or electric burner, and the water in the bottom is converted to steam. When steam issues freely through the vent, the vent is closed and the pressure allowed to build up to 15 lb/sq in. The pressure in the cooker must be regulated manually by adjusting the degree of heat. The main limitation of the home-style pressure cooker is that it does not have a large capacity. However, for most routine microbiological work, it is completely satisfactory and is easier and quicker to use than an autoclave. Its cost is only a small fraction of the cost of even a modest-sized autoclave.

FIGURE 3.14
Pressure cooker suitable for the sterilization of small items. *Courtesy of National Presto Industries, Inc., Eau Claire, Wisconsin.*

The autoclave or pressure cooker can also be used for sterilization of glassware, pipettes, surgical instruments, and other items. It is operated normally except that pressure need not be reduced slowly, since if no liquid is being sterilized, there is no danger of boiling over. Quick reduction in pressure as used for these items is called *fast exhaust*. There will usually be some moisture condensing within the items to be sterilized. Some autoclaves have a special drying cycle following fast exhaust to eliminate this residual moisture. However, it is often easier to sterilize glassware by dry heat, as described below.

Some items cannot be sterilized under pressure because of their sensitivity to the heat involved but can be sterilized by steaming, since that temperature is only 100°C. However, since the temperature is lower, the time required for sterilization by steaming is much longer than that required with an autoclave. To reduce the steaming time, an alternate procedure is used, called *fractional steam sterilization* (or *tyndallization* after John Tyndall, the scientist who developed it in the 1870s). The material to be sterilized is steamed for 30 to 60 minutes on each of 3 successive days. At the first heating, most microbes are killed, although bacterial endospores usually are not. During the intervening 24 hours, the spores present germinate, and on the second day of steaming, the vegetative cells derived from these spores are killed. The third steaming cycle is merely a precaution to make certain that the material is sterile. One of the most common ingredients sterilized by fractional steaming is elemental sulfur, used as an energy source by many sulfur bacteria. Elemental sulfur melts in the autoclave at 121°C but remains crystalline at the steaming temperature of 99 to 100°C.

DRY-HEAT STERILIZATION

The most convenient way of sterilizing objects such as empty test tubes, petri dishes, pipettes, hypodermic needles, or instruments is by use of dry heat. An ordinary kitchen oven is quite satisfactory, although most microbiology laboratories use a special larger oven with an automatic timing device. At least 90 minutes heating at 160 to 170°C is necessary for sterilization, although when bulky items or large loads are being sterilized, more time may be necessary. Since many items to be sterilized are wrapped in paper or closed with cotton, temperatures over 180°C must be avoided, as this will cause charring or burning.

GAS STERILIZATION

Objects that are sensitive to heat or that should not become damp, as happens in the autoclave, can often be sterilized by use of a gas. The most commonly used gas is ethylene oxide, although other gases such as methyl bromide, propylene oxide, or ozone may also be used. The items to be sterilized are placed in a gas-tight chamber, and the gas is admitted for the prescribed time. The chamber is then exhausted and the items aired for 8 to 10 days before use, since the gases are toxic to man as well as microbe. This method is finding increasing use in hospitals for the sterilization of plastics and other heat-sensitive materials (see Section 9.10). It is also used to some extent in the sterilization of certain foods, such as some grains, spices, nuts, and dried fruits, and for sterilizing soils.

INCINERATION

Microorganisms are easily destroyed by burning, a process called incineration. This is a quick and sure way of sterilizing an object but has limited application, since many materials are destroyed by such high temperatures. Incineration is used frequently in the laboratory, to sterilize the platinum and nichrome loops and needles used in transferring cultures. The contaminated instrument is heated in an open flame until it is red hot. Although tempered metal instruments such as forceps can also be sterilized in this manner, it is usually not desirable, as the heating discolors the metal and affects its temper. Incineration by burning in a furnace is used in hospitals and laboratories for the disposal of items that may have become contaminated with pathogenic organisms; such things as sputum cups, disposable paper sheets, and surgical dressings, as well as infected laboratory animals, are generally disposed of by incineration.

FILTRATION

An especially valuable technique for sterilizing aqueous solutions containing heat-sensitive materials is filtration. The filter used must have holes too small for the passage of contaminating microorganisms while still allowing the passage of the liquid (Figure 3.15). Common filter materials that have been used are asbestos (as in the commercially available Seitz filter), fritted glass (of a pore size labeled *UF*, ultrafilter), porcelain (as in the Chamberland filter), and diatomaceous earth (as in the Berkefeld filter). However, in recent years, most of these filters have been replaced by the *membrane filter*, which is a thin disc of cellulose acetate or cellulose nitrate. Membrane filters are available from a variety of commercial suppliers in various pore sizes and filter diameters. These were mentioned earlier (Section 3.3) for growing cells; in the former case, the cells on the filter were of interest; here, it is the cell-free filtrate that is collected.

Before sterilizing with a membrane filter, the filter apparatus and the receiving vacuum flask must be sterilized by autoclaving. The outlet of the vacuum flask must be closed with a cotton plug. After autoclaving, the filter apparatus is unwrapped, a sterile membrane filter disc is placed on the fritted glass platform of the base with a sterile forceps, and the top funnel part of the apparatus is quickly replaced. Since the funnel is exposed to the air, it of course soon becomes contaminated, but the portion of the filter apparatus below the filter and the flask itself remain sterile throughout. The liquid to be sterilized is poured into the top of the filter apparatus, and the flask outlet is connected to a vacuum supply. The liquid is pulled through the filter into the flask (Figure 3.16).

The volume of liquid that can be filtered through a single filter before clogging will occur depends on the pore size, on the type and size of filter, and on the initial clarity of the liquid. With typical clear culture media and standard membrane filters, a liter of medium can easily be filtered without clogging.

Once the liquid has been filtered, it is transferred from the

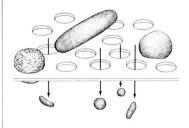

FIGURE 3.15
Diagram of a membrane filter; such filters retain cells and particles larger than the pore diameters.

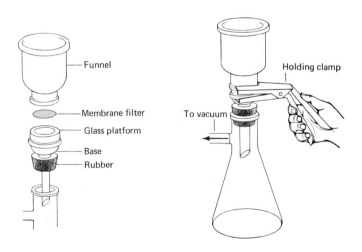

FIGURE 3.16
Diagram of membrane filter apparatus used for sterilization of culture media by filtration.

filter flask to sterile tubes, bottles, or flasks for storage and use. This transfer must of course be done aseptically to avoid any contamination of the now sterile liquid. It should be noted that filtration only removes all organisms larger than the pore size of the filter. With most filters, viruses and very tiny bacteria such as mycoplasmas easily pass through the filter; so sterilization by filtration is not so certain as sterilization by heat.

Any container in which the contents are to be kept sterile must have an effective closure. Not only must the closure be effective in maintaining sterility, it must also be easy to handle so that it can be removed and put on again quickly during manipulations. In culture vessels where aeration is needed, the closure must also permit passage of air.

The traditional closure for a culture vessel is the *cotton plug*, made by twisting a piece of cotton and forcing it into the neck of the tube or flask. Air can pass through the plug, but organisms cannot. Cotton molds to the shape of the vessel, permits reasonable passage of air, can be autoclaved, can be fairly easily manipulated, and is very inexpensive. It has the disadvantages that it does not always prevent passage of contaminants down the wall of the vessel; it frays, and fibers of cotton may fall into the medium; vitamins and other organic materials present even in purified cotton may leach into the culture medium and affect results of nutritional experiments; it becomes wet if used in water baths; it can easily become dislodged from the vessel, especially if vigorous aeration on a shaker is carried out; and plugs are tedious and time consuming to prepare.

A variety of alternatives to the cotton plug exist (Figure 3.17). If aeration is not a problem, screw-capped tubes, bottles, and flasks

FIGURE 3.17
Several kinds of closures used to keep tubes sterile. Left to right, these are cotton; plastic foam; screw cap; metal cap.

may be used. The screw cap should have a white rubber or Teflon liner rather than paper since the paper liners do not hold up well when wet (as with culture media).

Stainless steel caps are an excellent substitute for cotton for use with culture tubes. Plastic caps are also available and are probably the cheapest substitute for cotton, although they may not have so long a lifetime as stainless steel. These caps are very fast and easy to use, since they merely slip over the top of the tube. Air enters through the space between the cap and the wall of the tube; contaminants are unable to follow. The plastic foam plug is another substitute for cotton. This is a premolded plug that functions like a cotton plug, but avoids some of the disadvantages of cotton. Air moves down along the porelike channels of the foam, and the contaminants are filtered out. These plugs are easy to manipulate and can be used several times before wearing out.

Any manipulations of cultures, media, or sterile materials must be done in such a way as to avoid contamination. Airborne contaminants are the most common problem, since the air always has a population of microorganisms. When containers are opened, they must be handled in such a way that contaminant-laden air does not enter. This is best done by keeping the containers at an angle so that most of the opening is not exposed to the air vertically. Operations should be carried out in a room in which air currents are absent. A small room is best because it is likely to have less air movement, and all windows should be closed. The room should be kept clean of dust, since dust particles are laden with contaminants. In many laboratories, a special small room called a transfer room is used. Such a room can be fitted with germicidal ultraviolet lights that are left on when the room is not in use and turned off, to avoid eye or skin damage, when work is being done.

When transferring a culture with a loop or needle from one tube to another, the two tubes are held together side by side in one hand and the loop or needle holder in the other. First, the loop or needle is flamed to a red heat and allowed to cool briefly in the air. Then, with the fourth and fifth fingers, the closure is removed from the tube containing the culture and the loop or needle carefully inserted into the culture. The loop or needle is removed, the closure replaced, and the tube to receive the inoculum is opened. After inoculating, the tube is closed and the loop or needle reflamed before the holder is laid down. This last procedure is done so that organisms present on the loop or needle are not spread around the laboratory. In a minor variation of the foregoing procedure, the rims of the tubes are *flamed* by holding them in the burner flame for 1 or 2 seconds when the closures are removed, in order to kill any organisms on the rims of the tubes. If cotton plugs are used such flaming is usually desirable, since the outside of the tube is generally exposed to the air and will be contaminated; but if a metal or plastic cap is used that covers the outside of the tube, flaming is probably not necessary and may be detrimental, since during the time it is done the open tube has that much more chance of becoming contaminated.

When flaming the loop or needle, it is desirable to heat it slowly in order to avoid spattering; spattered liquid may contain living organisms, and it is of course highly undesirable to disperse them throughout the laboratory.

Transfers of liquid cultures are often made with sterile *pipettes*. Pipettes are usually sterilized by dry heat in metal cans that hold several pipettes, although they can also be individually wrapped and sterilized. If pipettes are sterilized in the autoclave, steam will inevitably condense inside, and the pipettes should be dried in an oven before use. For handling nonpathogenic cultures, the pipettes need not have cotton plugs in the end, but with pathogens, cotton plugs should always be used to avoid accidental sucking of cultures into the mouth. With the most virulent pathogens, the mouth should not be used at all; a special remote pipetting bulb should be used instead, of which several are marketed commercially.

To transfer a culture, the tubes are held as shown in Figure 3.18; a pipette is removed from a can or is unwrapped and held carefully in the air so that none of it touches anything other than the culture. After the culture is transferred and the tubes closed, the pipette should be discarded. If work with pathogens is involved, the pipette must be discarded directly into a jar or tray of antiseptic

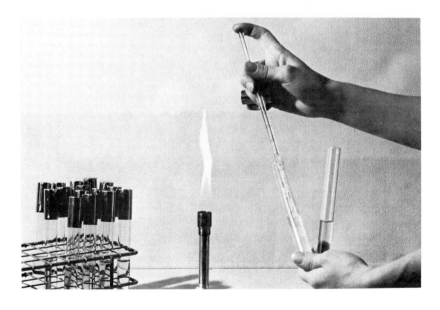

FIGURE 3.18
Aspetic transfer of a liquid culture using a sterile pipette.

solution, such as phenol or mercuric dichloride. With nonpathogens it is usually sufficient to discard the pipette into a jar of soapy water. Once the pipette has been used with a culture, care must be taken that the tip does not touch the laboratory bench or anything else, as this will result in transferring organisms to the object touched.

Working with agar in petri plates requires special precautions because when the lid of the plate is removed, the surface of the agar is exposed to the air and contamination is very easy. When inoculating a plate, the lid is raised on one side only and just high enough to allow the loop to enter. After streaking, the plates are incubated upside down so that the moisture that condenses on the surface of the lid will not drip onto the agar.

During manipulation of cultures, care must be taken to avoid the production of aerosols, since droplets of culture medium laden with microbes can be a serious source of contamination when carried into the air. This is especially important if pathogenic organisms are being studied.

Aerosols are created when containers with liquid medium are shaken, when pipettes are used, or if spattering occurs during flaming. They can be prevented by taking care not to make rapid movements, by avoiding rapid shaking, and by flaming carefully to avoid spattering. There are several commercial burners available that are designed to prevent spattering during flaming, the loop or needle being inserted into a tube at the base of which is the flame. An effective way of evaluating aerosol production is to carry out manipulations with an organism such as *Serratia marcescens*, which has colonies easily detectable because they are bright red. Agar plates with the lids off are placed at various locations around the working area, and after work is completed the plates are incubated. If red colonies appear, this is an indication that the organism was dispersed in aerosols and hence that faulty techniques were employed.

When working with pathogenic microorganisms, it is not sufficient merely to keep cultures uncontaminated. One must be constantly on the watch to avoid infecting oneself or others in the laboratory. Prevention of aerosols, just described, is only one precaution. For work with the most virulent pathogens, special bacteriological transfer hoods are available, inside of which all operations are carried out, thus confining the pathogen to a single location (Figure 3.19). The transfer hood is disinfected before and after use by means of a gaseous disinfectant or a germicidal lamp.

Naturally, all utensils, pipettes, glassware, and other items that have come in contact with the pathogen must be sterilized before

AEROSOLS AND THEIR PREVENTION

SPECIAL PRECAUTIONS WITH PATHOGENS

FIGURE 3.19
Special biohazard transfer hood
used for working with especially
virulent pathogens and other
dangerous materials. *Courtesy
of Contamination Control, Inc.,
Kulpsville, Pennsylvania.*

FIGURE 3.20
Special container to be used for
the shipping of pathogenic
cultures (approved by The
Center for Disease Control,
Atlanta, Georgia).

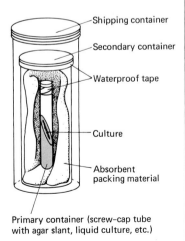

Shipping container

Secondary container

Waterproof tape

Culture

Absorbent
packing material

Primary container (screw-cap tube
with agar slant, liquid culture, etc.)

being discarded. Special disinfecting jars for pipettes and small
instruments are desirable. Pipettes should never be manipulated by
mouth but by remote control with a pipetting bulb. All cultures
containing the pathogen must be clearly marked, and separate
incubators should be used rather than general-use incubators.

When a large amount of work is to be done with pathogens,
special laboratories are often built in which the air flow can be
controlled. The outside of such a laboratory should be at a higher
air pressure than the inside so that air flows into the laboratory, not
out. The air-exit system should contain an incinerating device so that
any pathogens leaving the laboratory in the exhaust air are killed.

Most important of all, work with pathogens should be done
only by someone with proper training who is already skilled in their
study. Improper handling of pathogens by a laboratory worker can
endanger that individual, coworkers, and innocent bystanders. Fortunately, most microorganisms are nonpathogenic; so the student
can learn the proper procedures of aseptic technique in complete
safety.

It is important that the shipping and transfer of pathogenic
organisms be carried out properly. The Center for Disease Control,
Atlanta, Georgia, has recently approved the use of special containers
for shipping dangerous materials (Figure 3.20). Pathogenic cultures
or other harmful agents are to be placed inside a primary container

that is wrapped in absorbent material. If breakage occurs, any liquid will be absorbed and will not leak out. The primary container and absorbent packing are then placed inside a secondary container that is in turn put inside an outer shipping container. Appropriate warning labels must be used, and receipt of delivery should be acknowledged. The people who handle pathogenic materials in the mails cannot be expected to know their danger and must be protected.

Conclusion

In this chapter, we have seen that it is possible to culture microorganisms under controlled conditions in the laboratory, thus making possible a study of their properties and activities. All our advances in the science of microbiology have arisen in some way from laboratory study of microorganisms, and a knowledge of appropriate procedures and methods is an essential part of the skills of the microbiologist. Although laboratory culture requires skill and careful attention to detail, it is not an especially difficult undertaking. Culturing microorganisms is actually fun, since results are obtained quickly and successes are easily realized. It is considered an enjoyable challenge to attempt to isolate into culture a new microorganism. Occasionally this is a difficult undertaking, but often it is not. We shall see in subsequent chapters how widespread and useful a knowledge of the care and feeding of microbes can be.

Study questions

1 Name three types of energy source; give an example of each.
2 Distinguish between the terms *autotrophic* and *heterotrophic*.
3 A handy mnemonic device to remember the elemental requirements for growth is the following: *C HOPKiNS CaFe Mg* (read this "C. Hopkins Cafe, mighty good"). The order of importance is not quite accurate, and the *i* is just to make it sound better. Using this aid, list the elements necessary for growth, and indicate how they are obtained by the cell.
4 What are growth factors? Name at least three.
5 Contrast synthetic and complex media.
6 Why is agar so useful as a gelling agent in culture media?
7 Explain why the following are important in the culturing of microorganisms and how each is controlled: pH, temperature, oxygen.
8 What factors might account for the different heat sensitivities of various organisms?
9 Define *aerobe*; *anaerobe*. Distinguish between obligate and facultative anaerobes.
10 Define *anaerobic respiration*. List the substances that can replace O_2 in this process, and state the products formed from them.
11 What are the products formed when glucose is fermented by yeast under anaerobic conditions? State briefly the economic importance of each product.
12 What is an enrichment culture? Give two examples.
13 What is a pure culture? Describe briefly how a pure culture can be obtained and its purity verified.

14 What is a stock culture? a strain?
15 What is an autoclave? What is it used for in a microbiology laboratory?
16 List and briefly explain four different methods for sterilization of items used in a microbiology laboratory.
17 What do we mean by aseptic technique?

Suggested readings

Brock, T. D., *Biology of microorganisms*, 2nd ed., Prentice-Hall, Inc., Englewood Cliffs, N.J., 1974. Current textbook giving details of the nutrition and environmental requirements of microorganisms.

Olds, R. J., *Colour atlas of microbiology*, Wolfe Medical Books, London, 1975. Excellent photographs of bacterial and fungal colonies and stained cells, and of differential tests for identifying organisms.

Seeley, H. W., Jr., and P. J. Vandemark, *Microbes in action : a laboratory manual of microbiology*, 2nd ed., W. H. Freeman and Co., San Francisco, 1972. Good explanations of work with microorganisms in the laboratory.

Sistrom, W. R., *Microbial life*, 2nd ed., Holt, Rinehart and Winston, New York, 1969. Elementary discussion of microbial nutrition and biochemistry.

Stanier, R. Y., E. A. Adelberg, and J. L. Ingraham, *The microbial world*, 4th ed., Prentice-Hall, Inc., Englewood Cliffs, N.J., 1976. Extensive discussion of metabolism, nutrition, and biochemistry of microorganisms.

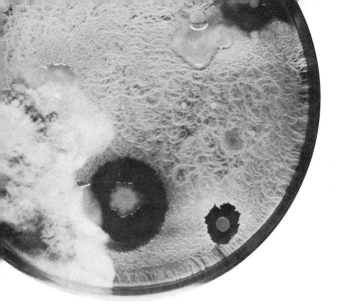

Chapter

4

Microbial growth and its control

A knowledge of the processes involved in the growth of microorganisms is essential if we are to predict their activities and control them for our own benefit. Microbiologists often measure microbial growth, either by counting numbers of cells or by measuring microbial mass. A characteristic feature of microbial growth is the rapidity with which the population increases; very large numbers are reached quickly, often with harmful consequences. In this chapter, we discuss the measurement of microbial growth and examine some of the chemical and physical agents that are used to control it. Among the latter, we shall discuss sterilization procedures, antiseptics and disinfectants, and antibiotics and other chemotherapeutic agents.

Growth is defined as an increase in the number of microbial cells or an increase in microbial mass. *Growth rate* is expressed as the amount of growth per unit time.

    In unicellular microorganisms, growth usually involves an increase in cell number. A single unicell continually increases in size until it is double its original size; then cell division occurs, resulting in the formation of two cells the size of the original cell. During this cell division cycle, all the structural and functional components of the cell double in amount. The interval for the formation of two

**4.1
Growth**

cells from one is called a *generation*, and its time is called *generation time*. The generation time is thus the time required for cell number to double. Because of this, the generation time is also sometimes called the *doubling time*. Note that during a single generation, both the cell number and cell mass have doubled. Generation times vary widely among organisms. Most bacteria have generation times of 1 to 3 hours, but a few are known that divide in as little as 10 minutes. At the other extreme, some slow-growing protozoa and algae have generation times of 24 hours or more.

In unicellular organisms, the progressive doubling of cell number results in a continually increasing rate of growth in the population, a process called exponential growth. Thus, one cell doubles to give rise to two cells, two cells give rise to four, four to eight, eight to sixteen, and so on. If this process continues unchecked, very quickly an enormous number of cells will be produced. For instance, beginning with a single cell with a doubling time of 30 minutes, at the end of 10 hours, there will be over 1,000,000 cells.

Because of the rapid exponential growth of many microorganisms, large populations of cells develop quickly, and one is often forced to deal with very large numbers. For instance, cell numbers in the millions, hundred millions, and even billions occur quite often. Since it is difficult to handle such large numbers with so many zeros, the microbiologist makes use of exponents of 10. Thus we express 1,000,000 (one million) as 10^6; 10,000,000 as 10^7; 100,000,000 as 10^8; and 1,000,000,000 (one billion) as 10^9. To express an uneven number, such as 5,000,000, the figure is written as the unit integer multiplied by the proper power of 10: thus $5,000,000 = 5.0 \times 10^6$; $25,000,000 = 2.5 \times 10^7$; and $700,000,000 = 7.0 \times 10^8$. Although it is initially somewhat difficult to learn to think in terms of exponents, the procedure is so useful in microbiology that learning it is worth the effort.

One of the characteristics of exponential growth is that the rate of increase in cell number is slow initially but increases at an ever faster rate (Figure 4.1). This results, in the later stages, in an explosive increase in cell numbers. A practical implication of exponential growth is that when a nonsterile product such as milk is allowed to stand under conditions such that microbial growth can occur, a few hours during the early stages of exponential growth are not detrimental, whereas standing for the same length of time during the later stages is disastrous.

EXPONENTIAL GROWTH

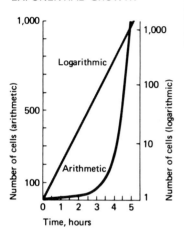

FIGURE 4.1
Exponential growth of a cell population from a single cell (generation time 30 seconds). Data below are plotted on both arithmetic and logarithmic scales.

Time hours	Number of cells
0	1
0.5	2
1	4
1.5	8
2	16
2.5	32
3	64
3.5	128
4	256
4.5	512
5	1,024
⋮	⋮
10	1,048,576

GROWTH CYCLE

A unicellular microorganism exhibits a characteristic growth cycle, which can be divided into several distinct phases: the *lag phase*, *exponential phase*, *stationary phase*, and *death phase* (Figure 4.2).

4.1 GROWTH

99

When a microorganism is inoculated into a fresh medium, growth usually does not begin immediately but only after a period of time. This interval is called the *lag phase*, and it may be brief or extended, depending on conditions. The lag usually occurs because the cells must become adjusted to the new medium before growth can begin. A lag can be avoided if an exponentially growing culture is transferred to the exact same culture medium at the same temperature. In this case, no adjustment by the cells is needed, and growth can continue at the same rate.

The *exponential phase* of growth has already been discussed. As noted, it is a consequence of the fact that each cell divides to form two cells, each of which also divides to form two more cells, and so on. Most unicellular microorganisms grow exponentially, but rates of exponential growth vary greatly. For instance, the typhoid organism, *Salmonella typhosa* grows very rapidly, with a generation time of 20 to 30 minutes; whereas the tubercle bacterium, *Mycobacterium tuberculosis*, grows slowly, with only one or two doublings a day. The rate of exponential growth is determined primarily by characteristics of the organism itself. In general, bacteria grow faster than eucaryotic microorganisms, and small eucaryotes grow faster than large ones. The growth rate of an organism is also greatly affected by culture medium and environmental conditions, especially temperature. In practice, one usually tries to choose a culture medium and a temperature that are optimum for the organism in question.

Exponential growth cannot occur indefinitely, else the world would be swamped with tons of microorganisms. One can calculate that a single bacterium with a generation time of 20 minutes would, if it continued to grow exponentially, produce at the end of 48 hours a population that weighed about 4,000 times the weight of the earth! Obviously, something must happen to limit growth long before this time. What generally happens is that either an essential nutrient of the culture medium is used up or some waste product of the organism builds up in the medium to an inhibitory level. Thus exponential growth ceases, and the population reaches the *stationary phase*.

The stationary phase may extend indefinitely, but it is sometimes followed by a *death phase*, in which the organisms in the population die. Death occurs either because the organisms undergo starvation or because some toxic product that accumulates kills them. In some cases, not only does the organism die, but also the cells disintegrate, a process called *lysis*. It is possible to maintain viable cultures and prevent onset of the death phase by removing cultures from the incubator shortly before exponential growth is over and placing them at a lower temperature, where growth and function are slowed. Death often occurs less rapidly if the organism is grown on a culture medium that is not too rich, since in a rich medium a high population density occurs and hence more waste products build up.

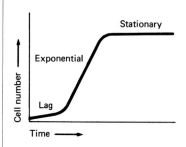

FIGURE 4.2
Typical growth curve for a bacterial population.

If the organism forms spores, the culture can be maintained more easily; the spores formed are usually quite stable and do not die readily.

Some very sensitive organisms die quickly during the stationary phase. The only way a culture of such organisms can be maintained is by periodic transfer to fresh medium. The growth of the culture must be watched, and as soon as the stationary phase is approached, a transfer to a fresh medium should be made. Experience will tell how often a specific culture must be transferred.

4.2
Measurement of growth

Growth is measured by following changes in number of cells or weight of cell mass. There are several methods for counting cell number or estimating cell mass, suited to different organisms or different problems.

TOTAL CELL COUNT

The number of cells in a population can be measured by counting under the microscope, a method called the *direct microscopic count*. Two kinds of direct microscopic counts are done, either on samples dried on slides or on samples in liquid. With liquid samples, special *counting chambers* must be used.

There are two types of chambers for counting cell number in liquid samples: the *hemocytometer*, or blood cell-counting chamber, for use with organisms 3 to 4 μm in diameter or larger; and the *Petroff-Hausser* counting chamber, for use primarily with bacteria. In both of these chambers, a special grid is marked on the surface of the glass slide, with squares of known small area. A special flat cover slip is placed on top of the grid, and a ridge on each side of the grid holds the cover slip off the grid by a precisely defined distance. Thus over each square on the grid is a volume of known size, very small but precisely measured. A sample of the suspension to be counted is allowed to flow under the cover slip and to fill the counting chamber. After the cells have settled in the chamber, the number per unit area of grid can be counted under the microscope, giving a measure of the number of cells per small chamber volume (Figure 4.3). Converting this value to number of cells per milliliter of suspension is easily done by multiplying by a conversion factor based on the volume of the chamber sample.

The *dry-slide method* for direct microscopic counting is used primarily in dairy work but can be applied in other studies also. A known small volume of sample is spread uniformly on a 1-sq-cm area marked on a slide. After drying and staining, the cells in several microscopic fields are counted. The area of the microscopic field is determined; then, knowing this and the volume of sample spread in the centimeter-square area, the number of cells per unit volume is

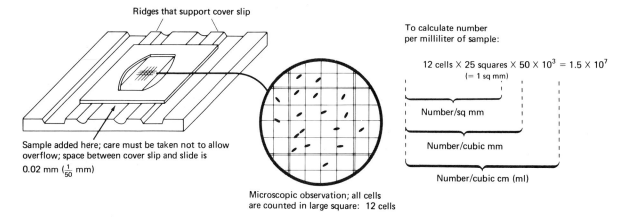

Ridges that support cover slip

To calculate number
per milliliter of sample:

$$12 \text{ cells} \times 25 \text{ squares} \times 50 \times 10^3 = 1.5 \times 10^7$$

(= 1 sq mm)

Number/sq mm

Number/cubic mm

Number/cubic cm (ml)

Sample added here; care must be taken not to allow
overflow; space between cover slip and slide is
0.02 mm ($\frac{1}{50}$ mm)

Microscopic observation; all cells
are counted in large square: 12 cells

calculated. A description of the method of counting bacteria in milk is given in Section 15.5.

Direct microscopic counting is tedious but is a quick and easy way of estimating microbial cell number. However, it has certain limitations: (1) Dead cells cannot usually be distinguished from living cells. (2) Small cells are difficult to see under the microscope, and some cells may be missed. (3) Precision is difficult to achieve. (4) The method is not suitable for cell suspensions of low density. With bacteria, if a cell suspension has less than 10^6 cells per milliliter (ml), few if any bacteria will be seen.

To avoid the tedium of direct microscopic counting, *electronic cell counters* have been invented. These counters were first devised for the counting of red blood cells in hospital laboratories but can also be used for counting microbial cells. In electronic counting, the electrical resistance of the fluid within a small hole is measured. As each cell passes through the hole, the resistance increases sharply, and the increase is recorded. A known small volume of liquid containing cells is allowed to flow through the hole, and each cell is counted as it passes. An electronic cell counter is accurate and quick but is fairly expensive. It cannot distinguish between an organism and an inert particle; hence debris and precipitates must be absent from the sample being counted. An electronic cell counter is only valuable if a very large number of similar samples are being counted.

In the methods just described, both living and dead cells are counted. In many cases we are interested in counting only live cells since these affect us most, and for this purpose viable cell counting methods have been developed. A viable cell is defined as one that is able to divide and form offspring, and the usual way to perform a viable count is to determine the number of cells in the sample

FIGURE 4.3
Direct microscopic count of cell number, using a Petroff–Hausser counting chamber. The total ruled grid has 25 large squares and measures 1 sq mm in area. When making microscopic counts, the practice is to count several squares, then average the numbers of cells counted.

VIABLE COUNT

capable of forming colonies on a suitable agar medium. For this reason, the viable count is often called the *plate count*, or *colony count*.

There are two ways of performing a plate count: the spread plate method and the pour plate method (Figure 4.4). With the *spread plate method*, a volume of culture no larger than 0.1 ml is spread over the agar surface, using a sterile glass spreader. The plate is then incubated until the colonies appear, and the number of colonies is counted. It is important that the surface of the plate be dry so that the liquid that is spread soaks in. Also, volumes greater than 0.1 ml should never be used, since the excess liquid will not soak in and may cause the colonies to coalesce as they form, making them difficult to count. In the *pour plate method*, a known volume of 0.1 to 1.0 ml of culture is pipetted into a sterile petri plate; melted agar medium is then added and mixed well by gently swirling the plate on the table top. Because the sample is mixed with the liquid agar medium, a larger volume can be used than with the spread plate; however, with the pour plate the organism to be counted must be able to withstand the temperature of melted agar, 45°C.

With both the spread plate and pour plate methods, it is important that the number of colonies developing on the plates not be too large, since on crowded plates some cells may not form colonies and the count will be erroneous. It is also essential that the number of colonies not be too small, or the statistical significance of the calculated count will be low. The usual practice, which is most valid statistically, is to count only those plates that have between 30 and 300 colonies. To obtain the appropriate colony number, the sample to be counted must thus usually be diluted. Since one rarely knows the approximate viable count ahead of time, it is usually

FIGURE 4.4
Two methods of performing a viable count (plate count).

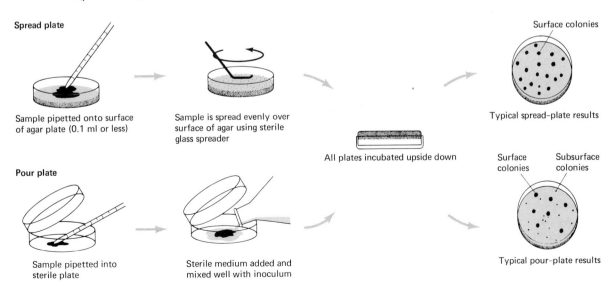

Spread plate

Sample pipetted onto surface of agar plate (0.1 ml or less)

Sample is spread evenly over surface of agar using sterile glass spreader

All plates incubated upside down

Surface colonies

Typical spread–plate results

Pour plate

Sample pipetted into sterile plate

Sterile medium added and mixed well with inoculum

Surface colonies   Subsurface colonies

Typical pour–plate results

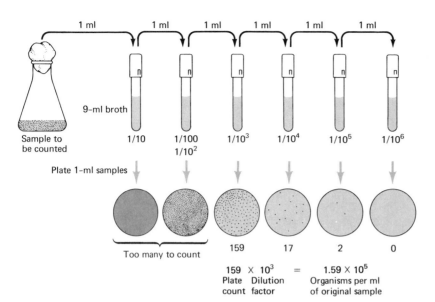

FIGURE 4.5
Procedure for viable count using serial dilutions of the sample.

In the figure:

1 ml   1 ml   1 ml   1 ml   1 ml   1 ml

9-ml broth

Sample to be counted

$1/10$ $1/100$ $1/10^3$ $1/10^4$ $1/10^5$ $1/10^6$
         $1/10^2$

Plate 1-ml samples

Too many to count   159   17   2   0

$$159 \times 10^3 = 1.59 \times 10^5$$

Plate count — Dilution factor = Organisms per ml of original sample

necessary to make more than one dilution. Several tenfold dilutions of the sample are commonly used (Figure 4.5). To make a tenfold dilution, one can mix 0.5 ml of sample with 4.5 ml of diluent, or 1.0 ml with 9.0 ml diluent. If a hundredfold dilution is needed, 0.05 ml can be mixed with 4.95 ml diluent, or 0.1 ml with 9.9 ml diluent, or of course two successive tenfold dilutions may be made. In most cases, such *serial dilutions* are needed to reach the final dilution desired. Thus, if a $1/10^6$ (1 to 1,000,000) dilution is needed, this can be achieved by making three successive $1/10^2$ (1 to 100) dilutions or six successive tenfold dilutions. The liquid used for making the dilutions is important. It is best if it is identical with the liquid medium used in making agar, although for economy, it is often possible to use a simple solution of inorganic salts or a phosphate buffer. Distilled, deionized, or tap water should never be used as a diluent, since organisms are often killed when placed in such liquids. The diluent should of course be sterile. The usual practice is to make up a large number of tubes of diluent at desired volumes and keep them on hand for use as needed. It is important when making dilutions that a separate sterile pipette be used for each dilution, even of the same sample. This is because in the initial sample, which contains the largest number of organisms, not all the organisms will be washed out of the pipette when its contents are expelled. Organisms sticking to the pipette will be washed out in later dilutions and can cause serious error in the final count obtained.

The number of colonies obtained on a plate will depend not only on the inoculum size but also on the suitability of the culture medium and the incubation conditions used; also, it will depend on the length of incubation. The cells deposited on the plate will not all develop into colonies at the same rate, and if a short incubation time is used, less than the maximum number of colonies will be obtained. Further, the size of colonies often varies. If some tiny colonies develop, they may be missed during the counting. It is usual to determine incubation conditions (medium, temperature, time) that will give the maximum number of colonies and then to use these conditions throughout. Viable counts are usually subject to large error, and if accurate counts are desired great care must be taken and many duplicate plates must be used.

Despite the difficulties in viable counting, the procedure is so useful in microbiology that it is widely used. In food, dairy, medical, and aquatic microbiology, viable counts are used routinely. Standard methods have been developed that are suitable for viable counts of different materials, and these methods are clearly described in appropriate manuals. In these fields of applied microbiology, identical methods are used in all laboratories, so that the results in one laboratory are comparable to those in another. Probably the most useful aspect of viable counting is the sensitivity of the method. Samples containing only a very few cells can be counted, thus permitting sensitive detection of contamination of products or materials. Some examples of dilution problems follow:

1 A urine sample is plated and gives 210 colonies on the plate receiving 1 ml of the 100,000-fold diluted sample. What is the count of bacteria per milliliter of urine?   **Answer:** 2.1×10^7
2 Three hundredfold serial dilutions of a soil sample are made: 1-ml samples are plated, and the count from the third of these is 54. How many cells per gram of soil (1 g of a solid is considered equivalent to 1 ml when preparing dilutions)?   **Answer:** 5.4×10^7
3 You wish to determine the viable count of a sample of milk (estimated count is 1.5×10^5 cells per milliliter). How would you propose to set up the assay?   **Answer:** Prepare serial dilutions and plate 1-ml samples from the $1/10^2$ to the $1/10^5$ dilutions. If the estimate is correct, the plate from the $1/10^3$ dilution will have between 30 to 300 colonies and will be used to determine the viable count. If the estimate is incorrect, one of the other plates will have a proper number of colonies.

For some work, *membrane filters* are used instead of agar plates, as mentioned in Chapter 3. The principle of counting is the same as in any viable count. An appropriately diluted sample is filtered, and the filter is placed on a suitable culture medium. After incubation, the colonies that develop on the filter are counted. Membrane filtration is most useful for counting when the sample contains only a very small number of organisms, since a relatively large volume of sample can be passed through the filter. The method is used in this way in aquatic microbiology, especially for obtaining counts of drinking water and other relatively unpolluted waters. It is also useful for counting organisms present in materials such as antiseptic or antibiotic solutions. Since the antiseptic or antibiotic would be inhibitory to organisms present if it were included in the sample added to the culture medium, it must be removed, and this can best be done by passing the sample through a membrane filter. The filter containing the organisms is washed free of the substance and then is placed on a nutrient medium, incubated, and counted.

Some organisms do not readily form colonies on agar plates or membrane filters but will initiate growth in liquid medium. To count such organisms, a technique called the *most probable number* (or MPN) method has been developed, which permits an estimate of viable numbers after incubation in liquid medium. With this method, the sample is diluted to the point at which some, but not all, portions contain a cell. If a series of tubes is inoculated with identical portions taken at this dilution, some will contain a cell whereas others will not; after incubation, some will show growth whereas others will not. By counting the fraction of tubes showing growth, one can estimate the viable count, using statistical tables that have been developed. As will be described in detail later (in Section 14.3), MPN methods are used most commonly in aquatic microbiology.

For many studies it is desirable to estimate the weight of cells rather than the number. Weight can be measured directly by taking a sample, centrifuging to sediment the cells, and then weighing the sedimented cells with a balance. Dry-weight measurements are made by drying the centrifuged cell mass before weighing, usually by placing it overnight in an oven at 100 to 105°C.

A simpler and very useful method for obtaining a relative estimate of cell weight is by use of *turbidity* measurements. A cell suspension looks turbid because each cell scatters light. The more cells present, the more the suspension scatters light and the more turbid it will be. Turbidity thus relates directly to the number and the weight of cells present. Turbidity can be estimated simply by comparing a tube of the turbid suspension with a series of tubes of varying known turbidities. More accurately, turbidity can be measured with an electrically operated device called a *colorimeter*, or *spectrophotometer*. With such a device, the turbidity is expressed in units of *optical density* (O.D.). In unicellular organisms, O.D. is proportional to cell number as well as cell weight, and turbidity readings can thus be used as a substitute for counting. To perform cell counts in this way, a standard curve must be prepared for each organism studied, relating cell number to O.D. Turbidity measurement is a much less sensitive way of measuring cell density than is viable counting but has the virtues that it is quick, easy, and does not destroy the sample. Such measurements are used widely to follow the rate of growth of cultures, since the same sample can be checked repeatedly.

In later chapters, we shall describe a number of situations in which a knowledge of how microorganisms grow can be put to good use. The rest of this chapter will be devoted to ways in which the growth of microorganisms can be controlled by use of physical and chemical agents.

4.3
Sterility testing

It is frequently necessary to determine whether a product is sterile. Sterility means the complete absence of all living microbes. Sterility tests are performed on foods and on many products used in medicine, such as injectable drugs and solutions, gauzes and bandages, and sutures. Sterility testing of products for human and veterinary use is done by specific procedures usually prescribed by government regulation. Tests are usually made for the presence of bacteria, fungi, and yeasts.

The container to be sampled must first be opened by aseptic procedures. The surface of the container should be cleansed and then disinfected to remove any surface contaminants. If the container is under vacuum, it must be opened in such a way that the air rushing into the container is sterilized. This can be done by inserting a hypodermic needle attached to a syringe barrel filled with sterile nonabsorbent cotton. More than one unit from any product should be tested, and each unit should be tested with more than one culture medium. Tests should be made for both anaerobic and aerobic bacteria, as well as for fungi and yeasts. The material to be tested is added to the medium and incubation is carried out at 25 to 30°C for 7 to 14 days. The length of incubation is important, since if only a small number of organisms is present or if the contaminant is slow-growing, a considerable time will be required before a population giving visible turbidity develops. Should the sterility testing be improperly done, the nonsterile product might wrongly be considered sterile. On the other hand, care must be taken that contamination is not introduced into the medium from the air or the outside of the product, else it might be concluded that a product is nonsterile when it is indeed sterile.

4.4
Control of microbial growth

The control of microbial growth is necessary in many practical situations, and some of the most important advances in medicine, agriculture, and food science have been made by applications of microbiological knowledge. Control can be effected either by *killing* organisms or by *inhibiting* their growth. The killing of organisms is generally called *sterilization* and is brought about by use of heat, radiation, or chemicals. An agent that kills microorganisms is sometimes called a *germicide*. Sterilization means the complete destruction of all microbial cells present; once a product is sterilized, it will remain sterile indefinitely if it is properly sealed.

One of the most important and widely used methods for sterilization is the use of heat, and an understanding of the principles of heat sterilization is important in microbiology. We have already discussed the use of heat to sterilize culture media in Section 3.8, and here we shall discuss some of the several principles and practices of heat sterilization. The lethal temperature varies among microorganisms: some are very sensitive and are killed at temperatures as low as 30°C, whereas others are quite resistant and require temperatures of boiling or higher. When a population is heated to a lethal temperature, death of all cells in the population does not occur at the same time, but a certain percentage of organisms dies during each given time period. However, the higher the temperature, the faster the rate of killing (Figure 4.6). The *thermal death time* is defined as the time required at a given temperature to kill *all* organisms in the population.

The thermal death time depends on the number of organisms present, since it obviously takes longer to kill a large number of organisms than a small number. Other factors to be considered are the species of microorganisms involved and the nature of the product being sterilized. In general, sterilization is quicker at low pH than at neutral or higher pH, because most microorganisms are sensitive to low pH. For this reason, acidic foods such as tomatoes, fruits, and pickles are much easier to sterilize than are more neutral foods such as corn and beans. Moisture also affects the thermal death time. Dry cells are usually more resistant to heat than moist ones; for this reason, heat sterilization of dry objects always requires higher temperatures and longer times than does the sterilization of moist objects. Sugary and salty products also are usually quite resistant to heat sterilization, probably because the sugar or salt has a dehydrating effect on the microbial cells.

The single most important factor affecting the rate of heat sterilization is the presence or absence of bacterial endospores. Because of the high heat resistance of bacterial spores, very much longer heating times are necessary to sterilize a product containing spores than one containing only vegetative cells. If only vegetative cells are present, most products can be sterilized by heating at 60 to 70°C for a few minutes, whereas if spores are present, temperatures as high as 121°C are usually required. To achieve temperatures this

HEAT STERILIZATION

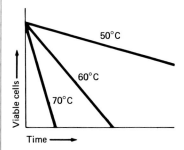

FIGURE 4.6
Effect of temperature on viability.

high, since they are above boiling, autoclaves or pressure cookers are required. The use of these devices was described in Section 3.8.

One of the most important practical processes involving heat sterilization is the *canning* of foods. Foods are sealed in airtight containers and heated to kill all of the organisms in the food. The resulting sterile product will keep nearly indefinitely. Canned foods have very long shelf lives and make possible the storage of foods for use at times or places when fresh foods are not available. The principles and practices of canning are discussed in detail in Section 15.4.

PASTEURIZATION

Pasteurization is a process using mild heat to reduce microbial populations in milk and other foods. The process is named for Louis Pasteur, who first used heat for controlling the spoilage of wine (see Section 1.5). Pasteurization is not synonymous with sterilization, since not all organisms are killed. Pasteurization is used to kill pathogenic microorganisms that might be present, since these are usually fairly sensitive to heat and rarely form endospores. It is also used to improve the keeping qualities of the product. Pasteurization is used commonly for milk, a rather heat-sensitive product, the flavor of which is altered greatly by heat sterilization but much less so by pasteurization. Beer, wine, cider, and other beverages are also pasteurized.

Pasteurization can be done by heating the substance at 63 to 66°C for 30 minutes and then quickly cooling. For pasteurizing large quantities of a liquid such as milk, a process called *flash pasteurization* is preferable. The product is passed continuously through a heater where its temperature is raised quickly to 71°C, held there for 15 seconds, and then quickly cooled. Flash pasteurization is more satisfactory than bulk pasteurization, since the product is heated for a shorter period of time and its flavor is therefore less altered. Flash pasteurization has the additional advantage that since it is carried out on a continuous basis, it is more adaptable to large-scale operation (see Figure 15.21).

INCINERATION

Incineration is a process of sterilization involving very· high temperatures, the contaminating organisms actually being ignited and burned to death. Incineration is chiefly used for sterilizing heat-resistant items such as metal inoculating loops and needles, and for disposing of burnable contaminated materials from hospitals and laboratories. Incineration is also a suitable way of sterilizing air, which is passed over copper coils or pipe heated to a high temperature. It is of course not suitable for sterilizing foods or other materials containing liquids or organic materials.

Since all microorganisms require water for growth, microbial growth can be controlled by removing water from the product. However, drying often does not kill organisms already present in the product but only prevents them from growing, so that it is essential that the drying process be used on products or materials that are free of undesirable organisms. We discussed the drying of cultures as a means of preserving them in Section 3.7.

Most microorganisms can grow only if water in the liquid state is present. Some fungi can grow under conditions of humid air even in the absence of liquid water. They apparently are able to extract water directly from the air. These fungi are common in the humid tropics and often cause serious damage (*mildew*) to cloth and leather goods, as well as to cameras and other optical instruments (Figure 4.7).

We stated earlier that spores formed by many microorganisms are not easily killed by drying. Vegetative cells of some microorganisms are also resistant; in general, small cells are more resistant to drying than large cells, and procaryotes are much more resistant than eucaryotes. Some disease-causing organisms resistant to drying are *Mycobacterium tuberculosis*, the causal agent of tuberculosis, and *Staphylococcus aureus*, a causal agent of pimples, boils, and pneumonia. On the other hand, the spirochete that causes syphilis, *Treponema pallidum*, is very sensitive to drying and dies almost instantly when exposed to air. For this reason, the organism is only transmitted by intimate contact, such as sexual intercourse.

Growth of organisms in foods can be prevented by drying the food material. Although all foods can be dried, some foods retain their flavor and texture upon drying better than do others. Milk, meats, fish, vegetables, fruits, and eggs are all preserved by drying. The practical aspects of preserving food in this way will be discussed in Section 15.4.

FIGURE 4.7
Development of mildew on leather stored in damp warm conditions. The boot on the left was stored for 6 months under humid tropical conditions. Compare with the boot on the right, stored under dry conditions. *Courtesy of National Bureau of Standards.*

Most organisms grow very little if at all in the cold; for this reason low temperatures are frequently used for the storage of perishable food products, thus slowing the rate of microbial growth and consequent spoilage. Only a few organisms are able to grow at refrigeration temperatures; these *psychrophiles* are able to grow to a limited extent, and some may cause spoilage.

Storage for much longer periods of time is possible at temperatures below freezing. Freezing greatly alters the physical structure of many food products, so that it cannot be used for many things; but it is widely and successfully used for the preservation of meats and many vegetables and fruits. Mechanical freezers providing temperatures of $-20°C$ are most commonly used for storing frozen products. At such temperatures, storage for weeks or months is possible, but even at these temperatures some microbial growth may occur, usually in pockets of liquid water trapped within the frozen mass. For very long-term storage, lower temperatures are necessary, such as $-70°C$ (dry ice temperature) or $-195°C$ (liquid nitrogen temperature), but maintenance of such low temperatures is expensive, and consequently they are not used for routine food storage.

It should be emphasized that many microorganisms survive freezing quite well. Thus foods to be frozen should be of the best quality and free of spoilage. If a contaminated product is frozen, its microbial load will be maintained, and when it is thawed many of these organisms will still be alive and can cause problems.

EFFECT OF RADIATION ON MICROORGANISMS

Radiations are invisible rays emanating from natural or artificial sources. Some types of radiation cause death of living organisms and are of value in the sterilization and control of microorganisms in various materials. We distinguish between two kinds of radiations: electromagnetic and ionizing. *Electromagnetic* radiations that have effects on living organisms include ultraviolet, visible, and infrared radiations. The latter two are usually beneficial and are used as sources of energy by photosynthetic organisms, whereas ultraviolet radiation is usually lethal. *Ionizing* radiations include X rays, cosmic rays, and emanations from radioactive materials (Figure 4.8).

Both intensity and wavelength can affect the killing power of *ultraviolet radiation*. Intense sources are more lethal than weak sources, and sources of short wavelength are more lethal than sources of long wavelength. Ultraviolet radiation from the sun that reaches the earth's surface is of longer wavelength and is only weakly lethal to microorganisms. Man-made ultraviolet sources, such as are found in germicidal lamps, are of shorter wavelength and are much more effective in killing microorganisms. The lethal effect of ultraviolet radiation is due to its effect on deoxyribonucleic acid (DNA). At low doses, ultraviolet radiation may not kill but only cause genetic changes called *mutations*, whereas at higher doses ultraviolet radiation is lethal.

Germicidal ultraviolet radiation passes very poorly through many kinds of glass and not at all through opaque objects, so that its practical uses are limited. It can be used to sterilize air or surfaces, such as toilet seats or counter tops. It is often used in hospital surgeries and in culture transfer rooms in microbiology laboratories, since it can be applied for a few hours or overnight before the room is used. Because radiation from germicidal lamps is very damaging

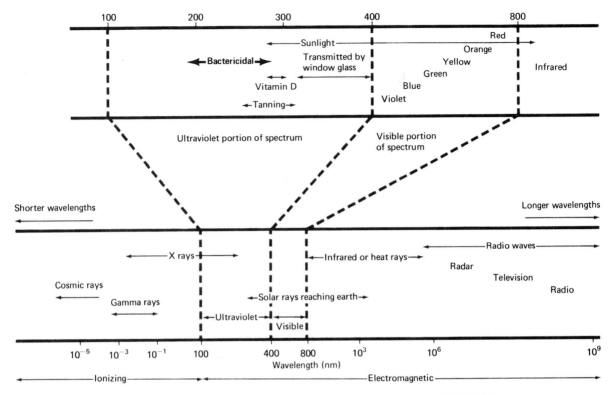

FIGURE 4.8
Wavelengths of radiation, showing regions of visible light and of bactericidal rays.

to the eyes, it must be used with caution and should never be allowed to enter the eyes, either directly or after reflection from surfaces.

Germicidal ultraviolet lamps are available commercially from various lamp manufacturers. They are supplied in tubes resembling fluorescent light tubes and are used in fixtures similar to those used for fluorescent lamps.

Visible light may also cause microbial death if it is of sufficient intensity. In contrast to ultraviolet radiation, visible light is only lethal if O_2 is present, the light causing a photooxidation of vital cell constituents. The lethal effects of sunlight on microorganisms are usually due to this photooxidation by visible light rather than to the ultraviolet radiation of sunlight. Some microorganisms have formed special pigments that protect them against photooxidation. These pigmented organisms are especially common in the air and in surface water exposed to bright light. Very few disease-causing bacteria survive long in bright sunlight, but *Staphylococcus aureus*, a yellow-pigmented organism, does. This pathogen is often airborne, and its pigment probably enables it to survive transport through the air.

Ionizing radiations do not kill by directly affecting cell constituents but by inducing in the water surrounding the cells the formation of highly reactive substances called *free radicals*; these react with and inactivate sensitive cell constituents. All kinds of cell constituents are affected, but death usually results from effects on DNA, the most vital cell constituent.

Ionizing radiation may be used to sterilize materials such as drugs, foods, and other items that are heat sensitive. In contrast to ultraviolet radiation, ionizing radiation penetrates well and hence can be used on products even after they have been packaged. Sources of ionizing radiation are X-ray machines, radioisotopes such as cobalt-60, and neutron piles in atomic energy installations. The most convenient commercial sources for radiation sterilization are cobalt-60 radioisotopes, which require no maintenance or input of energy once installed.

Strict precautions must be used in the handling of ionizing radiation, since it is highly lethal for humans as well as microorganisms. Protective lead shielding is essential, and the doses received by operators must be carefully monitored. Although ionizing radiation has been used experimentally for the sterilization of foods, especially meats, it has not as yet proved practical and economic for routine use. This is partly because of cost factors and partly because ionizing radiation sometimes produces off-flavors and discoloration in the product.

4.5 Antimicrobial agents

An antimicrobial agent is a chemical that kills or inhibits the growth of microorganisms. Such a substance may be either a synthetic chemical or a natural product.

Agents that kill organisms are often called "cidal" agents, with a prefix indicating the kind of organism killed. Thus we have *bactericidal*, *fungicidal*, and *algicidal* agents. A bactericidal agent, or bactericide, kills bacteria. It may or may not kill other kinds of microorganisms. Cidal agents with a broad spectrum of target organisms are usually called *germicides*. Germicides are sometimes conveniently divided into two groups: *antiseptics* and *disinfectants*. An antiseptic is a germicide that can kill most microorganisms and is sufficiently harmless that it can be applied to the skin or mucous membranes, although it is not necessarily safe enough to be taken internally. A disinfectant is an agent that kills microorganisms (but not necessarily their spores) and is distinguishable from an antiseptic by the fact that it is not safe for application to living tissue, and its use is restricted to inanimate objects, such as tables, floors, or dishes (Table 4.1). Disinfectants used for surfaces are generally applied as solutions, by swabbing with sponges or mops. Dairy equipment and medical instruments may be dipped or rinsed in solutions. Water supplies that are chlorinated are bubbled with the gas, which dis-

TABLE 4.1
Antimicrobial agents

Agent	Use
Disinfectants	
Mercuric dichloride	Tables, bench tops, floors
Copper sulfate	Algicide in swimming pools, water supplies
Iodine solution	Medical instruments
Chlorine gas	Purification of water supplies
Chlorine compounds	Dairy, food industry equipment
Phenolic compounds	Surfaces
Cationic detergents (quaternary ammonium compounds)	Medical instruments; food, dairy equipment
Antiseptics	
Organic mercurials	Skin
Silver nitrate	Eyes of newborn, to prevent gonorrhea
Iodine solution	Skin
Alcohol (70% ethanol in water)	Skin
Bis-phenols (hexachlorophene)	Soaps, lotions, body deodorants
Cationic detergents (quaternary ammonium compounds)	Soaps, lotions
Chemical food preservatives	
Sodium or calcium propionate	Bread
Sodium benzoate	Carbonated beverages, fruit juices, pickles, margarine, preserves
Sorbic acid	Citrus products, cheese, pickles, salads
Sulfur dioxide	Dried fruits and vegetables
Chlortetracycline	Chicken, fish
Formaldehyde (from food-smoking process)	Meat, fish

solves in the water. Antiseptics that are used on the skin may be swabbed on with cotton. Soaps containing antiseptics leave a residue of the compound on the skin after the soap is rinsed off, and this active residue may continue to affect microbial viability for some time, thus increasing the efficacy of the treatment.

Agents that do not kill but only inhibit growth are called "static" agents, and we can speak of *bacteriostatic*, *fungistatic*, and *algistatic* agents. The distinction between a static and a cidal agent is often arbitrary, since an agent that is cidal at high concentrations may only be static at lower concentrations. To be effective, a static agent must be continuously present with the product, and if it is removed or its activity neutralized the organisms present in the product should initiate growth. Static agents are often used as food preservatives, and since they must be present continuously to be effective, they remain in the food when eaten and hence must be nontoxic. Many drugs used in treatment of microbial infections are static agents and must be kept present for a period of time long enough for body defenses to destroy the infecting organism. Such drugs obviously must be nontoxic to the body. Although there is a wide variety of chemicals that are static agents, most of them are too toxic for use in foods or as drugs.

Many antimicrobial agents are effective in low concentrations, 1 to 10 parts per million (1 to $10 \mu g/ml$). This is of practical significance since it means that the agent will be active even after it is highly diluted, making it possible to apply the agent in effective concentration to animals by injection or to plants by spraying. Some agents not active at low concentrations can still have practical use, however, especially when applied to inanimate objects where high concentrations can be more effectively employed.

How do antiseptics and disinfectants act? The most common effect is the *denaturation* of the proteins of the microbial cells. The proteins are irreversibly altered so that they are no longer functional; without proteins no cell can live. Since human tissue also contains vital proteins, it is easy to see why so many compounds of both static and cidal type are not suitable for use on humans: proteins of both microbe and man are damaged. Another common effect of disinfectants is an attack on the cell membrane so as to cause loss of vital cell constituents, or lysis. Again, there is no specificity in this attack, both microbial and human cell membranes being similarly affected. Thus, an important feature of many antimicrobial agents is *selective toxicity*, which means that the agent is more active against the microbe than against the animal or plant host. Agents that act selectively on disease-causing organisms without affecting human tissue are of course of most medical value. Some antimicrobial agents act selectively against procaryotic organisms and are relatively harmless to eucaryotes. Since humans are eucaryotes and many disease-causing microorganisms are procaryotes, agents selective against procaryotes potentially have wide medical uses. A variety of such selective agents is known. There are also a few agents that act selectively on eucaryotic microorganisms (fungi, protozoa, algae) without affecting the eucaryotic cells of higher animals, including man, and such agents find important medical applications. A substance that selectively attacks microorganisms without harming human cells is called a *chemotherapeutic agent*.

DISPOSAL OF TOXIC AGENTS

It should be noted that disposal of some disinfectants and antiseptics, as well as various other laboratory chemicals, poses a considerable problem. Compounds containing mercury, copper, iron, arsenic, or cyanide as well as various other compounds such as detergents, acids, and organic solvents used to be indiscriminately discarded into sewer systems. Now, however, federal and state regulations impose certain restrictions on the disposal of waste products.

Many compounds formerly released into the environment may have caused widespread harm to aquatic life. In some cases the compounds were accumulated by organisms, interfering with food chains at various levels. To prevent such dispersal, some wastes must now be collected, treated, and in many cases stored for proper disposal, so as to avoid polluting public water supplies. Compounds with mercury, copper, iron, or zinc are precipitated, and the solids are stored for disposal. Arsenic and silver are precipitated and recycled for further use. Compounds such as acids must be neutralized before disposal, then diluted and flushed with excess water. Organic solvents, phenolic compounds, and certain other volatile compounds must be stored in safety cans until they are sent to facilities where they can be handled properly for disposal.

An antibiotic is a chemical substance produced by one microorganism that is able to kill or inhibit the growth of other microorganisms. Thousands of antibiotics have been discovered, but only a relative few have turned out to be of great practical value in medicine. However, the most important of these antibiotics have found widespread use in the treatment of many infectious diseases.

4.6 Antibiotics

The production of antibiotics by microorganisms is a common occurrence, and a variety of microorganisms produce antibiotics acting against other microorganisms (Figure 4.9). Antibiotic-producing microorganisms are especially common in the soil. Three groups of microorganisms are responsible for the production of most of the antibiotics used in medicine: (1) fungi, especially those of the genus *Penicillium*, which produce antibiotics such as penicillin and griseofulvin; (2) bacteria of the genus *Bacillus*, which produce antibiotics such as bacitracin and polymyxin; and (3) actinomycetes (filamentous bacteria) of the genus *Streptomyces*, which produce antibiotics such as streptomycin, chloramphenicol, tetracycline, and erythromycin. It is among members of the genus *Streptomyces*, a genus of organisms widespread in soil, that most of the antibiotics have been discovered.

Antibiotics are a diverse group of compounds, but they can be grouped in families with similar chemical structures. The antibiotics of one group usually have similar types of activity and find similar uses in practice. For instance, there is a large group of penicillins, most of which have similar activity.

The sensitivity of microorganisms to antibiotics varies. Gram-positive bacteria are usually more sensitive to antibiotics than are

FIGURE 4.9
Crowded colonies of soil microorganisms, showing antagonism. The clear zones around several colonies are due to antibiotic action. *Courtesy of Eli Lilly and Company.*

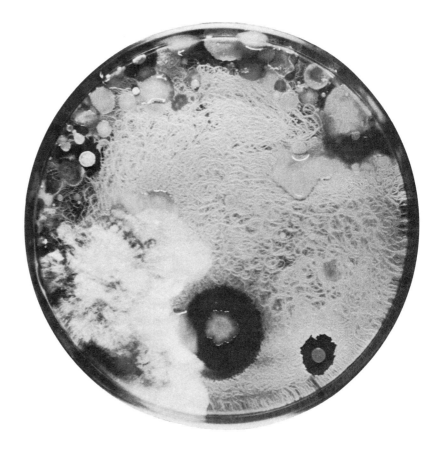

Gram-negative bacteria, although conversely, some antibiotics act only on Gram-negative bacteria. An antibiotic that acts upon both Gram-positive and Gram-negative bacteria is called a *broad-spectrum antibiotic*. In general, a broad-spectrum antibiotic will find wider medical usage than a narrow-spectrum one, although the latter may be quite valuable for the control of certain kinds of diseases (Figure 4.10). The physician needs and utilizes a wide variety of antibiotics and selects carefully the one needed for a particular patient with a particular kind of infection.

Although antibiotics are usually selective in their action, their toxicity for animals and man varies, and a knowledge of the severity of toxicity is essential for the wise use of antibiotics in medicine. Some antibiotics are so toxic that they can never be used in man, others have a limited toxicity and may be used with care, while others are essentially nontoxic. Some antibiotics cause intestinal disturbances or kidney damage, others cause deafness, and still others are harmful because they induce allergic reactions. Individual patients vary, and an antibiotic that may be perfectly harmless for one person may be quite toxic to another. This individual variability is most frequently seen with antibiotics that cause allergic reactions, such as penicillin.

Antimicrobial activity is measured by determining the smallest amount of agent needed to inhibit the growth of a test organism, a value called the *minimum inhibitory concentration* (MIC). A series of culture tubes is prepared, each tube containing medium with a different concentration of the agent, and all tubes of the series are inoculated. After incubation, the tubes in which growth does not occur (indicated by absence of visible turbidity) are noted, and the MIC is thus determined (Figure 4.11). This simple and effective procedure is often called the *tube dilution technique*. The MIC is not an absolute constant for a given agent, since it is affected by the kind of test organism used, the inoculum size, the composition of the culture medium, the incubation time, and the conditions of incubation, such as temperature, pH, and aeration. If all conditions are rigorously standardized, it is possible to compare different antibiotics and determine which is most effective against a given organism or to assess the activity of a single agent against a variety of organisms. Note that the tube dilution method does not distinguish between a cidal and a static agent, since the agent is present in the culture medium throughout the entire growth period.

Another commonly used procedure for studying antimicrobial action is the *agar diffusion method* (Figure 4.12). A petri plate containing an agar medium evenly inoculated with the test organism is prepared. In one procedure, filter-paper discs (containing defined amounts of the antimicrobial agent) are placed on the surface of the agar. During incubation the agent diffuses from the filter paper into the agar; the further it gets from the filter paper, the smaller is the

FIGURE 4.10
Range of activity of several antibiotics against various microbial groups, illustrating the idea of broad-spectrum antibiotics. Adapted from V. J. Cabasso, *BioScience* **17**, 796, 1967.

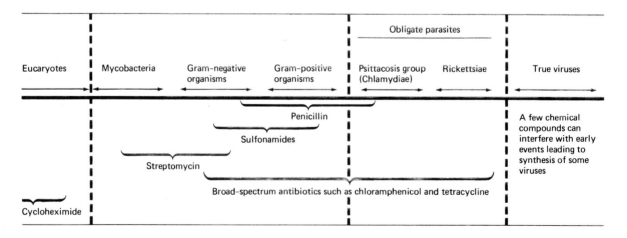

FIGURE 4.11
Antibiotic tube dilution series. A series of increasing concentrations of antibiotic is prepared in culture medium. Each tube is inoculated with the test organism and incubated. Growth decreases with increase in drug concentration (concentration of antibiotic increases left to right).

FIGURE 4.12
Steps in the agar diffusion method for assaying antibiotic activity.

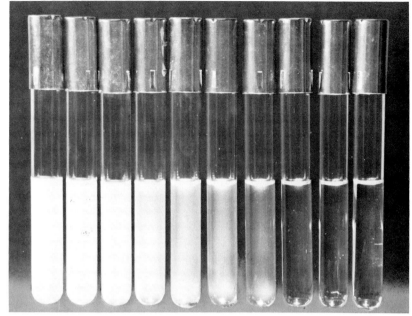

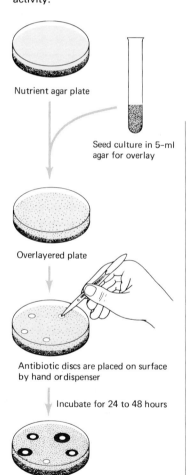

Nutrient agar plate

Seed culture in 5-ml agar for overlay

Overlayered plate

Antibiotic discs are placed on surface by hand or dispenser

Incubate for 24 to 48 hours

Test organism shows sensitivity to some antibiotics, indicated by zones of inhibition around discs

concentration of the agent. At some distance from the disc the MIC is reached. Past this point growth occurs, but closer to the disc growth is absent. An inhibition zone is thus created, and its size can be measured with a ruler. In another procedure, small stainless steel or porcelain cylinders 6.0 mm in diameter by 10.0 mm in length are placed on the surface of the agar. The solution to be assayed is pipetted into the cylinder, the plate incubated, and the zone size read.

The size of the zone is affected by the sensitivity of the organism, the culture medium and incubation conditions, the rate of diffusion of the agent, and the concentration of agent in the cylinder or on the filter-paper disc. The interpretation of the significance of wide and narrow inhibition zones is therefore not at all simple. An agent that produces a wide zone is not necessarily more active than an agent that produces a narrow zone, since the diffusion rate in agar of different agents varies widely. For comparing two agents, the tube dilution method is preferable. The agar diffusion method is simpler to set up, however, and is widely used for the assay of antibiotics and in clinical medicine.

Measuring the concentration of an antimicrobial agent that is present in unknown amount in a solution is done by comparing the activity of the unknown with known concentrations of the same agent. Either tube dilution or agar diffusion methods can be used, although the latter is more commonly used. First, a standard curve is prepared by determining the zone sizes obtained under standard assay conditions with different concentrations of the agent. This standard curve is displayed on graph paper, and if the logarithm of the concentration is plotted against zone diameter, a straight line is usually obtained. The concentration of the unknown is then obtained by noting the zone size that it elicits and reading its concentration from the standard curve (Figure 4.13). It is very important that the unknown and the standard be compared using the *same* test organism and assay conditions. The best procedure is to place both the unknown and the standard on the same agar plate so that their zones develop under identical conditions.

In assaying antibiotic preparations, a standard preparation is necessary, and the World Health Organization has established International Standards and Units. An International Standard is a preparation usually of pure antibiotic that has been defined as the standard and to which other preparations are compared. These International Standards are held and distributed by the International Laboratory for Biological Standards, National Institute for Medical Research, Mill Hill, London, N.W.7, England. Standard preparations are maintained for all of the antibiotics used in medical practice, at present more than 20 distinct antibiotics. In the United States, Standards are available from the Food and Drug Administration, Washington, D.C. For day-to-day use, a working standard is prepared. Its activity is determined by assay against the International Standard, and this working standard is then used in the assay of each unknown preparation.

Details of the standard assay procedure for each antibiotic are precisely specified, including method of assay (agar diffusion or tube dilution), test organism, culture medium, incubation time and temperature, and even the number of replicate assays and the statistical evaluation of results. Most antibiotics are assayed by the agar diffusion method, but in a few cases the tube method is used. Stock cultures of the test organisms used for antibiotic assay can be obtained from the recognized type culture collections.

In medical practice, the sensitivity to antimicrobial agents of cultures isolated from patients is often needed to help determine the treatment of the patient. The sensitivity of a culture can be most easily determined by an agar diffusion method. Federal regulations of the Food and Drug Administration (FDA) now control the procedures used for sensitivity testing in the United States. The recommended procedure is called the *Kirby–Bauer method*, after the workers who developed it (Figure 4.14). A plate of suitable culture medium is inoculated by spreading an aliquot of culture evenly across the agar surface. Filter-paper discs containing known concentrations of different agents are then placed on the plate. The concentration of each agent on the disc is specified so that zone diameters of approp-

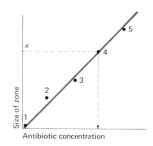

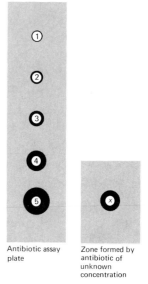

Antibiotic assay plate

Zone formed by antibiotic of unknown concentration

FIGURE 4.13
Procedure for determining unknown antibiotic concentration. To determine unknown concentration x, the zone size is located on the vertical axis, and the concentration is read off the horizontal axis.

TESTING CULTURES FOR ANTIBIOTIC SENSITIVITY

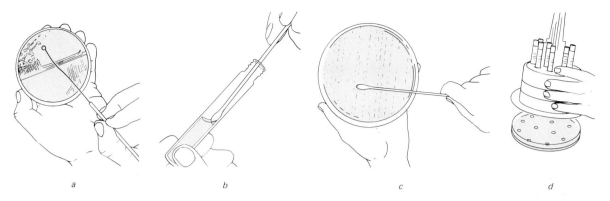

a b c d

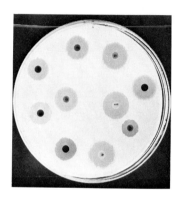

e

FIGURE 4.14

The Kirby–Bauer procedure for determining the sensitivity of an organism to antibiotics. *a* Isolates are selected from a plate streaked with patient's specimen and are grown to specified density in broth. *b, c* Swab dipped in broth culture (*b*) is spread evenly over a plate of the test medium (*c*). *d* Antibiotic discs are distributed on the plate and pressed gently onto the agar. *e* After incubation the zones of inhibition are measured. The susceptibility of the organism is determined by reference to the zone-size interpretive chart (see Table 4.2).

riate size will develop to indicate sensitivity or resistance. After incubation, the presence of inhibition zones around the discs of the different agents is noted. It should be emphasized that the relative sizes of the zones do not necessarily indicate the relative activities of the various agents, since as noted, zone size depends on rate of diffusion of the agent as well as on its potency. Table 4.2 presents typical zone sizes for several antibiotics. Zones observed on the plate are measured and compared to the figures given, to determine if the isolate is truly sensitive to a given antibiotic.

It is convenient in the sensitivity testing of large numbers of cultures, as in a hospital laboratory, to use special dispensers that will drop a number of separate discs onto a plate in a single operation, as shown in Figure 4.14. If small discs are used, as many as 12 can be placed on a single standard-sized petri plate. The discs are purchased commercially from suppliers with the antimicrobial agents already present in the dry discs at the desired concentrations.

KINDS OF ANTIBIOTICS AND THEIR ACTIONS

A number of antibiotics have been well-characterized and their practical uses defined. We discuss some of the most important antibiotics in the following pages.

CHLORAMPHENICOL   This broad-spectrum antibiotic, produced biologically by a *Streptomyces* species, was the first antibiotic also to be produced by chemical synthesis. It inhibits protein synthesis in both

Antibiotic	Amount on disc	Inhibition zone diameter (mm)		
		Resistant	Intermediate	Sensitive
Ampicillin[a]	10 μg	11 or less	12–13	14 or more
Ampicillin[b]	10 μg	20 or less	21–28	29 or more
Chloramphenicol	30 μg	12 or less	13–17	18 or more
Colistin	10 μg	8 or less	9–10	11 or more
Erythromycin	15 μg	13 or less	14–17	18 or more
Gentamicin	10 μg	12 or less		13 or more
Kanamycin	30 μg	13 or less	14–17	18 or more
Methicillin	5 μg	9 or less	10–13	14 or more
Neomycin	30 μg	12 or less	13–16	17 or more
Penicillin-G[c]	10 Units	20 or less	21–28	29 or more
Penicillin-G[d]	10 Units	11 or less	12–21	22 or more
Polymyxin-B	300 Units	8 or less	9–11	12 or more
Streptomycin	10 μg	11 or less	12–14	15 or more
Tetracycline	30 μg	14 or less	15–18	19 or more

TABLE 4.2
Typical zone sizes for some common antibiotics

[a] For Gram-negative organisms and enterococci.
[b] For staphylococci and highly penicillin-sensitive organisms.
[c] For staphylococci.
[d] For organisms other than staphylococci; includes some organisms, such as enterococci, that may cause systemic infections treatable by high doses of Penicillin-G.
Data based on "Rules and regulations, antibiotic susceptibility discs," *Fed. Regist.* (September 30, 1972), **37**, 20525–20529.

Gram-positive and Gram-negative procaryotes. Chloramphenicol does have some toxicity in man and has caused anemia and death when used for prolonged periods at high doses, although it is relatively safe for short-term uses. Its most common medical use today is in the treatment of typhoid fever and rickettsial infections, such as typhus and Rocky Mountain spotted fever.

CYCLOHEXIMIDE   Sometimes called *Acti-Dione*, this antibiotic is also produced by a *Streptomyces* species. It inhibits protein synthesis in eucaryotes but has no effect in procaryotes; hence it is of no use in treating bacterial infections. Although effective against fungi and therefore of potential use for fungal infections, it is quite toxic to man and has only been used medically in rare emergencies. It has found some use in agriculture in the control of fungus diseases and has been marketed especially for control of cherry diseases and turf diseases on golf courses.

ERYTHROMYCIN   This antibiotic is one of a series of related antibiotics called the *macrolides*, a group that also includes *spiramycin*, *carbomycin*, and *oleandomycin*. All are produced by *Streptomyces* and inhibit protein synthesis in procaryotes but not in eucaryotes. They are relatively nontoxic to man, but their antimicrobial activity is mostly limited to Gram-positive bacteria. These antibodies are used primarily for treatment of infections due to *Staphylococcus* and *Streptococcus*.

GRISEOFULVIN   This antibiotic is produced by a fungus, *Penicillium griseofulvum*, and is active against other fungi. It is used medically for the control of fungal infections of the skin and is noteworthy because of its low human toxicity, in contrast to most other antibiotics that inhibit fungi. However, griseofulvin is not effective when applied directly to the skin but must be given orally. The drug then enters the intestine, passes through the bloodstream, and accumulates in the skin tissues, thus providing protection from fungal attack. To cure a fungal infection of the skin, the drug must usually be taken over a long period of time at fairly high doses, a procedure possible only because of the low toxicity.

ISONICOTINIC ACID HYDRAZIDE   This drug, abbreviated INH (and also known as *isoniazid*), is a synthetic chemical rather than an antibiotic, although it is usually classified with the antibiotics because of its manner of use. It is one of the most effective drugs for the treatment of tuberculosis and is quite specific in its action against the causal agent, *Mycobacterium tuberculosis*, being ineffective against other microorganisms. It is given orally, usually for extended periods of time. Although it could be used alone, it is the usual practice to administer INH together with another antituberculosis drug, such as the antibiotic streptomycin or the synthetic chemical *p*-aminosalicylic acid.

NYSTATIN   This antibiotic is a member of a large group of antibiotics called *polyenes* that are produced by *Streptomyces* and are active against fungi. Nystatin was one of the first polyenes to be discovered, and it has found some use in the control of fungal infections of the intestine. It is too toxic to be given by injection, but when given orally it remains in the intestinal tract and hence does not cause undue toxicity. Other polyene antibiotics include *filipin*, *candicidin*, and *amphotericin*. These antibiotics act by combining with and destroying the integrity of the cell membrane. They act only against eucaryotes and have no effect on bacteria. In addition to its use in the control of fungal infections, nystatin is also used in combination with antibiotics active against bacteria, such as streptomycin or tetracycline, to treat intestinal infections caused by bacteria. The reason for this is that use of an antibiotic effective against bacteria often leads to fungal growth in the intestine, and if nystatin is present, the fungal growth can be prevented.

PENICILLIN   This was the first antibiotic to find widespread medical use and today is still one of the most useful antibiotics. The story of the discovery of penicillin by Fleming and the development of methods for its large-scale production was presented in Section 1.7.

Penicillin is not a single compound but a whole family of antibiotics, all related chemically but differing in detail and varying considerably in the range of organisms attacked. Some penicillins affect only Gram-positive organisms; others are active primarily against Gram-negative bacteria; while still others are active against both kinds of bacteria. Some penicillins are destroyed by stomach acids and can only be given by injection, whereas others are acid stable and can be taken orally. The core of the penicillin molecule is a substance called 6-*aminopenicillanic acid* (APA), and attached to this core are the side chains that confer specific activities. The APA core is produced by several kinds of fungi, usually with one or more side chains attached. The penicillin first discovered, now called penicillin G, is produced by the fungus *Penicillium chrysogenum*, and it was from the name of this genus that the antibiotic name was derived. Some other species of *Pencillium* and fungi of the genus *Cephalosporium* also produce penicillin antibiotics. The most commonly used penicillins include penicillin G, ampicillin, and methicillin.

All penicillins have a similar mode of action, affecting cell-wall synthesis in procaryotes. Eucaryotes, which have a cell-wall composition quite different from that of procaryotes, are unaffected by normal concentrations of penicillin; these antibiotics are therefore quite nontoxic to man and most animals. However, penicillins do cause allergic reactions in some people and hence are not completely harmless. Because they affect cell-wall synthesis, penicillins are bactericidal only to growing cells, since as cells grow they must continue to synthesize new cell wall in order to maintain cellular integrity. The cell with a weakened cell wall eventually bursts, a process called *lysis*. Penicillin is without serious effects on nongrowing cells, since new cell-wall material is not being synthesized. Because of this, nongrowing cells may remain alive for long periods in the presence of penicillin.

POLYMYXIN   This antibiotic, produced by a species of *Bacillus*, acts on cell membranes and is especially effective against Gram-negative bacteria. It is unfortunately somewhat toxic to humans and hence is only used medically after less toxic antibiotics have proved unsuccessful. The most common use for polymyxin is in the control of *Pseudomonas* infections, since this Gram-negative bacterium is often resistant to other antibiotics.

STREPTOMYCIN   This antibiotic is a member of a group of chemically related antibiotics called the *aminoglycosides*, which also includes *neomycin, gentamycin*, and *kanamycin*. They prevent protein synthesis by affecting the activity of the ribosomes of procaryotes, and they are ineffective against eucaryotic ribosomes, although they are not without toxicity to man and animals. The aminoglycosides are produced by members of the genus *Streptomyces*, and the name streptomycin derives from the genus name of the producing microorganism. Streptomycin is effective against a variety of Gram-positive and Gram-negative bacteria but has found its widest use in the treatment of tuberculosis. It is usually given in combination with another antituberculosis agent, such as INH (discussed earlier). When given orally streptomycin remains in the intestinal tract and does not enter the bloodstream, so that for use in tuberculosis therapy, the antibiotic must be administered by injection. When given for prolonged periods, it may cause deafness and hence must be used cautiously. All of the aminoglycosides are able to cause deafness, and some also cause kidney damage. These toxic effects are unrelated to the action of the antibiotic against ribosomes and protein synthesis.

SULFA DRUGS   Although not antibiotics, the sulfa drugs, or sulfonamides, a group of synthetic chemicals, are quite useful in chemotherapy and are often used instead of or in conjunction with antibiotics. The sulfa drugs were among the first chemotherapeutic agents discovered, as we discussed in Section 1.7. All sulfa drugs contain a basic sulfonamide structure plus chemical modifications, and various sulfa drugs find use for control of different Gram-positive or Gram-negative bacteria. They continue to find widespread medical use even since the discovery of antibiotics, and a wide variety of sulfa drugs is currently marketed. The sulfonamides are the best example of a kind of antimicrobial agent called a *growth factor analog*. As we have seen, many microorganisms require growth factors in their diet. A growth factor analog is a substance that resembles a growth factor chemically but is sufficiently different so that it is nonfunctional. The organism is "fooled" by the analog and takes it in place of the growth factor, and its growth is thus inhibited. The sulfonamides are analogs of the growth factor *p*-aminobenzoic acid (PABA). The latter is a necessary part of the vitamin folic acid, and sulfa drugs thus prevent synthesis by the bacteria of folic acid. However, if an organism does not synthesize folic acid from PABA but requires folic acid preformed in the medium, then its growth is not inhibited by sulfa drugs. Humans require preformed folic acid and are not affected by the sulfa drugs; hence these drugs can be used selectively in the control of human infections.

TETRACYCLINES   The tetracyclines are another family of antibiotics and include the parent antibiotic *tetracycline* and two derivatives, *chlortetracycline* and *oxytetracycline*. These antibiotics are produced by *Streptomyces* and act by preventing protein synthesis, although by a mechanism different from that of the aminoglycosides. The tetracyclines act only against procaryotes and are much less toxic to man than are the aminoglycosides. In addition, they are absorbed into the bloodstream when given orally and hence are easy to administer. The tetracyclines are broad-spectrum antibiotics and are useful against a wide variety of Gram-positive and Gram-negative bacteria, although in contrast to streptomycin, they are not useful in the treatment of tuberculosis. Despite their lack of human toxicity, the tetracyclines must still be administered cautiously, since when given orally they alter the normal bacteria of the intestine and may cause intestinal disturbances.

Many organisms are able to develop resistance to antibiotics through genetic changes called *mutations* (Figure 4.15). Antibiotic resistance is usually a quantitative phenomenon; the resistant organisms may still be affected but only if higher concentrations of the agent are used. Antibiotic resistance is thus expressed in terms of a change in the minimum inhibitory concentration, discussed earlier in this chapter. Development of resistance to one antibiotic generally does not result in concomitant resistance to other antibiotics unless they are closely related. If an organism is resistant to a second antibiotic as a result of being resistant to the first, this is called *cross resistance*.

Several mechanisms of antibiotic resistance are known: (1) The sensitive structure (such as the cell wall, enzyme, or ribosome) may be lacking in the resistant form. For example, this is seen in certain forms, called *L forms*, that have acquired resistance to penicillin by losing the need to synthesize a cell wall. (2) The cellular structure that is the antibiotic's target may undergo an alteration so that it no longer binds the antibiotic but is still able to carry out its normal function. (3) The resistant organism may be impermeable to the antibiotic. If this is so, then the antibiotic may still act on the sensitive structure if the permeability barrier is altered or destroyed. (4) The organism may be able to modify the antibiotic to an inactive form. This is very frequently the case in penicillin resistance. A group of enzymes (penicillinases) split off the side chain of the penicillin molecule rendering it inactive. Penicillinases are produced by a wide variety of bacteria and have definitely been shown to confer antibiotic resistance. Some penicillins are less susceptible to penicillinase activity than are others and thus are effective on otherwise penicillin-resistant bacteria. Penicillins that are resistant to penicillinase action have different side chains from those which are sensitive, and in recent years, many of these resistant types have been synthesized chemically and are now widely used in medicine. We discuss the genetics of antibiotic resistance in Chapter 7.

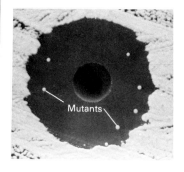

Mutants

FIGURE 4.15
Development of antibiotic-resistant mutants within the zone of inhibition. Close-up view of inhibition zone, showing mutants resistant to antibiotic.

Antibiotics were originally discovered by chance through studies on antagonism between various microorganisms, but during the past 20 to 30 years, most new antibiotics have been discovered through intense and careful search. Seeking new antibiotics has occupied the time of a large number of microbiologists in pharmaceutical companies. Many pharmaceutical firms have invested millions of dollars in the search for new antibiotics and have developed a variety of techniques for such searches. New antibiotics are still needed in part because antibiotic resistance to existing antibiotics develops and in part because a number of infectious diseases are still not effectively controlled by available techniques.

The most widely used procedure for searching for a new antibiotic is called the *screening approach*: A large number of possible antibiotic-producing organisms are isolated from natural environments, and each is tested against a variety of other organisms to see if antagonisms exist. The battery of test organisms is usually large and includes Gram-positive and Gram-negative bacteria, *Mycobacterium* (the causal agent of tuberculosis), yeasts and other fungi, and perhaps protozoa.

A convenient procedure in examining for antibacterial agents is to streak the possible antibiotic producer along one side of an agar plate and incubate the plate for a few days to allow the organism to grow and produce antibiotic. Then a series of test bacteria are streaked at right angles to the first streak, and the plate is reincubated and examined for zones of inhibition (Figure 4.16). Since different antibiotics act upon different groups of organisms, it is desirable that any antibiotic screening program use a wide range of test organisms to make certain that few antibiotic-producing organisms are missed. Since environment can greatly affect antibiotic production, obviously there must be appropriate medium and culture conditions permitting such production.

Once antibiotic activity is detected in a new isolate, the microbiologist will determine if the agent is new or identical with an existing one. With so many antibiotics already known, the chances are good that the antibiotic is not new. Often simple chemical methods will permit the characterization and identification of the antibiotic. If the agent appears to be new, larger amounts are produced, and the antibiotic is purified and sometimes crystallized. Finally, the antibiotic is tested for therapeutic activity—first in infected animals, later perhaps in man.

Most screening programs concentrate on antibiotics produced by the genus *Streptomyces*, a group of organisms that has yielded most of the medically useful antibiotics now known, including chloramphenicol, streptomycin, neomycin, erythromycin, cycloheximide, and tetracycline. *Streptomyces* occur predominantly in the soil, and most screening programs collect soil samples from a wide variety of geographical areas, under the assumption that soils from new places may yield new antibiotic producers. However, it does not necessarily follow that soils from distant parts will yield more new antibiotics than would local soils. In fact, since the wind-borne spores of

FIGURE 4.16
Method of testing an organism for the production of antibiotics, as used in antibiotic screening programs. The producer (*Streptomyces*) was streaked across one chord of the plate, and the plate was incubated. After good growth had been obtained, the test bacteria were streaked at right angles, and the plate was further incubated. The clear zones indicate production of antibiotic active against test organisms 2, 3, and 5.
Test organisms
1 *Escherichia coli*
2 *Bacillus subtilis*
3 *Staphylococcus epidermidis*
4 *Klebsiella pneumoniae*
5 *Mycobacterium smegmatis*

Streptomyces are widely dispersed, species producing the same antibiotics are found in similar kinds of soils throughout the world.

Organisms often produce more than one antibiotic and sometimes as many as five or six. Multiple antibiotic production presents special problems, as activity against a test organism may be due to the concerted action of the several agents. The different antibiotics may all be closely related chemicals, or they may differ markedly. They may all be active against a single test organism, or they may show different spectra of activity. The presence of multiple antibiotics presents further problems and challenges for the microbiologist and chemist: Which antibiotic in the mixture is likely to be medically useful? How can its production be favored over the others? And how can it be purified? Antibiotics research presents endless puzzles and fascination for the microbiologist. When a new antibiotic is discovered, the interesting work has only just begun.

Summary

In this chapter, we have seen some of the processes involved in microbial growth and its control. Unicellular microorganisms grow exponentially, and because of this, very large populations of cells can develop quickly. However, growth does not continue indefinitely. Eventually a stationary phase is reached, owing to either exhaustion from the medium of an essential nutrient or accumulation of a toxic waste product. The stationary phase is often followed by a death phase. Growth is monitored by measuring dry weight, estimating turbidity, or counting cell numbers. Both direct microscopic counts and viable counts can be performed, and each of these has advantages and disadvantages.

A number of processes or agents can be used to control microbial growth. When all organisms are killed, this is generally called sterilization. Heat is one of the most effective and widespread sterilization methods. Pasteurization is a heat process that kills some but not all organisms and is used mainly in the treatment of milk. Microbial growth can be prevented by use of low temperatures or drying, but these treatments do not result in sterilization. Sterilization can also be effected by ultraviolet radiation or ionizing radiation (X rays, nuclear energy).

A wide variety of chemicals, called *antimicrobial agents*, can be used to control microbial growth. Cidal agents kill microorganisms, resulting in sterilization, whereas static agents inhibit growth without causing death. An important feature of many antimicrobial agents is their *selective toxicity*, the ability to act more effectively against some organisms than others. Agents that are selective for microorganisms without affecting man can be used in control of infectious diseases. Antibiotics are an important class of antimicrobial agents. They are chemical substances produced by one microorganism that act upon others. *Broad-spectrum* antibiotics act on a wide variety of microorganisms; others have more restricted spectra of activity. Although hundreds of antibiotics are known, only a few are medically useful. Resistance to antibiotics can develop in sensitive microorganisms, presenting complications in the control of infection. The search for new antibiotics is carried out by large-scale testing of microorganisms isolated from soil. New antibiotics are still needed, in part because antibiotic resistance develops to existing antibiotics and in part because some infectious diseases are still not effectively controlled by available methods.

<table>
<tr><td>Study questions</td><td>

1 Define: *growth, growth rate, generation time.*

2 Draw a graph that describes the growth pattern of a unicellular microorganism after it is inoculated into a fresh medium. Name the various growth phases.

3 What is the practical significance of the fact that bacteria and other unicellular organisms grow at exponential rates?

4 Explain briefly the principle behind the direct microscopic method for counting cell number.

5 Explain briefly the principle behind the viable counting method.

6 Express the following numbers as exponents of 10: 1,000; 1,000,000; 1,000,000,000; 5,000,000.

7 Describe how turbidity measurements can be used to measure growth rate.

8 What is thermal death time? List several factors that affect this time, and indicate the practical significance of each.

9 Is the thermal death time longer or shorter if bacterial spores are present? Explain.

10 What is pasteurization? Describe two methods by which pasteurization is achieved. How is pasteurization used practically?

11 Explain why the following do not spoil when allowed to stand for long periods of time at room temperature: canned peas (unopened); strawberry preserves; powdered milk.

12 Explain why the following do spoil when allowed to stand for long periods of time at room temperature: canned peas (opened); fresh strawberries; pasteurized milk.

13 Define briefly: *bactericide, fungicide, germicide, antiseptic, disinfectant.*

14 How do a bactericidal and a bacteriostatic agent differ in their activity?

15 What is an antibiotic? a chemotherapeutic agent?

16 What is a broad-spectrum antibiotic?

17 Describe briefly the agar diffusion method for assaying antibiotic activity.

18 Describe briefly a method for determining the antibiotic sensitivity of a bacterial culture.

19 List four common antibiotics, and indicate how they are useful in medicine.

20 What is antibiotic resistance? Describe two mechanisms by which resistance arises.

21 How would you go about searching for a new antibiotic? How would you test its activity?

</td></tr>
</table>

<table>
<tr><td>Suggested readings</td><td>

Brock, T. D., *Biology of microorganisms,* 2nd ed., Prentice-Hall, Inc., Englewood Cliffs, N.J., 1974. Current textbook discussing the measurement of microbial growth and its control.

Olds, R. J., *Colour atlas of microbiology,* Wolfe Medical Books, London, 1975. Good photographs illustrating some of the actions of antibiotics and other chemotherapeutic agents.

Sistrom, W. R., *Microbial life,* 2nd ed., Holt, Rinehart and Winston, New York, 1969. Elementary discussion of microbial growth.

Stanier, R. Y., E. A. Adelberg, and J. L. Ingraham, *The microbial world,* 4th ed., Prentice-Hall, Inc., Englewood Cliffs, N.J., 1976. Good modern treatment of microbial physiology and growth.

</td></tr>
</table>

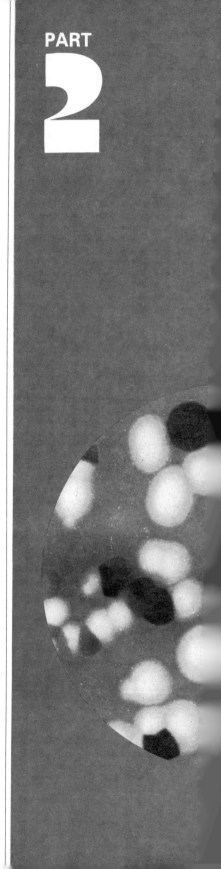

Molecular biology is that discipline that explains biological phenomena in terms of the molecules of cells. It attempts to provide chemical and physical explanations for cell structure and function. Molecular biology represents a merging of the more traditional fields of biochemistry and genetics. Although molecular biology is not restricted to a study of microorganisms, in actuality most of the knowledge in this field derives from studies with microbes. This has been because microbes are simple in structure and function and are likewise simple to study. However, the principles of molecular biology learned through a study of microorganisms are now being applied to higher organisms, with important consequences for an understanding of such important human problems as cancer. At the microbial level, there have also been many practical consequences of the study of molecular biology. In infectious disease, an understanding of the chemistry of the toxic products of microorganisms has greatly helped in devising appropriate immunization methods; in industrial micro-biology, new high-yielding microbial processes have been developed using the principles of molecular biology and genetics. The field of genetic engineering deals with procedures for constructing totally new

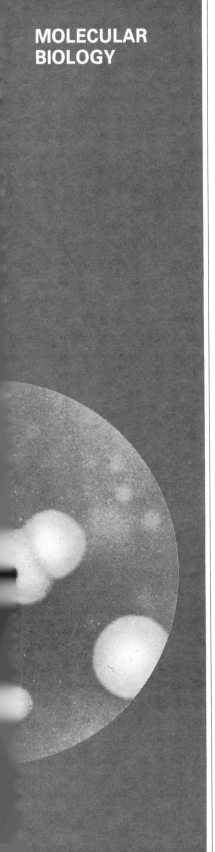

MOLECULAR BIOLOGY

organisms; this presents a real challenge, since useful organisms might be created, but continual watchfulness is necessary to ensure that harmful organisms are not created at the same time.

Because chemistry and physics provide the language of molecular biology, we begin this part with a brief chapter (Chapter 5) providing some of the necessary chemical and physical background. Chapter 6 then discusses the chemical processes by which microorganisms obtain energy and synthesize the vast array of chemical substances needed for normal cell function. Chapter 6 also discusses the synthesis of nucleic acids and proteins, providing some insight into how cellular growth processes are regulated. It concludes with a brief section on the modes of action of antibiotics, explaining how they work at the molecular level.

Chapter 7 then deals with microbial genetics, showing how microorganisms can undergo genetic change, either by mutation or by genetic recombination. The rudimentary sexual processes in bacteria are discussed here. The chapter concludes with a discussion of genetic engineering and its consequences and points out some of the problems of converting this technology into practice.

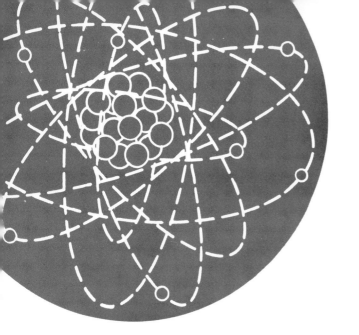

Chapter 5

Chemical and physical background

Matter is the substance of which physical objects are constructed. There are three states of matter: *solid*, *liquid*, and *gas*. Most substances can be converted from one state of matter to another by application of heat or cold. For instance, water is a liquid, but it can be converted to a solid, ice, by cold or into a gas, steam, by heat. When a liquid is converted into a gas, it *boils*; and when it is converted into a solid, it *freezes*. Some substances are more difficult to turn from one state of matter to another than water is. Iron metal, for instance, requires a great amount of heat to be converted into liquid (molten) metal, and this conversion is only done in special high-temperature furnaces. Lead metal, on the other hand, can be melted on an ordinary stove.

Substances are composed of elementary units called *atoms*. Atoms were first thought to be indivisible particles, constituting the smallest structures that substances could be divided into, but now we know that atoms themselves can be broken down into smaller particles. Over 100 different kinds of atoms are known.

A substance can be composed of just a single kind of atom or of more than one kind of atom. A substance that is composed of only one kind of atom is called an *element*. Examples of elements are iron, aluminum, mercury, and oxygen. Substances can also be formed of several different kinds of atoms. If the different kinds of

5.1
Matter

FIGURE 5.1
The structure of matter.

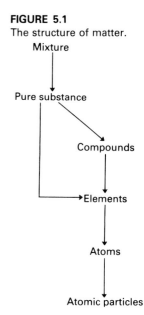

132

atoms making up the substance are present in definite proportions, then the substance is said to be a *compound*. Examples of compounds are table salt, water, and sugar. Many substances are composed of atoms in indefinite proportions; these are not compounds but *mixtures* (for instance, wood, sea water, a bacterial cell, air). By chemical procedures, it may be possible to separate out from a mixture the compounds or elements of which it is composed (Figure 5.1). This is the essential first step in determining the nature of a mixture. All studies on the nature of substances must be done on pure substances, with the constituent atoms present in definite proportions.

CHEMICAL SYMBOLS

TABLE 5.1
Some chemical symbols

Element	Symbol
Calcium	Ca
Carbon	C
Chlorine	Cl
Cobalt	Co
Copper	Cu
Hydrogen	H
Iron	Fe
Lead	Pb
Magnesium	Mg
Manganese	Mn
Mercury	Hg
Molybdenum	Mo
Nitrogen	N
Oxygen	O
Phosphorus	P
Potassium	K
Sodium	Na
Sulfur	S
Zinc	Zn

Chemists use a kind of shorthand to write about atoms and elements, and it is necessary for the student to have some familiarity with this shorthand. Each element is designated by a symbol of one or two letters, which is often derived from the name of the element. For instance, the symbol for oxygen is O, that for hydrogen, H, that for carbon, C, and so forth. Not all symbols are so easy to figure out as these. Some symbols are derived from the Latin names of elements, such as Fe for iron, from the Latin *ferrum*.

The chemical symbol for a compound is constructed by combining the symbols of its constituent elements in the proper proportions. Thus, water, composed of hydrogen and oxygen in the proportions 2 to 1, is symbolized H_2O; and table salt, composed of the elements sodium (symbol Na) and chlorine (symbol Cl) in equal proportions, is symbolized NaCl. Symbols for those elements of biological importance are given in Table 5.1. There is a brief discussion of the elements required for the growth of microorganisms in Section 3.1.

5.2
Structure of atoms

Atoms are composed of subatomic particles of a surprising variety and diversity. Modern high-energy physics has succeeded in breaking down the atom into ever smaller particles, and the end is not yet in sight. However, from the points of view of chemists and biologists, most of these particles are of little concern. For our purposes, we may consider the atom to be composed of only three kinds of particles: *protons*, *neutrons*, and *electrons* (Table 5.2). Protons and neutrons are associated together in the central core of the atom in a structure called the *atomic nucleus*. The electrons move in orbits around the nucleus. Electrons are much smaller in size than protons and neutrons; so the weight of an atom is determined almost solely by its complement of protons and neutrons. Protons and neutrons are of equivalent weight but differ electrically. The neutron is electrically neutral, whereas the proton is electrically positive. The electron has a negative charge that is exactly opposite the charge on the proton; thus, protons and electrons balance each other electrically.

The classic view of how these three kinds of atomic particles associate to form the atom is shown in Figure 5.2, which illustrates the so-called planetary model of the atom. (The nucleus is equivalent to the sun; the electrons to the planets.) Although the planetary model is graphic, it is not very accurate in showing how electrons really behave, and for most purposes, the model is no longer used. The positions of electrons are most accurately spoken of in terms of *energy levels*, which express the amount of energy it takes to move the electron. Electrons closer to the nucleus are harder to move than electrons that are farther from the nucleus. Energy levels are numbered, from one, the lowest, through seven, the highest. In the lowest energy level, only 2 electrons can exist; at the next higher energy level, 8 electrons can exist; and at the third level, 18 can exist. Each energy level can be viewed as a shell of electrons surrounding the nucleus.

Each element has a characteristic number of electrons. The number of electrons in the element are exactly balanced by the protons in the nucleus, so that the negative charges from the electrons are balanced by positive charges from the protons. Thus, the element as a whole has no charge.

The weight of an atom, its *atomic weight*, is determined by the number of protons and neutrons in its nucleus. The *atomic number* corresponds to the number of protons in the nucleus. The chemical properties of an atom are determined by the number of protons and electrons it has; neutrons add weight but do not change the chemical properties (although physical properties are affected by the number of neutrons). Beginning with the element hydrogen, which has 1 proton (atomic number 1) there is a step-wise increase in the number of protons from 1 element to the next. As the number of protons increases, the number of electrons increases correspondingly. The electrons are added to the shells surrounding the nucleus in a definite order. Table 5.3 gives the electronic configurations of the first 36 elements, beginning with hydrogen and ending with krypton. Most of the major biological elements in living organisms

Orbits of the electrons around the nucleus

The positively charged nucleus contains protons and neutrons

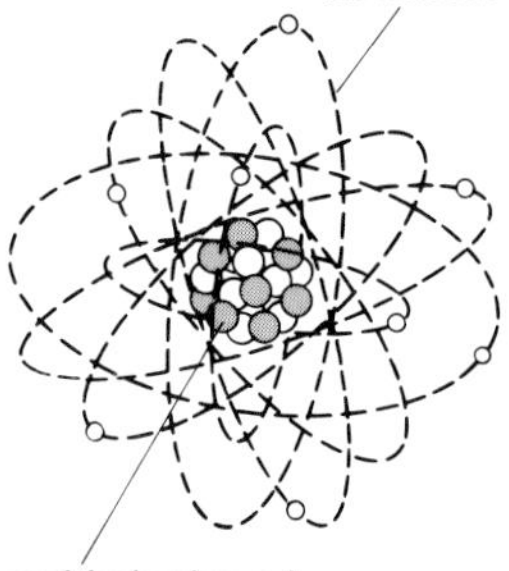

FIGURE 5.2
The structure of the atom.

Particle	Location	Mass[a]	Charge
Proton	Nucleus	1	Positive
Neutron	Nucleus	1	Neutral
Electron	Orbits	$\frac{1}{1,837}$	Negative

TABLE 5.2
Particles in the atom

[a] Mass given in atomic mass units (amu). One amu = 1.66×10^{-24} grams.

CHEMICAL AND PHYSICAL BACKGROUND

TABLE 5.3
Electronic configurations of some of the elements

Element	Atomic number (proton number)	Number of electrons in shells			
		Shell 1	Shell 2	Shell 3	Shell 4
Hydrogen	1	1			
Helium	2	2			
Lithium	3	2	1		
Beryllium	4	2	2		
Boron	5	2	3		
Carbon	6	2	4		
Nitrogen	7	2	5		
Oxygen	8	2	6		
Fluorine	9	2	7		
Neon	10	2	8		
Sodium	11	2	8	1	
Magnesium	12	2	8	2	
Aluminum	13	2	8	3	
Silicon	14	2	8	4	
Phosphorus	15	2	8	5	
Sulfur	16	2	8	6	
Chlorine	17	2	8	7	
Argon	18	2	8	8	
Potassium	19	2	8	8	1
Calcium	20	2	8	8	2
Scandium	21	2	8	9	2
Titanium	22	2	8	10	2
Vanadium	23	2	8	11	2
Chromium	24	2	8	13	1
Manganese	25	2	8	13	2
Iron	26	2	8	14	2
Cobalt	27	2	8	15	2
Nickel	28	2	8	16	2
Copper	29	2	8	17	2
Zinc	30	2	8	18	2
Gallium	31	2	8	18	3
Germanium	32	2	8	18	4
Arsenic	33	2	8	18	5
Selenium	34	2	8	18	6
Bromine	35	2	8	18	7
Krypton	36	2	8	18	8

Elements from Scandium (21) to Zinc (30) are bracketed as **Transition elements**.

are found in this list of 36 elements. Note that electrons are added to each shell until it is filled, after which electrons are added to the next higher shell. As we discussed, the shells correspond to energy levels. The third shell can actually accommodate 18 electrons, but after 8 electrons are added (element argon), further electrons are not added to this shell immediately, but additions to the fourth shell begin. After 2 electrons are added to the fourth shell, the addition of electrons returns to the third shell, with further additions to the specified 18 electrons. The elements in this series between atomic number 21 and 30 all have just 2 electrons in their outermost shell and because of this have some definite similarities in chemical properties. The elements in this series are called *transition elements*.

Table 5.3 illustrates the regular progression in addition of electrons to the electron shells. There are definite repeats of the electronic configurations of the outer shells of the elements, from one level to the next. Thus, lithium, sodium, and potassium have only one electron in their outer shell; oxygen, sulfur, and selenium have six electrons in their outer shell; and so forth. It has been found that elements such as these, with the same number of electrons in their outer shells, often have similar chemical properties. Elements with similar properties can be grouped in families.

Recognition of these families led to the development of the periodic table, which groups the elements in configurations that express both the periodic changes in electronic configurations of the outer shells and the grouping of elements by families. The periodic table of the elements is presented in Figure 5.3. Four families with distinctive properties are the *alkali metals*: lithium, sodium, potassium, rubidium, cesium, and francium; the *alkaline earth metals*: beryllium, magnesium, calcium, strontium, barium, and radium; the *halogens*: fluorine, chlorine, bromine, iodine, and astatine; and the *noble gases*: helium, neon, argon, krypton, xenon, and radon.

Although there are fixed relationships between the number of protons and the number of electrons, as shown in Table 5.3, the number of neutrons can vary. It is possible for two atoms to exist with exactly the same number of protons, and hence, with the same chemical properties, but with different numbers of neutrons. Two atoms with identical numbers of protons but different numbers of neutrons are called *isotopes*. Thus, oxygen, with 8 protons, generally has 8 neutrons as well, with an atomic weight of 16, but oxygen isotopes are known with 6, 7, 9, 10, and 11 neutrons, corresponding to atomic weights of 14, 15, 17, 18, and 19. The various isotopes of an element are symbolized by adding the atomic weight as a superscript: $O^{14}, O^{15}, O^{16}, O^{17}, O^{18}, O^{19}$.

Some isotopes of an element are *radioactive*, giving off emanations that can have biological effects on organisms. Oxygen isotopes that are radioactive are O^{14}, O^{15} and O^{19}. Those isotopes that are not radioactive are sometimes called *stable* isotopes.

Generally, one of the isotopes of an element is considerably more abundant in nature than the others. Thus, the most common oxygen isotope is O^{16}, which constitutes 99.759 percent of the oxygen of the atmosphere; O^{17} is found in 0.037 percent and O^{18} in 0.204 percent. The other isotopes of oxygen are known only from high-energy physics experiments and do not exist in nature.

Most atoms do not exist independently in nature but in combination with other atoms. A *molecule* is a specific association of atoms held together by strong forces called *chemical bonds*. Because of the

ISOTOPES

5.3
Molecules and ions
MOLECULES

FIGURE 5.3
The periodic table of the elements. The biologically important elements are shaded. The numbers are explained on the box for hydrogen, H.

strength of chemical bonds, the atoms of molecules remain tightly together and are separated only if sufficient energy is introduced to break the bonds. Chemical reactions occur only when chemical bonds are broken.

An example of a molecule is water, H_2O, which consists of one oxygen atom and two hydrogen atoms. Other examples of molecules are: hydrogen sulfide, H_2S; ammonia, NH_3; carbon dioxide, CO_2; methane, CH_4. Some molecules are composed of only one kind of atom. Thus, oxygen gas consists of two atoms of oxygen held together by chemical bonds, O_2. Likewise, nitrogen gas consists of two nitrogen atoms, N_2, and hydrogen gas of two hydrogen atoms, H_2.

An *ion* is a charged particle. It arises when an atom or molecule either gains or loses electrons. If a particle gains electrons, it acquires a negative charge, since electrons are negative. If a particle loses electrons, it acquires a positive charge, since it has lost a negative electron.

IONS

Some ions have special names. Thus, *chlorine* is the atom, but *chloride* is the ion. Iron has two ions, with different charges, named *ferrous*, Fe^{2+}, and *ferric*, Fe^{3+}. Most ions, however, do not have special names and are thus designated by their atomic names, such as calcium ion, sodium ion, magnesium ion. Some ions of importance in microbiology are listed in Table 5.4.

Ions can be moved by electrical currents. If two platinum wires are put into a solution of table salt (NaCl), one connected to the positive pole of a battery, the other to the negative pole, a current will flow through the solution. The sodium ions, being positively charged, will be attracted to the negative pole, and the chloride ions, being negatively charged, will be attracted to the positive pole. In electrochemistry, the negative pole is called the *cathode*, and the ions that accumulate at this pole are called *cations*. The positive pole is the *anode*, and negative ions are thus called *anions*. The words *cation* and *anion*, referring to positive and negative ions respectively, are often used in chemistry.

The ions discussed up to now have been derived from single atoms. Many ions contain more than one atom. Ions containing more than one kind of atom are called *polyatomic* ions. Some of the most common and important substances in microbiology consist of poly-atomic ions. A few examples are included in Table 5.4.

POLYATOMIC IONS

It is often necessary when discussing solutions and chemical reactions to express concentrations of dissolved substances. For many purposes, it is sufficient merely to state the weight of solute that is

5.4
Expressing concentrations

TABLE 5.4
Some ions of importance in
microbiology

Monoatomic ions

Hydrogen	H^+
Sodium	Na^+
Magnesium	Mg^{2+}
Sulfide	S^{2-}
Chloride	Cl^-
Potassium	K^+
Calcium	Ca^{2+}
Manganous	Mn^{2+}
Ferrous (iron)	Fe^{2+}
Ferric (iron)	Fe^{3+}

Polyatomic ions

Ammonium	NH_4^+
Sulfate	SO_4^{2-}
Carbonate	CO_3^{2-}
Nitrate	NO_3^-
Hydroxide	OH^-
Phosphate	PO_4^{3-}

dissolved in the solvent, either in grams or milligrams. However, because different chemical substances have different molecular weights, a gram of one substance does not have the same number of molecules as a gram of another substance. It is often desirable to express the number of molecules present, so that it can be determined if there are enough molecules of one substance present to react with another substance. The number of molecules is, in most cases, a very large number. To avoid large numbers but to express concentrations in terms of numbers of molecules, a quantity known as the *mole* is used. A mole of any one substance contains the same number of molecules as a mole of any other substance, 6.023×10^{23} molecules.[1]

A mole of an element is equal to its atomic weight in grams. Since carbon has an atomic weight of 12 (see Figure 5.3), a mole of carbon weighs 12 grams. Sodium has an atomic weight of 23, so that a mole of sodium weighs 23 grams. The number of atoms in 12 grams of carbon or 23 grams of sodium is the same.

For molecules, one mole is calculated by adding up the atomic weights of the atoms present. Thus, for NaCl, sodium chloride:

$$Na\ (23\ g) + Cl\ (35.5\ g) = NaCl\ (58.5\ g)$$

In practical terms, if we want to carry out a reaction between HCl, hydrochloric acid, and NaOH, sodium hydroxide (a strong base), so that the two exactly neutralize each other, we need to have present the same number of moles (or fractions of moles) of each substance:

$$HCl = 36.5\ g \quad H = 1 \quad Cl = 35.5$$
$$NaOH = 40\ g \quad Na = 23 \quad O = 16 \quad H = 1$$

Mix 36.5 grams of HCl and 40 grams of NaOH to obtain an exactly neutralized solution:

$$HCl + NaOH = NaCl + H_2O$$

For many biochemical purposes, a mole is too large a quantity to work with. Biochemists often use quantities such as $\frac{1}{1,000}$ of a mole, which are *millimoles*, or even $\frac{1}{1,000,000}$ of a mole, *micromoles*.

A *molar solution* contains one gram-molecular weight (or mole) of solute in a liter of solution. In the chemistry or biochemistry laboratory, concentrations are most often expressed in terms of *molarity*. The abbreviation for a molar solution is M, and is used in such expressions as: a $4\ M$ solution of sodium chloride, a $0.1\ M$ solution of silver nitrate. For the low concentrations often used in biochemistry, the words *millimolar* or *micromolar* are often used. Thus, a solution of penicillin containing $\frac{1}{1,000}$ of a mole in a liter

---

[1] This number is known as *Avogadro's number.*

would be called a 1-millimolar solution, and abbreviated 1 mM. Likewise, a solution of vitamin B_{12} might be prepared that is 3 micromolar, abbreviated 3 μM. This would contain 3 micromoles in a liter of water.

5.5
Acids and bases: the pH scale

As a first approximation, acids may be considered to be substances that yield hydrogen ions, H^+, when dissolved in water, and bases are then substances that yield hydroxyl ions, OH^-. Small amounts of H^+ and OH^- arise in pure water itself, because of the weak dissociation of water molecules:

$$H-O-H \rightleftharpoons H^+ + OH^-$$

Note that this is an equilibrium reaction, with water molecules dissociating into H^+ and OH^- and reforming into H_2O constantly. At equilibrium, the amount of H^+ and OH^- in pure water is very small. Since exactly equal numbers of H^+ and OH^- are present, if we give the concentration of one we automatically have the concentration of the other. By convention, concentrations of hydrogen ions are used. In pure water, this concentration is $10^{-7} M$, or 0.0000001 M. When equal numbers of hydrogen and hydroxyl ions are present, the solution is said to be *neutral*. If hydrogen ions are added, as from an acid, the solution becomes acidic, whereas if hydroxyl ions are added, as from a base, the solution becomes basic.

With a few exceptions, life processes occur primarily at neutrality, where concentrations of hydrogen ions are about $10^{-7} M$. It is frequently necessary to express the acidity or alkalinity of solutions near neutrality. Because the concentrations are low, it is inconvenient to use the actual molarities. Instead, the pH scale is used, which expresses the hydrogen ion concentration in terms of a logarithmic scale. A pH value is the logarithm of the reciprocal of the hydrogen ion concentration:

$$pH = \log_{10} 1/(H^+)$$

The relationship between pH and hydrogen (and hydroxyl) ion concentration is shown in Table 5.5. We discussed the importance of pH in relation to culture media in Section 3.2.

TABLE 5.5
The pH scale

H^+ concentration	1	10^{-1}	10^{-2}	10^{-3}	10^{-4}	10^{-5}	10^{-6}	10^{-7}	10^{-8}	10^{-9}	10^{-10}	10^{-11}	10^{-12}	10^{-13}	10^{-14}
OH^- concentration	10^{-14}	10^{-13}	10^{-12}	10^{-11}	10^{-10}	10^{-9}	10^{-8}	10^{-7}	10^{-6}	10^{-5}	10^{-4}	10^{-3}	10^{-2}	10^{-1}	1
pH	0	1	2	3	4	5	6	7	8	9	10	11	12	13	14

Concentrations are given in molarity. Note that the pH value is equivalent to the negative exponent of the hydrogen ion concentration.

CHEMICAL AND PHYSICAL BACKGROUND

As we noted in Section 5.3, the atoms in molecules are held together by chemical bonds. The strength of a chemical bond can be determined from the energy it takes to break it. Some chemical bonds are very strong, requiring large amounts of energy, whereas others are weak and are easily broken. Since chemical reactions involve breaking and rejoining bonds between different kinds of molecules, a knowledge of the forces holding chemical bonds together provides insight into how chemical reactions take place. Two principal types of chemical bonds are (1) *ionic bonds*, in which the atoms are held together by attraction of positive and negative charges, and (2) *covalent bonds*, which operate to hold together uncharged atoms.

IONIC BONDS

Ionic bonds arise because ions of opposite charge attract. Sodium ions, being positively charged, can form an ionic bond with negatively charged chloride ions to form sodium chloride, common table salt. A large number of other compounds can be formed in a similar manner by the attraction of positive and negative ions. Ionic bonds are not very specific, since any positive and negative ions will be attracted to each other.

Ionic bonds may form from noncharged atoms in the following way: Some atoms are more likely to gain than lose electrons, whereas other atoms usually lose rather than gain electrons. Whether an atom will gain or lose electrons depends upon its atomic structure, or position in the periodic table. Those elements that have one or a few electrons in their outermost shells will lose electrons on becoming ions, whereas those with many electrons will gain electrons (see Table 5.3). Thus, sodium, which has only one electron in its outermost shell, loses this electron and becomes the sodium ion:

$$Na \rightarrow Na^+ + e^-$$

Chlorine, on the other hand, has seven electrons in its outermost shell. It accepts another electron and becomes negatively charged:

$$Cl + e^- \rightarrow Cl^-$$

Note that when sodium loses an electron or chlorine gains an electron, they both achieve outer shells with eight electrons. An outer shell with eight electrons is the most stable configuration, so that sodium and chlorine tend to go toward this state. In fact, atomic sodium and atomic chlorine hardly exist in nature, whereas sodium and chloride ions are very common, as in crystals of salt.

When chlorine acquires an electron it has to come from somewhere. It can come from another atom that has a tendency to lose electrons. A logical source of an electron for chlorine would be sodium, since it tends to lose electrons readily. If sodium gives up its electron to chlorine, the two ions now have opposite charges and are attracted to each other:

$$Na \rightarrow Na^+ + e^-$$
$$Cl + e^- \rightarrow Cl^-$$
$$Na^+ + Cl^- \rightarrow Na^+Cl^-$$

An ionic bond is formed.

In crystals of common table salt, each sodium ion is surrounded by six chloride ions, and each chloride ion is surrounded by six sodium ions. Because there are equal numbers of negative and positive ions, the salt crystal is electrically neutral, even though it consists of charged particles.

When table salt is dissolved in water, the sodium and chloride ions go into solution but remain charged. Because there are still equal numbers of positive and negative ions, the salt solution is electrically neutral also. However, in solution, the ions are free to react with other ions that might also be added to the solution. Thus, chemical reactions can develop.

In the formation of the ionic bond, the electron of sodium was completely transferred to chlorine. Covalent bonds arise when electrons are only partially transferred to other atoms, so that they are *shared* between the two atoms. An example of a covalent bond is the molecule H_2, formed by the association of two hydrogen atoms. In such a molecule, ionic bonding is clearly not possible because the two atoms are alike and could not be of opposite charge.

The bond may form because one electron in the outer shell of each H atom *pairs* with an electron from the outer shell of the other H atom. This pairing of electrons is illustrated in Figure 5.4. The net effect is that the outer shell of each H atom has now acquired two electrons, the most stable configuration. The bond formed by this pairing of electrons is the covalent bond.

Covalent bonds are the kinds formed in most molecules of biological interest. In all the vast array of organic compounds found in living organisms, the atoms are held together by covalent bonds. A covalent bond between two atoms is usually written as a single straight line, thus: H—H; although sometimes it is written as two dots, to indicate the two electrons that are shared: H:H.

In many cases, the pairing of two atoms in the formation of a covalent bond occurs with the sharing of more than one pair of electrons. If two pairs of electrons are shared, the bond is called a *double bond*, symbolized as such: O=O, or O::O. A triple bond

COVALENT BONDS

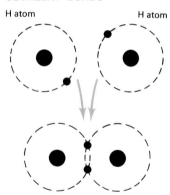

H atom   H atom

FIGURE 5.4
The formation of a covalent bond by the sharing of electrons between two H atoms.

MULTIPLE COVALENT BONDS

CHEMICAL AND PHYSICAL BACKGROUND

TABLE 5.6	Molecule	Formula	Structure
Covalent bonds of some simple molecules of microbiological interest	Hydrogen	H_2	H—H
	Oxygen	O_2	O=O
	Nitrogen	N_2	N≡N
	Water	H_2O	H H \\ / O
	Ammonia	NH_3	H—N—H | H
	Carbon dioxide	CO_2	O=C=O
	Hydrogen sulfide	H_2S	H H \\ / S
	Methane	CH_4	H | H—C—H | H

consists of one in which *three* pairs of electrons are shared, thus: N≡N, or N⋮⋮N. The reason that double or triple bonds form with some molecules is that the atoms involved have fewer electrons in their outer shells, so that they share more electrons in order to achieve a stable electron configuration.

Covalent bonds can, of course, also be formed between unlike atoms, and a single atom can form covalent bonds simultaneously with two or more atoms. The main requirement is that the sharing of electrons should lead to a proper number of electrons in the outer shell of each atom in the molecule. Some examples of common molecular substances with different bonding patterns are given in Table 5.6.

COVALENCE NUMBERS

Note in Table 5.6 that some elements can form more covalent bonds around a single atom than can other elements. Around the sulfur atom in hydrogen sulfide, only two single bonds to hydrogen form; whereas around the nitrogen atom in ammonia, three hydrogen atoms form bonds; and around the carbon atom in methane, there are four hydrogen atoms. Elements differ in their covalent-combining abilities, and these differences are expressed by *covalence numbers*. The basis of the covalence number is the number of electrons an element needs in order to achieve a stable electron configuration.

Some representative covalence numbers follow:

Hydrogen	Covalence 1	H—H
Carbon	Covalence 4	O=C=O
Nitrogen	Covalence 3	N≡N
Oxygen	Covalence 2	O=O
Chlorine	Covalence 1	Cl—Cl
Sulfur	Covalence 2	H—S—H

COVALENT VERSUS IONIC BONDS

A single element may form both covalent and ionic bonds, although most elements are more likely to form one kind of bond than the other. Chlorine, for instance, can form both covalent and ionic bonds: in Na^+Cl^- it is ionic, whereas in Cl_2 (chlorine gas) it is covalent, Cl—Cl. Sodium, on the other hand, forms only ionic bonds, as do potassium and rubidium. Calcium and magnesium primarily form ionic bonds but may also form covalent bonds under certain conditions. Carbon, on the other hand, forms only covalent bonds. Because carbon has a covalence of 4, it can form a vast array of compounds, since there are so many possibilities for bond formation. *Organic chemistry* is the chemistry of carbon compounds and has developed as an independent branch of chemistry because of the vast array of carbon compounds that can be formed. Organic chemistry is also of fundamental importance for an understanding of life processes. We shall discuss this area of chemistry more in Section 5.12.

BOND ENERGY

A chemical bond can be broken only if energy is used. The amount of energy necessary to break a chemical bond is called *bond energy* and expresses the strength of that chemical bond. Covalent bonds are strong bonds, requiring relatively large amounts of energy for breakage, whereas ionic bonds are weak bonds, being broken fairly easily. The weakness of an ionic bond is seen in how readily table salt, sodium chloride, dissolves in water. As the crystals dissolve, the bonds between the ions are broken, and the ions which were associated in the crystal are separated widely in the solution.

The strength of covalent bonds is seen in the difficulty of separating the atoms of covalent compounds. Covalent compounds such as those listed in Table 5.6 are quite stable, and the atoms remain associated when the compounds are placed in water, even after long periods of time.

5.7
The importance of weak bonds in biochemistry

Although covalently linked compounds are relatively stable, they still are able to interact with other compounds by way of weaker types of interactions. Weak bonds between molecules can form by several means and play important roles in life processes. Many of the key types of reactions occurring in living organisms are only possible through the making and breaking of weak bonds between organic molecules of cells.

HYDROGEN BONDS

An extremely important type of weak bond in living organisms is the hydrogen bond. A hydrogen bond arises when a hydrogen atom covalently linked to one atom with a partial positive charge interacts with an acceptor atom with a partial negative charge on another molecule. The simplest situation for visualizing hydrogen bonds involves water itself. Hydrogen bonds between water molecules result in the attraction of one water molecule to another (Figure 5.5). A network of water molecules forms, with hydrogen-bond formation occurring throughout the system. In liquid water, the network is not very orderly, and hydrogen bonds are continually forming and breaking, resulting in continuous movement of water molecules; but in ice, neighboring water molecules take up orderly positions. The water molecules in the ice crystal are thus held together by hydrogen bonds.

In biochemistry, the most important hydrogen bonds involve hydrogen atoms attached to oxygen (O—H) or nitrogen (N—H). The acceptor atom also is usually a nitrogen or oxygen atom. Some examples of hydrogen bonds formed between organic molecules are

FIGURE 5.5
a Formation of a hydrogen bond between two water molecules; *b* network of many water molecules.

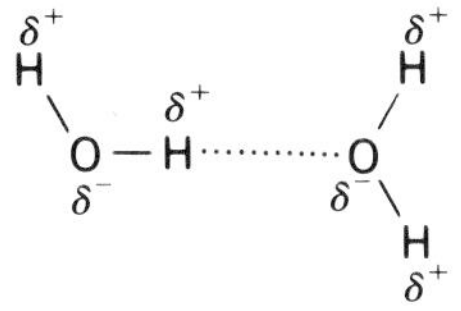

Partial positive charge, indicated by δ^+, arises because oxygen atom pulls one electron of the covalent bond towards it. Oxygen atom becomes partially negative in charge, indicated by δ^-.

Because of partial positive and negative charges, hydrogen atom is weakly attracted to oxygen atom, resulting in formation of a weak bond, the *hydrogen bond*.

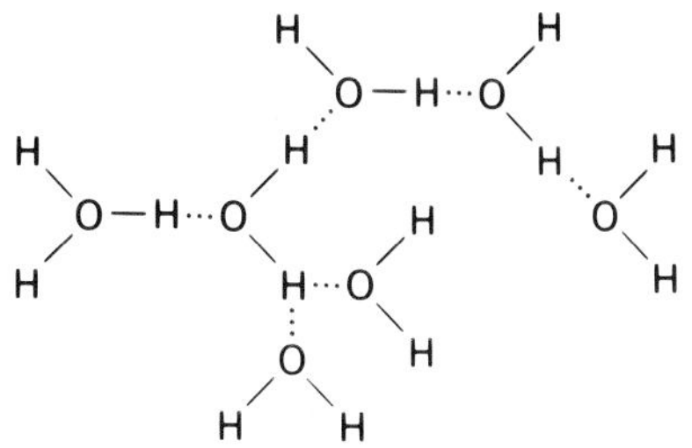

Network of water molecules due to hydrogen bond formation throughout the system.

FIGURE 5.6
Hydrogen bond formation
between organic compounds.

a

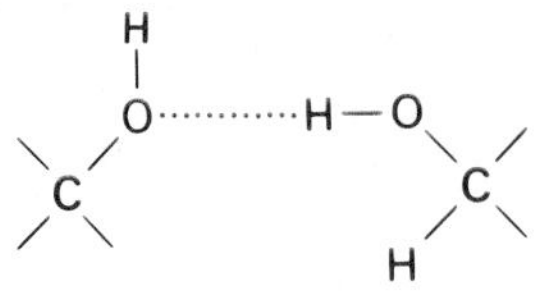

Hydrogen bond between two oxygen atoms

b

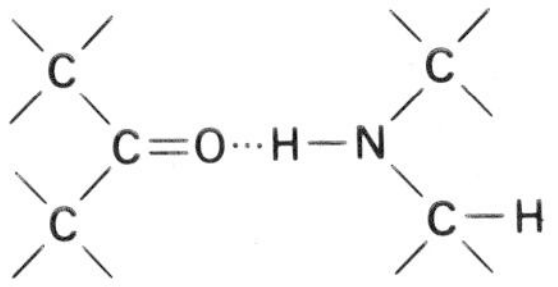

Hydrogen bond between oxygen and nitrogen atoms

shown in Figure 5.6. Hydrogen-bond formation is very important in the association of large organic molecules in living cells (see Section 5.12).

Molecules that will form hydrogen bonds with other molecules will also form them with water molecules. When a hydrogen-bond forming molecule is placed in water, some of the hydrogen bonds between water molecules are broken, and hydrogen bonds between water molecules and the added compounds are formed. Molecules that associate in this manner with water molecules are called *hydrophilic* (*hydro*, "water"; and *philic*, "loving"), and the atoms or groups of atoms in these molecules that undergo hydrogen-bond formation are often called *hydrophilic groups*. (They also are sometimes called *polar* groups, because the weak charges developed will cause them to accumulate to some extent at the poles of an electrical system.) It is because these molecules can form hydrogen bonds with water molecules that they are able to dissolve in water. Thus, water solubility goes together with hydrogen-bond formation.

Another type of bonding, even weaker and less specific, is called van der Waals bonding. It arises because whenever two molecules approach, they induce in each other movements of electrons that result in partial charge separations. This causes a very weak attraction between molecules, and bonding can thus occur only when molecules are very close.

Because van der Waals forces are weak, they are effective in holding molecules together only when several atoms in one molecule can interact with several atoms in another molecule. This means that two molecules can associate by van der Waals forces only if there is a precise fit between the atoms of each. This sort of precise fit between

VAN DER WAALS FORCES

CHEMICAL AND PHYSICAL BACKGROUND

molecules is crucial in a number of important biochemical situations, as we shall see.

HYDROPHOBIC BONDING

Substances such as sugars, which have many —OH groups, are hydrophilic and dissolve readily in water. On the other hand, many substances, such as cooking oils, gasoline, and kerosene, lack hydrophilic groups and are water insoluble. Atoms or groups of atoms in molecules that do not associate with water are called hydrophobic (literally, "water-fearing"). (They are also sometimes called *nonpolar* because they do not have any electrical character.) When a hydrophobic substance is placed in water, there is no tendency for hydrogen-bond formation between the substance and water. The water molecules remain firmly bonded together, and the hydrophobic groups associate with each other via van der Waals forces. This association between hydrophobic groups is in effect a type of bonding and has been called *hydrophobic bonding*. Hydrophobic bonding is very important in cells, because structures such as the cell membrane, high in fatty substances that are hydrophobic, are in essence held together by hydrophobic bonds.

COOPERATIVE ACTION OF WEAK BONDS

In the interactions between molecules, especially the large organic molecules found in living organisms, the various types of weak bonds act cooperatively to bring about interactions. Ionic bonds, hydrogen bonds, hydrophobic bonds, etc., all act together to bring about molecular associations. The precise alignment of atoms in adjacent molecules then becomes of great importance. It is a fascinating idea that heredity, growth, cell division, and cell behavior all depend ultimately on specific interactions between the large organic molecules of cells that arise as a result of the formation and breaking of weak chemical bonds.

5.8
Chemical reactions

When two or more kinds of molecules are mixed, there is the chance for reactions between them, leading to the formation of new products and the disappearance of the original substances. A variety of kinds of chemical reactions are known. The kind of chemical reaction that will occur depends on the reacting molecules and on the environment in which the reaction occurs. Chemical reactions can occur in either gaseous, liquid, or solid state, but in biological systems it is reactions in liquid (water) that are of prime importance. Reactions taking place in water involve molecules that are dissolved in the water. If a substance is insoluble in water, it is not likely to be involved to any great extent in chemical reactions in aqueous environments.

The simplest chemical reactions to visualize are those involving ions. Most ionic reactions involve the combination of positive and negative ions to form an insoluble substance. As an example of this sort of reaction, consider two solutions, one containing sodium chloride, the other silver nitrate. Both substances are soluble in water and form crystal-clear solutions. The solution of sodium chloride consists, really, of a mixture of sodium ions and chloride ions, whereas the silver nitrate solution consists of a mixture of silver ions and nitrate ions. We can visualize these solutions as having arisen as follows:

$$Na^+Cl^- \rightarrow Na^+ + Cl^-$$
$$Ag^+NO_3^- \rightarrow Ag^+ + NO_3^-$$

When these two clear solutions are mixed, the mixture turns cloudy white instantaneously, because of the formation of a precipitate of silver chloride, AgCl, a very insoluble substance:

$$Ag^+ + Cl^- \rightarrow AgCl \downarrow$$

(It is a chemical convention that whenever a precipitation reaction occurs, a vertical arrow pointing downward is written next to the substance precipitating.) Note that in this reaction, we write down only those substances reacting, Ag^+ and Cl^-. The Na^+ and the NO_3^- ions, which do not react, are, of course, present in the solution but need not be written.

Acid-base reactions are ionic reactions of special types, involving acids and bases. We discussed the nature of acids and bases in Section 5.5. When a solution of an acid such as hydrochloric acid and a base such as sodium hydroxide are mixed, the hydrogen and hydroxyl ions react to form water:

$$HCl + NaOH \rightarrow NaCl + H_2O$$

Note that in this reaction, two neutral products are formed: water and sodium chloride. Because the reaction of acid and base results in a neutral solution, this type of reaction is often called a *neutralization* reaction. In actuality, HCl is a solution of H^+ and Cl^- ions, and NaOH is a solution of Na^+ and OH^-. The actual reaction proceeds as follows:

$$H^+ + Cl^- + Na^+ + OH^- \rightarrow Na^+ + Cl^- + H_2O$$

Hydrochloric acid is a *strong acid*, and sodium hydroxide is a *strong base*. When they are dissolved in water they completely ionize, so that all the hydrogen ions and hydroxyl ions present are liberated into the solution. There are many acids that are called *weak acids* because they do not liberate all their hydrogen ions upon dissolving in water. Likewise, there are *weak bases*. Acetic acid, the principal component of vinegar, is an example of a weak acid. Its formula is CH_3COOH, and the right-hand hydrogen atom is the acid-forming one. When acetic acid dissolves in water, some of the hydrogen atoms are liberated and form hydrogen ions, but most of them are retained in the acetic acid molecule:

$$CH_3COOH \rightleftharpoons CH_3COO^- + H^+$$

The pH of a 1 M solution of acetic acid is not pH 0, as it would be for hydrochloric acid, but pH 4.7.

However, the fact that the hydrogen ions in acetic acid do not dissociate completely when the substance is dissolved in water does not mean that they are not free to react. If a base is added to a solution of acetic acid, the hydrogen ions will be pulled off the molecule, resulting in the formation of the ionized species, the acetate ion:

$$CH_3COOH + NaOH \rightarrow CH_3COO^- + Na^+ + H_2O$$

Note that in this reaction, one H of the water molecule comes from acetic acid, and the other from sodium hydroxide. If equimolar amounts of acetic acid and sodium hydroxide are used in the reaction, the solution formed is of neutral pH (the reaction is thus another example of a neutralization reaction). The negatively charged acetate ions are exactly balanced by the positively charged sodium ions.

There are many examples of weak acids in biology. Acetic, lactic, citric, propionic, butyric, succinic, fumaric, malic, pyruvic, and oxalacetic are all weak acids. Many of these play important roles in cellular metabolism, as discussed in Chapter 6. Note that because living organisms are generally at neutral pH, these substances do not exist as the free acids but as the ions. Because of this, they will generally be referred to as the ions; thus, acetate, lactate, citrate, propionate, and so forth.

In biological systems, the most common element involved in formation of *weak bases* is nitrogen, N, which is present in many of the important biochemicals of cells. Just as an acid is a substance that gives up H^+ when dissolved, a base can be viewed as a substance that *takes up* H^+ from water, leaving OH^- behind. The simplest N-containing base is ammonia, NH_3, which dissolves in water in the following way:

$$NH_3 + H_2O \rightarrow NH_4^+ + OH^-$$

When ammonia dissolves in water and takes up a H^+, it leaves behind the OH^-, and the solution becomes basic.

Ammonia is actually a fairly strong base, but in the organic substances in cells, the N is covalently bonded to carbon and has less affinity for H^+, so that the basicity of N-containing organic substances is weaker:

$$-\overset{|}{\underset{|}{C}}-NH_2 + H_2O \rightarrow -\overset{|}{\underset{|}{C}}-NH_3^+ + OH^-$$

The following is a simple neutralization reaction between acetate, a weak acid, and a N-containing base:

$$CH_3COO^- + -\overset{|}{\underset{|}{C}}-NH_3^+ \rightarrow CH_3COOH + -\overset{|}{\underset{|}{C}}-NH_2$$

Another important group of chemical reactions in living organisms involves the breaking of covalent bonds by water molecules. Because these reactions result in the degradation of larger molecules into smaller ones, they are called *hydrolysis* (literally, "lysis with water"; *lysis* means "breakdown" or "degradation"). A simple example of a hydrolysis reaction is the breakdown of common table sugar, sucrose. Sucrose is composed of two separate sugar molecules, glucose and fructose, connected by an oxygen bridge:

$$C_6H_{11}O_5-O-C_6H_{11}O_6 + H_2O \rightarrow C_6H_{12}O_6 + C_6H_{12}O_6$$

(Fructose)	(Glucose)		Fructose	Glucose
	Sucrose			

The hydrogens and oxygen of the water molecule are added to the sugars during the hydrolysis reaction, and the two separate sugars are liberated. The formation of sucrose can occur by the reverse of the foregoing reaction, by removing water from the two molecules as they are connected together. This reverse reaction is called *dehydration*, since water is removed. Dehydration reactions are very important in living organisms because they result in the *synthesis* of larger molecules. These reactions are very difficult to carry out, however, because most organisms live in an aqueous environment, and the high concentration of free water is contrary to the conditions for dehydration reactions. Organisms have thus had to evolve the capacity for special biochemical tricks in order to carry out dehydration reactions such as these.

Group-transfer reactions are those in which molecules exchange portions (or groups) of their structures. An important type of group-transfer reaction in biochemistry involves the transfer of phosphate from one compound to another:

$$A-P + X \rightarrow X-P + A$$

where A and X are two different molecules. The phosphate group has been transferred from one to the other.

In a sense, hydrolysis reactions are group transfers in which water serves as the second molecule:

$$A{-}B + H_2O \rightarrow A{-}OH + BH$$

OXIDATION-REDUCTION REACTIONS

Some of the most important chemical reactions in living organisms are *oxidation-reduction reactions*, sometimes called *redox* reactions. Oxidation-reduction reactions are of especial importance in energy generation by living organisms. A little confusion often arises at this point because oxidation seems to imply that oxygen is involved. Although oxygen often plays a major role in oxidation-reduction reactions in living organisms, many oxidation-reduction reactions occur in the complete absence of oxygen. Essentially, an oxidation-reduction reaction is one in which an electron is transferred from one molecule to another. We have already presented one example of an oxidation-reduction reaction in another context, in our discussion of the formation of ionic bonds in Section 5.6. We noted that when sodium and chlorine ionized, one of them gave up an electron and the other acquired an electron:

$$Na \rightarrow Na^+ + e^-$$
$$Cl + e^- \rightarrow Cl^-$$

In chemical terms, the reaction in which sodium gives up an electron is called an *oxidation*, and sodium is said to be *oxidized*. Likewise, the reaction in which chlorine accepts an electron is called a *reduction*, and chlorine is said to be *reduced*. The individual reactions are called *half-reactions*, but it should be emphasized that neither the oxidation nor the reduction half-reaction will occur by itself. The electron from sodium can only be pulled off if there is another substance, the *electron acceptor*, present. Thus, oxidation and reduction are really two sides of the same coin and must always be considered together. The terms involved in discussing oxidation-reduction reactions are summarized in Table 5.7.

Some other examples of oxidation-reduction reactions are:

$Fe^{2+} \rightarrow Fe^{3+} + e^-$	Iron (ferrous) gives up an electron and becomes oxidized to the ferric state.
$H_2 \rightarrow 2H^+ + 2e^-$	Hydrogen gas gives up two electrons and becomes oxidized to hydrogen ions.
$\frac{1}{2}O_2 + 2e^- \rightarrow O^=$	Oxygen gas accepts two electrons and becomes reduced.

The two half-reactions involving hydrogen and oxygen may be coupled as follows:

TABLE 5.7
Terminology for oxidation-reduction reactions

Electron donor
$$A \rightarrow A^+ + e^-$$
Reducing agent
Reductant
Becomes oxidized

Electron acceptor
$$B + e^- \rightarrow B^-$$
Oxidizing agent
Oxidant
Becomes reduced

$$H_2 \rightarrow 2H^+ + 2e^-$$
$$\tfrac{1}{2}O_2 + 2e^- \rightarrow O^=$$
$$2H^+ + O^= \rightarrow H_2O$$

This is the reaction that would occur if hydrogen gas were allowed to react with oxygen in air. The reaction is so favorable that it occurs with a violent explosion.

A wide variety of oxidation-reduction reactions are known, many of which are of importance only in chemistry, but there are a number of important biochemical oxidation-reduction reactions, several of which will be discussed in the next chapter.

Energy can be thought of as the ability to do work. Chemical reactions involve energy changes. Not all chemical reactions that we can think of will occur spontaneously. Some will occur only if energy is put into them. By understanding the energetics of chemical reactions, we can predict whether or not they will occur spontaneously and whether they can be harnessed to do useful work.

Energy can be expressed in a number of ways, but in biochemistry, the most useful expression of energy is the *calorie* or *kilocalorie*. A *calorie*, sometimes called a *gram calorie*, is the amount of energy necessary to raise the temperature of a gram of water from 14.5 to 15.5°C. In biochemistry, the kilocalorie is usually used. A *kilocalorie* (kcal) is the amount of energy it takes to raise the temperature of a kilogram of water from 14.5 to 15.5°C.

Chemical reactions that occur spontaneously are accompanied by the release of heat energy. A reaction in which heat energy is released is called an *exothermic* reaction. Chemical reactions that do not occur spontaneously may do so if energy is supplied to the system; such reactions are called *endothermic*. The amount of energy involved in a chemical reaction is expressed in terms of the gain or loss of heat energy from the system during the reaction, in kilocalories per mole of reactants.

There are two types of expression of the amount of energy released during a chemical reaction, abbreviated H and G. H, called *enthalpy*, expresses the total amount of energy released during a chemical reaction. However, some of the energy released is not available to do useful work. G, called *free energy*, is used to express the energy released that is available to do useful work. The change in free energy during a reaction is expressed as ΔG, where the symbol Δ (delta) should be read to mean "change in." If ΔG is negative, then free energy is released, and the reaction will occur spontaneously. If ΔG is positive, the reaction will not occur spontaneously.

5.9
Energy and chemical reactions

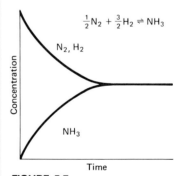

FIGURE 5.7
Chemical equilibrium. Changes in concentrations of reactants (N_2, H_2) and product (NH_3) during approach to equilibrium. At equilibrium, no further change occurs, although chemical reaction continues. The back reaction exactly balances the forward reaction.

When two chemical substances are placed together under conditions in which they can react, the concentration of the reactants is high at the beginning, while the concentration of products is zero. As the reaction proceeds, the concentration of products builds up, and the concentration of reactants decreases. Eventually, the concentration of products builds to the point where the reaction no longer proceeds. At this point, the reverse reaction, the conversion of products back to reactants, occurs. An equilibrium is reached in which the forward and reverse reactions are exactly balanced. At this point, although both reactions are continuing, no *measurable* reaction is seen. The trend toward equilibrium is shown graphically in Figure 5.7.

FREE ENERGY AND CHEMICAL REACTIONS

Returning to our discussion of energy release during chemical reaction, we can state that at equilibrium there is no change in energy, and the reaction will not proceed in either direction. If the reaction is exothermic, it will proceed spontaneously, whereas if the reaction is endothermic, it will not proceed spontaneously, but the reverse reaction will be spontaneous:

ΔG negative: reaction is spontaneous.
ΔG positive: reaction not spontaneous; reverse reaction spontaneous.
ΔG zero: reaction at equilibrium.

For example, in the reaction shown in Figure 5.7, the free energy change is -4.0 kcal under standard conditions:

$$\tfrac{1}{2}N_2 + \tfrac{3}{2}H_2 \rightleftharpoons NH_3 \qquad \Delta G = -4.0 \text{ kcal}$$

On the other hand, the following reaction, the formation of methane (CH_4) and O_2 from CO_2 and H_2O, has a positive ΔG:

$$CO_2 + 2H_2O \rightarrow CH_4 + 2O_2 \qquad \Delta G = +196 \text{ kcal}$$

Thus, the reaction to the right will not proceed spontaneously, but the reverse reaction will. Since the reverse reaction represents the burning of natural gas (methane), a clearly exothermic reaction, it is obvious that the reaction should go only in the direction to the left. If the reaction is now written in this form, it has a negative ΔG:

$$CH_4 + 2O_2 \rightarrow CO_2 + 2H_2O \qquad \Delta G = -196 \text{ kcal}$$

The energy yield in reactions such as this can be illustrated graphically as in Figure 5.8, which gives the free energy of the reactants and the free energy of the products; the difference between the two is the energy yield in the reaction.

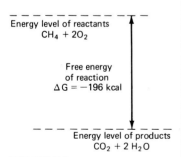

Energy level of reactants
$CH_4 + 2O_2$

Free energy
of reaction
$\Delta G = -196$ kcal

Energy level of products
$CO_2 + 2 H_2O$

FIGURE 5.8
Change in energy level during exothermic chemical reaction.

Even if the free energy yield of a given reaction is negative and the reaction would be predicted to occur spontaneously, the reactants may remain together indefinitely without reacting. This is because chemical reactions require that the bonds between atoms in molecules be broken or partially loosened if they are to react with each other. As we have seen, covalent bonds are strong bonds and will not break spontaneously. Therefore, for some reactions to be initiated, energy must be applied. The energy necessary to loosen up covalent bonds and get a reaction started is called *activation energy*. Once the reaction is initiated, it can continue spontaneously when the ΔG is negative. The concept of activation, which is extremely important in biochemistry, is illustrated in Figure 5.9a (compare with Figure 5.8).

A good example of the necessity of activation energy is shown by the reaction in which methane (natural gas) is burned. Methane can remain unchanged in air for long periods of time, provided no source of activation energy is present. If a spark or other small source of heat is added, some of the methane molecules are activated and react with oxygen in the air. The energy released from this initial reaction activates more methane molecules, leading to more reaction. Rapidly, the reaction rate builds up to a very fast rate and proceeds to completion. In fact, an explosion may occur, caused by the large amount of energy released in the reaction.

5.10
Activation energy and catalysts

FIGURE 5.9
a The concept of activation energy. Although the reaction should proceed spontaneously, it does not unless energy is added to initiate the reaction. Once the energy of the reactants has been raised to the activated state, the reaction proceeds spontaneously. The energy yields of the reaction is −196 kcal, rather than −216 kcal, because the activation energy must be deducted from the overall yield. *b* Catalysts reduce activation energies, making reactions occur more readily.

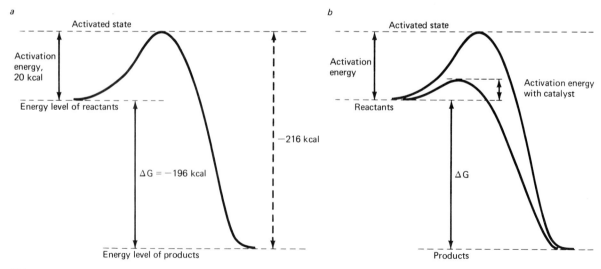

CHEMICAL AND PHYSICAL BACKGROUND

Catalysts are agents that reduce activation energies and thus make reactions occur more readily. A *catalyst* is defined as a substance that speeds up a chemical reaction but is not changed during the reaction. Good examples of catalysts used in chemistry are finely divided metals, such as nickel, palladium, and platinum. One of the most important chemical catalytic processes is the "catalytic cracking" of petroleum to produce gasoline from crude oil, in which a silica-alumina catalyst is used.

The manner in which a catalyst reduces the activation energy necessary for the initiation of a reaction is shown in Figure 5.9*b*. Because the necessary activation energy is reduced, the reaction proceeds more readily.

5.11
Enzymes as catalysts

Enzymes are the catalysts in living organisms. Without enzymes, life would be impossible, because most biochemical reactions have considerable activation energies and will not occur spontaneously. It is just as well that biochemical reactions do require catalysis, because if they occurred without catalysts, all reactants in cells would quickly proceed to the lowest energy levels of the products. It is only because of the presence of temporarily energy-rich compounds in cells that cell function is possible.

ENZYME ACTION

Enzymes are proteins. We shall present the molecular structure of proteins in Section 6.4. For the moment it is sufficient to know that enzymes are large, generally spherical molecules that have a number of reactive groups on the outside which interact with small molecules present in cells. These reactive groups on enzyme surfaces form weak bonds (hydrogen, ionic, and van der Waals bonds) with chemicals in the cell. Enzymes are very specific; a single enzyme will combine with only one or a very few kinds of molecules. The molecule with which an enzyme reacts is called its *substrate*. Binding of substrate to enzyme results in the formation of an *enzyme-substrate complex*. As a result of the formation of an enzyme-substrate complex, the substrate is activated, and chemical reaction can occur.

As an example of how an enzyme catalyzes a chemical reaction, let us consider the hydrolysis of common table sugar, sucrose, to its component sugars, glucose and fructose. We described this hydrolysis chemically in Section 5.8. Even though the free energy of hydrolysis of sucrose is negative and hence favorable, the reaction is extremely slow. This is shown by the fact that sucrose solutions in water (syrups) remain unchanged for long periods of time. An enzyme very common in yeast and other organisms called *invertase* (or *sucrase*) is able to catalyze this reaction and greatly speed it up. The manner in which invertase hydrolyzes sucrose is illustrated in Figure 5.10.

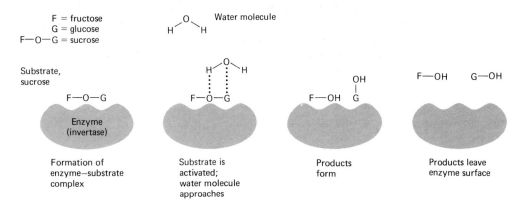

F = fructose
G = glucose
F—O—G = sucrose

Water molecule

Substrate, sucrose

F—O—G

Enzyme (invertase)

Formation of enzyme–substrate complex

Substrate is activated; water molecule approaches

Products form

Products leave enzyme surface

Several key aspects of enzymes deserve emphasis:

1 Enzymes are agents that very efficiently catalyze chemical reactions. They are active even when present in very low concentrations. It is usually difficult to detect the enzyme molecule directly in a cell, but it is readily detected through the chemical reaction that it catalyzes.

2 Enzymes are very specific. An enzyme catalyzes only a single or a very few chemical reactions. Thus, for the vast array of chemical reactions that occur in living organisms, a vast array of enzymes are necessary.

3 Enzymes, as other catalysts, are not changed as a result of the chemical reactions they catalyze. At the end of a reaction, the enzyme is still present to catalyze the same reaction again.

It is a fascinating thought that the basis of life function is the enzyme, an agent whose only role is to catalyze a chemical reaction. The differences between one organism and another, between plant and animal, bacterium and protozoan, man and beast, arise only because of the different enzymes that they possess.

Enzymes are generally named for the substrate they react with or the reaction that they carry out. The combining form *-ase* is added to the end to indicate that an enzyme is being named. Thus, an enzyme that hydrolyzes *cellulose* (the chemical substance of cotton) is called *cellulase*, the enzyme that hydrolyzes malt sugar, *maltose*, is called *maltase*, and so forth.

Many enzymes are named for the kinds of functions they carry out. Enzymes that remove hydrogen atoms from molecules are called *dehydrogenases*; the enzyme that removes hydrogens from glutamic acid is called glutamic dehydrogenase. Enzymes that introduce oxygen into molecules are called *oxidases*; the enzyme glucose oxidase introduces oxygen into glucose. Enzymes that catalyze the reduction of substances are called *reductases*; nitrate reductase causes the reduction of nitrate to nitrite. The enzymes that hydrolyze nucleic acids are called *nucleases*. If the nucleic acid is ribonucleic

FIGURE 5.10
Mechanism of enzyme catalysis. The enzyme invertase catalyzes the hydrolysis of sucrose into its component sugars, fructose and glucose.

NAMING OF ENZYMES

CHEMICAL AND PHYSICAL BACKGROUND

acid, the enzyme is a ribonuclease, sometimes abbreviated RNase; with deoxyribonucleic acid, the enzyme is a deoxyribonuclease, abbreviated DNase. Enzymes that hydrolyze proteins are called *proteases*. Because thousands of enzymes are known, thousands of enzyme names exist. Fortunately, we shall only have to concentrate on a few key names in this book.

5.12
Organic chemistry

Organic chemistry is the chemistry of *carbon* compounds. A single carbon atom has the ability to form *four* covalent bonds, and it can form bonds with a wide variety of other elements, such as H, O, N, S, and P. Further, carbon is unique among elements in its ability to form bonds not only with other elements but with itself (Figure 5.11). This capacity greatly increases the variety of compounds that can be formed. Over a million carbon-based compounds are known, and more are being discovered or synthesized each year. The vast array of compounds necessary for the functioning of a living cell are carbon compounds. In fact, it can be said that life has been possible only because carbon can form such a tremendous variety of compounds. For this reason, the chemistry of carbon compounds has come to be called *organic* chemistry, since the word *organic* refers to life or life processes.

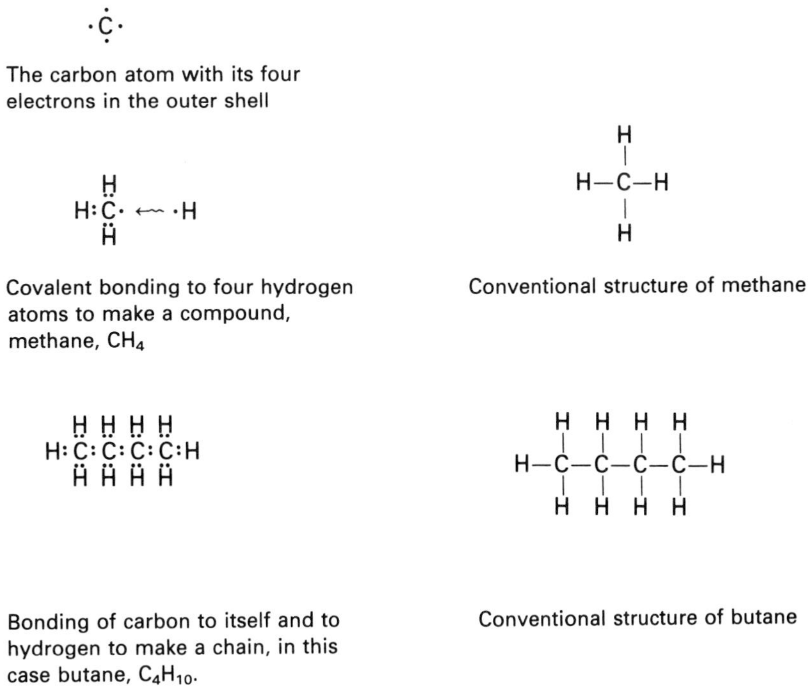

FIGURE 5.11
Covalent bonding of carbon to other atoms and to itself leads to the formation of organic compounds.

·Ċ·

The carbon atom with its four electrons in the outer shell

$$H:\overset{\cdot\cdot}{\underset{\cdot\cdot}{C}}\cdot \longleftarrow \cdot H$$

Covalent bonding to four hydrogen atoms to make a compound, methane, CH_4

$$H-\overset{\displaystyle H}{\underset{\displaystyle H}{C}}-H$$

Conventional structure of methane

$$H:\overset{\cdot\cdot}{\underset{\cdot\cdot}{C}}:\overset{\cdot\cdot}{\underset{\cdot\cdot}{C}}:\overset{\cdot\cdot}{\underset{\cdot\cdot}{C}}:\overset{\cdot\cdot}{\underset{\cdot\cdot}{C}}:H$$

Bonding of carbon to itself and to hydrogen to make a chain, in this case butane, C_4H_{10}.

$$H-\overset{\displaystyle H}{\underset{\displaystyle H}{C}}-\overset{\displaystyle H}{\underset{\displaystyle H}{C}}-\overset{\displaystyle H}{\underset{\displaystyle H}{C}}-\overset{\displaystyle H}{\underset{\displaystyle H}{C}}-H$$

Conventional structure of butane

Not all carbon-based compounds are found in living organisms. Most, in fact, are synthetic, having been made by chemists in laboratories. For this reason, the field of organic chemistry is much larger than merely the study of the chemicals of living organisms.

Because of the large number of possible compounds, and the large number of atoms that can be present in a single compound, writing formulas for organic compounds presents some difficulties. It is not sufficient merely to give the number of atoms of each kind present in an organic compound, as we would for other elements, because the same number of atoms may combine in a variety of ways to make a number of different compounds. The most accurate way of writing organic structures would be to indicate all bonds between atoms, but this would be cumbersome and space-consuming. Certain simplifications have been necessary in order to be able to write structures conveniently. In practice, what is done is to write condensed structures, in which key or important bonds are shown but other bonds are not. Figure 5.12 illustrates the way in which organic structures are written.

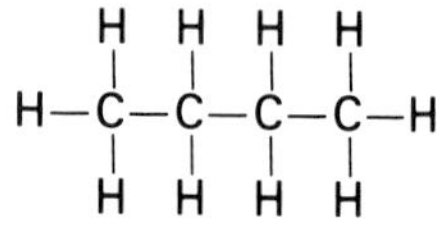

$$H_3C-CH_2-CH_2-CH_3$$

or

$$CH_3-CH_2-CH_2-CH_3$$

or

$$CH_3CH_2CH_2CH_3$$

Condensed structure of butane

Conventional structure of butane, C_4H_{10}

$CH_3CH_2CH_2CH_2NH_2$ Butylamine

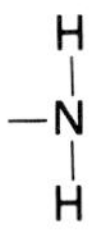

—N   Replaces one H in butane (with H above and H below)

$CH_3CH_2CH_2CH_2OH$ Butyl alcohol (butanol)

—O—H   Replaces one H

$CH_3CH=CHCH_3$ Butene

Double bond between two carbon atoms (2 H atoms removed)

Other condensed structures

FIGURE 5.12
Writing the structures of organic compounds.

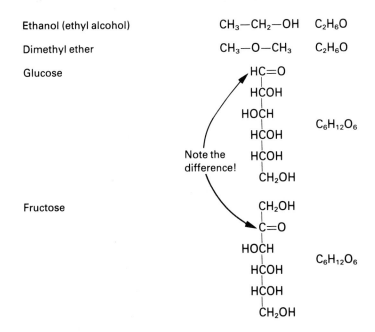

FIGURE 5.13
Isomers of organic compounds.

Ethanol (ethyl alcohol) $CH_3—CH_2—OH$ C_2H_6O

Dimethyl ether $CH_3—O—CH_3$ C_2H_6O

Glucose

Note the difference!

$C_6H_{12}O_6$

Fructose

$C_6H_{12}O_6$

ISOMERS

Isomers are chemical substances with the same overall formula but with different arrangements of the atoms, resulting in compounds with different properties. Formation of isomers is very common in organic chemistry because of the many possibilities for connecting carbon-to-carbon bonds. Examples of isomers of organic compounds are shown in Figure 5.13.

OPTICAL ISOMERS

Even more confusing, carbon compounds can also form isomers that appear to be exactly alike when the structures are written but differ in the way they affect polarized light. These compounds are called *optical isomers* and can only be differentiated when the three-dimensional structures of the compounds are made. Optical isomers are indicated by the letters D and L, to indicate the direction in which they rotate polarized light. D-compounds rotate polarized light to the right and are called *dextrorotary*, whereas L-compounds rotate polarized light to the left and are called *levorotary*.[2] In living organisms, the property of optical isomerism is very common. Indeed, the chemist Louis Pasteur (see Chapter 1) was first motivated to study living organisms after he had carried out extensive research on optical isomers and had learned how common they were in living organisms.

When compounds with optical activity are synthesized by chemists, equal amounts of both optical isomers are made simultaneously,

[2] This definition is oversimplified but is sufficient for our purposes.

so that the synthesized material is a mixture of D and L forms. (Such a mixture is called a *racemic* mixture.) In organisms, on the other hand, synthesis of a compound results in the formation of only one optical isomer, either D or L, depending on the compound.

It is convenient to simplify a discussion of organic chemistry by separating the various compounds into classes of similar types. In this way, it becomes easier to remember compounds and their structures. Some of the main classes of organic compounds and their characteristic structures follow:

1 *Hydrocarbons* are composed of only carbon and hydrogen atoms. Methane, CH_4, is the simplest hydrocarbon. The CH_3 group, which is found in many other organic compounds, is called a *methyl* group.
2 *Alcohols* contain the alcohol function, OH, joined to a carbon atom. Methyl alcohol (methanol) has the formula CH_3OH. Beverage alcohol, ethanol, has the formula CH_3CH_2OH.
3 *Ethers* contain an O atom, but it is connected differently than in alcohols. In ethers, the O atom is between two carbon atoms: CH_3-O-CH_3, dimethyl ether. The ether used in anaesthetics is diethyl ether, $CH_3CH_2-O-CH_2CH_3$.
4 *Aldehydes* contain a double-bonded oxygen atom at the end of the compound:

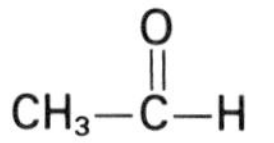

$$\begin{array}{c} O \\ \parallel \\ CH_3-C-H \end{array}$$

Acetaldehyde

5 *Ketones* contain a double-bonded O atom similar to an aldehyde, but it is in the center of a compound:

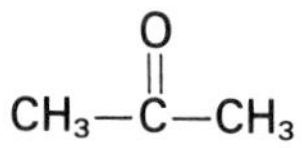

$$\begin{array}{c} O \\ \parallel \\ CH_3-C-CH_3 \end{array}$$

Acetone

6 *Carboxylic acids* contain the characteristic acid functional group:

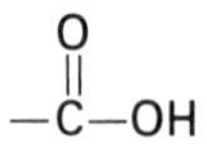

$$\begin{array}{c} O \\ \parallel \\ -C-OH \end{array}$$

which ionizes to form

$$\begin{array}{c} O \\ \parallel \\ -C-O^- + H^+ \end{array}$$

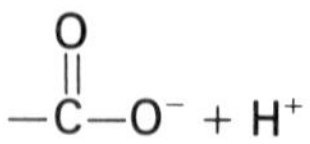

We have already mentioned carboxylic acids in Section 5.8. Carboxyl groups are often abbreviated COOH. Acetic acid, CH_3COOH, ionizes to form acetate ion, CH_3COO^-, and H^+.

7 *Esters* are combinations of alcohols and carboxylic acids, formed by dehydration:

$$\underset{\text{Acetic acid}}{CH_3\overset{\displaystyle O}{\overset{\|}{C}}-OH} \quad + \quad \underset{\text{Methanol}}{HOCH_3} \quad \rightarrow \quad \underset{\text{Methyl acetate}}{CH_3\overset{\displaystyle O}{\overset{\|}{C}}-O-CH_3} \quad + \quad H_2O$$

The OH of water comes from methanol, and the H from acetic acid.

8 *Amines* are N-containing compounds; they can be viewed as analogs of ammonia:

$$\underset{\text{Ammonia}}{NH_3} \qquad \underset{\text{Methylamine}}{CH_3-NH_2}$$

One of the H atoms of ammonia is replaced by a CH_3 grouping.

9 *Sulfhydryl* compounds (also called *mercaptans* or *thiols*) contain an S atom; they can be viewed as analogs of H_2S:

$$\underset{\text{Hydrogen sulfide}}{H_2S} \qquad \underset{\substack{\text{Methyl mercaptan} \\ \text{(methane thiol)}}}{CH_3SH}$$

One of the H atoms of H_2S is replaced by a methyl group.

10 *Aromatic* compounds contain six carbon atoms formed into a *ring*, with a hydrogen atom connected to each carbon atom. Aromatic rings are conventionally drawn as if there was a double bond between every other carbon atom, although the bonding between the carbon atoms is actually more complicated than that. The simplest aromatic compound is benzene, but because of the possibilities for substitution and addition, a huge array of aromatic compounds exists. Several important biochemicals are aromatic in character, and the aromatic ring is one of the main types of hydrophobic structures in living organisms:

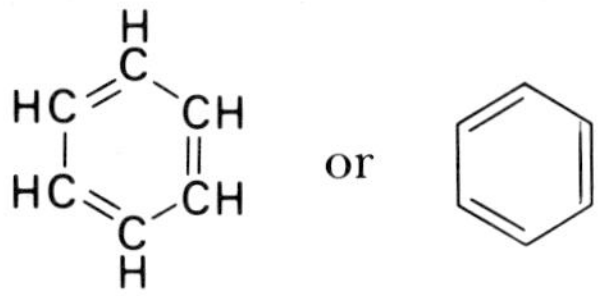

POLYMERS

Polymers are long molecular chains formed by connecting one or a few small molecules together in a repeating fashion. The repeating unit of which a polymer is constructed is called a *monomer*. Although some inorganic polymers are known, most polymers are organic. Many of the organic chemicals of living organisms are polymers, and there are also many synthetic polymers. For example, fibers such as Nylon, Orlon, and Dacron are synthetic polymers. Many useful plastics are also synthetic polymers, including polyethylene, polypropylene, Teflon, and polystyrene.

Polymer formation in organic chemistry is usually accomplished by allowing a reactive monomer to condense:

```
 H  H          H  H│H  H│H  H│H  H│H  H
 │  │          │  │││  │││  │││  │││  │
 C=C          —C—C┼C—C┼C—C┼C—C┼C┼C—C
 │  │          │  │││  │││  │││  │││  │
 H  H          H  H│H  H│H  H│H  H│H  H
 Ethylene              Polyethylene
```

The length of the polymer is determined by the time the reaction is allowed to proceed and the conditions used (temperature, pressure, etc.). From a single monomer, a variety of polymers can be formed, with differing lengths.

The molecular sizes of polymers may be very large. Because of this, polymers are often called *macromolecules*. This term is widely used to refer to the polymers of living organisms. We shall discuss the polymers of biological importance more fully in Chapter 6.

Suggested readings

Holum, J. R., *Elements of general and biological chemistry*, 4th ed., John Wiley & Sons, New York, 1975. A chemistry textbook that emphasizes biological and health-related topics.

Masterton, W. L., and E. J. Slowinski, *Chemical principles*, 3rd ed., William B. Saunders, Philadelphia, 1973. Elementary textbook of general chemistry.

Watson, J. D., *Molecular biology of the gene*, 3rd ed., W. A. Benjamin, Inc., Menlo Park, Calif., 1976. A standard reference to chemical aspects of living organisms. Good material on weak bond interactions.

Microbial metabolism and biosynthesis

Chapter 6

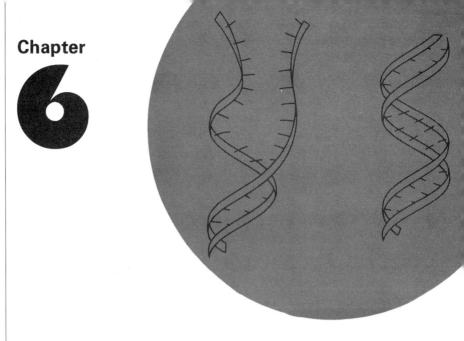

We have studied in earlier chapters the various factors that affect microbial growth; now we shall study growth itself. How do cells grow? What are the chemical reactions within the cell that result in growth? Only when we know something about these chemical processes can we really understand and *control* microbial growth.

All the reactions that occur within the cell during growth are grouped under the term *cell metabolism.* A cell has a very complex chemical composition and contains a wide variety of organic compounds, but each compound is present in a predictable and relatively constant amount for a given organism at a particular age growing in a particular medium. The chemical composition of a cell is by no means random!

Most culture media provide only very simple organic molecules that serve as carbon and energy sources. Inside the cell one finds many of these simple organic compounds, but in addition there is an enormous array of complex macromolecules called *polymers.* Polymers are large molecules built up from small subunits (called *monomers*) in much the same way that the words of our language are built up from the subunits we call letters (Figure 6.1). Each cell synthesizes many of these compounds; each time it divides, it must duplicate all of its own components to make a new cell. The sum of all these syntheses is called *biosynthesis,* and by carrying out these reactions, the cell is able to grow.

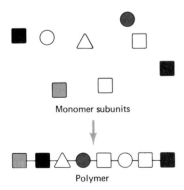

Monomer subunits

Polymer

FIGURE 6.1
Formation of a polymer from monomer subunits.

Biosynthesis requires an adequate supply of the subunits of macromolecules; these are obtained either from the medium or are formed in the cell from ingredients in the medium. Biosynthesis also requires energy, obtained by the breakdown of nutrient materials.

The processes by which nutrients are broken down leading to the release of usable energy are called *catabolism*; catabolism and biosynthesis (sometimes called *anabolism*) are opposing processes. We can represent with words the overall reaction of cell metabolism as follows; it is necessarily greatly simplified, but it shows the basic activities and requirements of a cell:

(1)	Nutrients	→ energy release + end products	Energy-yielding reactions (catabolism)
(2)	Nutrients + energy consumed	→ subunits of macromolecules	Biosynthesis (anabolism)
	Subunits + energy consumed	→ macromolecules	
Sum:	Nutrients	→ cell materials + end products	

Some of the energy released in catabolism is conserved in biosynthesis, but cells are only about 30 percent efficient, so that about 70 percent of the energy is not conserved; it is released as *heat*. Heat is wasted energy, and all cells produce considerable amounts of heat during catabolism. The reactions in (1) above can be termed *catabolic*, or *degradative reactions*; the reactions in (2) are *anabolic*, or *biosynthetic reactions*. We shall discuss each type of reaction in Sections 6.1 and 6.6.

In most organisms, the energy used to drive biosynthetic reactions is *chemical bond energy*. In the reaction below, for example:

$$C-C-C + C-C-C + energy \rightleftharpoons C-C-C-C-C-C$$

a new bond is formed between the third and fourth carbons of the product.[1] In biological systems, the energy required to make this new bond is provided in the form of *high-energy phosphate bonds*, shown as ~P (sometmes called "squiggle P"). A typical compound that provides such ~P bonds is *adenosine triphosphate*, ATP (Figure 6.2). There are two high-energy phosphate bonds in ATP, and when they are split away from the parent molecule, large amounts of energy are liberated that can be used to drive reactions. We shall see how ATP is used for the synthesis of proteins in a later section. In most reactions, only the outer high-energy phosphate bond is used, ATP then being converted into *adenosine diphosphate*, ADP.

By adding energy back to the ADP and phosphate, ATP can be regenerated.

**6.1
Obtaining energy: catabolism**

---

[1] The arrows pointing left and right in the foregoing equation signify a reversible reaction.

MICROBIAL METABOLISM AND BIOSYNTHESIS

$$ADP + P + energy \rightleftharpoons ATP$$

It is by cell oxidative reactions, the reactions of catabolism, that energy is released from nutrients to make the new high-energy phosphate bonds. The overall process is diagrammed simply as follows:

Nutrients $\xrightarrow{\text{catabolism}}$ products + energy

subunits waste products

~P

drives biosynthetic reactions

All these metabolic reactions occur in many small steps; we shall examine these pathways more carefully in the next sections. Because the processes are stepwise reactions under precise control, a maximum amount of the energy liberated is trapped to form high-energy phosphate bonds. Enzymes play two important roles: some allow the stepwise degradation of the substrate, and others help in the trapping of energy to make ~P bonds.

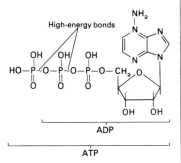

FIGURE 6.2
Structure of adenosine triphosphate, ATP, showing the location of the two high-energy phosphate bonds.

OXIDATION AND REDUCTION

In order to understand the reactions of metabolism, one must first have some understanding of how chemical reactions occur in cells. Metabolic reactions are simply chemical reactions that are usually catalyzed by enzymes. In energy metabolism there are two types of reactions: oxidations and reductions.

Oxidations are energy-yielding: a substrate is converted into oxidized products, and energy is released. Oxidation is accomplished by the addition of oxygen and/or the loss of hydrogen and electrons. In the reaction, glucose $(C_6H_{12}O_6) + O_2 \rightarrow 6CO_2 + 6H_2O$, oxygen has been added to each of the glucose carbons to form carbon dioxide (CO_2), while some of the electrons and all the hydrogen in the glucose molecule have been lost during the oxidation, being converted to water (H_2O).

Reductions are energy-requiring: energy must be added to drive the reaction, and the substrate is converted into a more reduced product. Reduction is accomplished by addition of electrons and the removal of oxygen and/or the addition of hydrogen. In the reaction, $CO_2 + H_2O \rightarrow C_6H_{12}O_6 + O_2$ (the reaction of photosynthesis), the CO_2 is reduced: the product glucose has less oxygen per carbon atom than does CO_2 (oxygen has been removed), and hydrogen atoms have been added.

Oxidation and reduction reactions also occur in inorganic systems. Ammonia (NH_3), for example, may be oxidized by certain bacteria to nitrite (NO_2^-), and nitrite may in turn be oxidized by other bacteria to nitrate (NO_3^-):

$$NH_3 \rightarrow NO_2^- \rightarrow NO_3^-$$

In the oxidation of ammonia to nitrite, hydrogen and electrons are lost from ammonia and oxygen is gained to form nitrite. When nitrite is further oxidized to nitrate, another two electrons are lost and an oxygen atom is added. If the reactions are reversed, so that nitrate is reduced to nitrite and then to ammonia, there is loss of oxygen and addition of electrons and hydrogen.

Oxidations and reductions are *always* paired. In the reaction showing the oxidation of glucose, above, the glucose was oxidized, and at the same time, the oxygen was reduced. The hydrogens liberated from the glucose oxidation were "used up" by reducing the oxygen to water.

The simplest way to think about these oxidation and reduction reactions is in terms of the donation and acceptance of electrons. The energy source donates electrons and becomes oxidized; it can be called the *electron donor*. In the same reaction, O_2 accepts electrons and becomes reduced; it is the *electron acceptor*. In every oxidation–reduction reaction there is an electron donor and an electron acceptor. Although O_2 is the most common electron acceptor, we shall learn later of several other electron acceptors that can be used in the absence of O_2.

Although we may often speak of electrons moving through the cell, electrons never move by themselves; they are always carried by specific *electron carriers*. The most common electron carrier is the substance *nicotinamide adenine dinucleotide*, NAD (Figure 6.3), which is capable of either accepting or donating two electrons. When it accepts two electrons, it becomes reduced (now written as NADH), whereas when it donates two electrons, it becomes oxidized and is

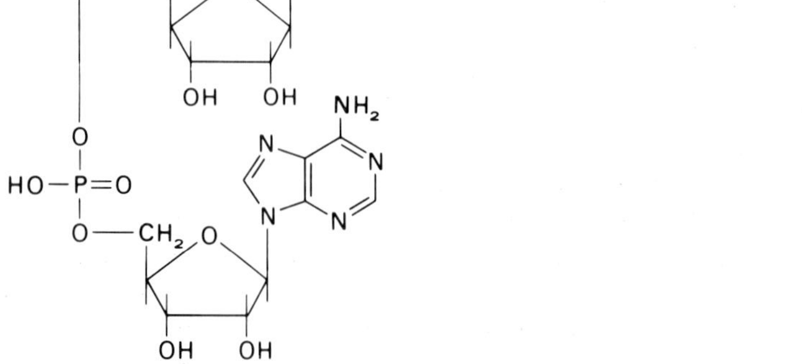

FIGURE 6.3
Structure of nicotinamide adenine dinucleotide, NAD, the major electron carrier in the cell.

MICROBIAL METABOLISM AND BIOSYNTHESIS

FIGURE 6.4
The electron transport system. Electrons removed from the organic substrate during its oxidation are transferred through the electron transport system to O_2. At three stages during this transfer, ADP is converted into ATP, leading to the synthesis of new high-energy phosphate bonds. Thus, for each pair of electrons transferred, three ATP molecules are synthesized.

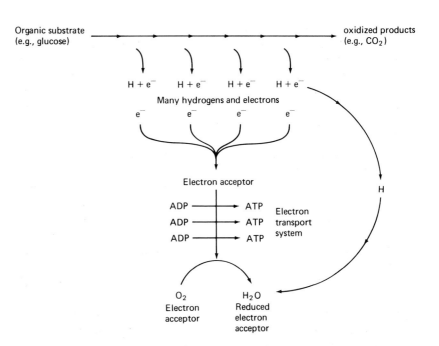

converted back into NAD. A related electron carrier is *nicotinamide adenine dinucleotide phosphate*, NADP. In general, NAD plays a role in catabolic reactions, and NADP acts in biosynthetic reactions.

SYNTHESIS OF HIGH-ENERGY PHOSPHATE BONDS (PHOSPHORYLATION)

Let us now consider how the energy released during oxidation is trapped by the cell to form ATP. There is a great deal of energy potentially available when electrons are removed during the oxidation of a substrate. There are three types of phosphorylation mechanisms by which this energy is conserved in high-energy phosphate bonds in ATP. These are called *electron-transport phosphorylation*, *substrate-level phosphorylation*, and *photophosphorylation*. In electron-transport phosphorylation, the electrons are passed directly to an *electron acceptor* and in turn are transferred to the other acceptors of the *electron-transport system* and finally to oxygen (Figure 6.4). The hydrogens released from the substrate combine with the reduced oxygen to form water, H_2O. The overall reaction is the sum of the oxidation and reduction:

$$\begin{aligned} \text{Glucose} &\rightarrow CO_2 + e^- + [H] \qquad \text{Oxidation} \\ O_2 + e^- + [H] &\rightarrow H_2O \qquad \qquad \qquad \text{Reduction} \\ \hline \text{Glucose} + O_2 &\rightarrow CO_2 + H_2O \end{aligned}$$

The key to the conservation of energy is the stepwise manner in which the electrons are transferred from glucose to oxygen by being passed along the electron-transport system. During the passage of electrons along the system, chemical bond energy is released that can be used to form molecules of ATP. The electron-transport system is coupled to a process called *electron-transport phosphorylation* (also sometimes called *oxidative phosphorylation*), in which ADP and inorganic phosphorus are made into ATP by the formation of a new ~P. Three high-energy phosphate bonds are usually synthesized for each pair of electrons transferred along the system to oxygen (see Figure 6.4).

Key components of electron-transport systems are flavoproteins and cytochromes, which act as electron carriers. *Flavoproteins* are proteins containing a derivative of riboflavin (one of the B vitamins); the flavin portion, (flavin adenine dinucleotide, FAD) which is bound to the protein, is alternately reduced as it accepts electrons and oxidized as the electrons are passed on to the cytochromes. The *cytochromes* are proteins containing a porphyrin ring (an iron-containing structure related to heme, a substance also found in the hemoglobin of red blood cells). It is the iron atom in the cytochrome that alternately becomes reduced and oxidized. In most electron-transport systems there are three cytochromes, and electrons are passed from one to the next. The final component in the electron-transport system is cytochrome oxidase, an enzyme that passes the electrons from a cytochrome to oxygen (Figure 6.5).

As noted, there are usually three high-energy phosphate bonds synthesized during the reduction of one oxygen atom to water. Each of these high-energy phosphate bonds is synthesized at a different place in the electron-transport system: one at the flavoprotein level and two at the cytochrome levels. By transport of electrons in a number of small steps through the electron-transport system, rather than in one large step, the energy released during oxidation-reduction is conserved in high-energy phosphate bonds, rather than being wasted in heat.

While most ATP is formed by electron-transport phosphoryla-

FIGURE 6.5
Steps in the electron transport system. NAD, nicotinamide adenine dinucleotide; FAD, flavin adenine dinucleotide; Cyt, cytochrome.

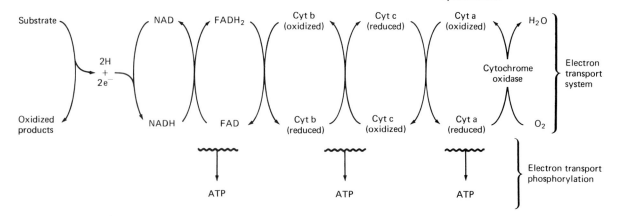

MICROBIAL METABOLISM AND BIOSYNTHESIS

tion, some is made by a process called *substrate-level phosphorylation*, which does not require the presence of oxygen. During catabolic reactions, several of the intermediates have high-energy phosphate bonds; via a group-transfer reaction (Section 5.8) this energy can be transferred to ADP to form ATP. One such intermediate is shown below:

$$
\begin{array}{ccc}
\underset{\substack{| \\ \text{COOH}}}{\overset{\substack{\text{CH}_2 \\ \| \\ \text{C}-\text{O}\sim\text{P}}}{}} + \text{ADP} \rightarrow \underset{\substack{| \\ \text{COOH}}}{\overset{\substack{\text{CH}_3 \\ | \\ \text{C}=\text{O}}}{}} + \text{ATP}
\end{array}
$$

Phosphoenol pyruvate · · · · · · · · · · · · · · · · Pyruvate

Other such substrate-level phosphorylations will be discussed later.

Photosynthetic cells have another way of generating high-energy phosphate bonds. They can carry out *photophosphorylation*, in which energy from sunlight is trapped by chlorophyll, liberating an electron that moves through an electron-transport system with cytochromes, resulting in the synthesis of a high-energy phosphate bond in ATP.

6.2 Biochemistry of energy generation

Now we shall look more carefully at the various reactions of catabolism. Just how are the substrate molecules degraded in order to enable the cell to produce the constant supply of the ATP it needs for biosynthesis? There are two major mechanisms for degradation: *fermentation* and *respiration*. Respiration requires O_2 and an electron-transport system; fermentation does not. By means of the metabolic reactions of fermentation and respiration, the cell can do several things: it can make compounds that serve as starting materials for the production of subunits that are necessary for polymer synthesis or as intermediates for other reactions; it can liberate electrons for the production of high energy phosphate bonds; and it can make end products.

FERMENTATION

In the absence of O_2, many organisms can still oxidize some organic compounds with the release of energy; part of the energy source is oxidized to CO_2, and the rest is reduced to some product that is excreted from the cell. The substance excreted, usually called a *fermentation product*, is ordinarily an acid or an alcohol. Common fermentation products include the alcohols ethanol and butanol, and lactic and acetic acids, all useful products for human beings.

The most common way in which a substrate such as glucose is fermented is by a process called *glycolysis*. Degradation of the substrate occurs by a series of enzymatic reactions, each one a small step carried out by a specific enzyme. The complete pathway for a

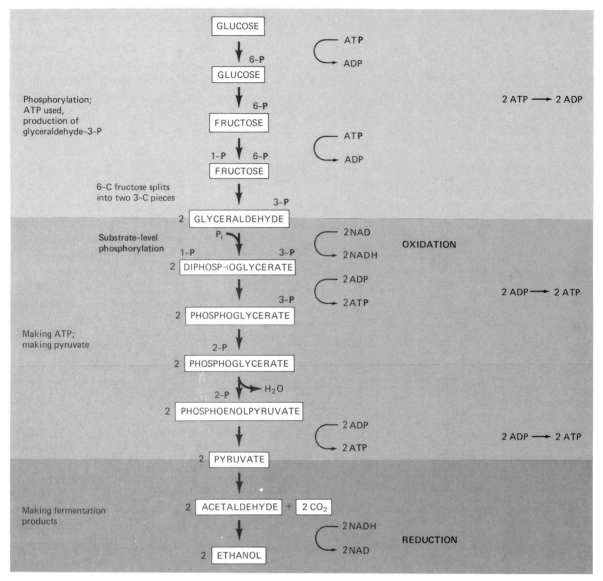

Phosphorylation;
ATP used,
production of
glyceraldehyde-3-P

GLUCOSE

ATP
ADP

6-P

GLUCOSE

6-P

2 ATP ⟶ 2 ADP

FRUCTOSE

ATP
ADP

1-P   6-P

FRUCTOSE

6-C fructose splits
into two 3-C pieces

3-P

2 GLYCERALDEHYDE

Substrate-level
phosphorylation

P_i

2 NAD
2 NADH

OXIDATION

1-P   3-P

2 DIPHOSPHOGLYCERATE

2 ADP
2 ATP

3-P

2 PHOSPHOGLYCERATE

2 ADP ⟶ 2 ATP

Making ATP;
making pyruvate

2-P

2 PHOSPHOGLYCERATE

2-P H_2O

2 PHOSPHOENOLPYRUVATE

2 ADP
2 ATP

2 ADP ⟶ 2 ATP

2 PYRUVATE

Making fermentation
products

2 ACETALDEHYDE + 2 CO_2

2 NADH
2 NAD

REDUCTION

2 ETHANOL

Overall reaction:

Glucose ⟶ 2 ethanol + 2 CO_2

2 ADP ⟶ 2 ATP

Note that there is no net change in the proportions of NAD and NADH

FIGURE 6.6
Glycolysis, the series of
biochemical steps by which
glucose is converted to ethanol
and CO_2 by yeast under
anaerobic conditions.

common fermentation, the production of alcohol from glucose by
yeast, is shown in Figure 6.6. The six-carbon glucose ring is broken
into two three-carbon pieces. Each reaction after that is in effect
doubled, since each three-carbon compound behaves in the same
way. The amount of ATP that is produced is not very large: two
ATP are used up to drive the early reactions; and then for each

three-carbon fragment, two more ATP are made. That is a total of four ATP produced, with two used up, or a net yield of two ATP for each glucose molecule used as substrate. Although two electrons are produced during the conversion of 3-P glyceraldehyde to 1, 3-di P glycerate (transferred to NAD), they are not available to go through the electron transport system since they are subsequently required for the reduction of acetaldehyde to ethanol. Note that there is no utilization of oxygen in the entire pathway; it is anaerobic.

Most anaerobic organisms metabolize substrates such as glucose by fermentation, using the glycolytic pathway, but the products are not always ethanol and CO_2. Some organisms make different products, such as lactic acid, acetic acid, or butyric acid, depending on what is done to pyruvic acid.

RESPIRATION

Most organisms grow aerobically; they utilize O_2 and oxidize their substrates by *respiration*. In respiration, a substrate such as glucose is metabolized as far as pyruvate by the glycolytic reactions just given, but the pyruvate is further metabolized aerobically by reactions known as the *tricarboxylic acid cycle* (TCA cycle). An overall scheme is given below, showing the alternative fates possible in the metabolism of glucose:

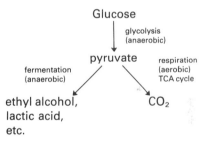

Pyruvate is clearly an important intermediate. For most organisms, aerobic or anaerobic, the reactions as far as pyruvate are the same and do not utilize oxygen. It is in the subsequent reactions of the TCA cycle, when the substrate is completely utilized, in respiration, that oxygen is necessary for the pathway to operate continuously and for ATP to be made.

For use in the TCA cycle, pyruvate must first be activated by being linked to coenzyme A, to form acetyl CoA and one molecule of CO_2. This activated two-carbon acetyl group can now enter the TCA cycle. The formation of acetyl CoA and all the reactions of the TCA cycle are shown in Figure 6.7. The name of the cycle derives from the first compounds, each of which has three carboxyl (—COOH) groups.

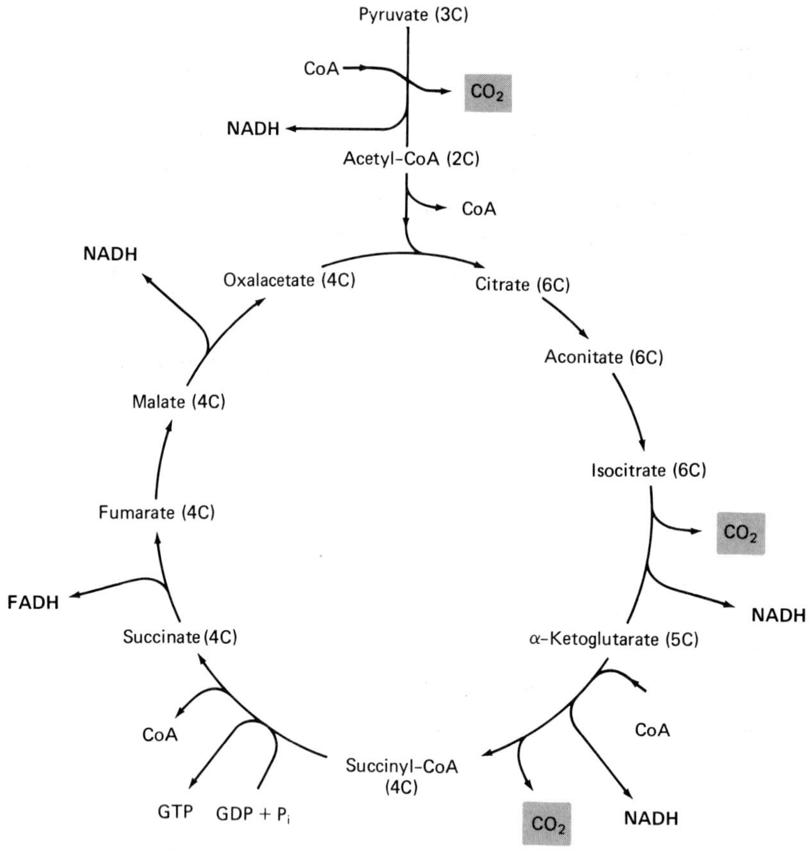

Pyruvate (3C)

CoA →

CO_2

NADH ←

Acetyl-CoA (2C)

→ CoA

NADH

Oxalacetate (4C)   Citrate (6C)

Aconitate (6C)

Malate (4C)

Isocitrate (6C)

Fumarate (4C)

CO_2

FADH

NADH

Succinate (4C)   α-Ketoglutarate (5C)

CoA   CoA

Succinyl-CoA (4C)

GTP   GDP + P$_i$

CO_2 NADH

Overall reaction: Pyruvate + 4NAD + FAD ⟶ 3CO$_2$ + 4NADH + FADH
GDP + P$_i$ ⟶ GTP
GTP + ADP ⟶ GDP + ATP
Electron transport 4NADH ≡ 12ATP
phosphorylation   FADH ≡ 2ATP   } 15ATP

FIGURE 6.7
The tricarboxylic acid cycle. The three-carbon compound pyruvate is oxidized to CO_2, and the energy released is conserved in NADH and FADH. Electron transport phosphorylation with these electron carriers leads to the synthesis of ATP. The cycle actually begins when the two-carbon compound acetyl-CoA condenses with the four-carbon compound oxalacetate to form the six-carbon compound citrate. Through a series of oxidations and transformations, this six-carbon compound is ultimately converted back to the four-carbon compound oxalacetate, which then passes through another cycle with the next molecule of acetyl-CoA. The overall balance sheet is shown at the bottom. The reducing equivalents formed (NADH and FADH) are shown in boldface type.

Three pairs of electrons are produced during each turn of the TCA cycle, and these go directly through the electron transport system. Each pair produces three high-energy phosphate bonds, and thus, the three pairs make $3 \times 3 = 9$ ATP. In addition, a fourth pair of electrons, from succinic acid, at a lower energy level, forms only 2 ATP. These 2 plus the 1 ATP from substrate level phosphorylation by succinyl CoA make $2 + 1 + 9 = 12$ ATP in all produced from the initial acetyl CoA that was condensed with oxalacetic acid to form citric acid. Another pair of electrons is produced during the conversion of pyruvate to acetyl CoA, and 3 more ATP are made. Thus a total of 15 ATP are produced from pyruvate:

$$\text{Pyruvate} + 2\tfrac{1}{2}O_2 + 15\,\text{ADP} \rightarrow 3CO_2 + 5H_2O + 15\,\text{ATP}$$

ENERGY YIELD IN CATABOLISM

Remember that there are two three-carbon fragments from each six-carbon glucose molecule; thus, the energy produced from both pyruvates derived from glucose is twice as great: a total of 30 ATP. Additional steps of oxidative phosphorylation, not shown in the illustrations, lead to a grand total of 38 ATP. The overall yields of energy from both respiration and fermentation are compared in Table 6.1. Respiration is obviously far more efficient than is fermentation. Aerobic organisms with their respiratory metabolism grow very efficiently, oxidizing substrate to CO_2 and deriving maximum energy from it. The oxygen is utilized in the removal of the pairs of electrons that are produced during the oxidation; it combines with hydrogen to form water. If the oxygen supply becomes limited, there can be no electron removal and the whole respiratory process backs up and stops. Products must be continuously removed for the process to operate.

Anaerobes using fermentative metabolism often do show good growth in culture, but they must use very large amounts of substrate since they have such a low energy yield per unit.

Many organisms are capable of using *either* respiration *or* fermentation, depending on whether oxygen is available. Yeast, for example, grows copiously on glucose when oxygen is available, producing large amounts of CO_2. If the same organism is grown without oxygen present, there is less cell yield, and alcohol is produced. It is this latter reaction, fermentation, that results in the production of beer and other alcoholic beverages. In the brewing industry, the amount of oxygen is carefully limited to allow maximal production of alcohol as a product. However, when yeast is used to make bread, CO_2 is the desired product; the gas causes the rising process that makes the desirably fluffy and airy bread.

The pathways of glycolysis and the TCA cycle are by no means the only ones operating in cell metabolism; they are very common, however, and a knowledge of their general operation is fundamental to understanding how a cell grows. There are a variety of other pathways and/or modifications or short cuts of these pathways that

TABLE 6.1 A comparison of fermentation and respiration	Fermentation	Respiration
Substrates	Sugars	Sugars or intermediate products
Products	Partially oxidized products	Completely oxidized product: CO_2
Oxygen requirements	None, anaerobic	O_2 required, aerobic
Pathway	Glycolysis	Tricarboxylic acid cycle (TCA cycle)
Amount of ATP	Low, 2 per glucose	High, 38 per glucose

are important in some cells. These will not be discussed in detail here.

Cells often do not utilize all their substrate at once; if it is present in excess, they may convert some of it into a storage polymer such as starch, glycogen, or poly-β-hydroxybutyric acid for use later, when substrate supplies dwindle. The biosynthesis of such polymers is discussed later in this chapter.

Although O_2 is the most common electron acceptor used by microorganisms, some organisms can use other electron acceptors. Oxidation with one of these alternate electron acceptors is called *anaerobic respiration*. The process is similar to respiration and uses an electron-transport system, yet oxygen is not involved, hence the name anaerobic respiration (see also Section 3.4).

A common anaerobic electron acceptor is *nitrate*, NO_3^-, which is converted into more reduced forms of nitrogen, such as nitrite NO_2^-, and nitrogen gas, N_2. The process of nitrate reduction to N_2 gas is called *denitrification* and when it occurs in agricultural soils, leads to the loss of useful nitrogen fertilizer from crop lands (see Section 16.5). Most bacteria that reduce nitrate also are able to use O_2 if it is present; they are thus called *facultative anaerobes*. It is always the case that if both O_2 and nitrate are present, the O_2 will be used first until it is mostly depleted before the organism will turn to nitrate. Less energy is obtained when nitrate is used as the electron acceptor instead of O_2; only two ATP are made in the electron transport chain instead of three.

Other electron acceptors used by some bacteria are sulfate, SO_4^{2-}, which is reduced to hydrogen sulfide, H_2S; carbon dioxide, CO_2, which is reduced to methane, CH_4; and ferric iron, Fe^{3+}, which is reduced to ferrous iron, Fe^{2+}. The bacteria that use sulfate and carbon dioxide are obligate anaerobes, not able to use O_2 at all. They have unique places in the cycles of matter in nature, as outlined in Sections 13.3 and 13.6.

We can divide the energy sources used by microorganisms into three groups: organic compounds, inorganic compounds, and light. We have already considered these briefly in Section 3.1.

Light energy is used by green plants, algae, and photosynthetic bacteria. No microorganisms of medical importance use light energy, although many algae and photosynthetic bacteria of significance in pollution control and water management do so. Most photosynthetic organisms have the ability to grow in the complete absence of organic compounds; they obtain the carbon they need for growth from carbon dioxide, CO_2. Organisms that are able to obtain all their needed carbon from carbon dioxide are called *autotrophs* (which roughly means "self-feeding"). These organisms have a specific means of synthesizing sugars from carbon dioxide, and from the sugar made they can then produce all the wide array of organic chemical substances needed for growth.

INORGANIC ENERGY SOURCES

Inorganic compounds are used by certain bacteria but by no eucaryotic microorganisms or higher organisms. Inorganic energy sources used by bacteria include sulfur compounds, such as hydrogen sulfide and elemental sulfur; the nitrogen compounds ammonia, NH_3, and nitrite, NO_2^-; and ferrous iron, Fe^{2+}. In all cases, the inorganic energy source is oxidized, the electrons being transported to an electron acceptor, generally oxygen. Organisms able to utilize inorganic energy sources are also autotrophs, as they are generally able to grow in the complete absence of organic compounds, obtaining the carbon they need for growth from CO_2.

ORGANIC COMPOUNDS

Organic substances are used as energy sources by all medically important microorganisms, as well as by the vast majority of microorganisms that live in the soil and water. We discussed the process of glycolysis, the means by which the sugar glucose is used as an energy source in Section 6.2. A variety of other sugars are also used by microorganisms and in fact sugars are the most common energy sources used by living organisms. The simplest sugar has three carbon atoms, but the most common sugars in living organisms have five or six carbon atoms. Five-carbon sugars are called *pentoses*, and six-carbon sugars are called *hexoses*. Polysaccharides are polymers formed from sugar monomers. As a group, sugars and polysaccharides are called *carbohydrates*.

The pentose sugars are most commonly associated with nucleic acids and will be discussed later in this chapter. The hexose sugars are the most common sugars used as energy sources by living organisms. The structure of the common hexose sugars glucose and fructose are given below:

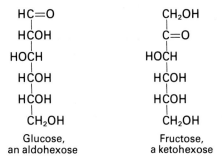

Glucose,
an aldohexose

Fructose,
a ketohexose

Although both sugars have the same molecular formula, $C_6H_{12}O_6$, they differ in that glucose has an aldehyde group on the carbon atom shown at the top in the preceding carbon-structure diagram, whereas fructose has an alcohol group on the top carbon atom and a keto group on the carbon atom that is second from the top. This difference affects the properties and reactivity of these two sugars.

Sucrose (common table sugar) is formed by connecting one molecule of glucose and one molecule of fructose together. Because sucrose contains two separate sugar molecules, it is called a *disaccharide*. Other disaccharides are *maltose* (malt sugar, common in beer making), which is formed from two glucose molecules, and *lactose* (milk sugar, the energy source for babies), which is formed from a molecule of glucose and a molecule of another hexose, galactose. Before these disaccharides are used as energy sources, they are hydrolyzed to their component sugars by specific enzymes.

Polysaccharides are long polymers formed from sugars. In most polysaccharides, only a single type of sugar is present; glucose is the most common polysaccharide-forming sugar and is the component in starch and cellulose. It is of interest that these two polymers are formed from the same monomer and yet have vastly different properties. Starch is readily digestible by many organisms and is a prime energy source. On the other hand, cellulose, the key constituent of plant cell walls, is more difficult to digest, and only a few kinds of organisms have the ability to use it as an energy source. Starch and cellulose differ only in the way in which their glucose molecules are connected; in starch, the molecular structure is open and easily attacked by enzymes, whereas in cellulose, the molecular structure is closed and less accessible to enzymatic attack. Vast amounts of cellulose are produced yearly by forests and grasslands, and this cellulose gets incorporated into the soil. Soil bacteria and fungi that have cellulase enzymes attack the cellulose and convert it into glucose, which is of course readily metabolized by many organisms. Another glucose polymer, glycogen, is the common polysaccharide of animal muscle and serves as an energy storage component. Many microorganisms also form glycogen as a storage material.

Other important organic energy sources utilized by microorganisms are the amino acids, derived from the breakdown of proteins, and fatty acids, derived from the breakdown of fats. Each of the 20 amino acids commonly found in proteins can be utilized as an energy source by one or more organisms. In most cases, the pathway for utilization involves the conversion to pyruvate or acetate, and the subsequent utilization of these substances via the TCA or similar cycles. Fatty acids are also utilized only after first being converted to acetate.

Some organic compounds can be used as energy sources either by fermentation or respiration. The sugars are in this category, as

MICROBIAL METABOLISM AND BIOSYNTHESIS

are certain of the amino acids. Acetate and the fatty acids, on the other hand, cannot be fermented but can be utilized only when respiratory processes are available. Thus, the utilization of these compounds depends on the availability of oxygen; under anaerobic conditions fatty acids will generally by quite resistant to breakdown. This fact is important in the operation of sewage-treatment systems, since the sludge digestors that carry out a major part of the treatment of the solid organic wastes of sewage operate anaerobically (see Section 14.10). Unless properly operated, sludge digestors are often poisoned by the build-up of acetic acid and certain fatty acids because the digestor bacteria are sensitive to the acidity that develops.

**6.4
Proteins**

Proteins are key constituents of organisms. Enzymes are proteins, as are a large number of other important cellular constituents. Proteins are polymers formed from amino acid monomers. Twenty amino acids are found in the proteins of living organisms, and a single protein usually contains all 20, connected together in specific sequences totalling 100 or more amino acid monomers. The characteristics of a given protein, and the enzyme reactions that it carries out are determined by the sequence of amino acids.

AMINO ACIDS AND PEPTIDE
BONDS

Amino acids have the following basic structure:

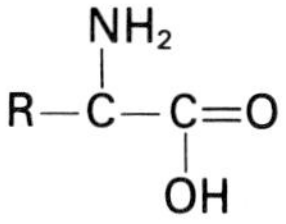

$$R-\overset{\overset{\displaystyle NH_2}{|}}{C}-\overset{\overset{\displaystyle}{}}{\underset{\underset{\displaystyle OH}{|}}{C}}=O$$

where R represents one of a number of side chains. Amino acids thus have both carboxyl groups, $-\underset{\underset{\displaystyle OH}{|}}{C}=O$, and amino groups,

$-NH_2$. The key point concerning the structure of an amino acid in relation to protein formation is that the presence of two functional groups on the same molecule makes possible the formation of polymers, by connection of the carboxyl group of one amino acid to the amino group of another by a *peptide bond*:

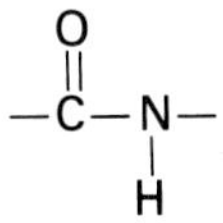

$$-\overset{\overset{\displaystyle O}{||}}{C}-\underset{\underset{\displaystyle H}{|}}{N}-$$

Peptide bond
structure

The peptide bond is formed by dehydration, the OH of the water molecule coming from —COOH and the H of the water molecule coming from —NH$_2$:

$$H_2N{-}R_1{-}\overset{|}{C}{-}\overset{|}{\underset{OH}{C}}{=}O + H_2N{-}\overset{R_2}{\underset{OH}{C}}{-}C{=}O \rightarrow R_1{-}\overset{H_2N}{C}{-}\overset{O}{C}{-}\underset{\underbrace{}_{\substack{Peptide\\bond}}}{N}{-}\overset{R_2}{\underset{OH}{C}}{-}C{=}O + H_2O$$

where R$_1$ and R$_2$ are different side chains.

A protein is a chain of amino acids held together by peptide bonds. At one end of the chain, there will be a free carboxyl group not involved in peptide-bond formation, and at the other end, there will be a free amino group. For purposes of designating amino acid sequences in proteins, it is necessary to distinguish between the amino acids at each end of the chain. These are called the *C-terminal* (for COOH) and *N-terminal* (for NH$_2$) amino acids.

When proteins are digested, water molecules are added, and the peptide bonds are broken, leading to the liberation of the individual amino acids. This is the process of hydrolysis and can be carried out chemically by heat and acid, but it is accomplished in living organisms at neutral pH and low temperature through the agency of enzymes. Enzymes that digest (hydrolyze) proteins are called *proteases.*

The peptide bonds connecting the amino acids in a protein are responsible for the overall structure of a protein molecule, but the specific properties of a single protein are determined by the *sequence* of amino acids. Once the sequence is formed, the protein molecule folds up into a fairly compact, generally spherical structure that is held together by weak bonds that form between the R groups of the amino acids.

As noted, there are about 20 amino acids commonly found in proteins. These amino acids differ in the side chains (the so-called R group) that they possess. Some of the side chains are acidic or basic and can form ionic or hydrogen bonds with other amino acid side chains. Some side chains are nonacidic but contain oxygen, sulfur, or other structures that make it possible for them to form hydrogen but not ionic bonds with other amino acid side chains. Some amino acids contain neither ionic- nor hydrogen-bond-forming side chains but contain methyl or other groups that are hydrophobic and form hydrophobic bonds. Structures of a few amino acids are shown in Figure 6.8, and a grouping of the amino acids by character of the side chain is given in Table 6.2.

MICROBIAL METABOLISM AND BIOSYNTHESIS

FIGURE 6.8
Structures of some amino acids. The R groups are circled.

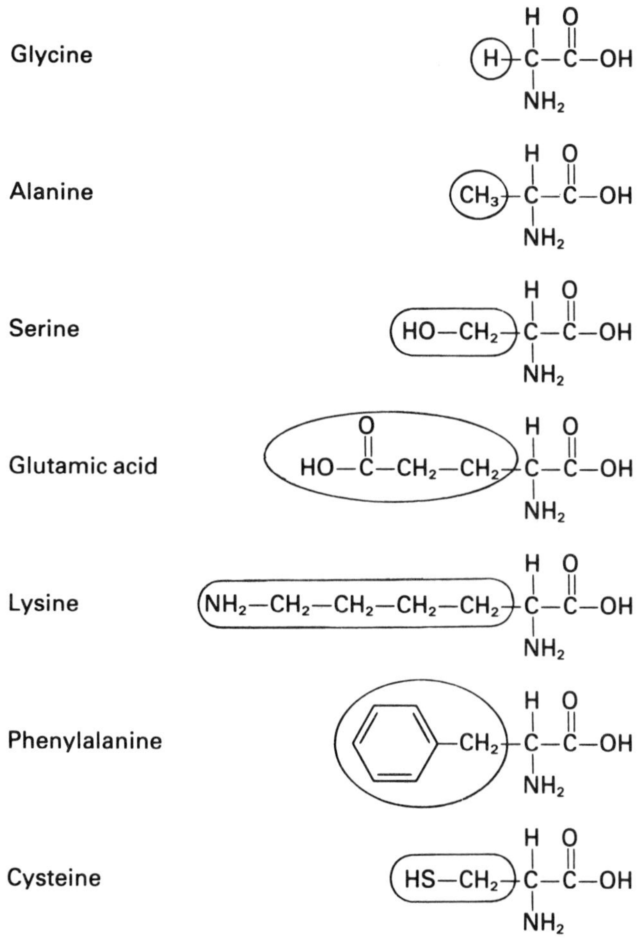

Glycine

Alanine

Serine

Glutamic acid

Lysine

Phenylalanine

Cysteine

TABLE 6.2
20 amino acids grouped by side-chain characteristics[a]

Ionizable side chains; form ionic and hydrogen bonds		Nonionizable side chains; form hydrogen bonds		Hydrophobic side chains	
Acid side chains		Glycine	gly	Valine	val
Aspartic acid	asp	Alanine	ala	Leucine	leu
Glutamic acid	glu	Serine	ser	Isoleucine	ile
Basic side chains		Cysteine	cys	Phenylalanine	phe
Lysine	lys	Threonine	thr	Tryptophan	trp
Arginine	arg	Tyrosine	tyr	Methionine	met
Histidine	his	Asparagine	asn		
		Proline	pro		
		Glutamine	gln		

[a] The abbreviations commonly used in writing protein structures are also shown.

As seen in Figure 6.8, the simplest amino acid is glycine. Its R group consists of a single H atom. Slightly more complicated, alanine has a CH_3 group for a side chain. In both these amino acids, the R group is weakly hydrophilic. Serine is similar to alanine but has an OH group substituting for one of the H atoms of the CH_3 group, enhancing the hydrophilic character of the R group. The next amino acid listed is glutamic acid, which has a carboxyl (acidic) function as part of its R group. Glutamic acid can thus form not only hydrogen bonds but also ionic bonds since the carboxyl group can ionize to an acid. The next amino acid in the list, lysine, has an NH_2 group at the end of its R group. Since an NH_2 group ionizes to form a base, the R group of lysine can form an ionic bond with an acidic R group, such as that of glutamic acid. Phenylalanine, the next amino acid in the list, has an aromatic (benzene) ring in its R group. Because aromatic rings are hydrophobic, phenylalanine forms hydrophobic bonds instead of hydrophilic bonds. The last amino acid listed is cysteine, which contains a sulfur atom. Cysteine can be viewed as an analog of serine in which the O of serine has been replaced by an S. The presence of cysteine in proteins is of great importance because one cysteine is able to form a *covalent* bond with another cysteine, by means of a disulfide bridge. The formation of such a disulfide bridge between two cysteines occurs as follows:

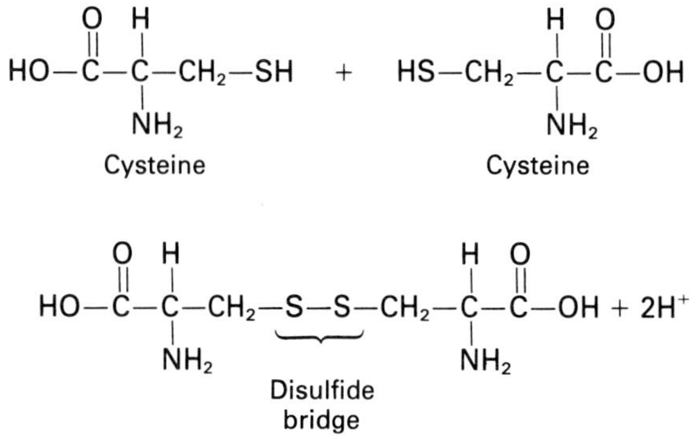

$$HO-\overset{\overset{\displaystyle O}{\|}}{C}-\overset{\overset{\displaystyle H}{|}}{\underset{\underset{\displaystyle NH_2}{|}}{C}}-CH_2-SH \quad + \quad HS-CH_2-\overset{\overset{\displaystyle H}{|}}{\underset{\underset{\displaystyle NH_2}{|}}{C}}-\overset{\overset{\displaystyle O}{\|}}{C}-OH$$

Cysteine   Cysteine

$$HO-\overset{\overset{\displaystyle O}{\|}}{C}-\overset{\overset{\displaystyle H}{|}}{\underset{\underset{\displaystyle NH_2}{|}}{C}}-CH_2-S-S-CH_2-\overset{\overset{\displaystyle H}{|}}{\underset{\underset{\displaystyle NH_2}{|}}{C}}-\overset{\overset{\displaystyle O}{\|}}{C}-OH + 2H^+$$

Disulfide bridge

Since covalent bonds are strong bonds, the disulfide bridge formed, if occurring between two cysteine residues in the same protein chain, results in a firm bond between the two parts of the chain. Such disulfide bridges are of great importance in stabilizing and conforming the structure of proteins.

The structures of proteins are determined first by their amino acid sequences, and then by the manner in which the long protein chains fold. Folding is determined to a great extent by the interactions between amino acids in different parts of the chain. A single protein may contain more than one chain.

N-terminal ends

A chain:
NH2 — Gly — Ile — Val — Glu — Gln — Cys — Cys—S—S—Cys — Ala — Ser — Val — Cys — Ser — Leu — Tyr — Gln — Leu — Glu — Asn — Tyr — Cys — Asn — COOH

B chain:
NH2 — Phe — Val — Asn — Gln — His — Leu — Cys — Gly — Ser — His — Leu — Val — Glu — Ala — Leu — Tyr — Leu — Val — Cys — Gly — Glu — Arg — Gly — Phe — Phe — Tyr — Thr — Pro — Lys — Ala — COOH

C-terminal ends

FIGURE 6.9
The amino acid sequence of a simple protein, the hormone insulin.

PROTEIN STRUCTURE

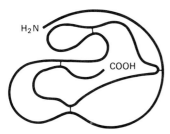

FIGURE 6.10
The folded, three-dimensional structure of the enzyme ribonuclease. This enzyme consists of a single chain with 124 amino acid residues. There are 8 cysteine residues, leading to the formation of 4 disulfide bridges, shown by short lines in the drawing.

Amino acid sequences for a large number of proteins are now known. Because proteins often have hundreds of amino acid residues, it becomes difficult to write the overall structure in simplified fashion. The structure of a simple protein, the hormone *insulin*, is shown in Figure 6.9. Note that this protein consists of two chains connected by disulfide bridges, and that there also is a disulfide bridge within the structure of chain A. The structure shown is called the *primary* structure of the protein. It shows the covalent bonds that are present but does not indicate how the chains are folded nor where any noncovalent (weak) bonds, such as hydrogen, ionic, or hydrophobic bonds, might exist. Since it is the folded protein that is the biologically active structure, the manner in which the protein folds is of considerable importance. An illustration of the folded, or three-dimensional, structure of the enzyme ribonuclease is shown in Figure 6.10. In recent years, the three-dimensional structures of a number of proteins have been determined, and considerable information about the types of bonding between portions of protein chains is available.

6.5
Nucleic acids

Nucleic acids are polymers in living organisms made up of sugar-phosphate chains to which special organic molecules called *nucleic acid bases* are attached. Two kinds of nucleic acids are recognized, based on the sugar present. Ribonucleic acids have the five-carbon sugar ribose, and deoxyribonucleic acids have deoxyribose, a sugar similar to ribose in which the oxygen on carbon two (see the following structural diagram) has been removed.

$$
\begin{array}{cc}
\text{CHO} & \text{CHO} \\
| & | \\
\text{HCOH} & \text{CH}_2 \longleftarrow \text{Oxygen missing} \\
| & | \\
\text{HCOH} & \text{HCOH} \\
| & | \\
\text{HCOH} & \text{HCOH} \\
| & | \\
\text{CH}_2\text{OH} & \text{CH}_2\text{OH} \\
\text{Ribose} & \text{2-Deoxyribose}
\end{array}
$$

NUCLEIC ACID BASES

The nucleic acid bases are N-containing molecules that are basic in nature. Two kinds of nucleic acid bases exist: *pyrimidines* and *purines*. Each pyrimidine contains a single ring containing two N atoms, with substitutions, and purines consist of two fused rings with four N atoms, also with substitutions. The two purines, adenine and guanine, and one pyrimidine, cytosine, are common to both RNA and DNA, whereas the second pyrimidine is different in these two polymers, RNA having uracil and DNA having thymine. (It is of interest that some nucleic acid bases are also key constituents of coenzymes, for instance, adenine in ATP and NAD.)

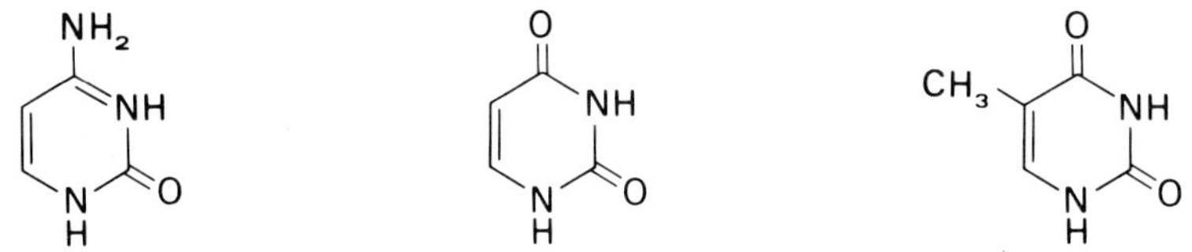

Cytosine   Uracil   Thymine

Pyrimidine bases

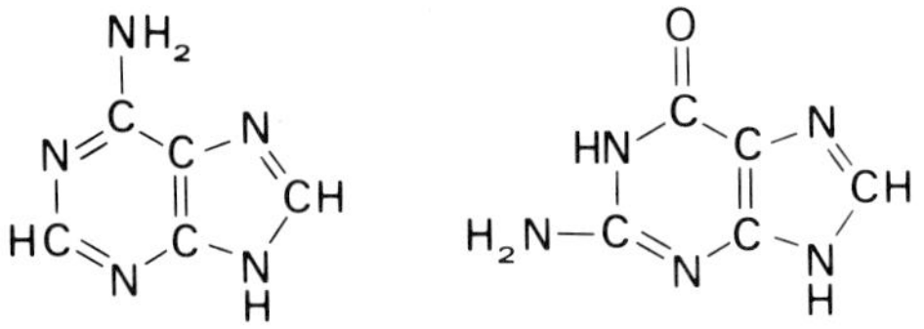

Adenine   Guanine

Purine bases

	Purine bases	Pyrimidine bases
RNA	Adenine, guanine	Cytosine, uracil
DNA	Adenine, guanine	Cytosine, thymine

The nucleic acid bases contain O and N side chains that are involved in hydrogen bonding and thus determine how the bases will interact with each other.

The nucleic acid bases are connected to the pentose sugars, ribose for RNA and deoxyribose for DNA. Connected to the sugar is a phosphate group. The combined base-sugar-phosphate is called a *nucleotide*; there is one nucleotide corresponding to each nucleic acid base. The structure of the uracil nucleotide is shown as an example:

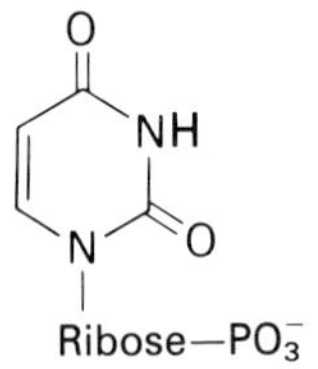

Ribose—PO_3^-

NUCLEOTIDES

The phosphate group has the ability to form a bond with two separate sugars, making the possibility for the formation of polymers:

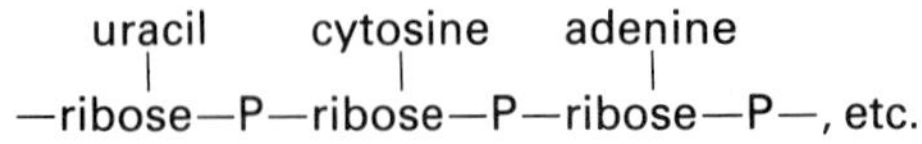

```
  uracil    cytosine   adenine
    |          |          |
—ribose—P—ribose—P—ribose—P—, etc.
```

In this way, long-chain polymers can be formed: the *nucleic acids*, RNA and DNA. The backbone of the RNA chain consists of alternating ribose and phosphate, and it is the same in all RNA molecules. The differences between RNA molecules, which determine their functional properties, reside in the *sequence* of bases connected to the backbone. With the four bases uracil, cytosine, adenine, and guanine in a long polymer, a vast number of sequences are possible. Similarly, the backbone of DNA chains consists of alternating deoxyribose and phosphate molecules, with the four bases thymine, cytosine, adenine, and guanine.

Of major importance for the overall structure and function of the nucleic acids are the specific types of hydrogen bonds that can be formed between different bases. In DNA, thymine has such a structure that its hydrogen-bond-forming groups line up exactly with the hydrogen-bond-forming groups of adenine. Cytosine, on the other hand, has such a structure that its hydrogen-bond-forming groups associate with those of guanine (Figure 6.11). Thus, thymine forms specific associations with adenine, whereas cytosine forms specific associations with guanine. This leads to the possibility for the formation of a double-stranded polymer, the two strands being held together by hydrogen bonds. However, such a double-stranded polymer will form only if the sequence of bases on one strand is exactly complementary to the sequence of bases on the other strand. *Complementary* means that wherever there is a thymine on one strand there will be an adenine on the other, and wherever there is a cytosine on one strand there will be a guanine on the other. Note that there are three hydrogen bonds formed between guanine and cytosine, whereas there are only two hydrogen bonds formed between adenine and thymine. This means that the guanine-cytosine pairs are stronger than the adenine-thymine pairs.

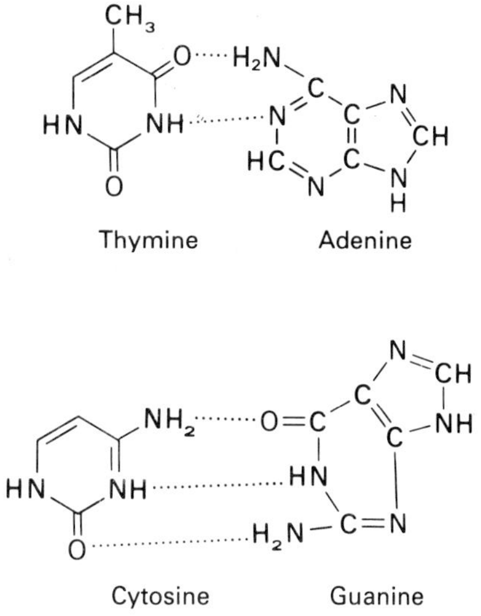

FIGURE 6.11
Complementarity of purine and pyrimidine bases in DNA and RNA. Cytosine always pairs with guanine, and adenine always pairs with thymine. In RNA, the same complementarity exists, except uracil replaces thymine. Complementarity also exists between DNA bases and RNA bases, so that the information in DNA can be transcribed into RNA. The hydrogen bonds are shown as dotted lines.

Thymine   Adenine

Cytosine   Guanine

The overall structure of the double-stranded DNA molecule is a *helix*, a screw-shaped structure that forms when two DNA strands which are complementary twist together in a specific configuration (Figure 6.12). The DNA molecule in a cell is generally present as a double helix. During duplication (replication) of the DNA, a new molecule forms along each single strand of the double helix, the double helix unwinding and nucleotides lining up opposite their complementary bases. Nucleic acid replication will be discussed later in this chapter.

Generally, RNA molecules do not exist in cells in double-stranded polymers, but as single strands. However, the bases of RNA can hydrogen-bond with the complementary bases on DNA. It is in this manner that RNA molecules are synthesized, using the DNA as a model, or *template*. The base sequence of an RNA molecule will thus be complementary to the DNA molecule that served as the template. Only one of the two strands of the DNA double helix usually serves as a template for RNA synthesis. The transfer of the base sequence of a DNA molecule into its complement in RNA is of importance in cell function, as this RNA molecule now contains the genetic code that will be used in the synthesis of protein, as described in the next section.

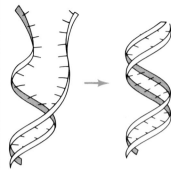

FIGURE 6.12
Formation of the DNA helix by specific hydrogen bond formation between complementary bases on opposite chains. The sugar-phosphate backbone defines the helix; the bases are shown as straight lines projecting from the backbone.

Let us examine what sorts of compounds the cell must synthesize. Table 6.3 shows some of the common types of polymers in a cell and the subunits from which each is built. The cell synthesizes these polymers by linking the subunits together. In addition, the cell must synthesize the various subunits themselves. Thus, we see that a vast array of synthetic reactions are required to produce the large number of polymers and other components found in a cell.

6.6
Biosynthesis: Anabolism

Let us consider the synthesis of protein. There are thousands of proteins in a cell, each with a unique structure and function. The structure of a protein is determined by the sequence in which its component amino acid subunits are connected (see Section 6.4). How is the cell able to ensure that for each separate protein, the proper amino acids are connected in the correct order? The amino acid sequence of the protein is specified by a gene, which is a portion of a DNA molecule, and it is the sequence of pyrimidine and purine bases within this gene that *codes* for the amino acid sequence of the protein. Microbial genetics will be discussed later, but the important point to understand here is that a sequence of *three* purine and pyrimidine bases codes for a single amino acid.

PROTEIN SYNTHESIS

TABLE 6.3
Summary of the key
macromolecules found in
living organisms

Macromolecule	Monomers	Formula	Linkage	Comments
Starch (a poly-saccharide)	Glucose	HC=O $\|$ HCOH $\|$ OHCH $\|$ HCOH $\|$ HCOH $\|$ HCOH $\|$ CH_2OH	Glycosidic α-1,4 $\|$ OCCCCCC OCCCCCC OCCCCCC $\|$ O—	1 kind of monomer; polymer of indefinite length; usually greater than 1,000 monomers.
Protein	Amino acids	R $\|$ $H_2N—C—C=O$ $\|$ OH R = specific side chain	Peptide H $\|$ —N—C— $\|\|$ O	20 different kinds of monomers; polymer usually of specific length and sequence; monomer number varies between 100 and 1,000.
RNA	Ribonucleotides	Purine-ribose-P (or pyrimidine-ribose-P)	Phosphodiester O $\|$ —C—O—P—O—C— $\|$ O	4 different kinds of monomers: adenine, uracil, cytosine, guanine; polymer usually of specific length and sequence; number of monomers often greater than 3,000.
DNA	Deoxyribo-nucleotides	Purine-deoxy-ribose-P (or pyrimidine-deoxyribose-P)	Phosphodiester O $\|$ —C—O—P—O—C— $\|$ O	4 different kinds of monomers: adenine, thymine, cytosine, guanine; polymer of specific length and sequence; number of monomers often greater than 10,000,000 but single genes within chain vary between 300 and 3,000 monomers.

The DNA is not directly involved in protein synthesis. The sequence of DNA purine and pyrimidine bases is first *transcribed* into an RNA molecule, and it is the RNA molecule that is directly involved in protein synthesis (Figure 6.13). During cell division, the sequence of DNA purines and pyrimidines is copied into a complementary molecule of DNA (see Section 6.5). Likewise, the sequence is also copied into RNA molecules in complementary order, with the additional specification that the pyrimidine uracil in RNA replaces the pyrimidine thymine in DNA.

The RNA molecule that represents the complement of the DNA specifying a particular protein is called a *messenger RNA* (mRNA). The mRNA interacts with ribosomes, as described below, to form a polyribosome, and it is on this polyribosome that protein synthesis occurs. How is the sequence of purines and pyrimidines in the messenger RNA *translated* into the amino acid sequence of the protein? Remember that the position of each amino acid in the protein is specified by a particular sequence of three purines and pyrimidines in the nucleic acid, the *triplet code*. The amino acids do not recognize this triplet code directly. Adapter molecules called *transfer RNA* (tRNA) carry the amino acids and recognize the triplet code.

With this background, how does protein synthesis occur? The cell must first synthesize the subunits, the amino acids (see Figure 6.8). Each of the 20 amino acids has a specific pathway for its synthesis. Once the amino acids are available, they are *activated* by the high-energy molecule adenosine triphosphate (ATP):

$$\text{aa} + \text{ATP} \xrightarrow{\text{enzyme}} \text{aa} \sim \text{AMP} + \text{P} \sim \text{P}$$

<table>
<tr><td>Amino
acid</td><td>Activated
amino acid</td><td>Inorganic
pyrophosphate</td></tr>
</table>

One of the high energy bonds (~P) from ATP is transferred to the amino acid, and inorganic pyrophosphate (also containing a high-

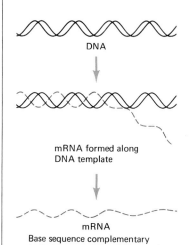

DNA

mRNA formed along
DNA template

mRNA
Base sequence complementary
to that in one of the DNA strands

FIGURE 6.13
Transcription, the transfer of the genetic information in DNA into a messenger RNA molecule, the first step in protein synthesis.

FIGURE 6.14
Synthesis of a protein.

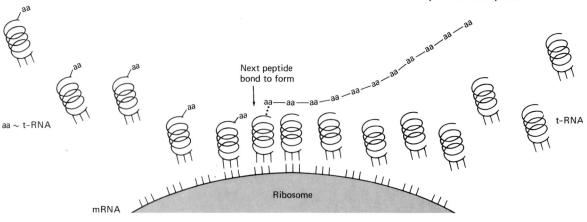

MICROBIAL METABOLISM AND BIOSYNTHESIS

energy bond) is split off (two high-energy phosphate bonds thus being consumed). Each of the 20 different amino acids is activated in this way, each requiring its own specific activating enzyme. Each amino acid is then transferred to a specific transfer RNA (tRNA):

$$aa \sim AMP + tRNA \rightarrow aa \sim tRNA + AMP$$

The amino acid-tRNA molecules are then polymerized; the overall reaction is shown below (where n represents the number of amino acids in the protein):

$$n(aa \sim tRNA) \rightarrow aa—aa—aa \cdots aa + n(tRNA)$$

But it is the *sequence* in which the amino acids occur that is important, for that is what determines which protein is formed. The sequence is determined on the surface of the ribosomes by a messenger RNA (mRNA) that serves as a *template*. This is diagrammed in Figure 6.14. Each aa-tRNA attaches briefly to the mRNA according to the specific nucleotide triplet code. Each time the proper triplets from tRNA and mRNA are paired, the amino acid corresponding to that triplet pair will be incorporated into the growing peptide chain. As soon as the amino acid chain is released from the ribosome, it is further twisted and folded into a configuration that is typical for each specific protein. This is the final form of the protein, which then can act in the cell, either as an enzyme or as a part of some cellular structure.

NUCLEIC ACID SYNTHESIS

The cell must also synthesize many other polymers in addition to proteins. The syntheses of the polynucleotides DNA and RNA are similar in many ways to that of protein. The cell first synthesizes the purine and pyrimidine bases, each of which is then linked to a sugar (deoxyribose in DNA and ribose in RNA) and phosphate to form a nucleotide (see Section 6.5). Then each nucleotide is *activated* by the addition of two high-energy phosphate bonds. In the presence of primer nucleic acid, the activated nucleotide subunits can then be polymerized into nucleic acid. The synthesis of DNA is diagrammed in Figure 6.15. The presence of primer DNA is of utmost importance, for it acts as a template to determine the sequence of the incorporation of the subunits, and it is this sequence that determines the properties of the polymer. The synthesis actually occurs along the DNA double helix; the helix splits, and the new chain is formed along the old template.

OTHER MACROMOLECULES

Polysaccharides play significant roles in cellular physiology. The cell wall structures of most eucaryotes consist of polysaccharide, as do

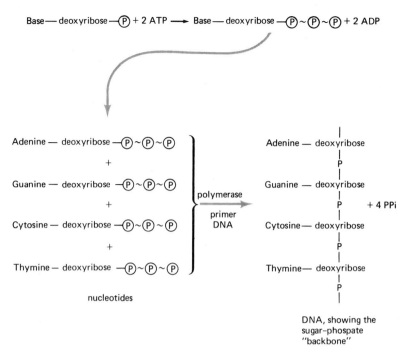

FIGURE 6.15
The synthesis of DNA.

the capsules and slimes of procaryotes. A variety of storage materials (starch, glycogen) are also polysaccharides. They are formed in a simpler way than are proteins and nucleic acids. The subunit sugars are first activated with high-energy phosphate groups and then are transferred to the end of the growing chain of the polysaccharide. No templates are involved; generally, only a single sugar is present in a polymer, so that the mechanism of biosynthesis need not be so complicated as for proteins and nucleic acids.

Bacterial walls have a rigid layer called *peptidoglycan*, which is, in a sense, a complex polysaccharide. It is a macromolecule composed of repeating units of two sugar derivatives, N-acetyl-glucosamine and N-acetylmuramic acid, and a small group of amino acids. The sugars and amino acids form a large sugar peptide which is then polymerized; the polymers are interconnected by peptide cross-links to form the peptidoglycan layer.

We can now see the metabolic events occurring in a cell from an overall viewpoint. We have only touched on a few of the reactions that occur in cells, but they should serve as an example of how a "typical" cell functions. A diagrammatic summary of cell metabolism is shown in Figure 6.16. There is a constant flow of nutrients into the cell; some nutrients are metabolized to end products so that ATP can be made, while other nutrients are converted into building blocks for the synthesis of cell components, using the ATP to drive these synthetic reactions.

6.7
Regulation of metabolism

How is the vast array of metabolic reactions regulated? The genetic capability of the cell determines much of what occurs; only certain nutrients can be used, only some substances can be synthesized, only certain energy sources used. Such factors as pH, temperature, concentration of substrate and products, and amount of enzyme present all interact to control the metabolic processes of the cell. But most important are several specific mechanisms for turning on or shutting off metabolic pathways.

From the great complexity of the metabolic reactions that we have been studying, we can see that the cell must have an enormous number of enzymes. Not all are necessarily present at all times. Many enzymes are found in a cell nearly all the time; these are called *constitutive* enzymes. The constitutive enzymes are those involved in the wide variety of interdependent reactions affecting nearly all processes; they are required if the cell is to grow under any conditions.

Other enzymes, however, are not needed all the time. The cell's work is made easier by the fact that it generally doesn't make enzymes that it doesn't need. For example, a cell may be genetically

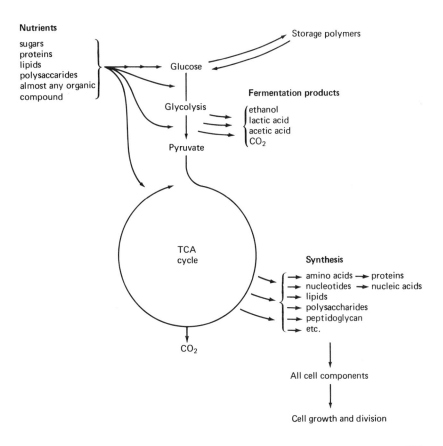

FIGURE 6.16
Overall view of cell metabolism.

capable of making a particular vitamin; however, if the vitamin is provided preformed in the medium, the cell will not make it. None of the enzymes of that biosynthetic pathway will be made.

To regulate and control the multienzyme systems of the cell so as to allow maximum efficiency in growth under a variety of conditions requires very precise control mechanisms in the cell. How are these controls on enzyme synthesis effected?

Two general mechanisms for control of enzyme synthesis are induction and repression of enzymes. A cell may have a genetic capability for making a particular enzyme but may still not synthesize it. In the case of inducible or repressible enzymes, expression of the cell's genetic capability depends on the medium. For an inducible enzyme to be made, an inducer substrate must be in the medium. For an enzyme that is being made to be repressed instead, a repressor substance must be present.

For example, when *Escherichia coli* is grown on a simple mineral medium containing an ammonium salt as a nitrogen source and glucose as a carbon source, the cells contain all of the normal constitutive glycolytic and TCA cycle enzymes. However, *E. coli* is capable of making many other enzymes as well, none of which will be present in the cells of this culture; these are *inducible* enzymes. One such inducible enzyme is β-galactosidase, which breaks down the sugar lactose, a β-galactoside. The enzyme is formed only when lactose is present in the medium. The differences between the constitutive enzymes (those for utilization of glucose), which are present in the cells at all times, and an inducible enzyme, such as β-galactosidase, are clear. The value of such a mechanism to the organism is great, since it does not have to make an enzyme until it is needed, that is, until the specific substrate is present.

A phenomenon related to induction is *repression*. The enzymes catalyzing the steps in the synthesis of a specific compound are often not present if the compound is present already in the medium. When it is absent, then the enzymes are synthesized, and the product is made by the cell. The presence of the product inhibits or represses the formation of the enzymes. The value to the organism of such a mechanism is clear; it does not synthesize enzymes when they are not needed.

What happens at the molecular level during induction or repression? Although the medium ingredients causing induction or repression seem to affect enzyme synthesis in opposite ways, they actually act quite similarly. To understand their action, it is necessary to recall the details of protein synthesis. When an inducer is added to a cell culture, it initiates enzyme synthesis by allowing formation of a new specific mRNA that codes for the particular enzyme. When a substance causing repression is added, it inhibits formation of a specific mRNA. How do these substances affect mRNA formation? They combine with specific proteins, called *repressors*, that in turn affect mRNA synthesis. In the case of repressible enzymes, it is thought that a specific repressor protein combines with the DNA at the point where synthesis of the mRNA is normally initiated. The synthesis of the mRNA is thus blocked, and the protein that is specified by this mRNA cannot be made (Figure 6.17).

For induction, the situation is reversed. The repressor protein is thought to be active at all times in the absence of the inducer, blocking the synthesis of the mRNA. When the inducer is added, for example, lactose, it combines with the repressor protein and inactivates it. The inhibition is overcome or derepressed, and mRNA can now be made (see Figure 6.17).

FEEDBACK INHIBITION

Another very specific control mechanism occurs in many cells, called *feedback inhibition*. This does not affect the synthesis of enzymes, but

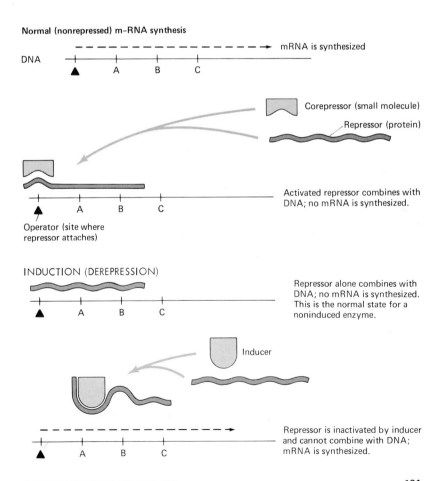

FIGURE 6.17
Manner by which induction and repression occur to regulate the synthesis of proteins in the cell.

their activity. In some metabolic reactions, the end product of a long series of reactions can regulate the activity of the pathway by inhibiting one of the enzymes acting at an early step in the pathway. This can be shown in the reaction below. Let B, C, D, and so on be a series of intermediates in a biosynthetic pathway leading from A, the substrate, to G, the final product (an amino acid, for instance). Each step from A to G is catalyzed by a specific enzyme, labeled *enz* in the diagram.

$$A \xrightarrow{enz} B \xrightarrow{enz} C \xrightarrow{enz} D \xrightarrow{enz} E \xrightarrow{enz} F \xrightarrow{enz} \text{product } G$$

If there is a large amount of product, the early reactions of the pathway will be severely inhibited, and the amount of product will be correspondingly lessened so that the cell will not make excess product. When the product is once again in short supply, the inhibition will be less, and synthesis will proceed. This type of control seems to be an important mechanism to regulate the amount of product formed in a variety of biosynthetic pathways. If pre-formed product is added to a culture, the cells stop making it and use the added source instead. This type of control is called feedback inhibition because the final product feeds back to the first step in the pathway (the word *feedback* was first used in electronics to refer to a type of control system widely used in high-fidelity and stereo sets). Note that feedback inhibition is an immediate type of control mechanism, acting on preexisting enzymes and leading to an immediate inhibition of the biosynthesis of the product. Induction and repression, on the other hand, do not act on preexisting enzymes but only on the synthesis of new enzymes; they are thus less immediate types of control mechanisms. Feedback inhibition can be viewed as a mechanism for finely regulating cell metabolism, whereas induction and repression are coarser, less precise mechan-isms. Working together, both these types of mechanism lead to an efficient regulation of cell metabolism, so that energy is not wasted by carrying out unnecessary reactions.

Antibiotics were discovered and exploited before the field of molecu-lar biology was established, but it required extensive developments in molecular biology to explain how antibiotics could be so selective and potent in their activities. We have already discussed the nature and testing of antibiotics in Section 4.6; we present here a brief discus-sion of how antibiotics act at the molecular level.

    Antibiotics are chemicals. They can only affect cells if they are able to bind to vital cell constituents and inhibit their functions. Four major cell structures are affected by one or more antibiotics: cell wall, membrane, nucleic acids, and ribosomes (Table 6.4). Since some of these structures differ between procaryotes and eucaryotes

6.8
Mode of action of antibiotics

TABLE 6.4
Modes of action of
antibiotics

	Specificity
Cell wall synthesis	
Penicillin	Procaryotes
Vancomycin	Procaryotes
Bacitracin	Procaryotes
Membrane structure	
Polymyxin	Procaryotes (eucaryotes less affected)
Gramicidin	Procaryotes (eucaryotes less affected)
Nystatin	Eucaryotes
Amphotericin	Eucaryotes
Protein synthesis	
Chloramphenicol	Procaryotes
Streptomycin	Procaryotes
Tetracycline	Procaryotes
Cycloheximide	Eucaryotes
Nucleic acid synthesis	
Actinomycin	Procaryotes, eucaryotes
Mitomycin	Procaryotes, eucaryotes
Rifamycin	Procaryotes (eucaryotes less affected)

(see Chapter 2), it is perhaps not surprising that certain antibiotics will act on procaryotic functions and others on eucaryotic functions. It is this differential activity that is one basis of selective toxicity, making possible the use of antibiotics in the therapy of infectious disease. The other basis of selectivity is permeability; some antibiotics penetrate more readily into certain kinds of organisms than into others. If an antibiotic does not bind or penetrate the cell, it cannot have any activity on this cell.

Certain antibiotics act by inhibiting synthesis of the *cell wall* in procaryotes. As noted in Section 6.7, the cell wall of procaryotes is chemically different from that of eucaryotes, and different enzymes are involved in its synthesis. A number of antibiotics inhibit cell wall synthesis in procaryotes (see Table 6.4), of which the most important is penicillin. Other antibiotics inhibiting cell wall synthesis are vancomycin and bacitracin. Although these latter antibiotics have much more restricted medical uses than penicillin, they do have roles to play in the treatment of certain diseases caused by organisms that have become resistant to penicillin.

Some antibiotics are able to combine with vital constituents of the *cell membrane* and either destroy the permeability barrier of the cell, causing cell lysis, or inhibit the function of key constituents in the cell membrane, such as the respiratory enzymes. Because procaryotes and eucaryotes have some differences in chemical composition of the cell membrane, certain of these antibiotics also have selective toxicity. Antibiotics affecting membranes in procaryotes include polymyxin and gramicidin; since these antibiotics also have some effects on eucaryotic membranes, they have a fair degree of toxicity and must be used cautiously in clinical practice. In recent years, only polymyxin has had any medical use. Another group of antibiotics affect membrane structure in eucaryotes, by acting on sterols that are specific components of the membranes of these organisms. These antibiotics are all in one chemical grouping called *polyenes*; specific antibiotics include nystatin and amphotericin. Nystatin, especially, has found considerable use as an agent for control of fungus infections, particularly of the intestine. (It can be used even though humans are eucaryotes, because it is not rapidly absorbed and remains localized in the intestine where its toxicity cannot be expressed.)

A number of medically useful antibiotics act by inhibiting *protein synthesis*. Recall that protein synthesis occurs on ribosomes. The ribosomes of procaryotes are different from the ribosomes of eucaryotes, and antibiotics are known that affect protein synthesis either in procaryotes or eucaryotes by selective action on ribosomal structures. Chloramphenicol, streptomycin, and the tetracyclines all act specifically to inhibit protein synthesis in procaryotes (see Table 6.4). These antibiotics have been some of the most widely used broad-spectrum antibiotics in human medicine. However, even these have some toxicity to humans under certain conditions, because the protein synthesis machinery in the mitochondria of eucaryotic cells is really procaryotic in nature. Chloramphenicol, especially, has shown considerable toxicity for newborns and infants and for certain adults and must be used cautiously, although it still is an important antibiotic for the treatment of typhoid fever. One antibiotic, cycloheximide, specifically inhibits protein synthesis in eucaryotes. This antibiotic is of little use for treating fungal infections because it is also quite toxic to humans; it does, however, find considerable use in research work when it is necessary to inhibit protein synthesis in eucaryotic microorganisms.

Several antibiotics inhibit *nucleic acid synthesis*. The antibiotic actinomycin inhibits both DNA and RNA synthesis, but RNA synthesis is inhibited at much lower doses. This antibiotic is actually more toxic to man and animals than to microorganisms and has therefore found no use in the therapy of infectious disease, although it has found wide use in research on mRNA synthesis. Mitomycin specifically inhibits DNA synthesis but is also active against both procaryotes and eucaryotes. The antibiotic rifamycin inhibits mRNA synthesis by action on the enzyme RNA polymerase; it is considerably more active against procaryotes than eucaryotes.

From the foregoing discussion, we can see that selective toxicity can manifest itself in various ways. At present we have a fair amount of knowledge concerning the molecular mechanisms of antimicrobial action, and we can see how these actions at the molecular level might lead to inhibition of growth and to cell death. An understanding of these matters may help in discovering new agents, or in redesigning (by chemical modification) existing agents. Finally, this knowledge may help physicians to use these agents more effectively in treating infectious diseases.

Summary

Metabolism includes all the biochemical reactions of the cell. The energy required to drive many of the cell's reactions is obtained from catabolic reactions. These are degradative reactions by which the cell breaks down the nutrients supplied in the culture medium. Energy is released during catabolic reactions and is trapped by the formation of high-energy phosphate bonds, such as in ATP. The reactions of catabolism are oxidative reactions and occur in many small steps, each step being mediated by an enzyme. Oxidations are accomplished by the addition of oxygen and/or the loss of hydrogen and electrons. Oxidations are always paired with reductions, which are accomplished by the addition of electrons and the removal of oxygen and/or the addition of hydrogen. Energy made available when electrons are removed during cell oxidations can be conserved by phosphorylation reactions: electron transport phosphorylation, substrate level phosphorylation, and photophosphorylation.

Substrate molecules are degraded by a variety of mechanisms during the catabolic reactions of a cell. Fermentation reactions occur in the absence of oxygen; glycolysis is the most common fermentation pathway and yields alcohols or acids as end products, with low energy yield. Respiration reactions are aerobic and are found in most organisms. Reactions of the TCA cycle require oxygen, produce CO_2 as the end product, and yield large amounts of energy in the form of ATP. Some organisms are capable of anaerobic respiration, using a compound other than oxygen (such as nitrate) as an electron acceptor. Energy sources are most frequently organic compounds but may also be inorganic compounds or light for certain specialized organisms. Cells may convert excess substrate into some storage material to store for later use.

The cell uses its biosynthetic reactions to make all of the compounds required to duplicate itself during growth. Proteins are synthesized according to directions derived from DNA and transcribed by RNA. The nucleic acid triplet code is translated into the amino acid sequence of the protein by the action of transfer RNA that aids in assembling activated amino acids into the polymer. Nucleic acids are synthesized from activated nucleotide subunits, using primer nucleic acid as a template for the sequence of the subunits.

Regulation of metabolism occurs primarily by the action of several specific mechanisms for turning on or shutting off pathways. By induction and repression of enzymes, the cell regulates the synthesis of enzymes so that only those needed at a particular time in a particular medium are made. Feedback inhibition operates in some cells to control the activity of enzymes; regulation is immediate and acts on preexisting enzymes.

Antibiotics act by binding vital cell constituents so as to inhibit their functions. Structures affected are typical for each antibiotic and may include cell wall, membrane, nucleic acids, and ribosomes. The value of antibiotics lies in their selective toxicity, their differential activity on man and microbe.

1 Define *metabolism, catabolism, anabolism.*
2 A cell is only 30 percent efficient. Explain this, and account for the other 70 percent of substrate energy.
3 What role does ATP play in cell metabolism?
4 Define *oxidation* and *reduction*, and show how they are complementary.
5 Tell whether the following reactions are oxidations or reductions, and name the electron donor or the electron acceptor, as appropriate:
  (a) $CO_2 \rightarrow C_6H_{12}O_6$
  (b) $NH_3 \rightarrow NO_2^-$
  (c) $C_6H_{12}O_6 \rightarrow CO_2 + H_2O$
  (d) $O_2 \rightarrow H_2O$
6 What are the three ways in which high-energy phosphate bonds are synthesized?
7 Show by diagram how electrons are passed through the electron transport system and finally to oxygen.
8 Compare and contrast fermentation and respiration with respect to pathways, oxygen requirements, energy yield, end products.
9 Why is aerobic growth so much more efficient than anaerobic?
10 What is anaerobic respiration? Show how nitrate can be used in this process.
11 What are the three types of energy sources? Give examples showing how each type is used by microorganisms. Which type is most commonly used?
12 Describe the process by which proteins are synthesized. Show how the following are involved: DNA, messenger RNA, transfer RNA, triplet code, ATP, amino acids.
13 Describe the synthesis of DNA, including the role played by the nucleotide bases, the primer nucleic acid, ATP.
14 Compare and contrast constitutive and inducible enzymes.
15 Compare and contrast induction and repression.
16 What is feedback inhibition? How does it differ from the induction–repression mechanism?
17 It is the selective toxicity of antibiotics that makes them useful in the therapy of infectious disease. Explain.
18 Explain the mechanism of action of penicillin, polymyxin, streptomycin.

Suggested readings

Brock, T. D., *Biology of microorganisms*, 2nd ed., Prentice-Hall, Inc., Englewood Cliffs, N.J., 1974. A fairly elementary treatment of microbial physiology and biochemistry is found in Chapters 4, 5, 6, and 7 of this text.

Lehninger, A. L., *Biochemistry*, 2nd ed., Worth Publishers, Inc., New York, 1975. Detailed discussion of all aspects of biochemistry and molecular biology, with emphasis on mammalian systems.

Stanier, R. Y., E. A. Adelberg, and J. L. Ingraham, *The microbial world*, 4th ed., Prentice-Hall, Inc., Englewood Cliffs, N.J., 1976. A standard advanced-level microbiology text, with good material on the physiology of bacteria.

Watson, J. D., *Molecular biology of the gene*, 3rd ed., W. A. Benjamin, Inc., Menlo Park, Calif., 1976. An excellent elementary treatment of molecular biology, with emphasis on chemical matters.

Genetics

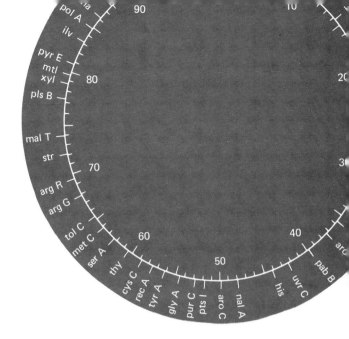

Microorganisms exhibit hereditary phenomena similar to higher organisms. Many eucaryotic microbes have male and female types. No male and female forms exist in procaryotic microbes, but these microorganisms have a variety of special mechanisms for passing on characteristics to succeeding generations. *Genetics* is the study of how organisms transmit characteristics to their offspring. Microbial genetics is an important field of microbiology, because it helps us explain how important characteristics of microorganisms, such as disease-causing properties, antibiotic resistance, formation of desirable substances such as penicillin, and so forth, are inherited. In this chapter, we present a brief discussion of our current understanding of microbial genetics with emphasis on bacteria.

As we discussed in earlier chapters, the bacterial nucleus consists of DNA arranged in one long circular strand. This is essentially the "chromosome" of the cell. Although it is quite different from chromosomes of eucaryotes, it is similar in that it determines the *genotype* of the cell; that is, it determines all the inherited properties of the cell.

Changes may occur in the genetic material of the cell, and when they do, they may be observed as changes in the hereditary properties of the organism. Genotypic changes can occur in various ways: by *mutation*, either spontaneous or induced; and by *genetic recombination*, the "crossing" of bacterial genomes with those of other bacteria or of viruses.

When the DNA of the cell replicates, during cell division, the existing double strand splits apart and each single strand acts as a template for the synthesis of a new complementary strand (Figure 7.1*a*). We discussed in Section 6.5 the complementarity between DNA bases: adenine pairs with thymine, and guanine pairs with cytosine. It is this complementarity that permits the precise copying of one DNA strand; wherever an adenine was in one strand, a thymine will be in the copy, and wherever there was a guanine in one strand, there will be a cytosine in the copy. When this complementary strand is itself copied, *its* complement will be exactly the same as the original DNA strand. The word *mutation* is used to refer to changes in the base sequence in the DNA that result in a DNA structure different from that of the original. In most cases, mutations that occur in the base sequence of the DNA lead to genetic changes in the organisms; these changes are mostly harmful, although beneficial changes do occur occasionally.

a Replication of DNA

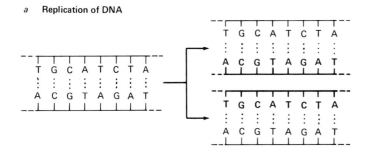

b Spontaneous mutation during replication

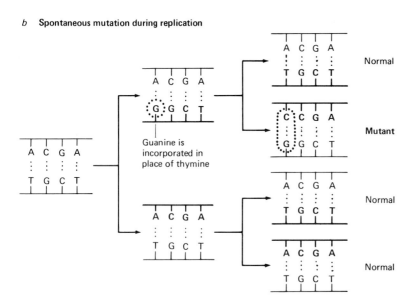

FIGURE 7.1
DNA replication and mutation. *a* Replication of DNA. The new strands are formed by complementary base-pairing with the pre-existing strands. *b* Spontaneous mutation during replication. A pairing error results in a substitution of the wrong base. Subsequent replication of this faulty strand results in the formation of a mutant.

Mutation can be either spontaneous or induced. *Spontaneous mutations* occur during replication, as a result of errors in the pairing of bases, leading to changes in the replicated DNA. In fact, spontaneous mutations occur in every 10^6 to 10^{10} replications, depending on which particular error is considered. Thus in a normal, fully grown culture of organisms having approximately 10^8 cells/ml, there will probably be some kind of mutant in each milliliter.

Figure 7.1*b* shows a typical way in which a spontaneous mutation can occur. The cell receiving the altered DNA is a mutant. What is the result? The DNA containing the error directs the synthesis of RNA containing the error, and this in turn is transmitted to the protein that uses this mRNA as a template. The triplet code that directs the insertion of the tRNA amino acids will be changed and will likely cause the wrong amino acid to be incorporated. If this wrong amino acid occurs at an important site in the protein, the properties will be different and there will be an observable mutation. In other cases, the change may not be so important and may not even cause a noticeable effect.

Some general properties of spontaneous mutation may be observed. Mutation is by nature a sudden and unpredictable event and is relatively independent of the cell's environment. Mutational changes in the cell genome are permanent but are quite rare events.

THE EFFECT OF MUTAGENS

Some agents, such as X rays, ultraviolet irradiation, and certain chemicals, can cause mutations. These *mutagens* cause a change in the DNA that is identical to that which occurs in spontaneous mutation, but the rate at which these mutational events occur is much higher. Thus, if a large number of mutants are desired for certain studies, they can be obtained reasonably quickly. The culture is treated with a mutagen and then grown in media that will select for the growth of the desired mutants.

Mutagens either act directly on bases in the DNA, changing them to other bases that are copied to produce altered DNA strands, or act in such a way that they increase the chance of an error during copying. In either way, the result is the same: the production of an altered (probably faulty) DNA. Since many of the mutagenic agents that are known also occur naturally, it is possible that many spontaneous mutations are actually induced by natural mutagens. For instance, ultraviolet radiation from the sun can cause some mutations, and cosmic rays from outer space are similar in their mutagenic effects to X rays. Although X rays do not exist naturally at the surface of the earth, this type of radiation is widely used in medical practice, and great care must be taken to ensure that individuals are adequately protected from X-ray overdoses. Radioactivity is also mutagenic, and both natural and man-made radioactive materials can cause mutations under appropriate conditions.

A wide variety of chemical mutagens are known, some of which are also natural products. One of the most effective mutagens is *nitrous acid*, which forms from nitrite (NO_2^-) in acid solutions. We already noted that a number of bacteria can produce nitrite by reduction of nitrate, and nitrite is also added to sausage and some other kinds of meat as a preservative. Nitrous acid reacts with adenine in DNA, converting it into the base hypoxanthine, which then pairs with cytosine during replication. Thus nitrous acid converts an adenine-thymine base pair into a cytosine-guanine pair. Other chemical mutagens include nitrosamines (formed in some foods by reaction of nitrite and amino acids), the nitrogen mustard gases (once used in warfare), the antibiotic mitomycin (another natural product), and the acridine dyes. Current research using bacteria as test agents is uncovering a wide variety of other mutagens, including many compounds present in the environment. Some of these environmental mutagens are probably natural in origin, but others (such as nitrous acid) are almost certainly from man-made sources as well.

Virtually every mutagen is also a carcinogen, capable of inducing cancer in animals or man. Consequently, it is useful to carry out preliminary screening of compounds for carcinogenicity by seeing whether they are mutagenic in bacteria. Because of their ease of culture and study, and because the presence of mutations can often be easily detected, bacteria make ideal test organisms for determining mutagenicity of actual or potential environmental chemicals. This has been a significant practical development of the study of mutation in bacterial systems.

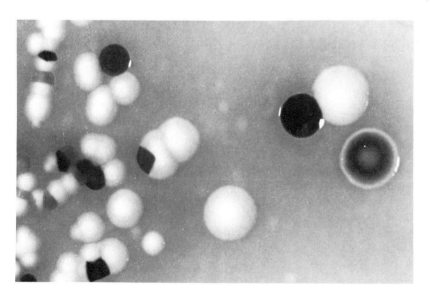

FIGURE 7.2
Origin of pigmented mutants in a nonpigmented strain of *Chromobacterium violaceum*. The pigmented sectors probably arose because of a mutation at an early stage in the development of the colony. From R. Sivendra, H. S. Lo, and K. T. Lim, *J. Gen. Microbiol.*, **90**, 21–31, 1975.

Typical mutations include loss of various synthetic capabilities (a requirement for the vitamin niacin, for example), loss of ability to utilize a certain sugar (unable to utilize lactose, for instance), and loss or gain of pigment (Figure 7.2). Mutations may affect an organism's ability to survive, which is of special importance in natural environments. For example, a mutation may affect the growth rate; faster growth will allow a mutant to become the dominant member of the population. Conversely, a mutation causing slower growth will result in the elimination of the organism from a population. Or a cell may become better able to resist extremes in pH or temperature, which also has obvious survival advantage. As noted earlier, development of antibiotic-resistant mutants has been of the most practical consequence.

7.2
Genetic recombination

Genetic recombination can be defined as a process in which a cell receives genetic information from each of two parents. While mutation usually changes a very small amount of a cell's DNA, resulting in a change in only one or a very few proteins, genetic recombination may involve either small or very large changes. In some ways, genetic recombination in bacteria resembles genetic recombination in higher organisms, but there are major differences.

In eucaryotes, genetic recombination is a result of the sexual reproduction process. Special cells called *gametes* are formed that contain only half the chromosome number of the parent. In animals, the male gametes are sperm, and the female gametes are eggs. The male sperm unites with the female egg to form a fertilized egg, restoring the chromosomes to the parental number. The fertilized egg then develops into a new individual, which has received half its genes from each parent (Figure 7.3).

In bacteria, distinct gametes are not formed, and the genetic recombination process is fragmentary. Usually, one bacterial cell, the *recipient*, retains most of its normal gene complement, and another bacterial cell, the *donor*, transfers a small part of its gene complement to the recipient. By convention, the recipient is sometimes called the *female*, and the donor the *male*, although these words obviously do not have the same meaning that they have in higher organisms. Thus, in bacterial recombination, a part of the donor cell's DNA is transferred into the cell of the recipient. During subsequent cell replication, genetic elements from each DNA are incorporated into the DNA of the new daughter cell; this recombined DNA is then the genetic material for all subsequent progeny (Figure 7.4). There are three major mechanisms of gene transfer in bacteria: transformation, conjugation, and transduction. Recombination of genetic material occurs in much the same way in each; it is the method of getting the donor DNA into the recipient cell that varies.

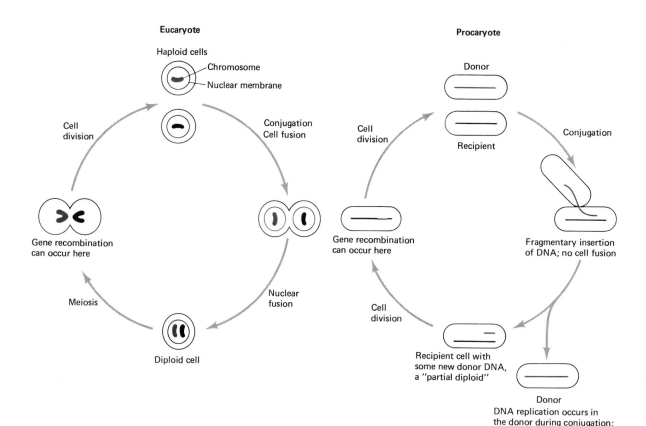

Eucaryote

Haploid cells

Chromosome
Nuclear membrane

Cell division

Conjugation
Cell fusion

Gene recombination can occur here

Meiosis

Nuclear fusion

Diploid cell

Procaryote

Donor

Recipient

Cell division

Conjugation

Fragmentary insertion of DNA; no cell fusion

Gene recombination can occur here

Cell division

Recipient cell with some new donor DNA, a "partial diploid"

Donor
DNA replication occurs in the donor during conjugation; the donor thus retains a complete genome and is viable

FIGURE 7.3
Contrasts between sexual reproduction and genetic recombination in eucaryotes and procaryotes.

FIGURE 7.4
The basic process of genetic recombination. Portions of the genetic material of the donor DNA are incorporated into the parent DNA, resulting in a genetic recombinant. The letters denote specific genes. For example, "A" and "a" might be lactose fermentation ("A" being capability of fermenting, "a" being lack of capability): "B, b" might be penicillin resistance or sensitivity ("B" being sensitivity, "b" being resistance); "C, c" might be methionine synthesis ("c" being a requirement for it, "C" being ability to synthesize it), and so on.

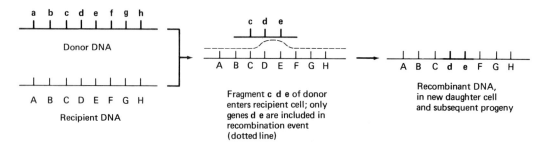

a b c d e f g h

Donor DNA

A B C D E F G H

Recipient DNA

c d e

A B C D E F G H

Fragment c d e of donor enters recipient cell; only genes d e are included in recombination event (dotted line)

A B C d e F G H

Recombinant DNA, in new daughter cell and subsequent progeny

GENETICS

7.3
Transformation

Transformation is a process by which free DNA is inserted directly into a recipient without the agency of any carrier. DNA fragments obtained from donor cells by lysis or extraction can be mixed with live recipient cells. With appropriate cells under proper conditions, the DNA fragments are taken up. The recipient cells grow, and if they express genes from the donor DNA, then transformation is known to have occurred. The process of transformation is diagramed in Figure 7.5. One prime requirement for transformation is of course that the donor DNA preparation be able to enter the recipient. Only cells from the same or closely related species will undergo transformation; in fact, only a very few organisms have been observed to do it at all. It is presumably quite difficult for a molecule as large as DNA to pass through the barrier presented by the cell wall and cell membrane.

What sorts of genes are transferred? Various ones are seen in the progeny: those specifying capsule formation, drug resistance, nutritional changes, pathogenicity, and so forth. The recombinant progeny must be detected in some way to know that recombination has indeed occurred. In the case of capsule or pigment formation, visual inspection of the recombinant colonies that arise after transformation may show some which are obviously different. However, in other cases selective media must be employed. For example, if recipient cells requiring arginine for growth are mixed with donor DNA from cells that are arginine independent (synthesizing their own arginine), then the recombinant progeny are arg^+ (synthesizing their own arginine). The way to find these arg^+ recombinants is to plate the transformed mixture onto a medium *without* arginine; only the recombinants (arg^+) are able to grow. This same method of observing recombinants works for a large number of nutritional markers. A similar method is useful in tracing the acquisition of resistance to a drug, such as penicillin. If the donor DNA comes from penicillin resistant (pen^r) cells, and the recipient is penicillin sensitive (pen^s), then after mixing and plating onto a medium containing penicillin, only the pen^r recombinants will be able to grow and form colonies. These examples of transformation are diagramed in Figure 7.6.

FIGURE 7.5
Transformation, the incorporation of free DNA into a cell and its integration into the cell genome.

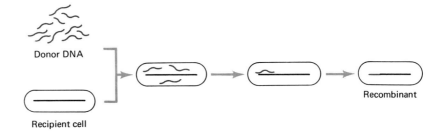

Donor DNA

Recipient cell

Recombinant

FIGURE 7.6
Laboratory procedure for the
detection of genetic
transformation. This procedure
is effective even if the efficiency
of transformation is low.

a Mix DNA from cells able to synthesize their own arginine
   (arg$^+$) with arginine-requiring recipient cells (arg$^-$)

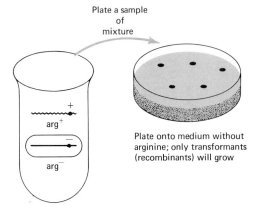

Plate a sample
of
mixture

arg$^+$

arg$^-$

Plate onto medium without
arginine; only transformants
(recombinants) will grow

Recombinant genotype:
arg$^+$

b Mix DNA from penicillin resistant organism (pen^r)
   with penicillin sensitive recipient cells (pen^s)

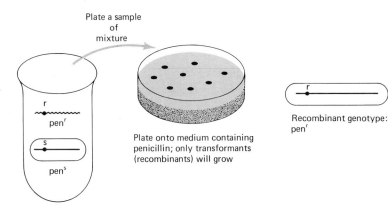

Plate a sample
of
mixture

pen^r

pen^s

Plate onto medium containing
penicillin; only transformants
(recombinants) will grow

Recombinant genotype:
pen^r

Genetic markers are transferred from one living cell to another in
bacterial conjugation. This is the closest that bacteria get to the
sexual reproduction common in higher forms. The requirements
necessary for conjugation include direct contact between cells, and
the cells must be closely related, although of opposite "sex" or
mating type.

Transfer of genes is one way only, from donor to recipient, in
an orderly, sequential manner. This is diagramed in Figure 7.7. In a
given strain, the genes always enter in the same sequence. Since
mating pairs do not remain attached indefinitely, it is more common
to see recombinants with the genes that come over early, and it is
very rare to see recombinants with late genes. The ability to function
as a donor is determined by a gene called F^+. In F^+ donors, the F

7.4
Conjugation

factor is not on the chromosome and transfers quite freely in the cytoplasm, but other genes are not transferred. If F⁺ becomes attached to the chromosome, transfer of the chromosome occurs at high frequency; such donors are called *Hfr* (for *H*igh *f*requency of *r*ecombination). With the F⁺ donors, the event of gene transfer itself is quite rare, in contrast to the Hfr donors.

Typical examples of conjugation are shown in Figure 7.8. Selection for recombinants is done by plating the mating mixture on media that allow growth of only the desired genotypes. For instance, in the first selection in Figure 7.8, the medium contains streptomycin; only those recombinants retaining the Str^r from the F⁻ recipient grow. Remaining viable donor cells (Str^s) are killed. The medium does not provide threonine or leucine; therefore only recombinants receiving both T⁺ and L⁺ from the Hfr donor grow.

When conjugation occurs, the circular DNA is broken at a point specific for each genotype; in the case of Hfr cells the F factor attaches to the end. Then, as shown in Figure 7.7, the donor DNA begins to enter the recipient cell and mating is underway. Knowing

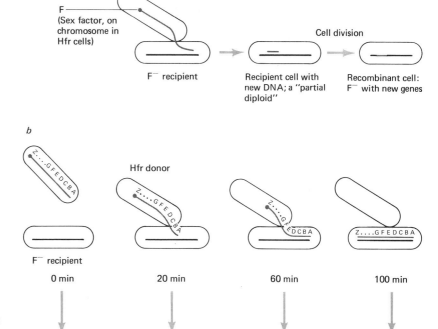

FIGURE 7.7
Bacterial conjugation, the transfer of genetic material from donor to recipient by cell-to-cell contact. *a* Incorporation of genetic material. *b* Sequential transfer of genes.

Genotypes of recombinants recovered by plating after conjugation

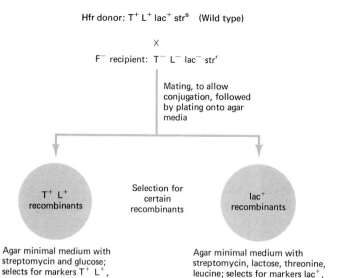

Hfr donor: T⁺ L⁺ lac⁺ strˢ   (Wild type)

X

F⁻ recipient: T⁻ L⁻ lac⁻ strʳ

Mating, to allow
conjugation, followed
by plating onto agar
media

T⁺ L⁺
recombinants

Selection for
certain
recombinants

lac⁺
recombinants

Agar minimal medium with
streptomycin and glucose;
selects for markers T⁺ L⁺,
does not select for lac

Agar minimal medium with
streptomycin, lactose, threonine,
leucine; selects for markers lac⁺,
does not select for T, L

FIGURE 7.8
Laboratory procedure for the
detection of genetic conjugation.
Symbols for genetic markers:
"T", threonine; "L", leucine;
"lac", lactose; "str",
streptomycin.

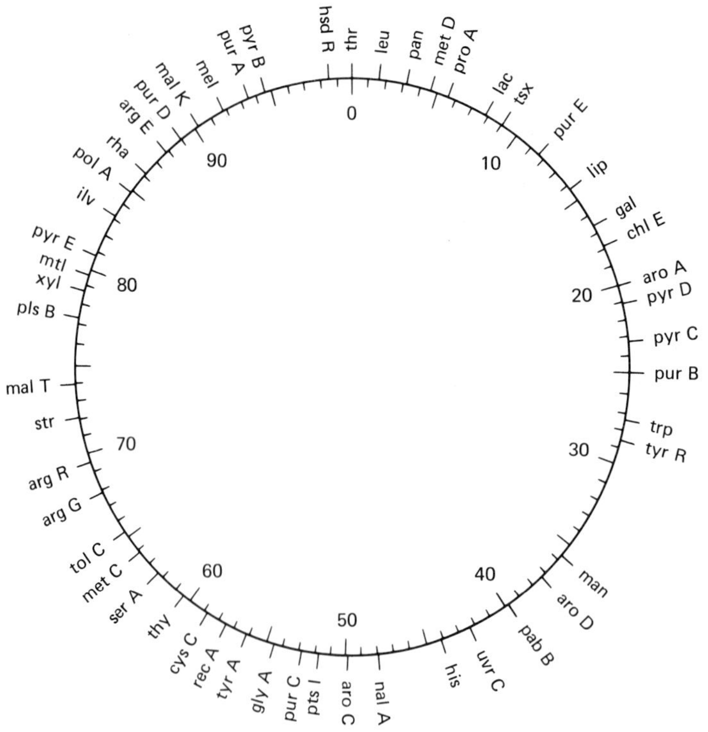

FIGURE 7.9
The genetic map of *Escherichia coli.* Only about 50 of the best known and most commonly used gene loci are shown, chosen from the more than 650 loci now known for this organism. From B. J. Bachmann, K. B. Low, and A. L. Taylor, *Bacteriol. Rev.,* **40**, 116–167, 1976.

GENETICS

that the transfer of genetic material is always sequential, one can determine the *order* in which the markers on the chromosome are transferred. If a large number of matings are carried out using different genotypes of an organism, a *genetic map* can be formulated. A simplified version of the genetic map for *Escherichia coli* is shown in Figure 7.9. This is a greatly abbreviated map, since there are actually more than 650 known genes that could be shown in a complete map.

7.5
Transduction

The third way in which genetic recombination can occur is by transduction, the virus-mediated transfer of genetic material from a donor cell to a recipient cell. The growth of animal viruses was mentioned in Section 2.7. Bacterial viruses, called *bacteriophages*, replicate in a way that is essentially similar, shown in Figure 7.10. The most common result of bacteriophage infection of a cell is lysis, with the release of large numbers of new bacteriophages. Viruses that cause cell lysis are called *virulent*. However, some viruses, called *temperate*, do not lyse the cell; instead of multiplying within the cell,

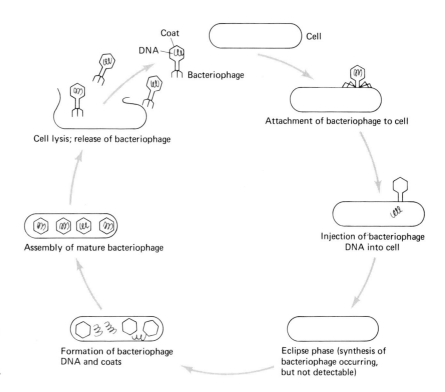

Coat
DNA
Bacteriophage
Cell

Attachment of bacteriophage to cell

Cell lysis; release of bacteriophage

Injection of bacteriophage DNA into cell

Assembly of mature bacteriophage

Formation of bacteriophage DNA and coats

Eclipse phase (synthesis of bacteriophage occurring, but not detectable)

FIGURE 7.10
The manner in which a bacterial virus replicates in its host. The example shown is for a virulent virus, which always kills its host.

their DNA becomes *integrated* into the genetic apparatus of the host and replicates with it. At a later time, the integrated virus may break from the cell chromosome and replicate, forming new virus particles and lysing the cell. An organism that contains a temperate virus is called *lysogenic*, and it is this type that is involved in transduction. Occasionally, when the virus breaks from the cell chromosome, it takes cell genes with it, which can then be transferred to a new cell upon infection by the virus.

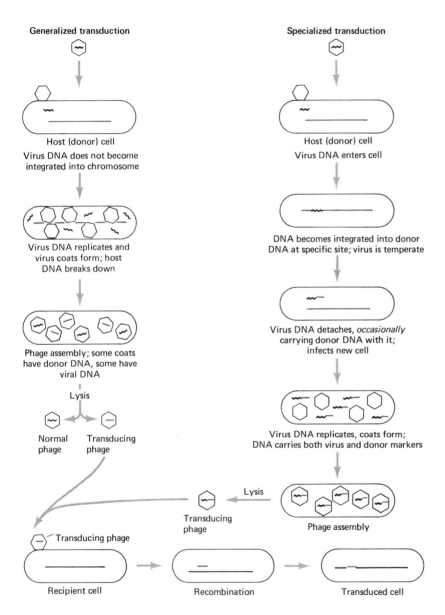

Generalized transduction

Host (donor) cell
Virus DNA does not become integrated into chromosome

Virus DNA replicates and virus coats form; host DNA breaks down

Phage assembly; some coats have donor DNA, some have viral DNA

Lysis

Normal phage   Transducing phage

Transducing phage

Recipient cell   Recombination   Transduced cell

Specialized transduction

Host (donor) cell
Virus DNA enters cell

DNA becomes integrated into donor DNA at specific site; virus is temperate

Virus DNA detaches, *occasionally* carrying donor DNA with it; infects new cell

Virus DNA replicates, coats form; DNA carries both virus and donor markers

Lysis   Transducing phage   Phage assembly

FIGURE 7.11
Transduction, the transfer of genetic material from donor to recipient via virus particles.

Genetic transfer of host genes by viruses can occur in two ways. In the first, called *specialized transduction*, a restricted group of host genes is integrated into the virus DNA and is transferred to the recipient during lysogenization. In the second, called *generalized transduction*, host genes from any part of the cell's chromosome are incorporated into the virus and then transferred to the recipient. The two types of transduction are diagrammed in Figure 7.11. They are similar in that both involve the acquisition of donor DNA by a phage particle, although the precise mechanism of that acquisition differs. The transducing phage, from either source, subsequently infects a recipient cell, and recombination of phage and recipient DNA result in a transduced cell that now bears genes obtained from the donor.

**7.6
Plasmids**

Genetic elements that occur in the cytoplasm of bacterial cells, unconnected with the chromosomal DNA, are called *plasmids*. Plasmid DNA molecules are much smaller than chromosomal DNA molecules and replicate independently. They can also be transferred by conjugation much more readily than chromosomal DNA; it is this property that has made their recognition possible. Plasmids contain genes that are nonessential for growth, so that under many conditions, they can be lost or gained without harm to the cell (Figure 7.12). Treatment of cells with certain drugs can result in loss of plasmids without any other genetic change (this is called *curing* (Figure 7.12c). Although they are nonessential genetic elements, plasmids do carry genes that confer selective advantage to cells under certain conditions. The most dramatic examples of the significance of plasmids are those plasmids that contain genes for antibiotic resistance. Bacteria containing antibiotic-resistance plasmids present some of the most serious problems in the antibiotic therapy of infectious disease. The fact that plasmids conferring antibiotic resistance are transferred at high frequency to other cells means that resistance can spread rapidly through the population. Often, the plasmids contain genes for resistance to several antibiotics; this is termed *multiple resistance*. The emergence of bacteria containing plasmid-mediated multiple antibiotic resistance correlated with the increasing use of antibiotics for the treatment of infectious diseases.

The F^+ particle, involved in bacterial conjugation, is itself a plasmid. When genes for antibiotic resistance become incorporated into the F^+ particle, the particle becomes an antibiotic-resistance plasmid. Other genes that can become incorporated into plasmids are the genes for resistance to heavy metals (mercury, copper, nickel, etc.) and genes for the production of toxins and other substances involved in pathogenicity. The fact that pathogenicity is often plasmid-mediated may sometimes explain attenuation of

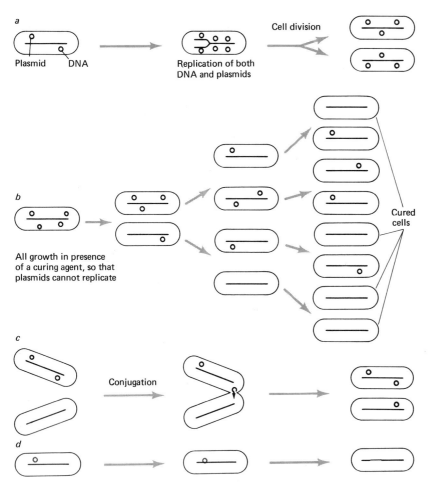

FIGURE 7.12
Plasmids, genetic elements in bacteria that are not integrated with the chromosome.
a Replication of a plasmid in the cell independent of the chromosome. *b* Curing of plasmid by drug treatment.
c Transfer of plasmid from cell to cell. *d* Integration of plasmid DNA into chromosome under rare conditions.

virulence such as Pasteur first discovered in fowl cholera (see Section 1.6). By curing a cell of a toxin-conferring plasmid, it is possible to derive a strain that resembles the parent in many ways but which no longer causes disease, and such a strain might be used in an immunization procedure.

Although most plasmids do not become integrated into the chromosome, certain ones can (Figure 7.12*d*); these are called *episomes*. The F$^+$ particle is thus really an episome, since it can become integrated into the chromosome, the cell then becoming an Hfr cell. In a certain sense, temperate viruses are also episomes. The various similarities and differences between plasmids, episomes, and viruses are summarized in Table 7.1.

One of the recent advances in bacterial genetics, which has arisen from detailed study of plasmids and from a knowledge of the

7.7
Genetic engineering

TABLE 7.1
Characteristics of plasmids, episomes, and viruses

	Plasmid	Episome	Temperate virus	Virulent virus
Independent replication	+	+	+	+
Transmission by cell-to-cell contact	+ or −	+ or −	+ or −	−
Integration into host genome	−	+	+	−
Acquisition of host genes	+	+	+	−

enzymes involved in DNA replication, is the development of the ability to manufacture genetic elements that can then be incorporated into bacterial cells. The techniques of manufacturing and manipulating genetic material to achieve certain desired results has been called *genetic engineering*. It permits the construction of organisms of desired genotype, organisms that may have medical, agricultural, or industrial significance.

The basic procedure involved in genetic engineering is as follows: DNA containing the desired genes is obtained from plasmid-bearing cells, the cells being lysed and the plasmid DNA purified. (Being small, plasmid DNA is easier to purify than chromosomal DNA.) Two plasmid DNAs are prepared, each containing genes that it is desired to incorporate together in a single organism. The plasmid DNAs are treated with an enzyme that breaks them partially but, most importantly, makes the ends of the DNAs sticky.[1] Then the two treated plasmid DNAs are mixed and treated with another enzyme (DNA ligase) that connects the sticky ends together. The connected DNA, a test-tube hybrid, is then mixed with appropriate recipient cells under proper conditions, and the DNA is allowed to become integrated into the recipient. If the plasmid DNA contains an antibiotic-resistance gene, it is a relatively simple matter to select a plasmid-containing strain from the mixture of plasmid-containing and plasmid-free cells by plating on an agar medium containing the antibiotic.

Although genetic engineering is a powerful tool for creating new organisms, it is not without dangers. If pathogens are involved and if antibiotic-resistance genes are part of the plasmid (as is usually the case, since they make selection easy), it could transpire that a new antibiotic-resistant pathogen would be created that might present serious troubles in medical practice. For this reason, scientists working in this area have agreed that any genetic engineering

---

[1] "Sticky" in this context means that the end of each DNA has a short single-strand fragment complementary to a single-strand end of the other DNA. The molecules thus specifically associate by their single-strand ends.

practices that might lead to harmful consequences should only be performed in laboratories where appropriate confinement facilities are available, so that the newly created organisms cannot escape and reach the general public.

One of the possible practical benefits of genetic engineering is the creation of organisms with desired properties that would not otherwise be possible. Thus, it has been possible to transfer the nitrogen-fixing genes to bacteria that ordinarily do not fix nitrogen (N_2), making these bacteria independent of sources of combined nitrogen in the medium. Such nitrogen-fixing organisms may have useful agricultural or industrial properties. Other types of genetic engineering procedures may be easily conceived; a whole new field of practical microbiology is thus opening up.

Summary

Microbial genetics deals with the way in which various characteristics are transmitted from microbes to their offspring. Bacterial DNA is arranged in one long circular strand. Changes in the DNA result in changes in the genetic makeup of the cell and may be shown as changes in the cell's progeny. A change in the base sequence of the DNA of the cell is called a mutation. Spontaneous mutations occur as a result of errors in the pairing of bases during DNA replication; the rate is quite low. The frequency of mutation can be increased greatly by inducing mutations with the use of mutagens. A mutation is observed in progeny cells as a change in some characteristic; for example, there may be a loss of some metabolic activity, a change in the growth rate, or a change in response to an environmental factor.

Genetic recombination is a process in which a cell receives genetic information from two parent cells. Usually, a larger amount of DNA is involved than in mutation. A recipient cell receives DNA from a donor cell; during subsequent cell replication, genes from each DNA are incorporated into the daughter cell, and this recombined DNA is the genetic material of all subsequent progeny. The DNA is transferred from the donor to the recipient in one of three ways: transformation, conjugation, or transduction. In transformation, the donor DNA is not cellular; it is obtained from donor cells by lysis or extraction and enters the recipient without the agency of any carrier. In conjugation, genetic material is transferred from one living cell to another by direct contact. The circular DNA strand opens and enters the recipient cell, transferring markers in sequence. Transduction is a bacteriophage-mediated transfer of genetic material from donor to recipient. Some bacteriophages are temperate and become integrated into the DNA of a host donor cell. At a later time, the virus may break from the cell chromosome taking some donor genes with it; the donor genes can be transferred to a new (recipient) cell upon infection by the virus. Transduction may be specialized, involving a restricted group of host genes, or generalized, involving genes from any part of the host chromosome.

Plasmids occur free in the cytoplasm, apart from the chromosomal DNA, and replicate independently. They are transferred readily during bacterial conjugation. Examples of genes transferred by plasmids are those for antibiotic resistance. Plasmids that become integrated into the chromosome are called episomes. Genetic elements can be synthesized in the laboratory and incorporated as episomes into cells to change their genetic character. Such genetic engineering has widespread practical significance.

Study questions

1 Define the following terms: *genetics, genotype, mutation.*
2 Draw a diagram of a DNA molecule; show how a mutation can occur.
3 What are mutagens? How do they act? Why are they used?
4 Why should special precautions be used to avoid overexposure to radiation during a chest X ray?
5 What sorts of mutants are seen? How are they recognized?
6 Compare genetic recombination in eucaryotes and procaryotes.
7 How does transformation occur?
8 Describe a method for isolating penicillin-resistant recombinants derived from the following transformation experiment: the recipient cells were sensitive to penicillin and the donor DNA had a penicillin-resistance marker.
9 Explain why the transfer of genes in conjugation is one way only. What is the fate of the donor cell?
10 In a mating experiment, cells of the following genotypes were allowed to conjugate: donor cell: arginine$^+$, lactose$^+$, streptomycin sensitive; recipient cell: arginine$^-$, lactose$^-$, streptomycin resistant. After mating, the cell mixture was plated on agar medium containing 100 μg/ml streptomycin, with lactose as the only carbon source and no arginine present. What will be the genotype of recombinants growing on such a medium? What prevents the remaining viable donor cells from growing? Why won't remaining recipient cells grow?
11 Why must bacteriophages be temperate in order to carry out transduction? Why aren't lytic phages involved?
12 Describe how a lysogenic cell donates genes to a donor cell by transduction.
13 Differentiate between specialized and generalized transduction.
14 What are plasmids? How are their genes transferred to cells?
15 Describe how a cell can become penicillin resistant through the action of plasmids.
16 Compare and contrast plasmids, episomes, and viruses.
17 Give some possible benefits to be gained from genetic engineering. What are some possible dangers?

Suggested readings

Brock, T. D., *Biology of microorganisms*, 2nd ed., Prentice-Hall, Inc., Englewood Cliffs, N.J., 1974. A fairly elementary treatment of microbial genetics of both procaryotes (see Chapter 10) and eucaryotes (see Chapter 11).

Stanier, R. Y., E. A. Adelberg, and J. L. Ingraham, *The microbial world*, 4th ed., Prentice-Hall, Inc., Englewood Cliffs, N.J., 1976. A standard advanced-level text in microbiology with extensive material on the genetics of bacteria.

Watson, J. D., *Molecular biology of the gene*, 3rd ed., W. A. Benjamin, Inc., Menlo Park, Calif., 1976. An excellent elementary treatment of the genetics of both procaryotes and eucaryotes.

The science of microbiology developed as a result of the need to understand the nature of infectious disease. Some of the most important advances in medicine and public health have occurred because of the knowledge gained by microbiologists and physicians who study the causes and cures for microbial diseases. We discussed in Chapter 1 some of the historical background for this development. In the following five chapters, we consider the current understanding of infectious disease. Chapter 8 presents some of the general principles of microbial infection, how microorganisms are able to invade the body, grow, and induce harmful changes leading to sickness and death. The chapter also deals with immunity, the ability of the body to resist microbial attack. Some kinds of immunity are innate, found in all healthy individuals, whereas others are acquired, often as a result of infection itself. Acquired immunity results primarily from the formation by the body of specific substances called antibodies, which combine with and neutralize invading microorganisms or their toxic products. In our discussion of antibodies, we shall also include the subject of immunization procedures, the techniques by which antibody formation is induced.

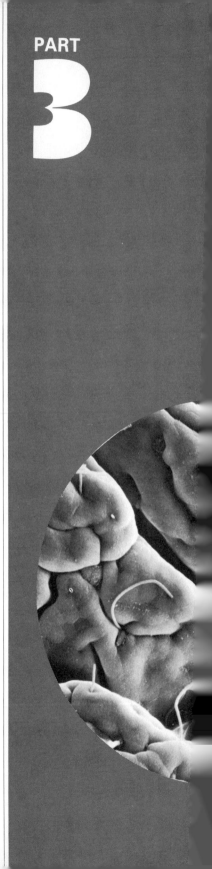

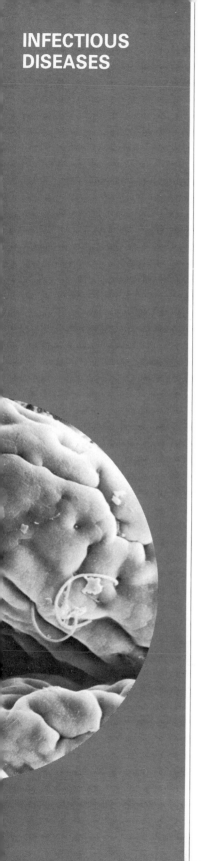

INFECTIOUS DISEASES

In Chapter 9, we shall consider how infectious diseases are spread through groups of individuals, a study often called epidemiology. Because the control of the spread of disease in hospital situations is so crucial for the protection of patients who are sick from other causes, a considerable amount of material in this chapter will deal with hospital sanitation. Finally, we consider the process of embalming and how it is used to provide for a sanitary disposal of the dead.

Chapter 10 considers specific bacterial diseases, and Chapter 11 discusses diseases caused by fungi and protozoa. Here we consider not only the vast diversity of agents able to cause specific types of infectious diseases, but also the ways they interact with hosts, how they are spread through populations, and how they are combated by drugs and public health procedures. We also consider clinical diagnostic methods used to identify various agents.

In Chapter 12, we discuss viruses and virus diseases. We consider some of the important differences between viruses and other infectious agents and how these differences are significant in the diagnosis and treatment of virus diseases. Discussion of the major virus diseases of humans is found in this chapter.

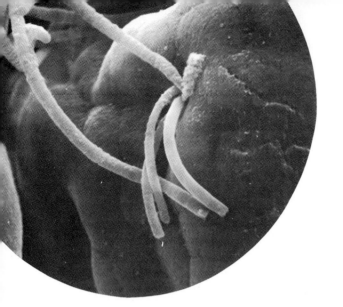

The ability to cause disease in higher organisms is one of the most dramatic and significant properties of microorganisms. Not all microorganisms cause disease; most are harmless, and some are beneficial or even essential for the life of an animal. Those microorganisms that live in and cause damage to other organisms are called *parasites*; the organism affected is called the *host*. A parasite that causes disease is called a *pathogen*, and the degree to which a pathogen affects a host is determined by its *virulence*. Highly virulent parasites cause more serious damage than do less virulent organisms. At the same time, the extent of damage is affected by the resistance of the host. If the host is very susceptible, even a weakly virulent parasite may cause damage, whereas a very resistant host may be affected only by a highly virulent pathogen. We discussed the history of infectious disease research in Chapter 1.

The growth of a microorganism in a host is called an *infection*. Infection is not the same as disease, since an organism may grow in a host without causing harm. Not all diseases are caused by microorganisms. They may also be caused by nutritional deficiencies, genetic defects, hormonal imbalances, and other factors. Diseases caused by microorganisms are often called *contagious*, or *infectious*, *diseases*, since the causal agent is transmissible from one individual to another.

8.1
The normal flora
of the body

Harmless microorganisms live in many parts of the body. Microorganisms are found on the skin, in the mouth, respiratory tract, intestinal tract, and genitourinary tract, all of which are regions of the body exposed directly or indirectly to the outside. The microorganisms that are commonly found in these places are called the *normal flora* of the body. Microorganisms are not normally found in the interior organs of the body or in the blood or lymph systems except in disease. The microorganisms of the normal flora sometimes are beneficial, but under certain conditions of unusual host susceptibility, they may be harmful. They are frequently sources of contamination of sterile goods and hence must be considered important in surgical and medical procedures.

SKIN

Much of the surface of the skin is not a favorable location for microbial growth as it is usually too dry. Most skin microorganisms are restricted to living in the hair follicles or in sweat glands, where moisture is higher and nutrients are present (Figure 8.1). Microorganisms that do live directly on the surface of the skin generally are found on the scalp, face, ears, underarm regions, urinary and anal regions, palms, and between the fingers and toes, where sufficient moisture is present. Underarm odor is caused by the action of bacteria on the secretions from the sweat glands. Underarm odor can be prevented either by destroying the bacteria present or by decreasing the flow of perspiration from the glands. Most body deodorants work in the latter way, by decreasing the flow. Skin antisepsis, an important medical practice, is discussed in Chapter 9.

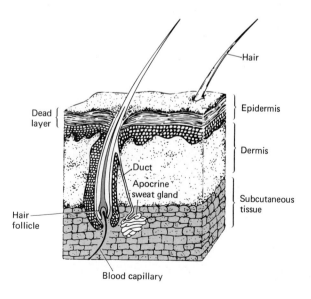

FIGURE 8.1
Anatomy of the human skin. Microbes are associated primarily with the sweat ducts and the hair follicles.

The pH of skin secretions in man is usually acidic, with pH between 4 and 6, and the microorganisms of the normal skin flora are hence usually mild acidophiles. The organisms are predominantly bacteria, including several species of *Staphylococcus* and *Corynebacterium*. The species *Corynebacterium acnes* is normally harmless but sometimes can cause *pimples* or *acne*. Most bacteria foreign to the skin die quickly when inoculated onto it. This is either caused by their inability to tolerate the dryness and low pH of the skin or because they cannot compete with the resident bacteria, which are better adapted to the skin environment.

The skin of the fetus is normally sterile but becomes infected at birth. The newborn baby is susceptible to infection by disease-causing organisms, such as the bacterium *Staphylococcus aureus* or the yeastlike fungus *Candida albicans*, because it lacks a normal flora that can compete with these pathogens. Within a week or two after birth, however, the infant has acquired its own normal skin flora, and its susceptibility to skin pathogens decreases.

The mouth is a favorable habitat for the growth of microorganisms. Saliva itself is not an especially good medium for the growth of

MOUTH

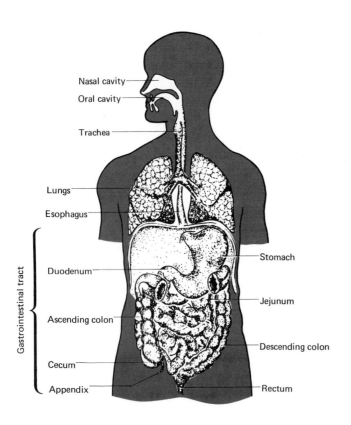

FIGURE 8.2
Anatomy of the human body. Microorganisms of the normal flora are associated primarily with the oral and nasal cavities and the gastrointestinal tract.

INFECTION AND IMMUNITY

microbes, but the surfaces of teeth are quite favorable, and very large numbers of microbes occur there. On the surface of the teeth, a thin coating develops that consists of organic materials and densely packed bacterial cells. This coating, called *dental plaque*, adheres tenaciously to the surface of the teeth and is not removed by ordinary brushing. The principal bacteria found in plaque are filamentous Gram-positive anaerobic bacteria, usually given the name *Leptotrichia buccalis*.

Associated with these filamentous bacteria are a variety of unicellular bacteria, including the important organisms *Streptococcus mutans* and *S. sanguis*, implicated as prime causal agents of tooth decay (*dental caries*). The crevices of the teeth, where food particles are retained, are the sites where tooth decay predominates. Diets high in sugars predispose toward tooth decay because the bacteria of the teeth produce lactic acid from the sugars; this acid attacks the hard surface of the tooth and causes removal of calcium from the tooth structure. Once the breakdown of the hard tissue has begun, microorganisms can penetrate more easily into the tooth and cause further breakdown. Susceptibility to tooth decay varies greatly among individuals and is affected by inherent traits in the individual as well as by diet, brushing of teeth, and other factors. The incorporation of fluoride into the tooth structure makes it more resistant to attack by acid. Addition of fluorides to drinking water or dentifrices therefore aids in controlling tooth decay.

GASTROINTESTINAL TRACT

The anatomy of the gastrointestinal tract is shown in Figure 8.2. Food digestion begins in the mouth, where the food is mixed with enzyme-containing saliva, and then continues in the stomach, where other enzymes present in the gastric juice are added. The gastric juice is quite acid, with pH about 2, and many bacteria that enter the stomach in the food are probably killed in this highly acidic environment. As the food passes into the intestine, the pH gradually rises, and the lower small intestine and the large intestine (or colon) are quite alkaline. Bacteria are present in large numbers on the walls of the stomach and intestine (Figure 8.3), and many are also found in the intestinal contents. Virtually all the bacteria of the intestine are either obligate or facultative anaerobes. The obligate anaerobic bacteria include many long, thin, Gram-negative rods (Figure 8.3*b*), which are as yet unnamed, as well as species of *Clostridium* and *Bacteroides*. Facultative anaerobes include *Escherichia coli* and *Streptococcus faecalis*.

In the lower small intestine and colon, water is gradually removed from the digesting mass, and the material is converted into *feces*, the excretory product. Bacteria, chiefly dead ones, make up about one-third the weight of feces.

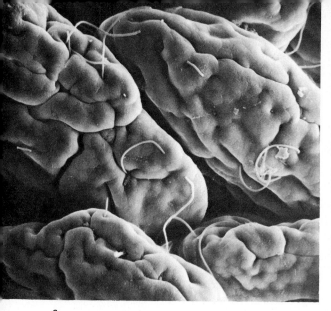

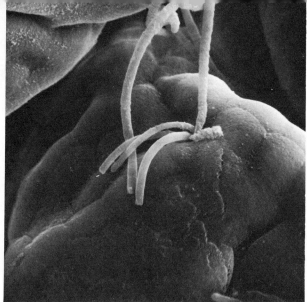

a

b

FIGURE 8.3
Microbial community of the small intestine of the mouse, as revealed by scanning electron microscopy. *a* Low magnification view of the vili of the intestinal epithelium, showing attached filamentous bacteria; magnification 400×. *b* High magnification, showing several bacteria attached end-on in a depression of a vilus; magnification 2,000×. From D. C. Savage and R. V. H. Blumershine, *Infection and Immunity*, **10**, 240–250, 1974.

The bacteria that grow in the large intestine are responsible for the gases and odors produced in and expelled from the large intestine. The gases are primarily nitrogen, carbon dioxide, methane, and hydrogen, and the odors are primarily due to hydrogen sulfide, indole, and skatole. A normal human expels about 450 to 650 ml of gas per day, composed of about 40 percent carbon dioxide, 50 percent N_2, and small amounts of methane and hydrogen. The N_2 comes from the air, but the CO_2, methane, and hydrogen all come from bacterial fermentation.

When some antibiotics are given orally, they may enter the intestine and prevent the growth of intestinal bacteria. As the intestinal contents move downward, the inhibited intestinal bacteria are expelled, leading to the near-sterilization of the intestinal tract. Antibiotics given orally are thus often used to sterilize the intestinal tract before bowel or intestinal surgery. In the absence of this normal flora, exotic bacteria such as antibiotic-resistant *Staphylococcus aureus* or the yeast *Candida albicans* may become established. Normally, these organisms do not grow in the intestine because they cannot compete with the normal flora, but with the normal flora eliminated, they can take over. Occasionally, establishment of these exotic organisms can lead to a harmful alteration in digestive function, but in the absence of continued antibiotic therapy, the normal flora will eventually become reestablished.

The composition of the intestinal flora is influenced by diet. Thus breast-fed infants have a flora consisting largely of the organism *Bifidobacterium bifidus*, which requires a growth factor found in human but not in cow's milk. Infants fed cow's milk usually have a more complex flora. A major constituent of milk is the sugar lactose, and many of the bacteria in the large intestine can utilize lactose for growth. In adults, whose milk consumption is low, lactose-utilizing bacteria represent only a small part of the normal flora.

OTHER REGIONS

Microorganisms enter the *nose* and *throat* in large numbers attached to dust particles from the air, but most of these organisms are trapped and expelled with nasal secretions. The resident organisms are primarily bacteria of the genera *Staphylococcus*, *Streptococcus*, *Corynebacterium*, and *Moraxella*. The *lower respiratory tract* (trachea, lungs, bronchi) is usually sterile because any organisms entering with the air inhaled are filtered out before they reach these regions.

The female *urethra* is usually sterile, whereas that of the male usually contains Gram-positive cocci and corynebacteria in the lower one-third. The *vagina* of the adult female usually contains a *Lactobacillus* species that produces lactic acid from a glycogen polysaccharide secreted by the vagina. Before puberty and after menopause, the vagina does not produce glycogen, and this *Lactobacillus* is absent.

8.2
Germ-free animals

As we have just learned, microorganisms are present in all normal animals. Animals without microorganisms, called *germ-free*, can only be obtained in the laboratory. A fetus is usually germ-free as long as it remains inside the mother, and only at birth does it become infected. To obtain a germ-free animal, the fetus is removed aseptically from the mother just before birth and placed in a sterile chamber called an *isolator*. All of the air, water, food, and other materials that go into the isolator must be sterile. The infant animal is fed sterile milk by hand from a bottle until it is old enough to feed itself. Careful attention must be given to be certain that the animals do not become contaminated (Figure 8.4). Although raising germ-free animals is a complicated business, it is now done routinely, and whole colonies of germ-free animals have been established in which germ-free parents are used to produce germ-free offspring. Chickens, mice, rats, guinea pigs, and rabbits have all been raised in the germ-free condition and colonies established. Large animals such as sheep, cattle, and man would of course be very difficult to raise germ-free, as very large isolators would be required. In some areas, germ-free mice can be purchased from laboratory animal supply companies. To prove that an animal is germ-free, it must be shown that no microbial growth is obtained when culture media are inoculated with feces, urine, skin, hair, or other parts of the animal. Since no one culture medium will support growth of all microorganisms, a wide variety of culture media should be used, and incubations should be carried out both aerobically and anaerobically.

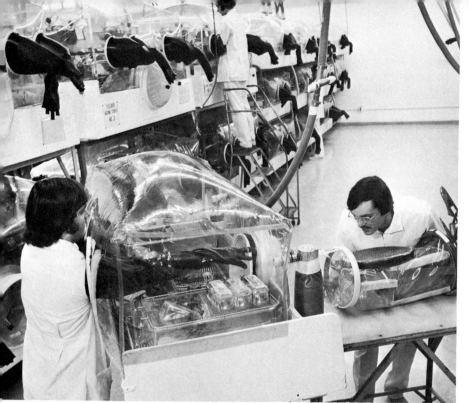

FIGURE 8.4
A germ-free isolator. *Courtesy of the Charles River Breeding Laboratories, Inc.*

Of what interest are germ-free animals? By studying the differences between germ-free and normal animals we can hope to find out whether the normal flora of the body is beneficial or harmful to the animal. The conclusion so far is that the normal flora can be both beneficial and harmful. Among benefits, bacteria of the intestine produce vitamin K, an essential vitamin for mammals. Germ-free animals must be given vitamin K in their diet in order to survive, whereas conventional animals can live without such vitamin K supplements. Another benefit is that the normal flora helps the animal resist infectious disease. Conventional animals are much more resistant to infection than are germ-free animals: the bacteria of the normal flora probably keep pathogenic bacteria from becoming established by virtue of competition for nutrients and growth habitats. Also, the immune processes are much better developed in the normal than in the germ-free animal. Therefore, although animals can live without their normal flora, they seem to be better off when bacteria are present. However, the normal flora can at times be harmful. For example, the bacterium *Clostridium perfringens*, sometimes a member of the normal flora, can produce toxic substances that may retard the growth of the animal. On balance, however, more benefit than harm is probably derived from the normal flora.

INFECTION AND IMMUNITY

Although the vast majority of organisms associated with the body are harmless, the very restricted variety of harmful organisms, called *pathogens*, assume dominant importance in our study of microbiology. We discussed in Section 1.3 the history of the germ theory of disease and emphasized the importance of Koch's postulates for proving that a particular infectious disease is caused by a particular organism.

Briefly, Koch's postulates require that it be shown that an organism suspected of causing disease is constantly present when the disease exists, is absent from healthy individuals, can be isolated in pure culture, and can be reinoculated into healthy individuals and cause disease again. By the application of these postulates, many bacterial pathogens have been identified and studied, and great advances in medicine and public health have resulted.

In some cases, Koch's postulates cannot be applied because the organism cannot be cultured. The causal agents of syphilis and leprosy have never been cultured away from living hosts, and of course, viruses can never be cultured in the absence of living cells. Despite inability to culture these agents, their causal role has been reasonably well established by virtue of observation of their constant association with the disease state and because drug therapy or other treatments that cure the disease lead to a disappearance of the causal agent.

In humans, intentional inoculation cannot be done since man is not an experimental animal, and hence one of Koch's postulates cannot be applied. One way in which suspected causality has been proved is through laboratory accidents involving research workers studying the organism. Despite considerable precaution, accidental infections do occur in the microbiology laboratory, and if infection is followed by disease, this is strong evidence for causality.

SPECIFICITY OF
CAUSAL ORGANISMS

In a number of cases a given microbial pathogen, when it infects a given host, causes the production of a certain set of symptoms that can be distinguished from all other sets of symptoms and which we recognize therefore as a specific disease. For example, virulent strains of *Corynebacterium diphtheriae*, when infecting susceptible humans, always give rise to the set of symptoms that we call *diphtheria*. Diphtheria can be caused by no other organism, and *Corynebacterium diphtheriae* can cause no other diseases. In this case, a specific disease is related to a specific pathogen. Other examples of organisms causing specific diseases are *Clostridium tetani* (tetanus), *Brucella abortus* (brucellosis), and *Treponema pallidum* (syphilis). In Chapters 10, 11, and 12, a number of such specific diseases will be discussed.

Infectious diseases are known in which less specificity is involved. For instance, the kidney may be infected by a number of

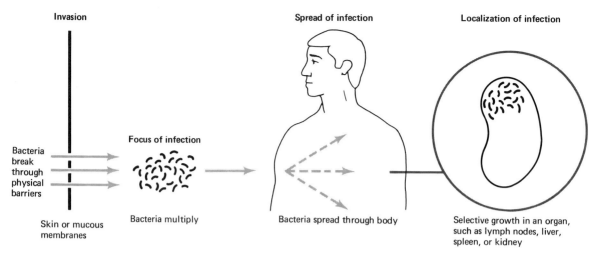

Invasion

Spread of infection

Localization of infection

Bacteria
break
through
physical
barriers

Skin or mucous
membranes

Focus of infection

Bacteria multiply

Bacteria spread through body

Selective growth in an organ,
such as lymph nodes, liver,
spleen, or kidney

FIGURE 8.5
Steps in infection of the body by
pathogenic bacteria.

species of bacteria, including *Escherichia coli, Staphylococcus aureus, Streptococcus faecalis*, and *Pseudomonas aeruginosa*, among others. All these organisms cause a disease in the kidney that is called *nephritis* (from *nephro*, referring to "kidney," and *itis*, "inflammation"). In this case, it is not possible from an analysis of the symptoms to make any good guess as to the causal organism of the infection in any particular host. Other diseases of this sort are endocarditis (infection of the endocardium of the heart), meningitis (infection of the membranes, called the *meninges*, surrounding the spinal chord), pneumonia (infection of the lungs), and peritonitis (infection of the peritoneal cavity). Many other situations of this sort might be described, and in each case, emphasis would be placed on the organ or location in the body that was infected rather than on the causal organism.

Between the two extremes of diseases caused by specific organisms and diseases caused by a variety of organisms, we have intermediate situations in which an individual organism may cause more than one clinical disease. Thus, *Mycobacterium tuberculosis* causes pulmonary tuberculosis, but it also may cause infections of skin, bones, or internal organs, although such infections are usually sufficiently characteristic of *M. tuberculosis* that they can be recognized and distinguished from infections of the same organs by other pathogens.

We can approach infectious diseases either through a discussion of individual pathogens and the diseases they cause or through a discussion of the disease conditions to which various organs of a host may succumb. The latter approach is followed by pathologists, who are interested primarily in the diseased organ and only indirectly in the organism causing the disease. The microbiologist, on the other hand, emphasizes the infectious agent and approaches infectious diseases through a study of the microorganisms involved.

INFECTION AND IMMUNITY

To be pathogenic, a microorganism must in some way cause harm to the host. In most cases, a definite sequence of events occurs during infection and disease:

1 Transfer of the parasite to the host (dispersal).
2 Entry of the parasite into the host's tissues and increase in numbers of the parasite (invasion).
3 Injury to the host (pathogenicity).
4 Response of the host to presence of the pathogen (immunity).

Dispersal, the first step in infection, is a complex process that depends on properties of the parasite, on the location in the body where infection occurs, and on the degree of incidence of the infection in the host population. We shall reserve a detailed discussion of mechanisms of dispersal until Chapter 9.

INVASION AND
GROWTH IN THE HOST

If the pathogen is to cause disease, it must penetrate the host tissues, a process called invasion (Figure 8.5). The skin is the major physical barrier to microbial invasion. Very few, if any, organisms gain entry through the unbroken skin. In burns where much skin is lost, a major cause of death is infection. Wounds, especially large dirty ones, are common sources of infection. In many wars, more soldiers have died as a result of wound infections than because of the direct effects of bullets and shells.

The mucous membranes that line the throat and respiratory tract also prevent entrance of microbes, although there are usually small breaks or lesions through which organisms may pass. Entry into the intestinal tract is generally via food particles carried through the stomach. Motility of the pathogen may help in invasion, although many pathogens are nonmotile.

Many pathogens have the ability to attach specifically to certain host cells or cell types. Streptococci attach to the mucosal epithelium of the upper respiratory tract, whereas other bacteria selectively colonize the gastrointestinal epithelium. In some cases, the pathogen can remain localized at the epithelium and initiate damage by liberating toxic substances (toxins are discussed in a later section), but in most cases, the pathogen penetrates the epithelium and either grows in the submucosa or spreads to other parts of the body where growth is initiated.

After initial entry the organism often becomes localized and multiplies to form a small *focus of infection* (Figure 8.6). In the skin, such foci are seen as boils, pimples, or carbuncles. These foci are usually filled with *pus*, a mixture of microorganisms, body fluids, and decomposing host cells. Infections that result in pus formation are called *pyogenic infections* (*pyo*, meaning "pus"). Access to the interior of the body generally occurs in those areas where lymph glands are near the surface, such as the nasopharyngeal region, the tonsils, or

FIGURE 8.6
A pathogenic organism in infected tissue. *Bacillus anthracis*, the causal agent of anthrax, is shown in a section of infected tissue; magnification 1,500×.

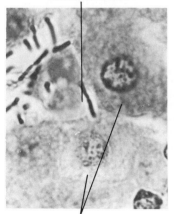

Bacillus anthracis

Tissue cells

the lymphoid follicles of the intestine. Inside the body, localization usually occurs in the lymph nodes, liver, spleen, or kidney. Only rarely is there extensive growth in a variety of organs or tissues. Microbes from foci of infection may pass into the bloodstream; when living bacteria are detected in the blood, the condition is usually called *bacteremia* or *septicemia*. Interestingly, blood itself is not an especially favorable medium for microbial growth, and when organisms are found there, they have usually come from some other place.

If the organism is to cause damage to the host, *growth* in the body must first take place. The organism must of course be able to grow at the temperature, pH, and osmotic conditions found in the body. It must be able to obtain its nutrients from body fluids and tissues; the body is a rich source of microbial nutrients in the form of carbohydrates, amino acids, nucleic acids, and lipids. Of some value to a pathogen are enzymes that can break down large molecules, such as polysaccharides or proteins; the components released can then be used as nutrients.

However, not all microbial nutrients may be plentiful at all times. Vitamins and growth factors are not necessarily in adequate supply in all tissues, and minerals may also be in short supply. Among the minerals, one frequently deficient is iron, a mineral required in fairly large amounts by most microorganisms. Ferric iron is highly insoluble at neutral pH values, and the host has a specific iron-binding protein called *transferrin*, which carries iron through the body. The affinity of this protein for iron is so high that there may be insufficient iron for microbial growth, and administration of a soluble iron salt to an infected animal can greatly increase the virulence of some pathogens. However, many bacteria produce specific iron-binding chemicals that enable them to grow even when concentrations of iron are low.

OVERCOMING HOST DEFENSE MECHANISMS

The host possesses a wide variety of defense mechanisms for counteracting microbial invasion, and the pathogen must be able to overcome these defenses if it is to become established. Briefly, host defenses include (1) cells called *phagocytes*, which engulf and remove invading microbes; (2) antimicrobial substances such as lysozyme and other enzymes, which kill and digest microbes; and (3) specific immune substances called *antibodies*, which affect specific microbes only. We shall discuss host defenses in some detail later.

Invading microbes can frequently avoid being engulfed by phagocytes (Figure 8.7). One way in which this is done is by the production of a *capsule* or *slime layer* that prevents the phagocyte from eating the cell. When such a surface layer is present, the phagocyte is unable to stick to and engulf the microbe. Some microbes produce substances, called *leukocidins*, that destroy phagocytes. When such a microbial cell is engulfed, its leukocidin kills

INFECTION AND IMMUNITY

FIGURE 8.7
Evasion of phagocytosis by pathogenic bacteria.

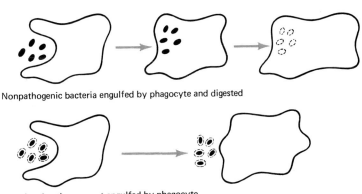

Nonpathogenic bacteria engulfed by phagocyte and digested

Capsulated pathogens not engulfed by phagocyte

Leukocidin-producing pathogens engulfed but destroy phagocyte

the phagocyte and the microbe leaves the phagocyte, still alive. Pathogens such as *Staphylococcus* and *Streptococcus* that cause pyogenic infections (boils, pimples) often produce leukocidins; the pus itself contains large numbers of dead phagocytes.

In some cases, pathogens are readily phagocytized but neither kill nor are killed by the phagocytes. These pathogens are able to remain alive within the phagocytes and even grow. Such pathogens set up *intracellular infections* and are able to maintain themselves for long periods of time. Within the phagocytes, such pathogens are well protected and are some of the most difficult organisms to eliminate from the body by drug treatment. Examples are *Mycobacterium tuberculosis* (causal agent of tuberculosis), *Salmonella typhi* (causal agent of typhoid fever), and *Brucella melitensis* (causal agent of undulant fever).

A common symptom of microbial infection is *fever*, an increase in body temperature. Fever is often caused by the release from pathogens of specific substances called *pyrogens*. Another cause of fever is the reaction of the body to foreign material, a process called *inflammation* (to be discussed later in this chapter).

Certain pathogens produce specific enzymes that are involved in invasion. One example is the enzyme *hyaluronidase*, which acts on *hyaluronic acid*, a polysaccharidelike substance that functions in the body as a tissue cement holding cells together. Pathogens are unable to spread easily through normal tissues, but if they produce the enzyme hyaluronidase they can spread more readily. Another cementing substance in the body is the simple protein *collagen*, and some pathogens produce the enzyme *collagenase*, which also makes spreading easier. In many infections, the focus of infection is walled off by the formation of a fibrin clot. Some invading bacteria, especially the streptococci, produce an enzyme called *streptokinase* that dissolves these clots. Other bacteria produce the enzyme *coagulase*, which, instead of dissolving clots, causes them to form; pathogens producing coagulase (mainly staphylococci) are able to cause localized infections such as boils and pimples in the region of the clotted serum.

Thus we see that pathogens produce a wide variety of substances that make possible their establishment and growth in the body. Different pathogens produce different substances, and the kinds of infections established are determined in part by the kinds of substances produced.

PATHOGENICITY

The mere presence of a microbe in the body rarely, if ever, causes damage to the host. The harm that is caused by pathogens is generally due to production of specific factors called *toxins*, which affect the host in some detrimental way. Two kinds of toxins are produced: (1) *exotoxins*, which are excreted or released from the pathogen; and (2) *endotoxins*, which generally remain bound to the cells and are released in significant amounts only when the pathogen dies and disintegrates (Figure 8.8).

Exotoxins are usually proteins. Since the exotoxin is released

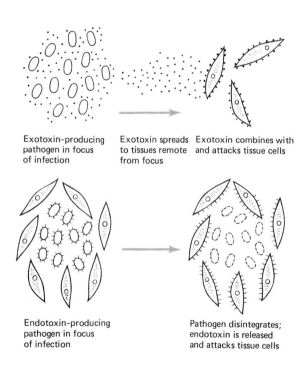

Exotoxin-producing pathogen in focus of infection

Exotoxin spreads to tissues remote from focus

Exotoxin combines with and attacks tissue cells

Endotoxin-producing pathogen in focus of infection

Pathogen disintegrates; endotoxin is released and attacks tissue cells

FIGURE 8.8
Differing actions of exotoxins and endotoxins.

TABLE 8.1	Bacterium	Disease	Toxin	Action
Some exotoxins produced by certain bacteria pathogenic for man	*Clostridium botulinum*	Botulism	Neurotoxin	Paralysis
	C. tetani	Tetanus	Neurotoxin	Paralysis
	C. perfringens	Gas gangrene	Alpha-toxin	Hemolysis
			Theta-toxin	Affects heart muscle
			Kappa-toxin	Digests collagen
			Lambda-toxin	Digests proteins
	Corynebacterium diphtheriae	Diphtheria	Diphtheria toxin	Inhibits protein synthesis
	Staphylococcus aureus	Pyogenic infections (boils, pimples), and food poisoning	Alpha-toxin	Hemolysis
			Leukocidin	Destroys leukocytes
			Enterotoxin	Induces vomiting, diarrhea
	Streptococcus pyogenes	Pyogenic infections, tonsillitis, and scarlet fever	Streptolysin O	Hemolysis
			Streptolysin S	Hemolysis
			Erythrogenic toxin	Causes scarlet-fever rash

from the pathogen, it can exert its action at sites distant from the pathogen itself. In fact, the toxin may even cause harm to the host after the organism is no longer present in the body. Table 8.1 lists some of the better known exotoxins produced by bacteria. The bacterial disease *diphtheria*, caused by *Corynebacterium diphtheriae*, can illustrate some of the principles of diseases involving exotoxins. The bacterium grows almost exclusively in the throat, never invading other parts of the body. The diphtheria toxin is released from this focus of infection and spreads throughout the body. All the symptoms of diphtheria are due to the action of the toxin, as shown by the fact that all of the disease symptoms, including death, can be induced in an experimental animal by injection of highly purified toxin. Further, if an antibody (see Section 8.7) specific to the toxin is injected, all the disease symptoms in an infected animal are neutralized. This has important consequences for therapy of diphtheria: elimination of the pathogen by treatment with antibiotics will not necessarily eliminate disease symptoms, if the toxin has already spread throughout the body. Thus, effective treatment of diphtheria involves injection not only of an appropriate antibiotic but also of a serum containing the specific antibody that will neutralize the toxin.

The disease *tetanus* is caused by another exotoxin-producing bacterium, *Clostridium tetani*. This organism is an obligate anaerobe and will not grow in the living tissues of the body, which are aerobic. However, where oxygen supply to tissues is cut off as a result of wounds, anaerobic conditions can develop, and *C. tetani* can grow.

The toxin released causes paralysis and eventually death. Another clostridium, *C. botulinum*, can cause death even without infecting the body. This organism, the cause of a very serious food poisoning (see Section 15.2), grows and produces its exotoxin in improperly canned foods. When the food is eaten, the toxin spreads through the body from the intestine and causes paralytic death.

Endotoxins are cell-bound toxins, usually part of the outer cell envelope. They are produced primarily by Gram-negative bacteria and are complex chemically, containing both lipid and polysaccharide. One current idea is that the lipid portion is responsible for toxicity, the polysaccharide acting mainly to make the lipid portion water soluble. However, research on the chemistry of endotoxins is still an active field, and many uncertainties exist on the exact nature of endotoxins. In general, their toxicity is lower than that of exotoxins, and much larger doses are required. However, endotoxins are of considerable medical significance and are responsible for many deaths. It is likely that small amounts of endotoxin are released even from living bacteria, but only when cells die and disintegrate are large amounts released. Symptoms induced by endotoxins include fever, diarrhea, hemorrhagic shock, and other tissue damage. Another important property is their ability to stimulate natural host defense mechanisms when present in the body in small amounts. Since certain bacteria of the normal flora, such as *Escherichia coli*, produce endotoxin in small amounts, natural stimulation of host defense mechanisms may occur more or less continuously, in a moderate way.

Allergy is also involved in the action of some pathogens. In many cases, disease symptoms are elicited in the host as a result of an allergic reaction of the body to the invading pathogen. We will discuss allergy and hypersensitivity later in Section 8.13. Diseases in which many or all the symptoms are caused by such allergic reactions include tuberculosis, rheumatic fever, and glomerular nephritis.

Virulence refers to the relative ability of a parasite to cause disease. Virulence is determined primarily by two properties of an organism: its *invasiveness* and its *toxigenicity*. A highly invasive organism is able to grow well in the body and to set up a widespread generalized infection. Even if it is only weakly toxic, it may cause host damage because of the large numbers of cells present. On the other hand, a weakly invasive organism, even though it may grow only poorly in the body, may still be highly virulent if it produces an extremely potent toxin (Figure 8.9).

An example of a weakly invasive organism that is highly virulent is *Corynebacterium diphtheriae*. Even though this organism infects and grows only in the throat, it can cause death because the toxin it produces is extremely potent and can spread throughout the body. An example of a weakly toxigenic organism that is still highly

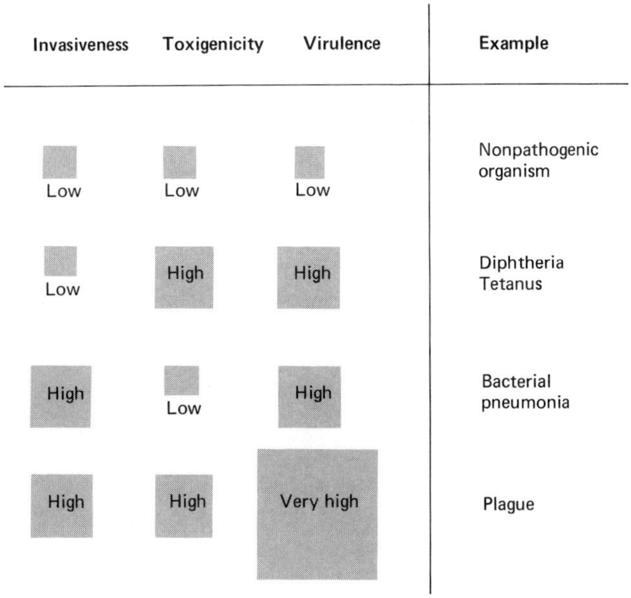

FIGURE 8.9

Interaction of invasiveness and toxigenicity in virulence. Both the invasive power of the organism and the severity of action of the toxin contribute to the virulence of the pathogen. Weakly invasive organisms may still be virulent if they produce potent toxins. Organisms producing weak toxins may be virulent if they are highly invasive.

Invasiveness	Toxigenicity	Virulence	Example
Low	Low	Low	Nonpathogenic organism
Low	High	High	Diphtheria Tetanus
High	Low	High	Bacterial pneumonia
High	High	Very high	Plague

virulent is *Streptococcus pneumoniae*, the main agent causing bacterial pneumonia. This organism is not known to produce a toxin yet is able to damage the host and even cause death because it is highly invasive in lung tissue and grows in the lungs in such large numbers that the functions of this organ are impaired.

ATTENUATION OF VIRULENCE

Pathogenicity is an inherited property of a parasite. A parasite may mutate and lose the ability to produce a capsule, toxin, or some other factor involved in invasiveness or virulence, thus becoming nonvirulent. When pathogens are kept in laboratory culture for long periods, their virulence may decrease or even disappear because of the selection of faster-growing nonvirulent strains. Such organisms are said to be *attenuated* and have all the properties of the source strain except its virulence. (We presented in Section 1.6 Pasteur's work on the attenuation of virulence.) Attenuated organisms are frequently useful as vaccines, as will be described later in this chapter.

Once an organism has become attenuated through laboratory culture, its virulence can sometimes be restored by inoculating it into an experimental animal (Figure 8.10). Although virulent strains may be present in the culture only in small numbers, their growth may be so favored in the host that they are selected. Virulence can often be greatly increased by successive passage of the parasite from one animal to another; after several passages, a highly virulent strain may be isolated. Of course, the same kind of selection process may occur naturally when one person becomes infected from another.

In terms of clinical symptoms, the course of disease can be conveniently divided into stages:

1 *Infection*, when the organism becomes lodged in the host.
2 *Incubation period*, the time between infection and the appearance of disease symptoms. Some diseases have short incubation periods; others, longer ones. The incubation period for a given disease is determined by inoculum size, virulence of pathogen, resistance of host, and distance of site of entrance from focus of infection.
3 *Prodromal period*, a short period sometimes following incubation in which the first symptoms, such as headache and feeling of illness, appear.
4 *Acute period*, when the disease is at its height, with such overt symptoms as fever and chills.
5 *Decline period*, during which disease symptoms are subsiding, the temperature falls, and a feeling of well-being develops. The

FIGURE 8.10
a Attenuation of pathogen. *b* Selection of a virulent strain from an attenuated culture.

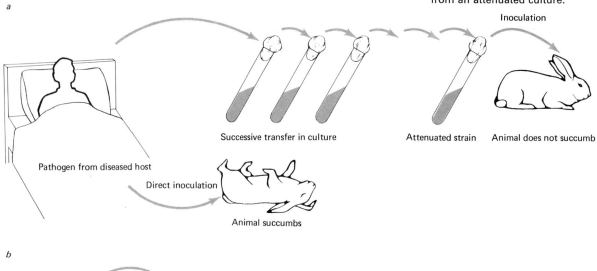

INFECTION AND IMMUNITY

decline may be rapid (within 1 day), in which case it is said to occur by *crisis*, or it may be slower, extending over several days, in which case it is said to be by *lysis*.

6 *Convalescent period*, during which the patient regains strength and returns to predisease status.

During the later stages of the infection cycle, the immune mechanisms of the host become increasingly important, and in most cases recovery from the disease requires the action of these immune mechanisms. We shall now describe the immune mechanisms of the host and how they operate to bring about recovery from infectious disease.

8.5 Immunity

We discussed the history of immunity and vaccination in Chapter 1. Immunity can be *innate*, that is, an inherent property of the host, or it can be *acquired*, that is, developing in the host as a result of previous infection or by immunization. After recovery from disease, the host is often immune to further infection by the same (and sometimes by related) pathogens. Immunization plays a major role in limiting the spread of epidemics and has been a major factor in increasing the general health level of the population. Immunity can be either *specific* or *nonspecific*. Mechanisms of nonspecific immunity act against any or all invading parasites, whereas specific immune mechanisms are highly selective, acting against single pathogens only. Nonspecific immunity is the initial defense against pathogens but is generally less effective than specific immunity, which often develops in response to infection.

BLOOD AND ITS COMPONENTS

Immunity develops to a great extent because of changes and activities of cells and substances found in the blood. Blood consists of cellular and noncellular components (Table 8.2a). The most numerous cells in the blood are the *red blood cells*, which carry oxygen from the lungs to the tissues. The *white blood cells*, or *leukocytes*, although much less numerous, play important roles in immunity, as will be discussed later. The fluid remaining after the cellular components are removed is called *plasma* (Table 8.2b). It consists primarily of water, in which a variety of salts and proteins are dissolved. Some of the plasma proteins play an important role in immunity, whereas others are involved in other functions. A plasma protein called *fibrinogen* is the clotting agent of the blood; when blood is removed from the body, fibrinogen undergoes a complex set of reactions and becomes *fibrin*, of which the clot is composed. Clotting can be prevented by adding *anticoagulants* such as potassium oxalate, potassium citrate, or heparin. When plasma is allowed to clot, the fluid components left behind, called *serum*, consist of all the proteins and

TABLE 8.2
Whole-blood components and blood-fraction terminology

a) Components of whole blood

Component	Proportion
Total blood volume, adult human	5 liters
Cellular components	45% of the blood volume
Red blood cells (erythrocytes)	5×10^9 cells/ml
White blood cells (leukocytes)	$0.005–0.01 \times 10^9$ cells/ml
Fluid components	55% of the blood volume
Plasma	Water, 90%; proteins,[a] 9%; salts, 0.9%; sugars and other organic compounds, traces

b) Terminology for blood fractions

	Cells	Fibrin	Fluid
Whole blood	+	+	+
Plasma	−	+	+
Serum	−	−	+

[a] Plasma proteins: albumin, globulins, and fibrinogen.

other dissolved materials of the plasma except fibrin. Since serum contains many of the substances involved in specific immunity, as will be described later, it is frequently prepared either for use in injections or for diagnostic studies.

Blood is carried within the circulatory system to various parts of the body (Figure 8.11). One function of the circulatory system is to ensure the distribution throughout the body of components of the immune system to combat infection. At the same time, the circulatory system facilitates the spread of pathogens to various parts of the body.

Lymph is a fluid similar to blood but lacks the red cell component. There is a separate circulation system for lymph, called the *lymphatic system*, within which lymph flows (see Figure 8.11). The *lymph nodes*, found at various locations throughout the system, function to filter out microorganisms and other particulate materials, but they may also be sites of infection, since organisms that are collected there by the filtering mechanisms may then proliferate. *Lymphocytes* are white blood cells involved in antibody formation; they are found in large numbers in the lymphatic system. The role of lymphocytes in antibody formation will be discussed later in this chapter.

A variety of properties and processes confer nonspecific immunity on the host. We can divide these into several groups: (1) phagocytosis, the action of host cells that engulf and kill microbes; (2) antimicrobial substances such as enzymes and other body substances that kill or prevent the growth of microbes; and (3) inflammatory responses to injury and infection.

8.6
Nonspecific immunity

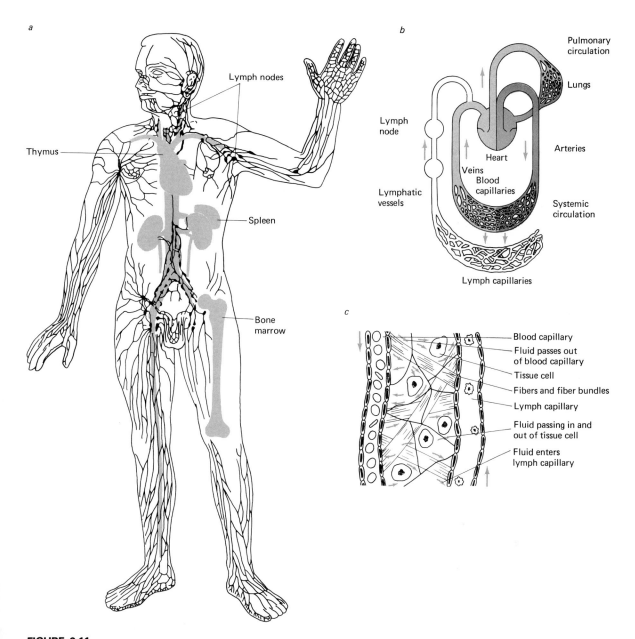

FIGURE 8.11

The blood and lymphatic systems. *a* Overall view of the major lymphatic systems, showing locations of major organs. The lymph nodes are heavily concentrated at the joints and trunk, where they filter out foreign particles and bacteria draining from the peripheries. *b* Diagramatic relationship between the lymphatic and blood systems. Blood flows from the veins to the heart, then to the lungs where it becomes oxygenated, then through the arteries to the tissues. Lymph flow rate is much slower than blood rate, as there is no organ such as the heart pumping the lymph fluid.
c Connection between the blood and lymphatic systems is shown microscopically. Both blood and lymph capillaries are closed vessels but are permeable to water and salts. Proteins of the blood also slowly move out of the capillaries. The tissue fluid is the means of transport between blood and lymph. Changes in capillary permeability, as a result of infection or other source of inflammation, can increase greatly the movement of fluid into the tissue spaces.

Phagocytes (literally, "cells that eat") are a prime line of defense against microbial invasion. Some of the leukocytes found in whole blood are phagocytes, and they are also found in various tissues and fluids of the body. Phagocytes are usually actively motile by amoeboid action. They are attracted to microbes, engulf them, and kill and digest them. Two types of phagocytes are *granulocytes*, which are actively motile cells containing distinctly visible granules, and *macrophages* (or *monocytes*), which are less granular and less motile (Figure 8.12). Granulocytes are short-lived cells found predominantly in the bloodstream and bone marrow; they appear in large numbers during the acute phase of an infection and, therefore, serve as an indicator of infection. Macrophages are of two types: *wandering cells*, which are found free in the blood and lymph; and fixed macrophages, or *histiocytes*, which are found embedded in various tissues and have only a limited mobility. Histiocytes constitute the active phagocytes of the reticuloendothelial system, which we shall discuss later.

A phagocyte is attracted chemically to an invading microbe, which it rapidly engulfs and digests (Figure 8.13). Phagocytes work best when they can trap microbial cells against surfaces such as blood vessel walls, blood clots, or connective tissue fibers. As we noted earlier, not all microbes are killed when engulfed by phagocytes. Some microbes produce toxins (leukocidins) that kill phagocytes, the microbes then being released unharmed, whereas other microbes do not actually kill phagocytes but are able to escape their destructive action and remain alive inside and even grow (see Figure 8.7). The action of phagocytes is not restricted to microbes; any foreign particulate material can be ingested and may be digested and thus eliminated from the body.

During many kinds of infection, an increased number of phagocytes are produced and disseminated through the bloodstream to the site of infection. Thus, infection may result in an increase in the number of phagocytic cells in the bloodstream, and a quantitative count of the number of such cells in a sample of blood can therefore be used as an indication of infection. Diseases often associated with an increased leukocyte count are appendicitis, meningitis, mononucleosis, pneumonia, gonorrhea, and pyogenic infections such as boils, in which the total white cell count may be doubled, tripled, or quadrupled. During the course of infection, there is a shift in the type of white cell that predominates. During the active stage of infection, granulocytes predominate; during subsidence, monocytes, which play a major role in ingesting dead or dying bacteria, are more common. Thus a *differential count* showing the proportions of various kinds of leukocytes aids in determining the progress of infection.

The liver, spleen, and lymph glands are unusually well supplied with fixed macrophages. As the blood and lymph pass through

FIGURE 8.12
Two major phagocytic cell types. *a* Granulocyte. *b* Macrophage. Magnifications 1,500×.

a

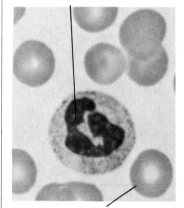

Granulocyte

Red blood cell

b

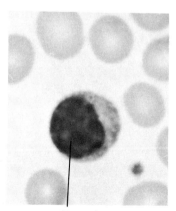

Macrophage (monocyte)

INFECTION AND IMMUNITY

FIGURE 8.13
Phagocytosis: engulfment and digestion of a large bacterial cell, *Bacillus megaterium*, by a human phagocyte. Magnification 1,350×, phase-contrast. From J. G. Hirsch, *J. Exptl. Med.*, **116**, 827, 1962.

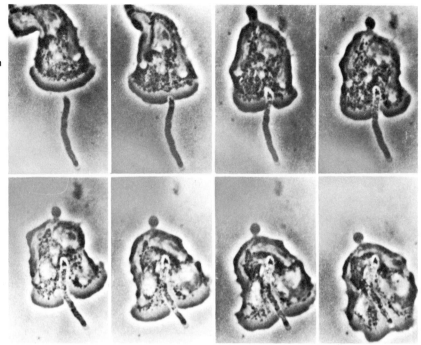

these organs, microorganisms present are filtered out and ingested. The system of fixed macrophages in these organs is called the *reticuloendothelial system* (RE system). The RE system is quite efficient in clearing foreign particles from the blood and lymph. Within hours after microbes enter, they are usually completely eliminated from the blood and can be found in the organs of the RE system, where they are usually quickly killed.

ANTIMICROBIAL SUBSTANCES

Several enzymes and chemical substances produced by the body are antimicrobial agents and kill some invading organisms. The most important is *lysozyme*, an enzyme found in tears, nasal secretions, saliva, mucus, and tissue fluids, which is able to digest the cell wall of bacteria, causing death and lysis of the cell. Lysozyme is most effective against Gram-positive bacteria but acts on Gram-negative bacteria under certain conditions. Saliva contains an additional antimicrobial defense, called the *lactoperoxidase system*: the enzyme lactoperoxidase functions together with two other substances found in saliva, thiocyanate (CNS^-) and hydrogen peroxide (H_2O_2), to kill a portion of the bacteria.

One other antimicrobial substance of considerable importance is the acid fluid of the stomach. The pH of the stomach is around 2, and many pathogens are rapidly killed at this pH; most organisms taken in with food thus cannot survive passage through the stomach unless protected inside food particles.

The host responds to foreign matter or trauma by a process called inflammation (Figure 8.14). Causes of inflammation include microbial infections, cuts, abrasions, burns, and sharp blows. The inflamed site becomes swollen and red and is usually painful to the touch. The symptoms of inflammation develop because the blood vessels near the site of the foreign body enlarge and the blood capillaries increase in permeability, allowing cells and fluids to escape from the bloodstream and enter the tissues. Phagocytes accumulate at the inflamed site, and if infection exists, the microbial cells present are ingested. A clot usually forms around the inflamed region and walls off and localizes the invading microbe. Thus, although inflammation is painful, it is beneficial because it leads to localization and destruction of the infecting organism.

One of the factors that start the inflammatory response is the chemical *histamine*, which is released by cells in the inflamed region and causes an increase in capillary permeability. Certain drugs, called *antihistamines*, counteract some of the effects of histamine and can be used to neutralize some of the pain associated with inflammation. However, antihistamines also prevent the benefits coming from inflammation; so that they must be used with caution.

Pathogens vary in the degree to which they bring about inflammation. Some pathogens that are not especially invasive induce considerable inflammation and in this way cause more host damage than would be predicted from their invasive powers.

CARE OF INFLAMED TISSUES   The care of an inflamed part of the body consists of rest, elevation, application of heat or cold, and sometimes incision and drainage. *Rest* of the inflamed part is essential because movement is not only painful but delays healing by reducing the walling-off process. Splints, bandages, or casts may be used to prevent movement. *Elevation* of the inflamed part permits the force of gravity to help drain the engorged blood vessels and tissue spaces, thus reducing swelling and pain. There is often almost instantaneous relief from pain when an inflamed part is elevated. *Cold* may be used in the treatment of recent inflammation; it causes a constriction of blood vessels in the inflamed region, which reduces the movement of fluid into the tissues. Cold may also slow down bacterial growth. *Heat* is more valuable in the later stages of inflammation, after swelling has decreased. It causes a relaxation of muscles, increases the blood supply to the inflamed area, and thus promotes healing. In seriously inflamed regions, *incision* and *drainage* may be used, resulting in immediate removal of fluid, phagocytes, and bacteria, thus reducing pain and eliminating the source of

INFLAMMATION

FIGURE 8.14
Events in inflammation. *a* Normal undilated capillary system. *b* Blood vessels and capillaries dilate and become filled with blood. *c* Phagocytes and plasma move out of capillaries into surrounding tissues. *d* Clot forms, and phagocytes ingest bacteria.

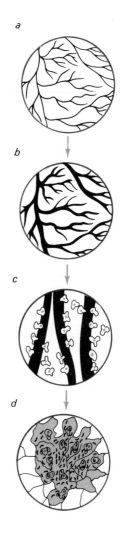

toxins. The fluids removed from an inflamed region may be highly infectious and should be disposed of carefully.

8.7
Specific immunity

In addition to the nonspecific types of immunity described above, higher animals and man produce *antibodies*, substances that act against specific organisms and prevent their growth in the body. Antibodies are specific proteins found in the globulin fraction of blood serum. They combine with and neutralize certain proteins or polysaccharides on the surface of microorganisms. The substance on the surface of the microorganism with which the antibody combines is called an *antigen*. Through this antigen-antibody reaction, the invading microorganism is often damaged and its pathogenicity neutralized.

ANTIGENS

Antigens are substances that induce antibody formation. They are usually proteins or polysaccharides, although some lipids and nucleic acids are also antigenic. Thus, many components of microbial cells are antigenic, so that there are many opportunities for the development of specific immunity. Antibodies do not form against the antigen as a whole but against certain parts of the antigen; those parts against which antibodies form are called the *antigenic determinants*. In addition to antibody formation against natural antigenic determinants, antibodies can form against artificial antigenic determinants. For instance, many drugs and other chemicals are able to combine with proteins, the combination then acting as an additional antigenic determinant, to which a specific antibody can be formed. Antibodies of this type are responsible for allergic reactions against drugs and other chemicals.

Antigens vary in the ease with which they induce antibody formation. Insoluble antigens are usually better than soluble ones. The effectiveness of a soluble antigen can usually be improved by mixing it with aluminum hydroxide (alum) or some other insoluble material. Substances such as alum that improve the effectiveness of antigens without actually being antigens themselves are called *adjuvants*.

ANTIBODIES

Antibodies are specific proteins found mainly in serum, but they are also present in other body fluids, as well as in milk. Those proteins of the serum that have antibody properties are called *immunoglobulins*. One part of the antibody molecule is the same in many antibodies, but the other part is specific and contains the antigen-combining portion (Figure 8.15).

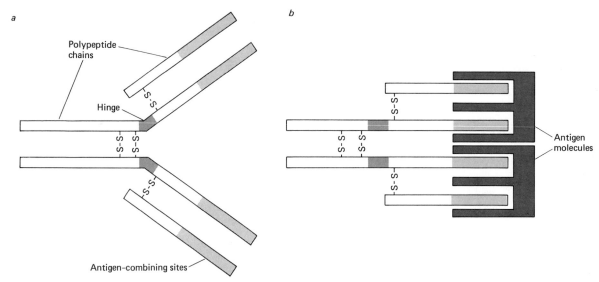

a

Polypeptide chains

Hinge

Antigen-combining sites

b

Antigen molecules

FIGURE 8.15
Structure of an antibody molecule and means by which antibody and antigen interact. *a* The antibody molecule, showing the four polypeptide chains of the protein, connected by —S—S— bridges. The longer central chains are flexible and can bend at the hinge. The antigen-combining sites are different in different antibodies, but the rest of the molecule is constant in structure. Thousands of different structures in the antigen-combining site are possible. *b* Manner of interaction of antibody and antigen. Two antigen molecules are shown, the antibody having bent at the hinge to conform to the structure of the antigen.

Antibodies are formed against all kinds of foreign proteins or polysaccharides as well as against those of pathogens. Antibodies are formed against antigens of pollen, spores, and various foods, and as we shall discuss later, allergies to these products develop because of antibody formation. However, antibodies do not form against the proteins or polysaccharides of the body's own cells. The specificity of antibodies is high but not absolute; thus an antibody may act against related antigens on different pathogens. In general, however, a single kind of antibody confers immunity against only a single disease.

The amount of a specific antibody in the blood is called the *antibody titer*. The antibody titer present initially will depend on the previous history of the host and can be greatly raised by immunization procedures involving injection of antigens. After an initial injection of antigen, there will be a period of several days during which no antibody will be present; then there is a slow rise in the antibody titer; and this is followed, in turn, by a gradual decline (Figure 8.16). If a second injection of antigen is given, the titer will rise rapidly to a level much greater than that achieved by the first injection. The response to the first injection is called the *primary response* and that to the second injection, the *secondary response*. The second injection is often called the "booster dose," since it boosts the antibody titer to a high level. Even after a booster the antibody level will eventually decrease, but it will return to a high level after another booster. Immunization schedules are designed to achieve and maintain high levels of antibody (see Table 8.6).

The primary and secondary responses will also occur as a result of microbial infection. Recovery from an infection in the absence of

INFECTION AND IMMUNITY

specific drug treatment is usually caused by antibody formation, probably as a result of a primary infection. If another infection with the same organism occurs subsequently, a secondary response can then take place. If the secondary response occurs sufficiently rapidly, the second infection may be quickly overcome and in fact may never lead to obvious symptoms.

The ability of the individual to form antibodies is influenced by nutritional state, age, hormonal balance, and general well-being. Very young persons and aged individuals are usually poor antibody producers. Certain drugs reduce the ability to produce antibody. This is especially true of anti-inflammatory agents such as cortisone and other steroids. Radiation (such as X rays or atomic radiation) also severely reduces ability to produce antibodies.

8.8
Kinds of antibodies and their properties

Studies of the chemistry of antibodies have shown that a variety of different antibody molecules are formed, each with its own function. The structures and functions of these various immunoglobulins are summarized in Table 8.3.

The antibody structure shown in Figure 8.15 is for the immunoglobulin called *IgG*. It is the major antibody found in the circulating body fluids, and plays a key role in antibody-antigen reactions in the blood, lymph, and tissue fluids. IgG molecules are *bivalent*, which means they have two antigen-combining sites, as illustrated in the figure. Immunoglobulin *IgM* has a much more complex structure than IgG and is much larger in size; it has five antigen-combining sites. Perhaps because of its size, IgM does not enter the tissue fluids but remains confined to the blood and lymph systems. IgM is the first antibody to appear after immunization; IgG appears later.

Immunoglobulin *IgA* is of interest because it is secreted into

FIGURE 8.16
Primary and secondary responses to the injection of an antigen. The amount of antibody is expressed as the antibody titer, usually assayed by determining the smallest amount of serum that will induce an antibody-antigen reaction in a test-tube system.

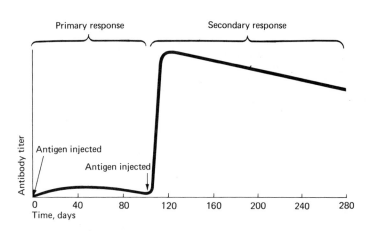

body fluids, such as the saliva, tears, breast milk and colostrum, gastrointestinal secretions, and mucous secretions of the respiratory and genitourinary tracts. IgA is also in the blood, but the IgA secreted into body fluids has a modified structure, with a protein-carbohydrate complex attached to it that apparently serves as the "secretory piece," protecting the molecule from digestion by proteolytic enzymes. The attachment of the secretory piece to the IgA molecule occurs in the mucosal cell through which the IgA passes on its way into the secretions. Because of its location in the secretions, IgA probably provides a first line of attack against bacterial invaders, which as we have seen, become established initially on tissue surfaces (generally bathed in secretions). A newborn sucking at its mother's breast receives IgA antibodies from both colostrum and milk and probably has fewer problems with bacterial infections than a newborn drinking cow's milk from a bottle.

Less is known about the other immunoglobulins, *IgE* and *IgD*. We do know that IgE is involved in allergic reactions, and its concentration rises during certain types of allergic responses (see Section 8.13).

TABLE 8.3
Properties of the immunoglobulins

Designation	Properties	Structure[a]	Proportion of total antibody (%)	Complement-fixing ability	Distribution
IgG	Major circulating antibody		80	Present	Extracellular fluids and blood and lymph
IgM	Complex structure; first antibody to appear after immunization		5–10	Present	Blood and lymph only
IgA	Secretory antibody	Secretory piece	10	Absent	Secretions; extracellular and blood fluids
IgD	Minor circulating antibody		1–3	Absent	Blood and lymph only
IgE	Involved in allergic reactions		0.05	Absent	Blood and lymph only

[a] See also Figure 8.16.
[b] Complement fixation is discussed in Section 8.10.

8.9
Antigen-antibody reactions

Many different kinds of antigen-antibody reactions occur. *Neutralization*, the simplest kind of reaction, occurs when an antibody combines with a soluble antigen, such as a toxin, thus blocking its toxicity (Figure 8.17). The antibody that combines with and neutralizes a toxin is sometimes called an *antitoxin*. Antibodies can also neutralize viruses by combining with them and preventing them from attaching to host cells.

Precipitation occurs when a number of molecules of antibody and antigen react together, producing a clump, or aggregate, of precipitate (Figure 8.18a). The mechanism of precipitation is illustrated in Figure 8.18b; when the antigen-combining sites on an antibody molecule react with separate antigen molecules, a large aggregate containing alternating units of antibody and antigen develops. Precipitation requires that approximately equal numbers of antibody and antigen molecules react. If either antibody or antigen is in great excess,

FIGURE 8.17
Antibody neutralization reaction. *a* Action of a toxin without neutralization. *b* Neutralization of a toxin by its specific antibody (antitoxin).

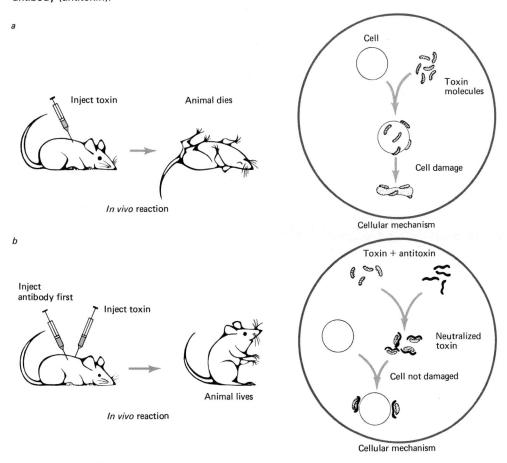

a

Inject toxin   Animal dies

In vivo reaction

Cell

Toxin molecules

Cell damage

Cellular mechanism

b

Inject antibody first   Inject toxin

Animal lives

In vivo reaction

Toxin + antitoxin

Neutralized toxin

Cell not damaged

Cellular mechanism

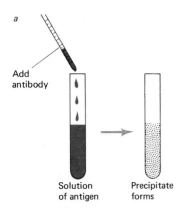

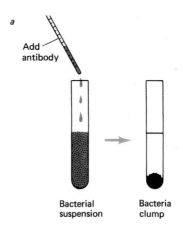

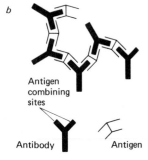

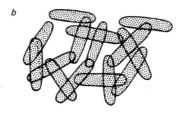

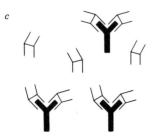

FIGURE 8.18
Antibody precipitation reaction.
a Demonstration of precipitation
in a test tube. *b* Mechanism of
precipitation; precipitate forms
when antibody and antigen are
at equivalence. *c* No
precipitation occurs when either
antibody or antigen is in excess.

FIGURE 8.19
Antibody agglutination reaction.
a Demonstration of
agglutination in a test tube. *b*
Appearance of agglutinated
bacterial cells when antibody is
directed against cell-envelope
antigens. *c* Appearance of
agglutinated bacteria when
antibody is directed against
flagella.

no precipitate will form, and the small antibody-antigen complexes will
be soluble (Figure 8.18*c*). Precipitates are phagocytized more readily
than are solutions, and precipitation thus leads to rapid clearing of
antigens from the blood or other fluids.

Agglutination is like precipitation except that the antigen is part of a cell (such as an endotoxin) rather than a soluble component; the antibody molecules cause the cells to stick together and form clumps (Figure 8.19*a*), which are more readily phagocytized than are individual cells. Agglutination reactions can occur either as a result of combination of the antibody with cell-envelope antigens, in which case a compact agglutinate forms (Figure 8.19*b*), or the antibody can combine with flagellar antigens, in which case a much looser agglutinate forms (Figure 8.19*c*). Antibodies against flagella can also cause *immobilization* of motile cells, since motility is caused by the action of flagella. Immobilization is one of the few antibody-antigen reactions that can be detected with single cells.

Antibodies can be made *fluorescent* by chemical combination to special fluorescent dyes. When a bacterial cell (containing antigen) is treated with fluorescent antibodies and viewed under a fluorescence microscope, the cell gives off light and appears colored upon a dark background (Figure 8.20). Fluorescent antibody preparations can be used to identify one organism, even in a mixture containing many kinds of organisms, since the antibody combines only with the organism to which it is directed. Thus fluorescent antibodies can be used in diagnosis of infections by permitting identification of the pathogen.

8.10
Complement and
complement fixation

A number of important antibody-antigen reactions require the participation of a group of substances called *complement*. Complement is not an antibody but is a series of enzymes found in normal blood serum that react with antigen-antibody complexes and inactivate or otherwise affect the antigen. These enzymes, found even in unimmunized individuals, are normally inactive but are activated when an antibody-antigen reaction occurs. One of the main functions of antibody is to recognize invading cells and activate the complement

a

b

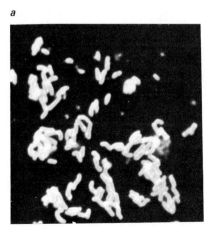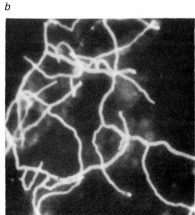

FIGURE 8.20
Fluorescent antibody reactions.
a Rhizobium japonicum,
magnification 1,000×. *Courtesy
of B. B. Bohlool and
E. L. Schmidt.*
b Filamentous bacterium,
magnification 850×. From B. B.
Bohlool and E. L. Schmidt, *Soil
Science,* **110**, 229–236, 1970.

system for attack. There is considerable economy in an arrangement such as this, since a wide variety of antibodies, each specific for a single antigen, can call into action the complement enzymatic machinery; thus the body does not need separate enzymes for the attack of each kind of invading agent.

Some reactions in which complement participates include: (1) bacterial lysis, especially in Gram-negative bacteria, when specific antibody combines with antigen on bacterial cells in the presence of complement; (2) microbial killing, even in the absence of lysis; and (3) phagocytosis, which may not occur during infection if the invading microorganism possesses a capsule or other surface structure that prevents the phagocyte from acting. When specific antibody combines with the cell in the presence of complement, the cell is changed in such a way that phagocytosis can occur. (This process in which antibody plus complement renders a cell susceptible to phagocytosis is sometimes called *opsonization*.)

Complement is a system of 9 proteins, designated C1, C2, C3, and so forth. Activation of complement occurs only by antibodies of the IgG and IgM types (see Table 8.3); when such antibodies combine with their respective antigens, they are altered in such a way that the first component of complement, C1 (which is really a complex of three subunits called C1q, C1r, and C1s), combines with the antibody-antigen aggregate (Figure 8.21). C4 then combines with the C1 complex and is converted by the enzymatic activity of C1 to an active fragment. This activated C4 binds C2, and the C2 molecule is then converted into an activated complex, C4,2, which is also an enzyme. The complement protein C3 is the substrate for the C4,2 complex, and when C3 becomes enzymatically activated, C5 can combine with it. Subsequent reactions with C5, C6, C7, C8, and C9 lead to the completion of the complement sequence of reactions, resulting in a variety of biological effects. The sequence and some of the resulting reactions are summarized in Figure 8.21. As seen in the figure, reactions at the C3 level result in chemotactic attraction of phagocytes to invading agents, and in phagocytosis (opsonization). Reaction at C5 also leads to leukocyte attraction. The terminal series of reactions from C5 through C9 results in cell lysis and death. Lysis itself results from destruction of the integrity of the cell membrane, leading to the formation of holes through which cytoplasm can leak.

The complement system is involved in other immunological reactions besides those leading to the destruction of invading organisms. Allergic and hypersensitivity reactions and certain aspects of the inflammatory response also involve actions of the complement system. C3 and C5 fragments cause release of histamine from leukocytes and from other cells that store this substance, and as we saw earlier in this chapter, histamine increases the permeability of capillaries, enabling leukocytes and fluid to escape into the tissue cells. One serious type of allergic reaction, *anaphylactic shock*, also

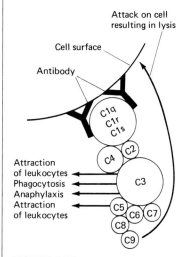

FIGURE 8.21
The complement system and its actions. After antibody has attached to the cell surface, the components of complement are added in sequence, as shown. The components are modified when they are added so that further additions can take place. Addition of C3 leads to chemical changes that lead to the attraction of leukocytes to the cells, and to phagocytosis (opsonization). In certain situations, addition of C3 can result in anaphylaxis, an allergic-type reaction. Addition of C5 through C9 leads to changes in the cell surface that result in leakage of cytoplasmic constituents and lysis. The cell is killed. In the absence of complement, the combination of antibody with the cell surface has only minimal effects on the cells, and killing and phagocytosis do not usually occur.

sometimes involves components of the complement system. We shall discuss allergy and hypersensitivity in Section 8.13.

In addition to the activation of the complement system beginning with C1, which was described above, there is another means of activating C3 that bypasses the C1 complex and C2 and C4. This is the *properdin* system. Properdin is a protein complex in normal blood serum (it may also be formed as a result of antigenic stimulation) that combines with bacterial polysaccharides or certain aggregated immunoglobulins to activate C3 and C5 directly, leading to opsonization and the production of inflammatory reactions. The importance of the properdin system is that it leads to defense against invading organisms without the production of specific antibodies; thus it plays a role in innate (nonspecific) immunity.

An important property of the complement system is that the components are enzymatically altered during reaction, so that they will no longer react in a new sequence of reactions. Complement thus appears to be used up during antibody-antigen reactions. This is called *complement fixation* and occurs whenever an IgG or IgM antibody reacts with antigen in the presence of complement, even if complement is not required in the reaction. Complement fixation is measured by assaying the concentration of complement after an antibody-antigen reaction has occurred, as illustrated in Figure 8.22. After the initial reaction to permit complement fixation, an indicator system is added, consisting of sheep red blood cells and antibody to these cells. In the absence of complement, lysis of the red cells will not take place, but if complement is present, the normal series of reactions leading to cell lysis occurs. Thus, if complement has *not* been fixed, lysis occurs, but if complement *has* been fixed, lysis will not occur. Sheep red cells are used since their lysis is readily observed visually, without a microscope. Appropriate controls must be set up to be sure that nothing in the system is inactivating complement nonspecifically.

By measuring complement fixation, one has a means of determining the occurrence of an antigen-antibody reaction even if precipitation or some other visible reaction has not occurred. Complement fixation is used in several diagnostic procedures, of which the most well known is probably the Wasserman test for syphilis (now no longer widely used).

8.11
Mechanism of antibody formation

How is it possible for higher animals to produce such a large variety of specific proteins, the antibodies, in response to invasion by foreign macromolecules? In considering the mechanism of antibody formation, we must first explore what happens to the antigen. The main sites of antigen localization in the body are the lymph nodes, the spleen, and the liver. It has been well established that antibodies

FIGURE 8.22
Demonstration of complement
fixation.

Stage 1

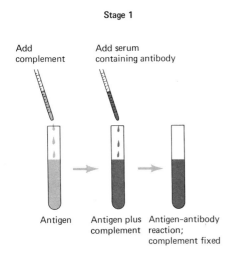

Add
complement

Add serum
containing antibody

Antigen

Antigen plus
complement

Antigen-antibody
reaction;
complement fixed

Add
complement

No antibody

Antigen

Antigen plus
complement

Complement
not fixed

Stage 2

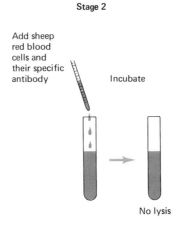

Add sheep
red blood
cells and
their specific
antibody

Incubate

No lysis

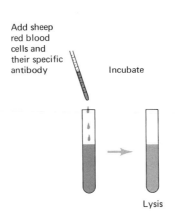

Add sheep
red blood
cells and
their specific
antibody

Incubate

Lysis

are formed in both the spleen and lymph nodes; the liver seems not
to be involved. If the antigen is injected intravenously, the spleen is
the site of greatest antibody formation, whereas subcutaneous,
intradermal, and intraperitoneal injections lead to antibody formation in lymph nodes. Fragments of lymph node or spleen from
immunized animals can continue to produce antibody when placed
in tissue culture or when injected into other, nonimmunized, animals.

The cells that produce immunoglobulins are special types of
lymphocyte cells called *plasma cells*, which are common in spleen and
lymph nodes. Since antibodies are protein molecules, it is assumed
that they are produced by the same biochemical mechanisms that
produce other proteins. This means that the structure of an individual antibody molecule is ultimately determined by a segment of
DNA (a gene). The puzzle arises when we try to explain the role that
the antigen plays in this process. The hypothesis which is best

INFECTION AND IMMUNITY

supported is that cells capable of forming specific antibody arise as a result of random mutation from a population of undifferentiated *stem cells*. Since antibody molecules of differing antigenic affinity have constant and variable regions in their polypeptide chains, it is likely that the variable regions are coded for by genes that are unusually prone to mutation; from the stem cell, a number of distinct cell lines may thus arise, each deriving from a separate mutational event. In the absence of antigenic stimulation, a cell line may persist only at a low population level; then, when an antigen complementary to the antibody produced by that cell line is introduced, the cells are stimulated to divide and a large population of antibody-producing cells develops. According to this theory, the differentiation of an antibody-producing population occurs by mutation of the variable genes, followed by antigen selection. The antigen is merely selective for a given antibody-producing cell type and is not required for the actual production of the antibody protein molecule. Thus, a population of antibody-producing cells can continue to function even in the absence of the antigen. Although in most cases an antigenic stimulus is necessary for development of the antibody-forming cell line, some antibodies can be formed by individuals that have never been in contact with the antigen.

The conversion of lymphocytes into antibody-producing plasma cells is of considerable interest. It is now known that there are two types of lymphocytes involved in immune responses, called B cells and T cells. Both are derived from stem cells in the bone marrow, and their subsequent differentiation is determined by the organ within which they become established (Figure 8.23a). It has

FIGURE 8.23
Role of B and T cell lymphocytes and macrophages in antibody formation. *a* Origin of B cells, T cells, and macrophages from stem cells derived from the bone marrow. *b*. Interaction between B and T cells and macrophages to produce a line of antibody-secreting plasma cells.

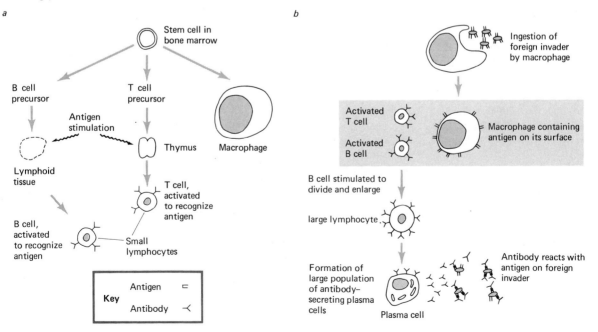

recently been discovered that T lymphocytes are differentiated in the thymus, an organ in mammals whose role has long been a puzzle. T cells have two kinds of functions in immune responses: they act as helper cells to B cells in production of antibody, and they play the prime role in cellular immune responses (delayed hypersentitivity) discussed in Section 8.13. B cells become differentiated into plasma cells and thus become the prime producers of immunoglobulins. In birds, it is known that B cells differentiate in an organ known as the *bursa of Fabricius*. A similar organ has not been found in mammals, and it is thought that B cells differentiate in lymph nodes or in lymphoid structures such as the appendix or tonsils. Some antibodies can be formed by antigenic stimulation of B cells alone, but for many antigens B cells do not form antibodies without the help of T cells. In addition to B and T cells, macrophages (phagocytic cells) are also involved in antibody formation.

How do B cells, T cells, and macrophages act together to produce antibodies? Macrophages act nonspecifically, phagocytizing antigens so that antigenic materials become attached to macrophage cell surfaces where they can interact with the other cell types (Figure 8.23*b*). Macrophages are very sticky and attach well to surfaces; their stickiness probably also promotes the attachment of B and T cells. B and T cells combine with macrophages to form a three-cell complex. T cells become immunologically responsive as a result of antigenic stimulation, and while in the three-cell complex, the T cell probably releases substances that stimulate the B cells to produce antibody. The details of these interactions are not known, but it *is* known that antibody production by B cells is greatly increased if they can interact with T cells and macrophages. It is important to note, however, that B cells can make IgM-type antibodies without interaction with T cells; it is the switch to making the more common IgG-type antibodies, leading to the secondary response (booster response), that requires the participation of T cells and macrophages.

T cells also can act in the absence of B cells to eliminate invading microorganisms and other foreign material by releasing large molecules called *lymphokines* that not only attack the antigen directly but attract macrophages which ingest the invading particle (Figure 8.24). Other affects of lymphokines resemble those previously described as inflammatory responses (Section 8.6). Because conventional antibodies are not involved in these reactions, they are generally classified under the category of *cell-mediated immune reactions*. Cellular immunity differs from antibody-mediated immunity in that the immune response cannot be transferred from animal to animal by antibodies or by serum containing antibodies but can be transfer-

8.12
Cellular immunity

INFECTION AND IMMUNITY

FIGURE 8.24
Cell-mediated immunity.
Antigen stimulation of
previously activated T cells
induces the release of
lymphokines, which have
various actions in attacking the
invading antigen.

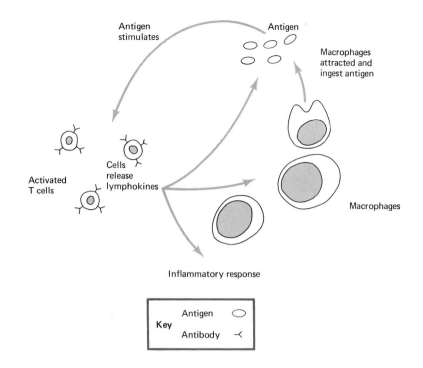

red by lymphocytes removed from the blood. The lymphocytes that function in the transfer of cellular immunity are T cells, which must have been activated by previous antigenic stimulation. Most of the lymphocytes circulating in the blood are T cells; the B cells do not generally circulate but are localized in lymphoid tissues.

Cellular immunity differs from antibody-mediated immunity in that it develops rather slowly, whereas antibody action is a rapid phenomenon, quickly observable after antigenic injection in an immune animal. For this reason cellular immunity is sometimes called *delayed-type immunity*.

The best example of a cellular immune response is the development of immunity to the causal agent of tuberculosis, *Mycobacterium tuberculosis*. This was first discovered by Robert Koch during his classical studies on tuberculosis (Section 1.3) and has been widely studied. Antigens derived from the bacterium, when injected subcutaneously into an animal previously immunized, elicit a characteristic skin reaction that develops only after a period of 24 to 48 hours. (In contrast, skin reactions to antibody-mediated responses, as seen in conventional allergic reactions, develop almost immediately after antigen injection.) T cells in the region of the injected antigen become stimulated by the antigen and release lymphokines that attract large numbers of macrophages. The macrophages are responsible for the ingestion and digestion of the invading antigen. The characteristic skin reaction seen at the site of injection is a result of an inflammatory response arising as a result of the release of lymphokines by the activated T cells. This skin response serves as the basis for the *tuberculin test* for determining prior infection with *M. tuberculosis* (see Section 10.7).

A number of microbial infections elicit cellular immune reactions. In addition to tuberculosis, these include leprosy, brucellosis, and psittacosis (caused by bacteria), mumps (caused by a virus), and coccidioidomycosis, histoplasmosis, and blastomycosis (caused by fungi). In all these cellular immune reactions, characteristic skin reactions are elicited upon injection of antigens derived from the pathogens, and the skin reactions can be used in diagnosis of prior exposure to the pathogen. In addition to pathogens, immunity to tumors, rejection of transplants, and some drug allergies also involve cell-mediated immune responses. We discuss drug allergies in the next section.

Although immunological responses are often beneficial to the body in protecting against infectious disease, some responses are harmful enough to cause severe symptoms or even death. These harmful responses are called hypersensitive reactions. They are also called *allergic reactions*, although not all allergies are due to immune responses.

Hypersensitivity can result either from conventional antibody-antigen reactions or from cell-mediated immune reactions. The former results in an almost immediate response to the antigen, and in some cases it has been shown that the reaction is connected with the formation and action of immunoglobulin IgE (see Table 8.3). This type of hypersensitivity is elicited by many pollen and food antigens. The delayed type of cell-mediated hypersensitivity has been shown to occur with poison ivy and with certain drugs and synthetic chemicals.

The most dramatic example of hypersensitivity is the phenomenon known as *anaphylactic shock*. If an animal that has previously been sensitized with small amounts of a foreign protein is injected a few weeks later with a large amount of the same protein, within a few minutes the animal shows signs of respiratory distress because of contraction of the bronchial muscles, which prevents the exhalation of air. The animal has an acute attack of asthma and usually dies within minutes. Anaphylaxis in humans is signaled by the same acute asthma and is accompanied by flushing and itching, circulatory collapse, acute emphysema, and (on recovery) a rash on the skin. The basis of anaphylactic shock is thought to be as follows: the initial sensitization induces antibody formation, but the antibody remains fixed to cells rather than circulating in the blood stream. When the challenge dose of antigen is injected, the antigen com-

8.13
Hypersensitivity (allergy)

bines with antibody on the cells and induces the release of chemicals such as histamine, that cause the contraction of smooth muscles, leading to the characteristic symptoms. Other chemicals released include serotonin and kinins, which also act on smooth muscle. Antihistamines are drugs that block the action of histamine and thus neutralize some of the effects of anaphylaxis.

Anaphylactic shock is of most concern with drug treatments. Injection of drugs such as penicillin into unusually allergic individuals may cause death by anaphylaxis. Because of the seriousness of this drug reaction, it is extremely important to know whether an individual is allergic to such drugs so that their use can be avoided. Fortunately, only a small percentage of people show these unusual reactions.

A number of other hypersensitivity reactions have a similar mechanism to anaphylaxis but are less severe and dramatic. *Serum sickness* results when a relatively large amount of foreign serum (horse serum most commonly causes this problem) is injected and antibodies form against the foreign-serum proteins. The symptoms are usually a rash, enlargement of lymph glands, and swelling of body tissues, all related to increased permeability of the capillaries, another manifestation of histamine action.

Allergies to drugs, foods, pollen, and spores are quite common. In the case of *drug allergies*, the drug itself is not an antigen but becomes an antigenic determinant when it combines with serum protein. Allergens in foods and pollen are usually proteins or polysaccharides. Allergies to pollen and spores are usually called *hay fever*; they tend to show pronounced seasonal incidence, developing when the particular pollen or spore is being released. Both drug allergy and hay fever are types of hypersensitivity that seem to have hereditary factors. The symptoms usually result from contraction of smooth muscle, and they vary in severity, depending on the individual and the route by which the antigen enters.

Allergies only develop after an initial sensitization has led to the production of antibody. A later experience with the same antigen then leads to an antigen-antibody reaction and hence to the development of allergic symptoms. The sensitivity of an individual to an allergen can often be determined by a skin test. A small amount of extract of the suspected allergen is injected under the skin; an allergic reaction will be seen as a reddening and swelling of the skin, usually within a few minutes. Desensitization to an allergen can sometimes be achieved by carefully injecting gradually larger doses of allergen, resulting in a neutralization of existing antibody or its release from smooth muscle.

Allergic reactions are responsible for some of the symptoms of many microbial infections. Upon initial infection, antigenic components of the microbes induce antibody formation. Subsequent infections then cause allergic-type reactions in the infected tissue or elsewhere in the body. Usually, the organisms that elicit allergic reactions cause chronic rather than acute infections. The most important diseases with allergic manifestations are tuberculosis and rheumatic fever.

The transplantation of tissues or organs from one individual to another is an increasingly important medical practice. However, in many cases, the transplanted tissue or organ is rejected and deteriorates and dies. Immune phenomena are generally responsible for rejection in tissue and organ transplantation; antigens in the transplanted tissue are foreign proteins and induce immunological reactions, leading to the destruction of the transplanted material. If it is essential to the life of a patient to carry out a transplantation, then the immune mechanisms of the patient must be neutralized by use of radiation or immunosuppressive drugs. Of course, this makes the patient unusually susceptible to microbial infection, so that antibiotics must be used as chemoprophylactics to combat possible microbial infection.

Homografts, or transplantation of tissue from one part of the body to another, can of course be performed without any immunological complications, since the body does not react immunologically against itself; hence, skin grafts can be done. Transplantation can also be done between identical twins, since their antigenic makeup is the same. One of the main factors delaying routine organ transplantation is the lack of knowledge of how to control the immune mechanisms responsible for transplant rejection.

Red blood cells contain antigens on their surfaces that can induce antibody formation, and an understanding of blood cell antigens is important for blood transfusions and certain other medical concerns. Since individuals do not form antibodies against their own body substances, the antigenicity of red cells is only a problem during transfusions or other situations in which foreign red cells enter the body. There are at least 12 different red blood cell antigen systems.

8.14
Blood groups

The most important system is that called the *ABO system*; it is the one that most frequently enters into transfusion problems. It is this system that is used most often in blood matching of individuals. The key factors in the ABO system are the two blood group antigens designated as A and B. The antigenic determinants of these two antigens are carbohydrates. The ability of a person to produce these antigens is inherited, and a person can produce antigen A, antigen

TABLE 8.4
Antigens and antibodies of
the ABO system

Blood group	Blood group antigen on red cells	Natural anti-bodies present in serum	Approximate incidence in Caucasian population (%)
A	A	Anti-B	40
B	B	Anti-A	10
AB	Both A and B	Neither	5
O	Neither	Anti-A, Anti-B	45

B, both, or neither. When both antigens are produced, the blood group is designated AB; when neither is produced, the blood group is designated O (Table 8.4).

Antibodies against the A and B antigens will be formed if blood containing these antigens is transfused into an individual lacking them. These antibodies, which are present in the serum, are called anti-A and anti-B. The same antibodies are produced in most people even without transfusion, since the antigenic determinants of A and B are also found in certain common foods that will almost inevitably be eaten. Such antibodies, produced without transfusion or intentional immunization, are called *natural antibodies*. An individual does not produce natural antibodies against his own red cell antigens, of course, but does produce them against red cell antigens he does not possess (see Table 8.4).

The reason for blood grouping, or typing, before blood transfusion is to prevent the blood cell agglutination that occurs if blood containing a particular antigen is transfused into a person containing an antibody against this antigen (Table 8.5). If agglutination occurred, the clumps of cells could lodge in blood vessels or arteries and block the flow of blood, causing serious illness or death. Transfusions are always given using blood that will contain no antibodies against the donor blood.

TABLE 8.5
Agglutination reactions[a]
when bloods of different
groups are mixed

Red cells from blood group	Serum from blood group			
	A	B	AB	O
A	−	+	−	+
B	+	−	−	+
AB	+	+	−	+
O	−	−	−	−

[a] KEY: +, agglutinates; −, does not agglutinate

Because the ability to form the A and B blood group substances is inherited, the frequency of occurrence of the various blood groups can vary in different populations. The approximate incidence of blood groups in Caucasian populations is shown in Table 8.4. In this population, which is of mixed ethnic background, blood groups A and O are the most common. Black African populations are generally high in incidence of B blood group, whereas American Indians completely lack B blood group and are predominantly group O. Frequencies of the ABO blood groups in various populations have been used in studying processes involved in human genetics.

One other blood group system, the Rh system, also has considerable medical importance. There are a number of Rh antigens, but for present purposes these can be simplified to Rh positive and Rh negative. An individual who possesses the Rh antigen is called Rh positive; such a person does not produce antibodies against Rh, since one does not make antibodies against one's own antigens. An Rh-negative individual does not possess the Rh antigen and therefore will produce antibodies against Rh upon coming into contact with it, as in a misdirected transfusion. A serious blood disease (*erythroblastosis fetalis*) can occur in a newborn if it is Rh positive and its mother is Rh negative. Red blood cells from the fetus can pass to the mother and induce her (since she is Rh negative) to produce anti-Rh antibodies. These antibodies pass back into the fetus and react with the fetal Rh-positive cells, destroying them and causing anemia. Because the ability to produce Rh antigen is inherited, the condition arises when an Rh-negative woman conceives by an Rh-positive man. Usually, the first pregnancy is normal, antibody to the child's Rh antigen not being formed in large amounts in the mother. It is the second or subsequent pregnancy which can lead to difficulty, since in this case there is the possibility of a secondary response, which as we have seen leads to considerably higher antibody titers. Not all pregnancies of an Rh-negative woman by an Rh-positive man lead to difficulty, but it is wise to know that the possibility exists so that corrective measures can be taken. One treatment procedure is to replace at birth all the blood of the infant with blood lacking Rh antibody. Rh disease can also be prevented by passive immunization of the mother with anti-Rh serum at delivery. This removes the Rh antigens obtained from her infant; subsequent infants will thus not be harmed by Rh antibodies in the mother.

Blood group is determined by mixing serum containing known antibody with blood of the person to be typed and then observing

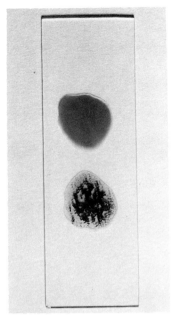

FIGURE 8.25
Procedure for blood typing. The blood used was group B. Anti-A serum was placed in the drop at the top, anti-B in the drop at the bottom. The agglutination of the red cells by anti-B results in the clump.

whether agglutination takes place. ABO typing is usually done on microscope slides (Figure 8.25). Two drops of blood are placed on a slide: to one a drop of anti-A serum is added; to the other, a drop of anti-B serum. After mixing for a few moments, the presence or absence of agglutination is observed. The presence of Rh antigen can be determined similarly with a third drop of blood receiving a drop of anti-Rh serum.

Before using even a blood of the same blood group for transfusion, it is important to cross-match it with that of the recipient to ensure that any unforeseen incompatibilities are absent. Cross-matching is done by mixing a drop of donor serum with a drop of recipient cells and observing whether agglutination occurs.

In emergencies, certain deviations from the rules of matching may be made. Since group O blood lacks both A and B antigens, it should not agglutinate with either anti-A or anti-B (see Table 8.5). A group O individual is sometimes called a *universal donor* because of this. Individuals of group AB lack both anti-A and anti-B antibodies in their serums and hence should not agglutinate blood of either group. Such individuals are sometimes called *universal recipients*. In practice, however, blood of the identical group should be used to avoid any unforeseen difficulties.

Because the ability to produce blood cell antigens is inherited, blood typing may be used to determine whether or not a child could *possibly* be the offspring of a particular set of parents, and hence blood cell typing is used in parentage determinations in courts of law. Although blood typing can never be used to prove *positively* the identity of the parents, in many cases it is possible to state that an individual could *not* be a parent of the child.

8.15
Immunization against disease

Knowledge of immune mechanisms has led to considerable advances in methods for the immunization against disease. Some of the major diseases that affect mankind have been controlled by appropriate immunization procedures. We discussed the history of immunization in Section 1.6, considering Jenner's pioneering work on smallpox and the progress Pasteur made in the preparation of vaccines.

A person may be immunized in either of two ways: by injection of an antigen that induces the formation of antibodies, or by the injection of antibodies themselves. The former is called *active immunity*, since the immunized individual is active in producing it, and the latter is called *passive immunity*. Active immunity, in general, does little to affect a disease in progress but wards against future attacks, whereas passive immunity is used to control a current attack of the disease. We discuss each type of immunization procedure below.

Passive immunity is conferred when a person is given an injection of serum containing antibodies taken from another person. The person donating the serum is usually *hyperimmune*, which means that he has a very high antibody titer. Hyperimmune serums are also taken occasionally from animals and injected into people. For instance, horses are frequently used for the production of serums containing antibodies against diphtheria, botulinum, and tetanus toxins (Figure 8.26). A serum containing antibodies is called an *antiserum*, and an immune serum active against a toxin is called an *antitoxin*. An immune serum is used to treat a person who is already suffering from a given disease and hence is used as a therapeutic agent. The

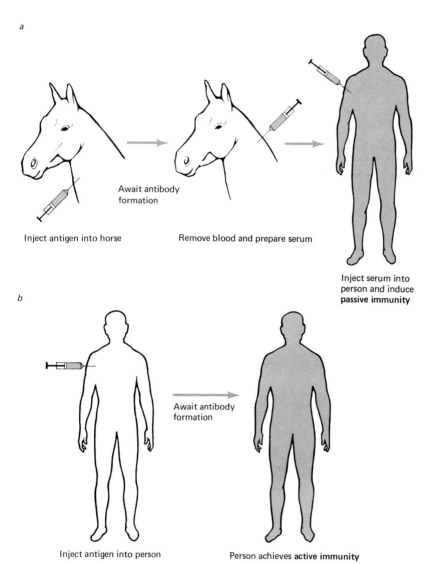

a

Inject antigen into horse

Await antibody formation

Remove blood and prepare serum

Inject serum into person and induce **passive immunity**

b

Inject antigen into person

Await antibody formation

Person achieves **active immunity**

FIGURE 8.26
a Active immunity. *b* Passive immunity.

immunity is called *passive* since the person did not produce the antibodies himself. Long-term immunity does not result, as the injected antibodies gradually disappear from the blood. However, the antibodies present will probably be able to inactivate the disease agent and result in a cure. If a subsequent infection occurs, the person will generally be as sensitive to the disease as an unimmunized individual. Repeated use of immune serum must be avoided, as the individual will begin to develop antibodies against the proteins of the serum and exhibit a reaction against the serum. This type of reaction is sometimes called *serum sickness* (see Section 8.13).

Natural passive immunity develops in the fetus carried within its mother's womb. Antibodies pass across the placenta from mother to child and are present in the newborn infant. These antibodies confer immunity to many infections and help the infant avoid infection during the early months of life. However, these antibodies gradually disappear and are usually gone by the age of 6 months, at which time the susceptibility of the infant to infection increases. Eventually, as a result of immunization or natural infection, the child produces its own antibodies.

ACTIVE IMMUNITY

Active immunity results when an individual is induced to form his own antibodies in response to injection of the appropriate antigen (see Figure 8.26). The material used to induce active immunity is known as a *vaccine* (see Section 1.6). Vaccines consist of harmless products containing antigens from the pathogen or its products. Some vaccines are derived from toxins produced by the pathogen. It is obvious that a toxin itself could not be injected, as this would cause damage to the host; rather, a chemically modified toxin called a *toxoid* is used. The toxoid still contains the appropriate antigenic determinants but is no longer toxic. One of the common ways of converting a toxin into a toxoid is by treating it with formaldehyde. If immunization requires the use of the microbe itself, the organism must obviously be inactivated in some way; usually, it is killed by treatment with heat, formaldehyde, or phenol. Vaccines against whooping cough and typhoid are routinely made from killed cultures of the causal agents. Killed cells are not as effective antigens as live cells, and in some cases, it has been possible to develop an *attenuated* strain of the pathogen, still effective antigenically but no longer able to initiate infection (see Sections 1.6 and 8.4). Vaccines made from attenuated strains are used in immunization against tuberculosis, anthrax, smallpox, polio, and measles.

An important distinction between active and passive immunity is that in active immunity the immunized individual is changed fundamentally in that he will be able to continue to make the antibody in question and will exhibit a secondary or booster response if he later receives another challenge with the antigen. Active immunity may remain throughout life. However, active immunity is *not* used to control a disease in progress but to prevent a future attack. It is hence a *prophylactic* measure. The effects of passive immunity, on the other hand, are more immediate but of shorter duration.

Some common immunization practices are summarized in Table 8.6. As discussed earlier, infants possess antibodies derived from their mothers and are relatively immune to infectious disease during the first 6 months of life. It is desirable to immunize infants for key infectious diseases as soon as possible, so that their own active immunity can replace the passive immunity received from the mother. However, infants have a rather poorly developed ability to form antibodies, so that immunization is not begun until a few months after birth. As noted in Figure 8.16, a single injection of antigen does not lead to a high antibody titer; it is therefore desirable to use a series of injections, so that a high titer of antibody is developed. The immunization schedule outlined in Table 8.6 provides for this spacing, as well as for periodic booster injections throughout life.

What happens inside the body during the time period that the injections are given? As noted in Section 8.11, antigen stimulation leads to the production of a line of activated lymphocytes, and once this line of cells has been produced, continual cell division will maintain a small but significant population of these cells for considerable periods of time (usually several years). Periodic booster injections will stimulate further division of the activated lymphocytes, thus making it possible to maintain an active population of cells throughout adult life. A natural infection will also result in a rapid booster response, leading both to a further increase in activated cells and to production of antibodies that will attack the invading pathogen. If booster injections or natural infections do not take place, eventually the population of activated lymphocytes will disappear, and susceptibility to infection will return. It is not really known how long immunity lasts in the absence of antigenic stimulation; the length probably varies considerably from person to person.

Immunization procedures such as listed in Table 8.6 are not only beneficial to the individual but are also effective public-health procedures, since disease spreads poorly through an immunized population (see Section 9.6). It is of some interest that the disease smallpox has been essentially eradicated from the world as the result of widespread and persistent immunization practices. Only in a few small areas of the world is smallpox still present. Because the procedure itself is not completely without risk, smallpox immunization has been abandoned in current medical practice in the United States.

INFECTION AND IMMUNITY

TABLE 8.6	Age	Immunization	Comments
Some recommended immunization procedures for infants and children[a]	During the first year of life	Diphtheria, pertussis, tetanus (DPT)	Given at 2, 4, and 6 months
		Oral polio (OP)	Given at 2, 4, and 6 months
	During the second year	Mumps, measles, rubella (MMR)	Given at 15–16 months
		DPT booster	Given at 18 months
		OP booster	Given at 18 months
	At 4 years or school entry	DPT booster, OP booster	May be given at 3 years on entering nursery school
	Between 10 and 13 years	Tuberculosis (BCG)	Not done in many countries
	At 15–19 years or on leaving school	DT booster	Given at 10-year intervals; pertussis omitted

[a] These are recommended practices in some areas. Actual practice may vary in the details. Smallpox vaccination is no longer recommended because the disease is virtually eradicated worldwide.

However, it is essential to immunize for smallpox if travel is intended to an area of the world where there is any possibility of contracting the disease.

IMMUNIZATION OF TRAVELERS

In Western Europe and North America, certain diseases are so uncommon that immunization against them is inadvisable. However, these diseases are still common in many other parts of the world, and in these days of quick and frequent travel, it is desirable for travelers to be immunized against diseases that are prevalent in the parts of the world to be visited. Table 8.7 summarizes immunization practices that are advised for travelers. The effectiveness of these immunization procedures varies, so that immunization does not automatically confer resistance. Further, some of these immunization procedures can cause side effects (sore arms, mild fever), so they should not be used unless necessary.

In some countries, entry may not be permitted unless immunization for certain diseases has been carried out. Air carriers generally control these situations by not permitting unimmunized persons on planes destined for locations where they will not be admitted. However, it is highly desirable to check ahead of time, either with the air carrier or the country, to be certain that all prescribed immunization procedures have been carried out. Otherwise, considerable inconvenience may be encountered during travel to distant lands.

Agent	Schedule	Destination of traveler
Cholera	Three doses over 6 months	India, Middle East, Southeast Asia
Plague	Three doses over 12 weeks	Areas where plague is occurring and rats are infected (parts of South America, Africa, Asia)
Smallpox	Every 3 years	Areas where smallpox is endemic[a]
Tuberculosis	Two months before exposure	Areas where disease is prevalent (children only)
Typhoid	Two doses 4 weeks apart	Most developing countries especially tropical
Typhus	Two doses 4 weeks apart	Remote areas of South America, Africa, and Asia where epidemics exist
Yellow fever	Every 10 years	South and Central America, Africa

TABLE 8.7
Immunizations for travelers

[a] Immunization schedule currently undergoing revision, as smallpox has been virtually eradicated worldwide (see Section 9.7).

8.16
Antibiotics and other drugs in the treatment of infectious disease

The development, chemical nature, and assay of antibiotics was discussed in Section 4.6, and the mode of action of antibiotics was discussed in Section 6.8. Antibiotics have had a major impact on the management of patients with infectious disease. Currently, a wide variety of antibiotic agents are available, a result of the intensive efforts in research and development over the past 30 years. Very effective agents are available for virtually all bacterial diseases, and a few agents effective against fungi and protozoa are also known. So far, control of viral infections has not been possible with antibiotics, although several chemicals have been discovered that inhibit the action of certain viruses (see Section 12.2). We shall discuss in Chapter 10 the antibiotics that are used in the therapy of specific bacterial diseases. Here we discuss a few general principles on the use of these agents in the treatment of infectious diseases. The use of antibiotics or other chemicals in the treatment of infectious diseases is called *chemotherapy*.

No one drug is effective against all pathogens, so that care must be taken to select the appropriate drug. Extensive studies are made with each new antibiotic before it is used in medicine, and these studies provide information needed to decide whether a particular antibiotic should be used for treating a particular disease. However, since pathogens may develop resistance to antimicrobial agents, it is essential to test the antibiotic sensitivity of the pathogen isolated from each patient. We have already described the methods by which antibiotic sensitivity is determined in Section 4.6.

INFECTION AND IMMUNITY

In many cases, treatment of an infectious disease requires more than the use of drugs. A drug will not affect toxin that has been produced by the organism, and since symptoms are usually due to the action of toxins, treatment with a drug alone will not be enough. Neutralization of toxin by use of passive immunization procedures is often required. Relief of symptoms can often be effected by surgical procedures such as cutting and removal of pus, by forcing fluids, or by injecting fluids into the bloodstream. It should definitely not be assumed that a drug must always be used to cure an infection. Drugs themselves have toxic side effects, and if a cure can be effected without drug treatment, that is preferable. In very few infections is the drug alone responsible for a cure. Most of the defense and immunity mechanisms described earlier in this chapter are essential to bring about a cure, even when a highly effective drug is used. Many drugs do not kill pathogens but only prevent their growth; it is the host's defenses that eliminate the pathogens from the body.

Many chemicals are inhibitory to pathogens in culture but are ineffective against the same pathogen in an infected host. The reasons for this difference vary. Sometimes the chemical is destroyed or inactivated by the host. In other cases, the chemical is poorly absorbed or is excreted too rapidly; only if a sufficiently high concentration can be maintained at the site of infection can a cure be effected. In some cases, the pathogen in the host is alive at some site of the body where the drug cannot penetrate; intracellular parasites, such as *Mycobacterium tuberculosis*, are especially resistant to action by many antibiotics.

In addition to use in curing diseases, drugs are sometimes used to prevent future infections in people who may be unusually susceptible to them. Such use is called *chemoprophylaxis*. The best example of chemoprophylaxis is the use of penicillin to prevent streptococcal sore throats in rheumatic fever patients, since these streptococcal infections often lead to a recurrence of rheumatic fever symptoms. Sometimes antibiotics are given to patients who have just undergone surgery, since the resistance of postoperative surgical patients to infectious disease is often low and the inoculum of foreign organisms during surgery may be high. However, such use of antibiotics should never supplant well-performed aseptic surgery.

Although modern medicine could not operate without antibiotics and other chemotherapeutic agents, these drugs also have certain disadvantages: (1) Some of them have serious toxic side effects or cause allergic reactions. Patients have actually died from the side effects of penicillin and chloramphenicol. (2) The pathogen may become drug resistant and may then be passed on to other people. New drugs must then be sought that will act on this newly resistant pathogen. (3) The drug may destroy or reduce the normal flora of the body. As we saw earlier, the normal flora plays important beneficial roles. (4) Drug treatment may permit *superinfection*, a condition in which a natural pathogen, usually held in check by the normal flora, is able to flourish once the normal flora is eliminated. (5) The ability of the body to develop immunity to a pathogen may be reduced if the pathogen is rapidly eliminated by drug treatment. Thus, if a later infection with the same organism occurs, drug treatment has to be repeated. In such cases, if the original infection is minor, it may be better to let it run its course so that immunity can build up, thus preventing subsequent reinfection.

Despite the obvious difficulties in using antimicrobial agents, it should be emphasized that they are in general highly effective and have played an enormous role in the elimination of infectious diseases as major causes of death. Before the availability of effective antimicrobial agents, diseases such as tuberculosis, pneumonia, typhoid fever, meningitis, and syphilis were leading causes of death in humans. Today these diseases are virtually absent from human populations where antimicrobial therapy is available. The practice of medicine has been completely changed since the widespread availability of antibiotics and other antimicrobial agents (see Figure 1.1).

The most important role of the microbiologist in medicine is in the isolation and identification of the causal agents of infectious disease.

8.17
Clinical microbiology

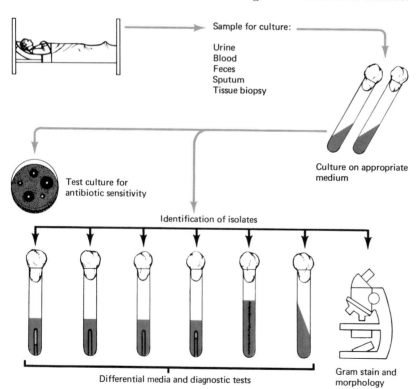

Sample for culture:

Urine
Blood
Feces
Sputum
Tissue biopsy

Culture on appropriate medium

Test culture for antibiotic sensitivity

Identification of isolates

Differential media and diagnostic tests

Gram stain and morphology

FIGURE 8.27
Flow sheet of diagnostic procedure for isolation and identification of a human pathogen.

INFECTION AND IMMUNITY

a

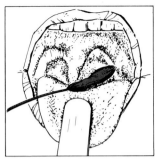

b

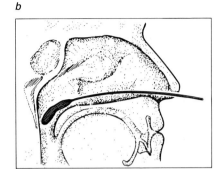

c

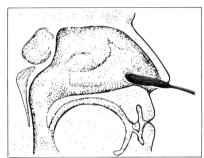

FIGURE 8.28
Methods for obtaining specimens from the upper respiratory tract. *a* Throat swab. *b.* Nasopharyngeal swab passed through the nose. *c.* Swabbing anterior nares.

This is a major area of microbiology called clinical microbiology and has greatly expanded in recent years because of the increasing awareness of the importance of identification of the pathogen for proper treatment of the disease. The days when the physician can substitute antibiotics for diagnosis are over; the microbiologist is a major force in ensuring proper diagnosis.

The physician, on the basis of careful examination of the patient, may decide that an infectious disease is present. Samples of infected tissues or fluids are then collected for microbiological analysis (Figure 8.27). Depending on the kind of infection, materials collected may include blood, urine, feces, sputum, cerebrospinal fluid, or pus. A sterile swab may be passed across a suspected infected area (Figure 8.28). Small bits of living tissue may be removed (a procedure known as biopsy). The sample must be carefully taken under aseptic conditions so that contamination is avoided. Once taken, the sample is analyzed as soon as possible. If it cannot be analyzed immediately, it is usually refrigerated to slow down deterioration.

Collection of urine for microbiological analysis requires special precautions to ensure that the specimen is not contaminated with organisms from the urethral opening. Urine samples can be collected by insertion of a sterile rubber tube up the urinary canal into the bladder (*catheterization*), but this procedure in itself presents some dangers of introducing infection and is also often painful. An alternative is to collect an uncontaminated sample voided by the patient. In the male, the penis is cleansed by repeated washing with soap and water, followed by sponging with an antiseptic solution. The first portion of urine voided is discarded, and the next portion (so-called midstream urine) is collected directly into a sterile tube or bottle. In the female, the vulva is cleansed three or four times with soapy water, always wiping back toward the anus. The container is covered with a sterile cap as soon as the sample is collected and is transported quickly to the laboratory. When the sample cannot be analyzed immediately, it must be refrigerated.

Blood for microbiological analysis is usually taken from a convenient vein with a sterile hypodermic needle and syringe. The site of puncture is first disinfected so that bacteria of the skin will not be carried into the bloodstream as the puncture is made. The blood is transferred immediately from the syringe into appropriate sterile media, which must also contain a substance such as sodium polyanethol sulfonate (SPS) which functions as a stabilizing agent and anticoagulant. Blood is normally sterile; hence any indication of the presence of bacteria is presumptive evidence for infection. For this reason, samples must be taken with full precautions to avoid contamination, since contaminants would mislead the diagnostician.

Feces for microbiological analysis are usually not collected aseptically. The patient voids into a bed pan, and a sample of feces is transferred to a sterile container for transport to the laboratory.

The removal of *spinal fluid* for diagnostic studies requires strict aseptic conditions, not only to avoid contaminating the sample but, more important, to avoid introducing contaminating bacteria into the spinal column, since this region is very easily infected. Meningitis, which is an infection of the membranes covering the spinal cord and brain, often leads to the appearance of infecting bacteria in the spinal fluid. Since the spinal fluid is normally sterile, the presence of bacteria is almost always an indication of infection.

The causal agent of pulmonary tuberculosis (*Mycobacterium tuberculosis*) grows in the lungs. Organisms are carried with sputum into the throat, and *sputum cultures* are often prepared for diagnosis of tuberculosis. The patient is requested to cough deeply, and the sputum is deposited in a sterile conical glass tube. The specimen should be examined to determine if the secretions are indeed sputum rather than saliva. Sputum, being heavier, will settle under saliva in the tube. Since some sputum is swallowed, the organism can also often be found in stomach fluids, and for some diagnostic purposes these are preferable to sputum, since they are less likely to be contaminated with normal organisms from the respiratory tract. *Stomach fluid* is obtained by inserting a rubber or plastic tube into the stomach via the mouth or nose and removing a sample by suction. The region surrounding the lungs is called the *pleura*, and *pleural fluid* may occasionally be removed for diagnosis. This is done by entering the pleura with a long, specially designed needle, through an incision in the patient's back. Although the sputum, stomach, and pleural fluids are most commonly examined for *M. tuberculosis*, diagnosis for other organisms infecting the lungs may also be done.

Many patients currently are given antibiotics as soon as infection is suspected. However, if microbiological analysis is to be carried out, samples should be taken before antibiotic therapy is begun so that growth of the pathogen is not prevented and the organism's viability is not reduced by antibiotic present in the sample.

Once the sample is collected, it must be analyzed to determine the numbers and kinds of organisms present. For certain diseases, it is of value to make a *direct microscopic examination* of the sample. This procedure is useful: (1) when the causal agent has distinctive structure or staining properties that make it quite different microscopically from the normal flora of the body; (2) when it is present in relatively large numbers; or (3) when it is present in a part of the body that is normally sterile.

Cultural studies are very important in diagnosis but must be performed and interpreted carefully. Successful culture requires selection of the appropriate culture medium and conditions of growth. Rich culture media containing blood are often used, as they provide relatively nonselective conditions in which a variety of pathogens will grow. If positive cultures are obtained from normally sterile fluids such as blood and spinal fluid, this usually indicates infectious disease; but cultures from areas of the body with an extensive normal flora, such as the skin, throat, or intestine, will always yield organisms even in a healthy person. The microbiologist must be able to distinguish members of the normal flora from potential disease agents. In many cases, *selective* and *differential* media, as described in Section 3.6, are used to culture specific pathogens suspected to be present. Once the organism is isolated, it may be obtained in pure culture and characterized taxonomically by means of morphological, physiological, and other more specialized tests. One of the most frequently performed procedures in clinical microbiology is determination of the antibiotic sensitivity of a culture (described in Section 4.6). Knowledge of the culture's antibiotic sensitivity permits the physician to select the proper antibiotic for treatment of the infectious disease.

One note of caution is necessary regarding the collection and handling of samples, and of the cultures derived from them. Any sample or culture is potentially infectious, presenting a certain danger for the clinical microbiologist. The fact that a diagnostic procedure is being conducted implies that the physician has some reason to suspect that a pathogen might be present. Consequently, great care must be taken in handling samples and cultures to ensure that laboratory and hospital personnel are not infected. Once samples have been processed, they should be sterilized before discard. Cultures present the greatest hazards, since the number of microbial cells has been built up to a high level, thus increasing the possibility of accidental infection. Continuous attention to all aspects of aseptic technique, as outlined in Section 3.8, is essential in all phases of clinical microbiology.

Summary

In this chapter, we have learned about the growth of microorganisms in both the healthy and the diseased body. Harmless microorganisms live in many parts of the body and are often beneficial to it; they are the normal flora. Specific microorganisms cause specific diseases. The proof that a microorganism causes a disease makes use of Koch's postulates, which place great emphasis on the isolation and culture of the pathogen. A variety of microbiological methods that permit culture and study of pathogens are available, and these are frequently used in diagnosis of infectious disease. The steps in the sequence of microbial infection and disease are: (1) passage of the parasite to the host, (2) invasion of the body, (3) growth of the parasite in the body, (4) production of disease symptoms, and (5) dispersal of the parasite to other hosts. Invasion of the host requires penetration through surface barriers such as the skin and mucous membranes and growth of the parasite in the body. Some microorganisms produce enzymes or other substances that play major roles in invasion. Harm to the host generally occurs as the result of the action of toxins produced by the pathogen. Exotoxins are excreted or released from the pathogen, whereas endotoxins generally remain bound to the cells and are released only at the time of death and disintegration of the pathogen. Virulence refers to the relative ability of a parasite to cause disease and is determined both by the invasiveness of the organism and by its toxigenicity.

The host resists infection by a variety of mechanisms, which can generally be divided into nonspecific and specific defense systems. Non-specific immunity consists of various properties and processes in the host that act indiscriminately against many or all pathogens. These include phagocytosis, host antimicrobial substances, and inflammatory responses. Specific immunity develops as a result of the host's production of antibodies, which then combine with antigens of the pathogen. Antigen–antibody reactions eliminate pathogens in a variety of ways, including agglutination, lysis, death, and opsonization. Individuals can be immunized against disease either by injecting into them antibodies produced by another individual (passive immunity) or by injecting antigens, which will induce individuals to produce their own antibodies (active immunity). Passive immunity is used in the treatment of a disease in progress, whereas active immunity is used as a prophylactic measure to prevent a future case of the disease. Although antibody formation is in general beneficial, under some conditions it can be harmful, resulting in allergic reactions. Antibodies also may cause difficulties in blood transfusions, and appropriate matching of the blood of donors and recipients is necessary. Antibody formation is responsible for rejection in tissue and organ transplantation.

The treatment of infectious diseases can often be carried out by use of antibiotics and other antimicrobial agents to which the pathogen is sensitive. To select the proper antimicrobial agent, a sensitivity test may be made on a culture isolated from the infected person. The availability of effective antimicrobial agents has virtually eliminated infectious diseases as causes of death in humans and has greatly altered the practice of medicine.

Study questions

1 An organism may infect a host and yet not cause disease. Explain.
2 In what regions of the body are the normal flora found? What areas of the body are generally sterile?
3 Explain in detail how Koch's postulates should be applied to determine

INFECTION AND IMMUNITY

the cause of an epidemic occurring, for example, in the mouse colony of a research laboratory.

4 What is the sequence of events occurring during infection and disease?

5 What defense mechanisms does the host have to counteract microbial invasion?

6 How does plasma differ from serum? from whole blood?

7 What are leukocytes? List three different types studied in this chapter.

8 What properties of the parasite enhance its virulence? Give several examples.

9 Immunity may be innate or acquired. Explain.

10 A host's nonspecific immune reactions include phagocytosis, production of antimicrobial substances, and inflammation. Describe each, showing how the host benefits.

11 Define: *specific immunity, antigen, antibody, immunoglobulin.*

12 What is the general structure of immunoglobulin? How do the structures of IgG and IgM differ?

13 Where are the antigen-combining sites on immunoglobulins? Do all immunoglobulins have the same number of antigen-combining sites?

14 What specific roles do IgA and IgE have in immunological reactions?

15 How do immunoglobulins interact with antigens to cause precipitation?

16 How will the antibody titer change after one injection of antigen? after a second injection?

17 Two ways in which bacterial cells are affected by antibodies are agglutination and immobilization. Describe briefly how each of these two types of reactions is brought about.

18 What are B cells, T cells? What is their origin, and how do they become differentiated in the body?

19 How do B and T cells interact in the formation of antibodies?

20 What roles do phagocytic cells (macrophages) have in antibody formation?

21 What is complement? How do the complement components interact with immunoglobulin and bacterial cells to cause death of the bacteria?

22 What is opsonization? What factors are necessary for it to take place? Why is it important in immunity to disease?

23 How do cell-mediated immune responses differ from conventional antibody-mediated immune responses in respect to: time for reaction to occur; body components involved; manner by which transfer to another individual can be effected?

24 What is meant by complement fixation? How is it used in diagnostic laboratories?

25 Compare and contrast passive immunity and active immunity.

26 Discuss allergy in immunological terms: symptoms, causes, mechanism.

27 Why is a homograft (transplantation of tissue from one part of the body to another) more likely to be successful than is transplantation of tissue from one person to another?

28 Why is a person with blood group O called a universal donor, while a person with blood group AB is a universal recipient? Explain.

29 Why is the Rh factor of importance to a fetus only if the mother is Rh negative and the father is Rh positive?

30 Why is careful blood typing so important before a blood transfusion is administered?

31 There are obvious advantages to the use of drugs for the treatment of disease, but there may be certain disadvantages as well. Explain, giving examples.

32 Collection of samples for the diagnostic laboratory requires extreme care

and strict aseptic technique. Why is this important? Describe how asepsis is maintained during collection of blood and spinal fluid.

33 Under what circumstances is the direct microscopic examination of a clinical sample useful? When would cultural studies be required instead for the identification of a disease organism?

Suggested readings

Cooper, M. D., and A. R. Lawton, "The development of the immune system," *Sci. Am.*, Nov., 1974, pp. 59–72. Discusses the nature, origin, and action of B and T cells.

Davis, B. D., R. Dulbecco, H. N. Eisen, H. S. Ginsberg, and W. B. Wood, *Microbiology, including immunology and molecular genetics*, 2nd ed., Harper & Row, New York, 1973. Excellent detailed treatment of the principles of infection and immunity. The section on immunology is especially recommended as a reference source.

Jerne, N. K., "The immune system," *Sci. Am.*, July, 1973, pp. 52–60. Immunoglobulin structure and ideas on how antibodies of differing specificities are formed.

Lennette, E. H., E. H. Spaulding, and J. H. Truant, eds., *Manual of clinical microbiology*, 2nd ed., American Society for Microbiology, Washington, D.C., 1974. Step-by-step descriptions of methods used for the culture and identification of pathogens from clinical materials.

Mayer, M. M., "The complement system," *Sci. Am.*, Nov., 1973, pp. 54–66. Explains in a clear manner the complexities of complement action.

The spread of disease

Chapter

9

In the previous chapter, we have discussed the processes involved in the infection of an individual, and in this chapter, we consider the prevalence and control of infectious diseases in populations. If an infectious agent is to maintain itself, it must be transmitted from the infected host to a susceptible host. Obviously, if the host dies before the parasite is transmitted, the parasite itself will very likely die. Thus, all parasites have evolved mechanisms to ensure their transmission from host to host. A knowledge of how parasites are transmitted helps us to control the spread of the parasite and thus to reduce disease.

The whole field of public health and preventive medicine is based on knowledge of the mechanisms of parasite transmission, and great success has been achieved in reducing or even completely eliminating many infectious diseases. Equally important has been the development of our understanding of the spread of infectious disease in hospitals, where there are of course many infectious individuals and where many other patients are unusually susceptible to infection. Thus an important part of the chapter is a discussion of many aspects of hospital sanitation.

9.1
Stages in the infection process

Let us review briefly the stages in the process by which infectious disease is initiated, as discussed in Section 8.4. The manner by which the pathogen is established and initiates infection, the locus in the body where infection takes place, and the immune responses of the body all influence markedly the extent and manner of spread of a pathogen through a population.

First, the pathogen must invade and grow in the body. The major sites of invasion are the skin and mucous membranes of the respiratory and the gastrointestinal tracts. Very few if any organisms are able to penetrate the unbroken skin. Wounds, especially large, dirty ones, are common sites of infection. Penetration through mucous membranes requires, first, that the pathogen be able to attach to the cells of the mucous membrane and start to grow; growth at the initial site of attachment is usually essential for invasion.

STEPS IN INFECTION

Second, the pathogen must be able to overcome host defense mechanisms, both specific and nonspecific. Nonspecific mechanisms include phagocytosis and the action of various chemical substances produced by the host. Specific mechanisms constitute the immune systems of the body, which attack invaders via antibodies and cell-mediated immunity. Pathogens affect hosts most commonly through the production of toxins, both exotoxins and endotoxins. Some host damage can also occur as a result of allergic reactions of the host to the invading microbe.

The *incubation period* is the time between the initiation of the infection and the first appearance of disease symptoms. Some diseases have short incubation periods; others have long incubation periods. From the point of view of spread through a population, the length of the incubation period is of considerable significance, because during this period the individual is infectious but is not aware of it and hence may infect numbers of other hosts.

EXIT FROM THE HOST

To spread through the population, the pathogen must leave the body and infect others. Only in this way is the pathogen able to maintain itself in the population. If it cannot leave the body in which it starts an infection, then of course it dies out and is of no significance in health and disease. All pathogens must have mechanisms for leaving the body and surviving the journey to other hosts. The pathogen can only spread in a significant manner from an infected host if it has grown sufficiently in that host to produce a large excess population of virulent organisms. The most common routes of exit from the body are the respiratory and gastrointestinal tracts. Organisms living in the respiratory tract exit as a result of coughing or sneezing, which expel droplets containing mucus particles within which the pathogen is entrapped. Many pathogenic organisms are excreted with the feces; this includes both those that infect primarily the gastrointestinal tract and others that infect interior organs of the body and spread via the blood stream to the intestines. Another important route of exit, used by just a few pathogens, is the genital tract, the organisms being emitted in secretions from the penis or vagina. Diseases in which the agents are transmitted in this manner are usually called *venereal diseases*. Organisms causing wound infections are generally transmitted by contact with others, either directly by touching or indirectly via infected objects. Finally, a number of pathogens are transmitted by the bites of animals. Insects and other arthropods are the most important animals transmitting pathogens by biting, but dogs, rodents, bats, and other mammals can also transmit pathogens.

9.2
Dispersal

FIGURE 9.1
Mechanisms of dispersal.

How does the parasite spread from one person to another? The mechanism of dispersal depends first on where in the body the parasite is living. Many parasites are dispersed through air, water, and food, via inanimate objects, or by animals, whereas some parasites are only dispersed by direct contact between people (Figure 9.1).

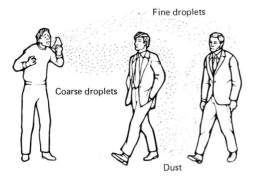

Fine droplets

Coarse droplets

Dust

Dispersal through air and dust

Dispersal by direct contact

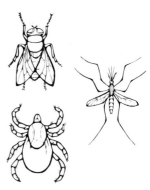

Dispersal by insects, mites, and ticks

Transmission of animal disease to human (zoonoses)

Dispersal by food

Dispersal by objects

Sewage

Stream

Dispersal by water

It is essential that the pathogen be able to maintain viability during the time that it is outside the body. The exterior environment is much more rigorous than the host in many ways, and pathogens can usually survive in it only for fairly restricted periods of time. Air especially is a harsh environment, and pathogens transmitted through this medium must be able to resist drying and the bactericidal action of bright sunlight if they are to be viable when they reach susceptible hosts. Water is somewhat more favorable as a dispersal agent, but the pathogen must contend with the lack of suitable nutrients for growth and with the presence of protozoa and other microbes that may eat it. In general, transmittal can be over greater distances via water than by air, although only a few pathogens are transmitted strictly by the water route. Some pathogens are so sensitive to the environment outside the host that they cannot survive even for a few minutes. Those pathogens transmitted by the genital tract are usually in this category, and it is for this reason that they are transmitted only by direct intimate contact between the two hosts so that the pathogen is never exposed to the outside world.

Not all pathogens are restricted to living only in the host. A few pathogens grow primarily in soil or water and only incidentally infect hosts under special conditions. Diseases in this category are tetanus and gas gangrene, in which the causal agents live primarily in soil. Of course, in such situations, the pathogen need not be transmitted to another host in order to maintain itself. These agents are really only incidental pathogens, causing diseases almost by accident. The organism that causes the kind of food poisoning known as botulism is even more restricted, as it does not grow in the body at all, and hence is not even an infectious agent. It spends its life in the soil and can become inoculated into foods where, under unusual circumstances, it can grow and produce toxin that is harmful if ingested.

Many parasites spread through the air, usually attached to some sort of particle. A wide variety of microorganisms are found in the air but usually only in low numbers. The organisms found in the air outdoors are primarily harmless soil organisms, but indoors the organisms may be those commonly found in the human respiratory tract. During coughing and sneezing, or even talking, numerous droplets of moisture are expelled from the mouth. During a sneeze (Figure 9.2), droplets laden with bacteria leave the mouth at a speed of over 200 miles per hour; the number of bacteria in a single sneeze may be as many as 10,000 to 100,000. In the air, the moisture in these droplets evaporates, leaving behind small particles of mucus to which the bacterial cells are attached. The bacteria in these lightweight particles come from the respiratory tract. Such bacteria are relatively resistant to drying and can remain alive for long periods of

AIR DISPERSAL

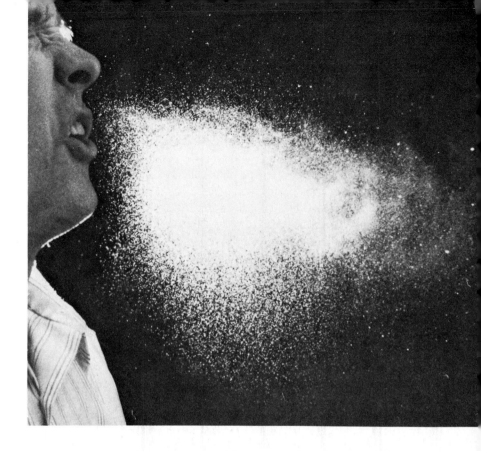

FIGURE 9.2
High-speed photograph of an unstifled sneeze.

time attached to mucus or dust particles. Thus infection can arise from inhalation of contaminated dust; this has often been a serious source of infection in hospitals. However, infection by direct transmission from one person to another is more common. This sort of direct transmission is more frequent when people are crowded, and this is one explanation for increased respiratory infections in winter, when people remain indoors and have more direct contact with each other.

It is often useful to have some idea of the approximate number of microorganisms in the air in different locations; this can be determined in a number of ways, with varying degrees of accuracy. The simplest and least expensive way is the *settle-plate method*, in which petri plates containing a suitable nutrient agar are exposed in areas of interest and the numbers of colonies appearing after incubation are counted. In highly contaminated or very windy areas, exposure for a few minutes may be sufficient to result in the development of many colonies, whereas in clean or quiet areas, exposure times of an hour or more may be necessary. Since heavier particles settle faster than light ones, this method is biased toward large microorganisms or those attached to particles of dust or soil. The method also does not measure the concentration of microorganisms per unit volume of air, since no definite volume of air is sampled. In spite of these limitations, the method may be very useful in locations such as hospitals, where an idea of the effect of hospital practices on airborne spread of pathogens may be desired.

More sophisticated air-sampling methods, used in research work or in special practical applications such as studies on microbial aerosols, involve *impaction samplers*. Known volumes of air are pumped through devices containing petri plates of nutrient agar; the equipment is so constructed that the microorganisms in the air sample are collected on the agar surface. The plates are then incubated and the colonies counted. With a properly designed impaction sampler, the volume of air sampled is known, and all the microorganisms in the air are taken out for counting. Thus more precise estimates of airborne microorganisms are obtained, although at considerably greater expense than with the settle-plate method.

The average human breathes several million cubic feet of air in a lifetime, much of it containing microbe-laden dust. The speed at which air moves through the respiratory tract varies; it is fast in the upper respiratory tract and quite slow in the lower respiratory tract. As the air slows down, particles in it stop moving and settle, the larger particles first and the smaller ones later, as shown in Figure 9.3. In the tiny bronchioles, only particles below 3 μm can penetrate. As also shown in Figure 9.3, different organisms reach different levels in the tract, thus accounting in part for the differences in the kinds of infections that occur in the upper and lower respiratory tract.

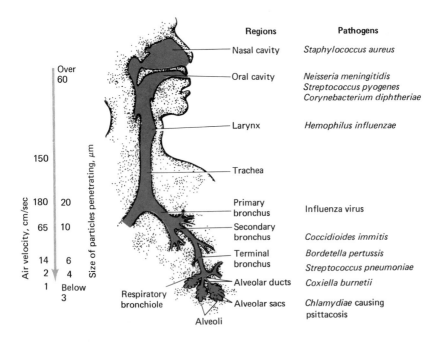

FIGURE 9.3
Characteristics of the human respiratory system and locations at which various organisms generally initiate infections. Modified from R. I. Mitchell, *Am. Rev. Resp. Dis.*, **82**, 630, 1960; and P. K. C. Austwick in M. F. Madelin, ed., *The fungus spore* Butterworth, London, 1966; and R. L. Dimmick, A. B. Akers, R. J. Heckly, and H. Wolochow, *An introduction to aerobiology*, Wiley-Interscience, New York, 1969.

THE SPREAD OF DISEASE

The kinds of infections in which the causal organism is transmitted through the air include diphtheria, pneumonia, meningitis, psittacosis, the common cold, influenza, streptococcal sore throat, histoplasmosis, and tuberculosis. In all these diseases, the pathogen lives in the respiratory tract and is expelled into the air by the host during the period of infection. Many of these diseases will be discussed in Chapters 10, 11, and 12.

To reduce the danger of spreading infectious organisms by coughing or sneezing, a face mask can cover the nose and mouth. Such masks are worn routinely by surgeons and nurses in operating rooms. In some parts of the world, face masks are even worn by lay individuals suffering from respiratory infections when they must go about in places where they might be a source of infection to others. The mask, which acts as an air filter, should consist of at least six layers of gauze but should not offer unusual resistance to air movement, as this would cause the wearer to rebreathe expired air. The mask should fit the face closely to prevent the escape of air around the sides.

In certain respiratory infections, such as whooping cough, the patient will cough violently, resulting in the spread into the environment of large numbers of infectious organisms. Face masks are virtually impossible to use in such cases, and the spread of the pathogen is best prevented by isolating the patient.

WATER AND FOOD DISPERSAL

Many harmless microorganisms are found in lakes, rivers, and oceans. In water polluted by sewage, however, pathogens are frequently also present. Pathogens found in water usually come from feces of individuals with intestinal infections. These pathogens live and grow in the intestinal tract and are often passed out of the body in large numbers. The methods of water testing and of water treatment and purification are designed to prevent the spread of these organisms and are described in Chapter 14. Pathogens transmitted by polluted water include the causal agents of typhoid fever (*Salmonella typhi*), cholera (*Vibrio cholerae*), amoebic dysentery (*Entamoeba histolytica*), and poliomyelitis (polio virus). Intestinal pathogens are also transmitted in foods that have become contaminated by infected food handlers. Chronically infected individuals who continuously release pathogens are called *carriers* and are the most serious sources of contamination of foods. Pathogenic microorganisms in foods are discussed further in Section 15.2.

DIRECT CONTACT

Some pathogens die quickly outside the body because they are very sensitive to drying, light, or other environmental factors. These pathogens are usually transmitted only by direct contact between people, as during sexual intercourse or kissing. When the causal agent is transmitted via the genital organs, the disease is called a *venereal disease*. Examples are syphilis and gonorrhea. Nonvenereal diseases in which the causal agent is transmitted by direct contact include pimples and boils, caused usually by *Staphylococcus aureus*.

Such things as pencils, toys, clothing, dishes, tables, or bedding can become contaminated by an infected individual. Pathogens that resist drying, such as *Staphylococcus aureus* and *Mycobacterium tuberculosis*, will remain on such objects and are the ones most frequently transmitted in this way.

The contamination of surfaces of objects can be assessed by a variety of techniques. In the simplest procedure, the *swab method*, a sterile swab is scrubbed over an area in standardized fashion; then the swab is rinsed in a sterile diluent, and the rinse liquid is diluted and plated for counting. This method, described in detail in Section 15.8 is the most common method used in testing restaurant sanitation. The swab method has the advantages that a high recovery of all organisms on the surface is possible, and cracks, crevices, and irregular surfaces can be sampled. It has the disadvantages that it is time-consuming and requires considerable laboratory work after sampling and that the location and manner of distribution of organisms over the area of surface sampled are not determined.

To determine the distribution of organisms on the surface, a variety of *agar-contact methods* have been developed. In all of these methods, an impression of the surface is made using a slab of sterile nutrient agar; contaminants from the surface are transferred to the agar. The agar slab is then incubated and colonies counted. This method reveals clearly the spatial distribution of organisms over the surface. It is also an easy method to carry out, as no transfers or dilutions are necessary. It has the disadvantage that only flat surfaces may be sampled, and if contamination is high, crowded plates may result that cannot be counted easily. Modifications of the agar-contact method can be made that permit evaluation of nonflat surfaces.

Sampling of bedding (sheets, blankets, pillows, and such) for microbial contamination requires different procedures, since most of the organisms are trapped within the fabric and must be dislodged for counting. One procedure is to force sterile air through a section of bedding and catch the organisms that are dislodged in an air sampler. Another method is to cut a section of bedding with sterile scissors and homogenize it in sterile diluent in a high-speed blender (12,000 revolutions per minute for 3 minutes is satisfactory); then viable counts of the homogenate are performed. This method is more efficient but of course damages the bedding and hence cannot be used for routine surveys.

OBJECTS

FIGURE 9.4
Zoonoses, diseases of animals transmitted to humans. Major zoonoses are:
Bacterial (leptospirosis, anthrax, brucellosis, psittacosis, tularemia, tuberculosis, typhus, spotted fever)
Fungal (ringworm, histoplasmosis, San Joaquin Valley fever)
Protozoal (toxoplasmosis, leishmaniasis, trypanosomiasis)
Viral (encephalitis, cat-scratch fever, rabies)

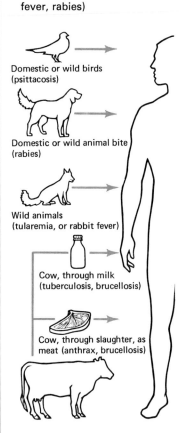

Domestic or wild birds (psittacosis)

Domestic or wild animal bite (rabies)

Wild animals (tularemia, or rabbit fever)

Cow, through milk (tuberculosis, brucellosis)

Cow, through slaughter, as meat (anthrax, brucellosis)

THE SPREAD OF DISEASE

Most organisms cannot penetrate the unbroken skin but can infect a wound. The organism may come from another person or from the soil. Examples of wound infections include pus-forming infections due to staphylococci and streptococci, gas gangrene, caused by *Clostridium perfringens*, and tetanus, caused by *Clostridium tetani*. The most dangerous wounds are those in which contaminated material is driven deep into tissue that has been extensively destroyed. In such dead tissue, oxygen supply is cut off, anaerobic conditions develop, and obligate anaerobes can grow.

Some pathogens are primarily animal parasites and only occasionally are transmitted to man. Diseases of this type, called *zoonoses*, may be passed to man as a result of an animal bite or through the intermediary of an insect or flea (Figure 9.4). Examples of zoonoses are rabies, caused by the rabies virus; plague, caused by *Yersinia pestis*; anthrax, caused by *Bacillus anthracis*; and rabbit fever (tularemia), caused by *Franciscella tularensis*. In some zoonoses of cattle, the pathogen is transmitted to man in infected milk. These include brucellosis and bovine tuberculosis. The control of a zoonosis is difficult because it requires the discovery and immunization or elimination of all infected animals. Such control is nearly impossible in the wild. For instance, rabies is widespread in foxes, skunks, squirrels, bats, and other animals, and the virus is transmitted from them to dogs and from dogs to man. One can reduce the incidence of rabies in man by immunizing dogs, but this does not eliminate the virus from wild animal populations.

9.3
Health of the population

If we are to understand and control disease, we must have some idea of its prevalance in the population. The prevalence of disease is determined by obtaining statistics of illness and death, and from these data, a picture of the public health in that population can be obtained. Public health varies from region to region and has varied with time, so that a picture of the public health at a given moment provides only an instantaneous picture of the situation. By continuing to examine health statistics for many years' duration, it is possible to assess the value of various public health protective measures in influencing the incidence of disease.

Mortality expresses the incidence of death in the population. Data for leading causes of death in the United States are given in Figure 1.1; it is noted there that infectious diseases were once the major causes of death, whereas currently, they are of much less significance, and diseases such as heart disease and cancer are of greater importance. However, this situation could change rapidly if there were a breakdown in public-health measures.

Data on causes of death are obtained by examining *death certificates.* For each death, a death certificate must be filled out by a physician, and these certificates are on file. The death certificate primarily establishes the fact and date of death for insurance and estate purposes, but an important feature of the certificate is the section where the cause of death must be entered. The validity of the statistics obtained depend, of course, on the accuracy and care with which the physician fills out the form. It should also be noted that the immediate cause of death may not be the same as the cause of the illness leading to death. For many chronic conditions, including cancer, termination of life may actually be caused by infectious disease to which a generally weakened patient is highly susceptible. In such cases, the physician would probably enter the chronic condition rather than the infectious disease as the cause of death.

The medical information requested on the death certificate is based on recommendations of the World Health Assembly meeting in Geneva, Switzerland, in 1948. This standard death certificate is used throughout the United States and in many other countries of the world. By use of the standard death certificate, a uniform reporting procedure is established, and comparative data from different parts of the world can be obtained.

There are marked influences of age, sex, and race on mortality. Mortality is high in infants and newborns, drops sharply after the first year of life, remains low throughout childhood and early adult life, and rises again in old age. Both infant and elderly mortality have been markedly influenced by medical practices, and the rates are much lower than formerly, but these two segments of the population still show the greatest mortality.

Female mortality is lower than male mortality at all ages. This difference is apparently due to inherent differences in susceptibility of the two sexes, since even newborn males have a higher death rate than females.

In the United States, death rates in black populations are considerably higher than the rates in the white population. This difference appears to be due to social, economic, and environmental factors, and inherent biological differences in susceptibility may not be involved. A genetic component of disease susceptibility may explain the fact that there is a greater concentration of disease susceptibility in certain familes than in others. Interestingly, mortality is greater in unmarried than in married individuals of the same age. This difference occurs irrespective of sex or race when all causes of death are included. There is also a high correlation between the

FACTORS INFLUENCING MORTALITY

length of life of husbands and wives, and spouses also tend to die from the same or similar causes. Occupation also has a marked influence on mortality, both on age at death and cause of death.

In most cases, the correlations among the various factors just listed and mortality have not been given any precise explanation. A large amount of research is necessary to explain in detail why mortality varies with age, sex, race, and so forth.

Morbidity refers to the incidence of disease in populations, and includes both fatal and nonfatal diseases. Clearly, morbidity statistics more precisely define the health of the population than do mortality statistics, since many diseases that affect health in important ways have only a low mortality (the common cold, for instance). Unfortunately, it is much more difficult to obtain meaningful and accurate statistics of morbidity, because there is no registration procedure such as the death certificate. A restricted number of diseases are reportable to public-health authorities (see Section 9.4), but in most cases, data on morbidity must be obtained from other statistical sources, such as prepaid health-program statistics, hospital and clinic records, absenteeism records in schools and industry, and routine physical examinations. For research purposes, detailed surveys have been made of the health of populations, based on census procedures. In 1956, the U.S. Congress authorized a National Health Survey, from which considerable data on morbidity have been obtained.

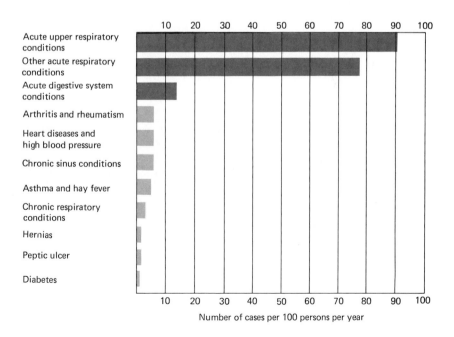

FIGURE 9.5
Leading causes of illness in the United States. Illness refers to any condition, acute or chronic, that affects the well-being of the individual. The three leading causes of illness are all the result of infectious diseases. Illness due to injuries has been excluded. From U.S. National Health Survey reports.

Figure 9.5 presents data on the leading causes of illness in the United States, obtained from the National Health Survey. As seen, the major causes of illness are quite different from the major causes of death given in Figure 1.1. The major illnesses are acute respiratory diseases or acute digestive system conditions, which are almost certainly due to infectious causes. These statistics confirm experience, since everyone is aware that the greatest incidence of sickness results from common colds, influenza, and communicable intestinal disturbances.

The data in Figure 9.5 do not express the severity of illness. When a person is too ill to go to work, or a child is too ill to go to school, the illness could be considered severe. Again, statistics show that the most important cause of lost time in work or school is acute respiratory illness.

So far, we have discussed the health of the U.S. population. To a considerable extent, the United States can be considered typical of those countries where public-health protection is highly developed. Other countries with similar characteristics include Japan, Australia, New Zealand, Israel, and the European countries. In quite another category as far as infectious disease is concerned are the so-called developing countries, a category that includes most of the countries in Africa, Central and South America, and Asia. In these countries, infectious diseases are still major causes of death. As shown in Table 9.1, certain diseases such as dysentery, typhoid fever, and diphtheria, which are virtually never causes of death in developed countries, are still significant causes of death in the less-developed and undeveloped countries. This is partly because of a generally lower health level in individuals in undeveloped countries, so that death from infection is more likely, and partly because of lower levels of public-health protection, making infection more likely in the first place. Interestingly, differences in mortality rates for respiratory infections are not very great between developed and undeveloped countries. This is reasonable when it is remembered that there are no effective public-health measures for the control of the spread of most respiratory infections.

As one means for controlling the spread of infectious disease, reporting of new cases to local public health officials is required. Diseases for which reporting is required are called reportable diseases. Several categories of reportable diseases have been established, the distinctions being based on the seriousness of the disease and the extent to which its communicability can be controlled by public-health means. Reporting requirements vary with the local

WORLD HEALTH

**9.4
Reportable diseases**

TABLE 9.1
Mortality rates (per 100,000 population) for selected diseases in various categories of countries for the years 1967 and 1968

Disease	Developed countries	Less-developed countries	Undeveloped countries
Respiratory infections (influenza, pneumonia, bronchitis)	60.3	62.1	141.5
Tuberculosis	6.8	15.3	34.3
Dysentery	0.05	0.1	3.8
Typhoid fever	0.01	0.1	1.6
Diphtheria	0.01	0.1	1.0
Whooping cough	0.02	0.1	6.0
Measles	0.2	0.5	11.3

Data from Cockburn, W. C., and F. Assaad, "Some observations on the communicable diseases as public health problems," World Health Organization *Bulletin*, **49**, 1–12, 1973.

jurisdiction, but certain general requirements can be stated. The report is usually directed to the local public-health official by the attending physician, although anyone who has knowledge that a communicable disease exists (which may not have been reported) is required to report it. This includes parents or guardians, teaching personnel, nurses, managers of hotels and lodging houses, officials of hospitals or clinics, dairy personnel, and licensed embalmers. The report should include not only the name of the disease, but the name, address, age, race, and sex of the person. Reporting may be by postcard, telephone, or telegraph, and the report is required within 24 hours of recognition.

KINDS OF REPORTABLE DISEASES

Some infectious diseases are more communicable than others, so that a series of categories of reportable diseases has been established.

CLASS 1 DISEASES   These are diseases in which a case report is universally required by international sanitary regulations. This class includes the six diseases in which quarantine (see Section 9.6) is required throughout the world. The diseases are cholera, plague, louse-borne relapsing fever, smallpox, louse-borne typhus fever, and yellow fever. The local authority receiving the report is required to transmit the report to the next superior jurisdiction immediately, if it is the first recognized case in the local area, otherwise weekly by mail for new cases.

CLASS 2 DISEASES   These are diseases in which a case report is regularly required wherever the disease occurs. Two subclasses are recognized, based on the relative urgency for investigation of outbreaks. In the first, required for the diseases typhoid fever and diphtheria, reporting is required by telephone, telegraph, or other rapid means. In the second, which includes brucellosis and leprosy, reporting is required by the most practicable means.

CLASS 3 DISEASES   These are diseases that are selectively reportable in areas where the disease is known to be generally present in the population, and local jurisdictions may require reporting because of undue frequency or severity in an area. Diseases in this category include tularemia, scrub typhus, and coccidioidomycosis.

CLASS 4 DISEASES   This category includes those diseases in which individual cases are not reportable but outbreaks are. These include food infections (salmonellosis), food poisonings (staphylococcal form and botulism), and influenza. Pertinent data reported are the number of cases, the time period of the outbreak, the approximate size of population involved, and the apparent mode of spread.

CLASS 5 DISEASES   Into this category are placed diseases for which an official report is not ordinarily required, either because the diseases are not generally transmissable from man to man (blastomycosis) or are of such a nature that no practical means for control exists (common cold).

It is important that the number of reportable diseases be restricted to those for which some definite public health measures are available. Otherwise the reports are merely a collection of statistics and serve no practical purpose. In such situations, the level of reporting deteriorates, as local reporting units lose interest in making the effort required to transmit the information. Thus, although it would be of interest to gather statistics for a larger number of infectious diseases than those listed above, it is preferable to restrict reporting to those diseases for which control services are available. We discuss the study of epidemics and their control in Section 9.5; the information for the existence of epidemics is derived primarily from case reports.

A disease is said to be *epidemic* when it occurs in an unusual number of individuals in a community at the same time. An *endemic* disease is one that is constantly present in a population. An epidemic can be caused by one of several factors. For example, the population may be highly susceptible to a pathogen that is normally absent from the area. If the pathogen is then introduced, an explosive epidemic can occur (Figure 9.6). Eventually, however, immunity develops as patients recover, and the incidence of disease declines again. An epidemic may also develop even in an immune population if the pathogen changes (by genetic mutation) to a new antigenic type against which the antibodies of the immune population are not

**9.5
Epidemics**

FIGURE 9.6

An example of an explosive epidemic caused by the introduction of a virulent pathogen into a susceptible population. Incidence of colds throughout the year in an isolated community on Spitsbergen, in the Arctic Ocean. Note the sharp rise in the number of cases soon after the arrival of the first boat of the summer. From C. H. Andrewes, *The natural history of viruses,* W. W. Norton and Co., New York, 1966.

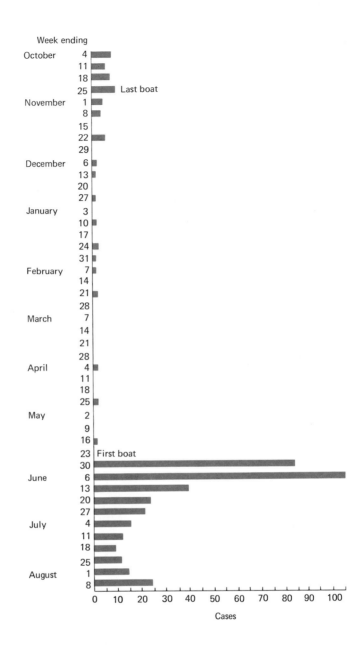

active. Likewise, when a group of susceptible individuals moves into an area where a disease is endemic, they may become infected, resulting in a small epidemic. Note that this situation meets our definition of an epidemic, even if only a small number of people are infected, since the disease is found in an unusual number of individuals in the community.

An epidemic may develop if a weakly virulent pathogen changes and becomes more virulent. The pathogen in question may have been widespread in the population for some time but caused only inapparent infection; if the new, more virulent form causes obvious disease symptoms, then the population will soon be aware that it is experiencing an epidemic. In this regard, it might be noted that a successful parasite is not necessarily a highly virulent one. From this point of view, success means the ability of the parasite to reproduce, grow, and be transmitted from individual to individual, permitting long-term survival of the parasite. A highly virulent parasite, which kills its host quickly, may not be transmitted and hence may die with its host.

A decrease in the nonspecific resistance of a population may result in an epidemic. As we have noted, resistance to infectious disease can be influenced by the nutritional status and general well-being of an individual. Deterioration of living standards, whether due to depressed economic conditions, warfare, natural disasters, or other causes, can lead to decreased resistance of a whole population and hence to increases in incidence of disease. Certain diseases arise more readily in response to such causes than others. Tuberculosis is one of the best examples of a disease the incidence of which rises as living standards decline.

Changes in behavior and personal habits may also lead to epidemics. Customs and habits influence the efficiency of transfer of pathogens. Venereal disease incidence increases drastically when sexual promiscuity increases. Intestinal infections are greatly affected by procedures for sewage disposal, water purification, hand washing, and other aspects of personal hygiene. Transmission of respiratory pathogens is affected by habits regarding coughing, sneezing, and spitting.

In an endemic situation, the pathogen may not be highly virulent or the majority of individuals may be immune, so that the proportion of individuals in the population who have the disease is low. However, as long as the endemic situation lasts, there will be a small proportion of individuals who can infect susceptible persons.

Certain diseases occur in cycles. These cycles develop because of changes in the proportion of individuals in the population with immunity to the disease agent. When virtually the whole population is immune, epidemic conditions cannot develop and the pathogen may disappear completely from the population. This then permits the development of a nonimmune population, perhaps through births or immigration. At a later date, if the pathogen is accidentally reintroduced into the population, the nonimmune individuals become infected and an epidemic ensues. If recovery of infected individuals results in immunity, the disease incidence subsides and the cycle is complete. A good example of a cyclic disease is measles in the days before a vaccine was available (Figure 9.7). This childhood disease

CYCLES OF DISEASE

FIGURE 9.7

Cycles of disease, as illustrated by measles. Reported cases of measles by 4-week periods in the United States. Note the sharp rise and fall of disease incidence and the marked decline in incidence after the introduction of immunization. From J. J. Witte, "Recent advances in public health," *Am. J. Pub. Hlth.*, **64**, 939–944, 1974.

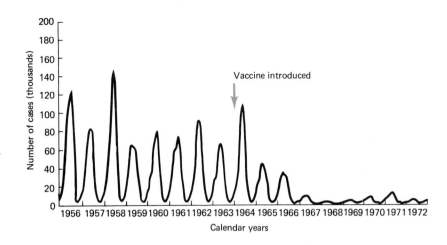

is caused by a virus transmitted efficiently from person to person by the respiratory route. Epidemics usually developed in school children, where opportunity for respiratory transmission is high. For a susceptible school population, introduction of the virus results in an explosive epidemic, which gradually subsides as the infected individuals form antibodies and become immune. After a while virtually all children have had the disease and are immune, and there is a quiescent period until a new susceptible population builds up as nonimmune children enter the lowest grades. Then another introduction of virus into the school results in another epidemic.

THE CARRIER STATE

Carriers are individuals who contain the virulent pathogen but do not exhibit symptoms of disease. They may be people in the incubation or convalescent stages of a disease who are about to have or have already had clinical attack, or they may be carriers who show no symptoms, either because they are immune or because they are only mildly infected. The carrier thus serves as a *reservoir* of the parasite, ensuring the maintenance and survival of the parasite in nature.

Carriers may be either temporary or chronic. Chronic carriers may have marked influences on public health, as they can serve as sources for the infection of large numbers of people over long periods of time. The classic example of a chronic carrier was the woman known as "Typhoid Mary," a cook in the New York–Long Island area in the early part of this century. Typhoid Mary (her real name was Mary Mallon) was employed in a number of households and institutions, and as a cook she was in a central position to infect large numbers of people. Eventually, she was tracked down after an extensive epidemiological investigation of a number of typhoid outbreaks revealed that she was the likely source of contamination. When her feces were examined bacteriologically, it was found that she had practically a pure culture of the typhoid bacterium, *Salmonella typhi*. She remained a carrier for many years, probably because her gall bladder was infected, and organisms were continuously excreted from there into her intestine. Public-health authorities offered to remove her gall bladder, but she refused the operation. In order to prevent her from continuing to serve as a source of infection, she was imprisoned. After almost 3 years in prison, she was released on the pledge that she would not cook or handle food for others and that she was to report to the health department every 3 months. She promptly disappeared, changed her name, and cooked in hotels, restaurants, and sanitoria, leaving a wake of typhoid fever. After another 5 years, she was captured as a result of the investigation of an epidemic at a New York hospital. She was again arrested and imprisoned and remained in prison for 23 years. She died in prison in 1938, 32 years after her first discovery as a typhoid carrier. This case shows the importance of carriers and the long period of time that they can remain infective.

Carriers can be identified by routine surveys of populations, using cultural, radiological (chest X ray), or immunological techniques. In general, carriers are only sought among groups of individuals who may be sources of infection for the public at large, such as food handlers in restaurants, groceries, or processing plants. Two diseases in which carriers have been of most significance are typhoid fever and tuberculosis, and routine surveys of food handlers for inapparent cases of these diseases are sometimes made. (Food-borne diseases will be considered in Section 15.2).

Control of epidemics can never be an individual matter but must involve participation of the entire community. The control of epidemics falls under the general heading of public health and preventive medicine. Because governmental units vary in desire and ability to carry out public health services, it is not possible to present a general picture of what is done. Instead, we indicate here a reasonably idealized account of what can be done (Table 9.2).

RECOGNITION OF AN EPIDEMIC

Recognition begins with proper diagnosis of the disease in individuals by physicians and reporting of the disease to some centralized body, such as a city, state, or national board of health. We have discussed reportable diseases and the manner by which reporting is effected in Section 9.4. As statistics accumulate, public-health authorities recognize that an epidemic is in progress. Notification of the public and the medical profession via newspapers, radio, and

TABLE 9.2

Epidemic diseases and their control

Disease	Infective organism	Infectious sources	Entry site	Method of spread	Incubation period	Prevention of epidemics
Bacillary dysentery	*Shigella* group	Contaminated water and food	Gastrointestinal tract	Patients and carriers; fecal-oral route	24–48 hours	Detection and control of carriers; inspection of food handlers; decontamination of water supplies
Brucellosis	*Brucella melitensis* and related organisms	Milk or meat from infected cattle, goats, and pigs	Gastrointestinal tract	Oral ingestion of infective material	6–14 days	Milk pasteurization; control of infection in animals
Diphtheria	*Coryne-bacterium diphtheriae*	Human cases and carriers; food; objects	Respiratory route	Nasal and oral secretions; respiratory droplets	2–5 days	Active immunization with diphtheria toxoid or toxin-antitoxin mixture; case quarantine; disinfection of carriers
German measles (rubella)	Virus	Human cases	Probably respiratory	Probably respiratory droplets	10–22 days (average 18)	Patient isolation when pregnant woman is in household; rubella vaccine
Gonorrhea	*Neisseria gonorrhoeae*	Urethral and vaginal secretions	Urethral or vaginal mucosa	Sexual intercourse	3–8 days	Chemotherapy of carriers and potential contacts; case-finding and treatment of patients
Hepatitis, epidemic	Virus	Contaminated food or water	Gastrointestinal tract	Fecal-oral route	2–6 weeks	Sanitary precautions applied to infected cases; passive immunization with gamma globulin
Influenza	Virus	Human cases	Respiratory tract	Respiratory	18–36 hours	Specific virus vaccine
Malaria	Protozoa, *Plasmodium vivax*	Human cases	Skin	Mosquitoes	2 weeks	Coordinated measures for wide-scale mosquito control; prompt detection and effective treatment of cases
Measles	Virus	Human cases	Respiratory	Nasal secretions	11–14 days	Measles vaccine
Meningococcal meningitis	*Neisseria meningitidis*	Human cases and carriers	Respiratory	Respiratory droplets	Variable	Group chemotherapy with sulfadiazine (when strain is sensitive to sulfonamides)
Paratyphoid fever	*Salmonella paratyphi*	Contaminated food and water, rectal tubes, and barium enemas	Gastrointestinal tract	Infected urine and feces	7–24 days	Control of public water sources, food vendors, and food handlers; treatment of carriers; individual vaccination with *S. paratyphi* A and B vaccine
Pneumococcal pneumonia	*Diplococcus pneumoniae*	Human carriers; patient's own pharynx	Respiratory	Respiratory droplets	Variable	Control of upper respiratory infections; avoidance of alcoholic intoxication; communicable disease precautions applied to cases
Poliomyelitis	Poliovirus	Human cases and carriers	Gastrointestinal tract	Infected feces and pharyngeal secretions	4–7 days	Wide-scale application of parenteral (Salk) and oral (Sabin) poliovirus vaccines; case isolation
Rocky Mountain spotted fever	*Rickettsia rickettsii*	Infected wild rodents, dogs, wood ticks, and dog ticks	Skin	Tick bites	3–12 days	Avoidance of tick-infected areas, or wearing of protective clothing in such areas; frequent search for and prompt removal of ticks from body; specific vaccination of exposed persons

TABLE 9.2
Epidemic diseases and their
control—*continued*

Disease	Infective organism	Infectious sources	Entry site	Method of spread	Incubation period	Prevention of epidemics
Syphilis	*Treponema pallidum*	Infected exudate or blood	External genitalia; cervix; mucosal surfaces; placenta	Sexual intercourse, contact with open lesions; blood transfusion; transplacental inoculation	10–90 days	Case-finding by means of routine serologic testing and other methods; and adequate treatment of infected individuals
Tuberculosis	*Mycobacterium tuberculosis*	Sputum from human cases; milk from infected cows	Respiratory or gastrointestinal mucosa	Sputum; respiratory droplets; infected milk	Variable	Early discovery and adequate treatment of active cases; milk pasteurization
Typhoid fever	*Salmonella typhi*	Contaminated food and water	Gastrointestinal tract	Infected urine and feces	5–14 days	Decontamination of water sources; milk pasteurization; individual vaccination; control of carriers
Whooping cough (pertussis)	*Hemophilus pertussis*	Human cases	Respiratory tract	Infected bronchial secretions	12–20 days	Active immunization with *H. pertussis* vaccine; case isolation

television is then carried out. Detailed information on epidemics is accumulated in the United States by the Center for Disease Control in Atlanta, Georgia, an agency of the Department of Health, Education, and Welfare. This agency publishes a weekly magazine, *Morbidity and Mortality*, which presents up-to-date statistics for epidemic diseases.

QUARANTINE

Quarantine involves the limitation of the freedom of movement of individuals with active infections, to prevent the spread of disease to other members of the population. The time limit of quarantine is the longest period of communicability of the disease. Quarantine must be done in such a manner that effective contact of the infected individual with those not exposed is prevented. Quarantine is not the same as strict isolation, used for unusually infectious diseases in hospital situations (see Section 9.11).

At one time, quarantine was required for a number of infectious diseases of childhood, such as measles, chicken pox, and mumps, and residences in which quarantined children were housed had placards affixed to the outside. Such measures were found to have little public-health significance in the control of the spread of these diseases, and quarantine is no longer required, although it is still advisable to prevent the contact of infected children with other (possibly susceptible) children.

Currently, quarantine is only required for smallpox, a disease that has almost been eradicated from many countries. Introduction of an individual infected with smallpox into many areas could have disastrous effects, since immunization for smallpox is no longer widely practiced. Thus, it is essential to quarantine an infected individual during the period of communicability.

PREVENTION OF AN EPIDEMIC

Prevention is best done before the epidemic begins rather than after it is in progress. Prevention requires an understanding of all the factors involved in the complex of relationships between host and parasite. For some diseases, prevention can best be accomplished by control of the spread of the parasite, whereas others can best be controlled by conferring immunity upon the host. *Water-borne diseases*, such as cholera and typhoid, are controlled by sanitary treatment of water and sewage. *Food-borne diseases*, such as typhoid, are controlled by restaurant and grocery store sanitation, survey of food handlers for carriers, and regulations regarding storage of foods. *Diseases of dairy animals* transmitted to humans, such as tuberculosis and brucellosis, are controlled by elimination of diseased animals in dairy herds. *Respiratory diseases* such as diphtheria, measles, and influenza are controlled by immunization of the population, since it is impossible to attempt to prevent transmission of pathogens via the respiratory route. Such immunization programs have been effective in practically eliminating epidemics of these diseases. However, it should be emphasized that a small reservoir of cases always remains; the potential for a renewed epidemic is therefore constantly present. If new generations of susceptible children are not immunized or if existing immunizations are allowed to lapse without boosters, then the proportion of immunized individuals in the population will fall too low to confer protection from an epidemic. It is almost certain that a new epidemic will occur. Recent increases in the incidence of measles in some areas are directly attributable to decreases in the number of persons receiving immunizations.

In a disease such as *tuberculosis*, which can be transmitted by a variety of means, control cannot be by a single procedure. Tuberculosis in food handlers is recognized by X ray or tuberculin test, and infected individuals are prevented from working until cured. Pasteurization of milk prevents transmission of the tuberculosis pathogen from cow to person, and elimination of tubercular cows provides an additional safety factor. Regulations prohibiting spitting in public places are designed to prevent transmission of the bacterium by dust and dirt. Finally, routine surveys of populations either by tuberculin test or by X ray permit the recognition of infected individuals, who can then be treated to control the disease.

Insect-borne diseases such as malaria and yellow fever are controlled by eliminating the insect vector. The control of such diseases in tropical countries has been greatly improved in recent years through the use of persistent and highly effective insecticides such as DDT, which keep the insect vector in check.

Among the most difficult diseases to control are the *venereal diseases*, of which gonorrhea and syphilis are the most prevalent. Incidence of these diseases depends greatly on the degree of sexual promiscuity in the population. Even though several highly effective antibiotics are available for treatment of both diseases, a worldwide resurgence, particularly of gonorrhea, has occurred, due in part to the difficulty of identifying those who require such treatment. Extensive education campaigns have been used occasionally in the attempt to control venereal diseases in certain populations, such as the military, but it is not certain that these campaigns have been effective. One problem associated with venereal diseases is that physicians do not always report cases, in deference to the sensitivities of their patients. As a result, individuals who may be foci of infection are not identified and their sexual contacts are not located and treated. Venereal diseases are thus social as well as medical problems.

SCHOOLS AND PUBLIC HEALTH

One of the most effective ways of controlling many diseases is through programs of immunization, diagnosis, and education in the schools. School health programs will be of greatest value to those children who receive little if any health guidance in the home, but the school can improve the health of all pupils. A variety of services may be offered: routine immunizations, screening of pupils for disease, studies of physical growth, diagnostic services aimed at detecting cases of clinical disease, vision and hearing tests, and so on. A large school usually has a full-time school nurse and a part-time school physician. Careful records are kept, and parents are contacted whenever anything serious is suspected.

STATISTICS

Statistics of disease incidence are kept by boards of health and are transmitted to a central agency for compilation in various reports. Statistics of this sort will always have a margin of error because of inaccurate reporting. Statistics of long-term changes in incidence of disease are especially unreliable, since the ability of physicians to diagnose specific diseases depends on their education and awareness of these diseases and on the availability to them of laboratory and other diagnostic aids. Thus a greater reported incidence of a disease today compared to a former time does not necessarily mean that the disease is more prevalent; it may merely be better recognized. Collection of statistics is still a valuable effort, however, since it helps to identify long-term trends and also serves to locate regions or segments of the population with unusually high disease incidence.

**9.7
Eradication of disease**

The overall goal of public-health measures should be the complete eradication of disease from the population. Worldwide eradication has not been achieved for any disease, but regional eradication has been accomplished for a number of diseases. If worldwide eradication of a disease could be achieved, further public-health measures for control of that disease would not be necessary. With regional eradication, continuous watchfulness to prevent new introduction of the pathogen is necessary.

Eradication can be achieved by any means that will prevent the establishment and spread of the pathogen. In practice, eradication has been achieved in one of three ways: (1) immunization, (2) elimination of insects responsible for transmission of the pathogen, or (3) drug treatment. The first two approaches have been most amenable to intense public-health effort, and some successes have been achieved.

IMMUNIZATION

Theoretically, if all individuals in a population were immune, the pathogen could not become established. Actually, immunization of *every* individual is not necessary; procedures that achieve a high level of immunity will suffice, because once immunity is conferred on most people, the chance of the pathogen's spreading through the remaining few people is greatly reduced, and the pathogen probably will die out. This approach has been most successful in the eradication of smallpox. We discussed the history of smallpox immunization in Section 1.6. Smallpox is an ideal target for eradication on a world scale because the disease is so obviously infectious that even the layman can understand its danger; as a result, immunization measures receive backing even in primitive areas. Vaccination confers complete immunity for 3 to 5 years, so that there is time to vaccinate on a large scale. The vaccine is easy to make, even under relatively primitive conditions, and is sufficiently stable so that it can be transported without refrigeration. The vaccination procedure is simple to carry out and can be done even by nontechnical people once they have been given brief training. The success of the smallpox eradication program has been so great that the disease is no longer a problem in most parts of the world. As we noted in Section 8.15, the disease is essentially nonexistent in the United States, and immunization of children is no longer prescribed because the dangers from immunization are greater than the dangers from infection.

Other diseases that have been eradicated or virtually eradicated by immunization include poliomyelitis, measles, diphtheria, and whooping cough. Rabies has been eliminated in England, although not in the United States. Rabies eradication in England was achieved by immunization of dogs and by strict quarantine on dogs entering the country. Eradication was helped by the fact that large wild areas where rabies could remain established in rodents or other animals do not exist in England, so that infection of dogs from this source does not occur. In the United States, despite immunization of dogs, rabies is still a problem because it is endemic in wild animal populations. Eradication is considerably easier to achieve in islands (such as England) than in large continental areas because control of the movement of populations is easier.

Insect-borne diseases can be eradicated by the elimination of the insect *vector*. This is most effective when only a single insect species or group of related species are responsible for the transmission of the pathogen. Eradication of the insect is most effectively achieved by the use of an insecticide, and the introduction of the insecticide DDT has had great impact on eradication measures for malaria and yellow fever. Elimination of insect vectors without insecticides can be achieved by drainage of swamps and elimination of other insect-breeding places or by introduction of biological agents that are pathogenic for the insects themselves (*biological control*).

ERADICATION OF INSECT VECTORS

The *malaria* parasite is transmitted by the mosquito *Anopheles*, and elimination of this mosquito can lead to eradication of the disease. Malaria has been eliminated from the United States for many years, and because the vector is absent, infection of the population will not occur even if infected individuals enter the country (for instance, returning military personnel). *Yellow fever* is transmitted to humans by another mosquito, *Aedes aegypti*. The elimination of this mosquito from the United States was achieved many years ago, the elimination made easier by the fact that the mosquito lives primarily in warm-climate areas so that it had a restricted habitat in this country. Complete elimination of yellow fever in jungle areas is impossible because the monkey population serves as a reservoir of infection; the virus is transmitted from monkey to monkey by mosquitoes other than *A. aegypti*.

In principle, drugs such as antibiotics can be used to eradicate diseases. If all infected individuals were treated, the transmission of the pathogen through the population would be quickly stopped. The difficulty with this approach is that the location of infected individuals is often difficult, especially if the disease is not rapidly fatal or debilitating. It is also essential, of course, that a highly effective drug be available. Some success with penicillin has been achieved in eradicating *yaws*, a bacterial disease caused by a spirochete. Penicillin is highly effective against the pathogen, and a single dose usually leads to cure. Yaws is a skin disease transmitted primarily by direct contact, and scanty clothing, poor personal and community hygiene, and the presence of minor skin lesions favor infection. The disease is present primarily in countries with hot and humid climates and is found mostly in lower socioeconomic groups. The ease and effective-

DRUG TREATMENT

ness of penicillin treatment has made this approach feasible in the eradication of yaws. Mass treatment campaigns have been initiated in those communities where the prevalence of the disease is 10 percent or more. In such campaigns, *every* individual in the community is given a single large dose of penicillin. In communities where the incidence is between 5 and 10 percent, all children under 15 years of age and the contacts of infectious cases are treated. For populations with rates less than 5 percent, more selective treatment is recommended, and only immediate family and other obvious contacts of infected individuals are treated. Eradication of yaws has been attempted in Jamaica, but it would be difficult to initiate a program in large countries, because even though the antibiotic is inexpensive, considerable expense is incurred in operating a mass treatment program.

9.8
Hospital sanitation

The hospital may not only be a place where sick people get well, it may also be a place where sick people can get sicker. The fact is that cross infection from patient to patient or from hospital personnel to patients presents a constant hazard. Sometimes, hospital infections are categorized under a special heading, *nosocomial* infections (*nosocomium* is the Latin word for "hospital").

Hospitals are especially hazardous for the following reasons: (1) Many patients have weakened resistance to infectious disease because of their illnesses. (2) Hospitals must of necessity treat patients suffering from infectious disease, and these patients may be reservoirs of highly virulent pathogens. (3) The crowding of patients in rooms and wards increases the chance of cross-infection. (4) There is much movement of hospital personnel from patient to patient, increasing the probability of transfer of pathogens. (5) Many hospital procedures, such as catheterization, hypodermic injection, spinal puncture, and removal of samples of tissues or fluids for diagnosis, carry with them the risk of introducing pathogens to the patient. (6) In maternity wards of hospitals, newborn infants are unusually susceptible to certain kinds of infection. (7) Surgical procedures (Figure 9.8) are a major hazard, since not only are highly susceptible parts of the body exposed to sources of contamination but surgery often diminishes the resistance of the patient to infection. (8) Many drugs used in hospitals increase susceptibility to infection. (9) Use of antibiotics to control infection carries with it the risk of selecting antibiotic-resistant organisms, which then cannot be controlled if they cause further infection.

Because of these widely acknowledged hazards of hospitals, it is of great importance that effective programs of hospital sanitation be established and that hospital personnel be trained in proper preventive measures. In any hospital, there should be a department or

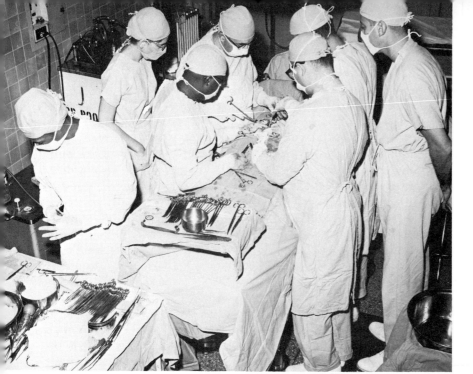

FIGURE 9.8
An operation in progress.

section of hospital sanitation, the size of this department to depend on the size of the hospital. This department establishes routines and procedures specifically related to hospital sanitation, trains and supervises personnel, exercises surveillance over sanitation procedures, investigates outbreaks of hospital infection, and keeps records and statistics on the sanitation program. An adequate bacteriology laboratory should be available to service this department. In a small hospital, the diagnostic bacteriology laboratory may be able to handle this task, but in a large hospital, a special environmental bacteriology laboratory is desirable. The extra cost of setting up a proper hospital sanitation program will, in the long run, be greatly offset by financial savings to both patients and hospital.

The broad area of hospital sanitation includes all aspects of operations within the hospital; the aim is to control the dispersal of organisms so as to eliminate infections arising within the hospital. Such control is accomplished in several ways; proper nursing practices and housekeeping procedures within the hospital facilities are perhaps the most important and must be maintained at a high level of quality at all times. Other procedures that are very important in more specialized areas of the hospital include the disinfection of skin before surgery, the preparation and sterilization of various supplies, solutions, and surgical equipment, and the isolation of highly infectious patients. These areas will be discussed in detail in the following pages.

ORGANISMS INVOLVED IN HOSPITAL INFECTIONS

Certain kinds of microorganisms seem to be most commonly involved in hospital infections. Perhaps the most important and widespread agent involved in hospital infections is *Staphylococcus aureus*. Certain strains of this species, of unusual virulence and frequently resistant to common antibiotics, have been so widely associated with hospital infections that they are sometimes designated "hospital staphylococci." The habitat of these staphylococci is the upper respiratory tract, usually the nasal passages, and they often become established as "normal flora" in hospital personnel. In such healthy personnel the organism may cause no disease, but these symptomless carriers may be a source of infection for susceptible patients. Since staphylococci are resistant to drying, they survive for long periods on dust particles and other inanimate objects and can subsequently infect patients. An infected patient may also be a serious source of infection for others. Because of the potential seriousness of infection with hospital staphylococci, and because of the difficulty of control if the organisms are antibiotic resistant, careful application of the principles of hospital sanitation is necessary.

Pseudomonas aeruginosa is a common hospital pathogen, causing infection of wounds, burns, and the urinary tract. It is a bacterium that is naturally resistant to many antibiotics and is adaptable to a wide variety of human habitats. Certain strains of *Escherichia coli* are causes of epidemics of gastrointestinal infection in newborn babies in maternity wards. Most newborns are unusually susceptible to intestinal infection, and an *E. coli* strain that is harmless to adults may spread rapidly through a newborn nursery, usually being transferred from infant to infant on the hands of nurses and other hospital personnel. Other organisms that occasionally cause outbreaks in hospitals include *Serratia*, *Salmonella*, *Clostridium*, certain viruses causing enteric infections, and herpes virus.

**9.9
Nursing and hospital practices**

Because of the considerable danger of infection in hospitals, a number of procedures and practices have been devised that help to ensure patient safety. Many of these practices have been specifically prescribed in manuals by groups involved with hospital operation; one is published by the American Hospital Association (see *Suggested Readings* at the end of this chapter). These procedures can be conveniently divided into those that are carried out by skilled, medically trained personnel and those routine hospital houseclean-

Strict Isolation

Visitors—Report to Nurses' Station Before Entering Room

1. **Private Room**—*necessary*; door must be kept closed.
2. **Gowns**—must be worn by all persons entering room.
3. **Masks**—must be worn by all persons entering room.
4. **Hands**—must be washed on entering and leaving room.
5. **Gloves**—must be worn by all persons entering room.
6. **Articles**—must be discarded, or wrapped before being sent to Central Supply for disinfection or sterilization.

Diseases requiring strict isolation:

1. Anthrax
2. Burns, extensive, infected with *Staphylococcus aureus* or Group A streptococcus
3. Diphtheria
4. Eczmea vaccinatum
5. Melioidosis
6. Neonatal vesicular disease (Herpes simplex)
7. Plague
8. Rabies
9. Congenital rubella syndrome
10. Smallpox
11. Staphylococcal enterocolitis
12. Staphylococcal pneumonia
13. Streptococcal pneumonia
14. Vaccinia, generalized and progressive

ing practices that are carried out predominantly by housekeeping personnel.

The nurse's duties cover a very broad spectrum of activities; some of these are directly related to preventing the spread of one patient's disease to other individuals. Some items are used frequently in the care of a patient and may be used in the care of several patients during a day. Clinical thermometers should be cleaned and sterilized after each use. The best sterilizing agent is a 0.5 to 1.0 percent solution of iodine in 70 percent alcohol, and the thermometer should be left in contact with the agent for at least 10 minutes. Bedside equipment such as glasses, pitchers, and washbasins should be kept clean. Bedpans, urinals, and equipment for administering enemas should be cleaned and sterilized after each use. The proper disposal of contaminated bedding, dishes and eating utensils, gowns and masks used in isolation areas, and infected secretions and excretions is of course important in preventing the spread of a patient's disease. The procedures should be part of normal hospital practice. Incineration is, of course, the most effective means of disinfecting contaminated materials and is used routinely for disposable items.

Dressing of infected wounds presents considerable risk. Masks and gowns should be worn. The soiled dressing should never be touched with the hands. Rather, it should be removed with a forceps and discarded to a special receptacle. The exposed wound should then be disinfected, using a swab, and the fresh dressing installed in such a way that the wound is not touched. Finally, the hands are washed with disinfectant detergent and dried.

NURSING PRACTICES

Careful housekeeping procedures, such as proper mopping of floors, cleaning of walls, ceilings, and windows, and emptying of trash, are essential to safe hospital operation. To accomplish this, the hospital should adopt standard and efficient practices for cleaning and then instruct all housekeeping personnel. Frequent surveillance of housekeeping personnel to be sure they are following prescribed methods is essential. The greatest impact that good housekeeping methods will have is in dust control, an important item, since as we have seen hospital staphylococci often survive for long periods on dust particles. Addition of germicidal solutions to water used for mopping floors will kill bacteria carried into the mop bucket from the floor. Dry mopping should never be done, as it merely redistributes and stirs up the dust. Bedding must not be shaken in rooms, since this greatly increases the dust level. Floors, walls, and ceilings should be constructed of a hard material and be free of cracks and crevices where dust can lodge. These surfaces should be well maintained and should be repaired promptly when damage occurs. Particular attention must be paid to housekeeping in high-risk areas such as isolation units, surgeries, nurseries, autopsy rooms, bathrooms and showers, and utility and treatment rooms. Recommended housekeeping practices for low- and high-risk areas have been established (Table 9.3).

Control of vermin such as rodents, roaches, and flies is of course an essential part of good hospital-housekeeping, as these animals may be agents for the transfer of infectious organisms.

SURGICAL SKIN ANTISEPSIS

During surgery, the patient is unusually susceptible to internal infection, and even organisms from the skin, not ordinarily pathogenic, may

TABLE 9.3
Minimum recommended housekeeping practices in hospitals

Item	High-risk areas	Low-risk areas
Floors	Scrubbed at least twice daily	Daily scrubbing
Walls	In isolation areas, complete washing when patient is discharged; in other areas, walls washed monthly or bimonthly	Semiannual washing
Ceilings	Washed on discharge of patient or quarterly; if acoustical tile, vacuum at least once a month	Semiannual washing
Toilet seats	Disinfected daily if used by a single patient; otherwise, after each use by a patient with gastrointestinal or staphylococcal infection	Disinfect twice daily
Toilet bowls	As toilet seats	

Isolation notice used in hospitals

Diseases requiring respiratory isolation:
1. Chickenpox
2. Herpes zoster
3. Measles (rubeola)
4. Meningococcal meningitis
5. Meningococcemia
6. Mumps
7. Pertussis (whooping cough)
8. Rubella (German measles)
9. Tuberculosis, pulmonary—sputum-positive (or suspect)
10. Venezuelan equine encephalomyelitis

initiate infection. For this reason, surgeons and other personnel involved in surgery must carry out careful skin antisepsis. Simple washing does not kill organisms and may actually cause an increase in bacterial numbers on the surface by bringing to the surface organisms embedded in the hair follicles. The best preparations for skin antisepsis in medical practice contain hexachlorophene, iodophors, or quaternary ammonium compounds mixed with foam wetting agents. Such an antiseptic cleansing agent must act both to remove all dirt and to kill organisms. The value of handwashing procedures for reducing the microbial load on the skin is illustrated in Figure 9.9.

In medical *handwashing procedures*, a foot-operated water faucet is preferred, so that the hands need not come in contact with the faucet after being treated. For general handwashing, the hands are wetted, and a small amount of antiseptic added and spread thoroughly over and around the hands for 15 to 20 seconds. If a hand-operated water faucet is used, the water should be left running during the entire operation. The hands are then rinsed in the full force of the water, care being taken to avoid touching either the body or the sink. The hands are dried with a sterile paper towel, and if a hand-operated faucet is used, the faucet should be turned off with the towel. The *surgical scrub* is a more elaborate procedure, in which both the hands and forearms are washed, and the length of washing is prolonged. A preliminary wash is done first, the hands and forearms being wetted, and a few drops of antiseptic added and spread over them. The arms and forearms are then rinsed by being held under the faucet so that they do not touch anything. This preliminary washing is designed to remove dirt, which would reduce the action of the antiseptic. Next, the nails are cleaned using a small wooden stick, after which the hands and arms are wetted again, and

FIGURE 9.9

The value of handwashing procedures for reducing the microbial load on the skin. *a* Dirty fingers touch agar. *b* After 24-hours incubation. There is now so much microbial growth that a mass of colonies develops. *c* After a 20-second rinse with cold water. Some large dirt particles are removed, but there is little reduction in bacterial numbers. *d* After 20-second wash with soap and water. Considerable reduction in bacterial numbers. *e* After an additional 20-second wash with soap and water. Further reduction has occurred. *f* After use of a sanitizing solution (50 ppm iodophor). No bacterial colonies seen. The hands are not really sterile, however, since bacteria are embedded in the crevices of the skin where they escape contact with the sanitizing solution. After vigorous rubbing, some of these bacteria would probably be brought to the surface and would be found viable. *Courtesy of Dr. George A. Schuler, Extension Food Scientist, University of Georgia, Athens.*

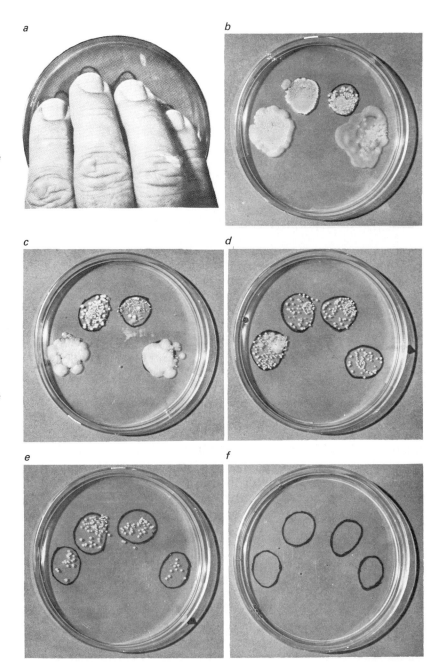

fresh antiseptic spread over them, preferably by use of a small brush. The hands and arms are thoroughly scrubbed with the brush so that dirt hidden in crevices is brought to the surface. The hands and arms should then be rubbed for 2 to 4 minutes, during which time the antiseptic is killing the bacteria. Finally, the hands and arms are rinsed thoroughly and dried with a sterile paper towel. The

Protective Isolation
Visitors—Report to Nurses' Station
Before Entering Room

1. **Private Room**—*necessary*; door must be kept closed.
2. **Gowns**—must be worn by all persons entering room
3. **Masks**—must be worn by all persons entering room.
4. **Hands**—must be washed on entering and leaving room.
5. **Gloves**—must be worn by all persons having direct contact with patient.
6. **Articles**—*see* manual text.

Conditions requiring protective isolation:
1. Agranulocytosis
2. Severe and extensive dermatitis
3. Certain patients receiving immunosuppressive therapy
4. Certain patients with lymphomas and leukemia

hands are dried first, then the arms. (Of course, rings, watches, and all other foreign objects should have been removed from the hands and arms prior to the scrub.)

The antiseptic agents used in handwashing procedures are not without some danger in themselves. They may cause allergic, toxic, or other undesirable side effects in some individuals. There is good evidence that hexachlorophene is toxic to newborn infants. Thus, although the use of antiseptics is essential in procedures where heavy contamination is present or where any contamination is undesirable, there are probably no advantages to their use for routine handwashing procedures under nonepidemic conditions. It is now recommended that for such routine use soap or detergent be used without antiseptic additives.

For many surgical procedures, sterile rubber or plastic gloves are worn, both to protect the patient from contamination by the physician or nurse and to protect the latter from contamination by an infectious patient. Gloves are usually packaged in a folded cloth wrapper and are sterilized together with a small packet of powder. The glove package is first opened so that the gloves and powder are accessible; then the hands are washed and dried as described above. After this they are powdered and the gloves put on, care being taken to avoid contaminating the glove surfaces by touching them to the outside of the package, to the body, or to anything else in the environment.

Preoperative skin antisepsis of the patient is necessary for any surgical operation involving cutting of the skin. The usual practice is to sterilize a region much larger than that of the immediate incision. The region to be sterilized is first washed thoroughly and then shaved to remove all hair. The skin is then cleansed with antiseptic, as described above. For abdominal surgery, care must be taken to cleanse the umbilicus, as inevitably there will be dirt lodged here. A cotton swab saturated with ether or hydrogen peroxide will help to remove the dirt in the umbilicus, and further cleaning may be done with a cotton swab saturated in soap or disinfectant.

Injections are frequently administered, and precautions to avoid carrying organisms from the skin into the body are necessary. Intramuscular and subcutaneous injections are less serious than intravenous injections, since organisms are not carried into the bloodstream, but local infections as a result of the former types of injections can occur. The injection site should be cleansed with an antiseptic, usually soaked into a cotton ball, the skin stretched taut, and the needle inserted. After the injection is complete, the needle is removed, while the cotton ball soaked with antiseptic is pressed on the area. Since the hole formed by the needle closes quickly, further precautions to avoid infection are usually not necessary.

9.10
Hospital sterilization procedures

The largest user of sterilized items is the hospital. Bandages, dressings, surgical suturing materials, hypodermic syringes and needles, tubing, glassware, scissors and other surgical instruments, catheters, forceps, and a wide variety of other items are sterilized and used in great number. Items either may be sterilized individually or, more commonly, may be prepared in *sterile packs* containing the necessary equipment, wraps, gauzes, and swabs for a single type of dressing or operation. To ensure delivery of properly sterilized items to the patient's bed or surgery, hospitals usually employ a central sterile-supply department, where cleaning, packaging, and sterilization occur, although small hospitals may purchase presterilized diposable goods, which they use and discard.

To ensure against sterilization failure, *sterility test strips* are recommended. These are paper strips containing spores of a very heat-resistant bacterium, *Bacillus stearothermophilus*. The strips are placed in the most inaccessible locations of the materials to be sterilized, such as the centers of large packs or within the fingers of gloves. After sterilization has been completed, the test strips are placed in culture medium and incubated, to determine whether any viable spores remain. Such test strips should be used periodically, at least once a month, to check on the proper functioning of the autoclave equipment. In addition to these bacteriological checks on sterilization, there are a variety of commercial devices on the market that provide a quick visual check on sterilization. These are paper strips that contain a chemical substance which becomes colored when it has been subjected to sterilization temperature for the proper length of time. One of these strips is placed in the very center of the load of material being sterilized, and its color change indicates that proper heating conditions existed throughout the autoclave. One of these strips should be used with each load being sterilized. Although they do not provide a direct bacteriological indication of sterility, these chemical sterility indicators are simple to use and hence are useful in routine work.

A variety of methods for sterilization are available, depending on the kinds of materials to be treated. Heat sterilization is the most widely used, but gas sterilization and chemical sterilization are also used for certain kinds of objects.

STEAM STERILIZATION

This is the most widely used method in hospitals, suitable for most items that are not heat sensitive. The recommended procedure is to use a temperature of 121 to 123°C for 15 to 45 minutes, depending on the parcel size and wrapping. It is essential that all the air be removed from within and around the packs, by allowing steam to flow through the system before the pressure is allowed to build up. Improper packing or positioning of the contents in the autoclave may lead to sterilization failure, due to the fact that not all the air is replaced. In such situations, even if the temperature and pressure gauges on the autoclave show proper values, sterilization may not be complete. Items sterilized by steam sterilization include syringes, needles, forceps, scissors, and other metal items, as well as dressings and other cloth goods. A new procedure, *high-vacuum sterilization*, lessens the problems of trapped air and considerably reduces the sterilization time. High-vacuum sterilizers develop a high vacuum once or twice before steam is introduced, to remove air, and again after sterilization, to remove moisture. The items are moistened for a much shorter period of time and emerge from the autoclave dry. The equipment is expensive and complicated to maintain, so that it has not replaced conventional autoclaving in most hospitals.

GAS STERILIZATION

The method of gas sterilization is used to sterilize heat-sensitive items, such as rubber goods and delicate instruments. The gas used is ethylene oxide, which penetrates wrappers fairly well and is highly germicidal. Because pure ethylene oxide is explosive, it is usually used in a mixture of 10 to 12 percent with an inert gas, such as Freon or carbon dioxide. Special gas-sterilizer units are used. The air is first removed from the sterilizer, to create a vacuum, and then the gas is introduced. To be effective, the gas concentration within the chamber must be at least 450 mg (milligrams) per liter, and the treatment time must be 3 to 12 hours. After sterilization, it is usually necessary to aerate the wrapped materials for several days to permit any residual gas to escape, since the gas is irritating to the skin and mucous membranes. Careful control of the process is necessary, and after any changes in wrapping materials or methods of packaging, special tests should be run to be certain that sterilization has not been affected.

CHEMICAL STERILIZATION

Chemical sterilization is sometimes called *cold sterilization* and is used for certain items, such as thermometers, lensed instruments, polyethylene tubing and catheters, inhalation and anaesthesia equipment, and smooth, hard objects of various sorts. The object to be sterilized is completely immersed in the solution of the chemical agent for the desired length of time, and then removed and used. Because most chemical disinfectants are neutralized by dirt and organic matter, the objects to be sterilized must be clean. In many cases, chemical treatment is not used to sterilize but to disinfect, reducing the total microbial load or eliminating possible pathogenic organisms present on the material. Chemical agents used include ethyl alcohol, iodine, formaldehyde, iodophors, quaternary ammonium compounds, phenol, sodium hypochlorite, and glutaraldehyde. Some specific recommendations for cold-sterilization treatments for various materials found in hospitals are given in Table 9.4. Since many of these sterilization agents are irritating or toxic, it is important to rinse treated materials that are to be used in the body in order to remove chemical residues.

ORGANIZATION OF THE STERILE-SUPPLY DEPARTMENT

There are three categories of supplies: dirty (or contaminated), clean, and sterile. Dirty items are those that have been used, generally in contact with a patient. Before reuse, they must first be

---

TABLE 9.4
Recommendations for chemical disinfection and sterilization of instruments

Equipment	Disinfecting treatment[a]	Sterilizing treatment[b]
Smooth, hard objects	Alcohol, 15 min Iodophor, 20 min Glutaraldehyde, 15 min	Glutaraldehyde, 10 hr
Rubber tubing and catheters, completely filled	Iodophor, 20 min	(Use gas sterilization)
Polyethylene tubing and catheters, completely filled	Alcohol, 15 min Iodophor, 20 min Glutaraldehyde, 15 min	Glutaraldehyde, 10 hr
Lensed instruments	Glutaraldehyde, 15 min	Glutaraldehyde, 10 hr
Thermometers, oral and rectal	Iodine, 15 min	Glutaraldehyde, 10 hr
Hinged instruments	Alcohol, 20 min Iodophor, 30 min Glutaraldehyde, 20 min	Glutaraldehyde, 10 hr
Inhalation and anesthesia equipment	Alcohol, 20 min Glutaraldehyde, 20 min	Glutaraldehyde, 10 hr

Chemical agents: alcohol, 70 percent ethyl alcohol; iodophor, 500 parts per million available iodine; iodine, 0.2 percent iodine in 70 percent ethyl alcohol; glutaraldehyde, 2 percent aqueous solution.
[a] Disinfecting treatment: kills vegetative cells and some viruses.
[b] Sterilizing treatment: kills vegetative cells, spores, viruses.
Adapted from *Infection control in the hospital*, 3rd ed., American Hospital Association, 1974.

cleaned, then wrapped and sterilized. An important principle is that the three categories of items must be kept carefully separated so that there is no chance of mixup and so that dirty items will not contaminate sterile ones. Thus, the same cart that is used to dispense sterile items should not be used to pick up dirty ones, and the sterile-supply department should be so laid out that confusion is not possible. Dirty supplies must be carefully segregated from clean and sterile ones, and a definite traffic pattern is required (Figure 9.10).

The items to be sterilized must be wrapped in paper that is rugged and has sufficient wet strength so that it can withstand the sterilization process; yet it must allow efficient penetration of steam. The paper should provide an effective barrier to the penetration of bacteria and should be conformable and drape around the items snugly. Some paper manufacturers may supply special grades of paper specifically for sterilization; otherwise, a good grade of kraft paper is suitable. It might be noted that paper is a better bacterial barrier than is cotton toweling, although the latter drapes better and is less likely to tear and hence may be more suitable for large packs. The wrapping should be so done that there are no openings or crevices leading to the interior of the pack. The wrapped pack is sealed with tape or tied with stout string. There are several brands of tape on the market that change color when they have been through a sterilization cycle. Employing one of these may help to control use of sterile goods, although this indicator tape should not be relied upon absolutely.

In principle, a well-wrapped sterile pack should remain sterile indefinitely. In practice, it is preferable to discard or resterilize it if it has not been used within a prescribed arbitrary number of days. A shelf life of 28 days can be safely adopted as the maximum. Actually, in most hospitals, sterile items are used sufficiently quickly (within 2 to 3 days) so that considerations of shelf life are not relevant. A reasonable precaution is to be sure that the most-recently sterilized items are used only after less-recently sterilized items have been used.

There is a considerable need for *sterile water and sterile solutions* in the hospital. These may usually be purchased presterilized from pharmaceutical firms, but the principles of sterilization should still be understood. Sterilization of water and solutions in bottles is more complicated than is sterilization of dry goods: during the autoclaving process some of the liquid evaporates to steam and leaves the bottle; then during the cooling process after autoclaving, a vacuum is created inside the bottle that may suck contaminants into it from the air. To circumvent both these problems, a specially designed seal is used (Figure 9.11).

One means to ensure the effectiveness of hospital practices is to carry out control procedures for microbiological contamination.

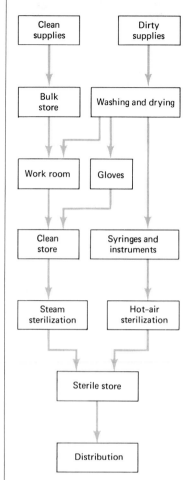

FIGURE 9.10
Flow of clean and dirty items through the sterile-supply department of a hospital.

MICROBIOLOGICAL SAMPLING IN THE HOSPITAL

THE SPREAD OF DISEASE

This involves the use of culture techniques that permit the counting of the microbiological load of objects or surfaces within the hospital. General plate-counting procedures may be used that enumerate the total microbiological populations, both pathogenic and non-pathogenic, or specific procedures may be used that monitor certain pathogenic organisms suspected of causing problems in the hospital. Routine sampling programs have been initiated in some hospitals, in which air, linens, surfaces in rooms, carpets, objects of various sorts, and so forth have been monitored on a regular basis in patients rooms, surgical suites, and nurseries, without regard to whether any specific microbiological problem exists. Such routine sampling programs are not recommended in general practice, because there is no evidence that routine environmental sampling contributes to a significant reduction in hospital infection. Hospitals are large and very complicated places, and routine sampling is unlikely to spot specific problem areas, since only a small number of areas are sampled in any routine survey.

Of greater value is the use of microbiological sampling as an adjunct to infection-control programs related to the investigation of specific hospital epidemics. If a series of similar cases or a significant epidemic occurs in a hospital and epidemiological evidence suggests that some group of articles, such as sterile solutions, nebulizer bottles, catheters, sterile instruments, and so on, are involved in the outbreak, intensive microbiological sampling of these items should be initiated, with the aim of identifying the contaminated items responsible for the epidemic and determining how contamination is occurring.

Similarly, if an outbreak of infection in patients occurs (for instance, caused by *Staphylococcus*, *Streptococcus*, or *Salmonella*, all common causes of hospital infection), then routine microbiological monitoring of hospital personnel is valuable. Such screening of personnel should be guided by the epidemiological findings in the case. For instance, if the outbreak is restricted to one floor or wing of the hospital, sampling should concentrate on personnel in that part of the hospital.

Microbiological control of sterilization procedures in hospitals is of course desirable on a routine basis, to make certain that sterilizers are working properly. Additionally, certain types of equipment used in hospitals are so prone to gross contamination that they should be checked routinely rather than only when an outbreak occurs. In this category are nebulizers, which are used to direct humid air directly to patients. The reservoirs of nebulizers contain water and are excellent places for microbial growth. Nebulizer reservoirs should be sampled routinely to be certain that their microbial loads are low.

Finally, it must be emphasized that microbiological sampling is only an adjunct to an infection-control program and should never

FIGURE 9.11
Special closure used to ensure sterility of a solution for hospital use. A rubber stopper allows injection of additives to the solution, while the metal rim prevents entry of microbes at the edge.

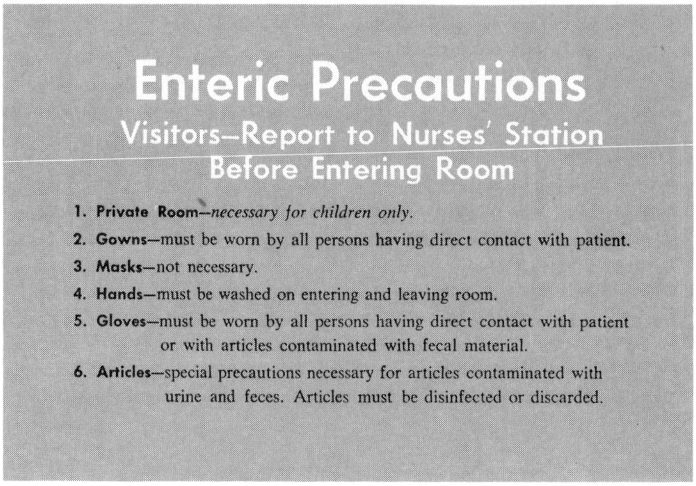

Enteric Precautions
Visitors—Report to Nurses' Station Before Entering Room

1. **Private Room**—*necessary for children only.*
2. **Gowns**—must be worn by all persons having direct contact with patient.
3. **Masks**—not necessary.
4. **Hands**—must be washed on entering and leaving room.
5. **Gloves**—must be worn by all persons having direct contact with patient or with articles contaminated with fecal material.
6. **Articles**—special precautions necessary for articles contaminated with urine and feces. Articles must be disinfected or discarded.

FIGURE 9.12
Isolation notices used in hospitals are shown on pages 298, 300, 302, 308, and 310. These notices are affixed as needed to the doors of patients' rooms.

Isolation notice used in hospitals

Diseases requiring enteric precautions:
1. Cholera
2. *Escherichia coli* gastroenteritis
3. Hepatitis, viral (infectious or serum)
4. Salmonellosis (including typhoid fever)
5. Shigellosis

replace the establishment and implementation of careful hospital practices. A trained hospital epidemiologist can more readily identify problem areas by visual examination than by microbiological sampling, except in very unusual situations. Dirty areas and sloppy practices must not be tolerated even if it can be shown by microbiological sampling that contamination is low.

9.11
Hospital isolation procedures

When there is great danger that a hospital patient may infect others, he or she may may be placed in isolation. The extent of isolation precautions depends upon the severity of the disease and its degree of communicability. The patient is placed in a private room and kept out of contact with all personnel except those specifically involved in his care. The room should be free of all unnecessary furniture and equipment; everything needed for the care of the patient is kept in the room, and the room has its own toilet facilities. Attendants of the patient are kept to a minimum; those present wear masks and gowns and discard them immediately after leaving the unit. All food is transferred to a tray kept permanently in the patient's room, and uneaten food or other solid waste is collected in paper bags for incineration. Bedding and towels used by the patient are placed in a special bag labeled CONTAMINATED and sterilized before routine cleaning. Feces, urine, and other body excretions may be poured directly in the toilet, but if this is not convenient they should be placed in a container with a solution of chlorinated lime and left standing for 4 hours before discarding. Books, magazines, and papers must be burned following contamination. Letters may be read or signed without contamination by placing them between sheets of newspaper or paper towels, with only the writing showing, so that the patient does not touch the letter.

There may be rare cases when even the foregoing procedures are not sufficient and the patient must be placed in a room with specially filtered air to avoid exit of pathogens into the outside air. Because of the expense of such extreme isolation, it may be preferable for the hospital to construct a special *isolation unit*, either in a separate building or in a separate wing, where several nursing units and the necessary laundry and cleaning facilities are available separate from the rest of the hospital. The isolation unit is entered through airlocks with positive air pressure from filtered air. Personnel change clothes within the unit and, at the end of the working day, bathe completely and discard all working clothes. Each room in the unit is designed so that it has no connecting ducts or channels to other rooms, and personnel cannot move directly from one room to another.

Infants born prematurely are unusually susceptible to infection and may be placed in isolators similar to those used for maintaining germ-free animals (see Section 8.2). The chance of contaminating pathogens reaching the interior of such isolators is minimal, and the survival rate of such infants has been raised considerably over the past several years.

Isolation procedures vary with the nature of the infection, and specific recommendations have been made by the U.S. Public Health Service. Special self-adhesive cards are available that can be affixed to doors of patients rooms, to call attention to the isolation procedures in force. These cards are illustrated in Figure 9.12, which also lists infectious diseases for which the specific isolation procedures are suggested.

PROTECTIVE ISOLATION

Protective isolation is also called *reverse isolation* and is used to protect patients with unusual susceptibility to infection. Patients in this category include those with extensive burns, premature infants, patients with immune-deficient diseases, patients who have received transplants and because of this have been given drugs to suppress the immune system, and patients who have received intensive radiation or corticosteroid therapy. Such patients may succumb to infections by organisms not generally considered pathogenic for normal individuals. Any source of contamination is potentially dangerous, so that the patient must be maintained in an environment where the level of asepsis is comparable to that in an operating room. All persons with suspected infection must be excluded from the room, and all personnel entering the room must wear clean caps and sterile masks and gowns. When it is necessary to touch the patient, sterile gloves must be used, and precautions are taken to ensure that all objects the patient touches are sterile. This type of protective isolation is a continuous, 24-hour responsibility and requires the use of specially trained personnel who are aware of the requirements.

Wound & Skin Precautions

Visitors—Report to Nurses' Station Before Entering Room

1. **Private Room**—desirable.
2. **Gowns**—must be worn by all persons having direct contact with patient.
3. **Masks**—not necessary except during dressing changes.
4. **Hands**—must be washed on entering and leaving room.
5. **Gloves**—must be worn by all persons having direct contact with infected area.
6. **Articles**—special precautions necessary for instruments, dressings, and linen.

NOTE: *See* manual for Special Dressing Techniques to be used when changing dressings.

Isolation notice used in hospitals

Diseases requiring wound and skin precautions:

1. Burns, extensive, not infected with *Staphyloccus aureus* or Group A streptococcus
2. Gas gangrene
3. Impetigo
4. Staphylococcal skin and wound infections
5. Streptococcal skin infection
6. Wound infection, extensive

Although procedures for disposing of the dead were based originally on superstitious customs or religious practices, there are good scientific reasons dictating that safe and sanitary disposal procedures be followed. Corpses may be important sources of infectious organisms and must be so treated that they are rendered sanitary. In addition, corpses rapidly putrefy, and if they are not to be buried immediately, they must be properly preserved.

The living body has tremendous powers to resist bacterial growth, but once death occurs, resistance is quickly lost, and bacteria that were not harmful during life multiply and spread rapidly. Even if death has been due to bacterial infection, it is not these pathogenic bacteria that attack the corpse but putrefactive bacteria found in the normal flora of the body or in the environment. The most important putrefactive bacteria are the coliform organisms, which are present in large numbers in the intestinal tract and can quickly spread throughout the dead body. Another important bacterium is *Clostridium perfringens*, also a member of the normal flora, which produces enzymes that rapidly digest tissues. The putrefactive bacteria produce gases such as hydrogen sulfide, hydrogen phosphide, mercaptans, methane, carbon dioxide, ammonia, and hydrogen. Several of these gases are quite odoriferous, so that the putrefying body exudes an awful stench.

The rate of decomposition of the body is determined by innate and external factors. Stillborn infants decompose very slowly because they lack a normal intestinal flora. Obese persons decompose more rapidly than thin ones, since they retain body heat longer and have a higher moisture content. Females have a higher moisture content than males and hence may decompose faster. If the tissues are

9.12
Embalming and disposal of the dead

dehydrated, as in death preceded by severe vomiting and diarrhea, putrefaction is delayed. The most important external factor influencing the decomposition rate is temperature. Most of the normal flora involved in putrefaction have optima around 37°C; therefore, at cold temperatures, such as may prevail in winter, putrefaction is greatly delayed. A body decomposes more slowly immersed in water than when exposed to air because in water free access of oxygen to the body is decreased.

After burial in soil, the rate of decomposition will be influenced by the soil moisture level; it is more rapid in moist than in dry soil. Decomposition is more rapid in porous, sandy soil than in heavy clay soil and is more rapid in shallow than in deep burial. Bodies buried in caskets decompose more slowly than those buried in direct contact with the soil.

Embalming is the process by which the dead body is treated to preserve it and to disinfect it. There are two reasons for embalming: (1) to preserve the body in a state as near natural as possible, in order that it may be viewed prior to the funeral without undue disturbance to the viewer; and (2) to disinfect the body, thus destroying any disease-producing organisms present. Embalming is accomplished by injecting into the body an embalming fluid that keeps the tissues from being decomposed and holds them in a nearly natural state. The embalming fluid is usually injected into the body through the arterial system, generally under considerable pressure in order to force it into all parts of the body. The active ingredient of the embalming solution is usually formaldehyde, which is a potent antimicrobial agent and one that kills rapidly on contact. Various additives are usually put in the aqueous formaldehyde solution, including anticoagulants, wetting agents, dyes, humectants (to help retain moisture in the tissues), and perfumes. The details of embalming for the individual corpse are highly variable, and a considerable knowledge is required on the part of the embalmer to produce a well-preserved corpse for the funeral.

If an embalmed corpse is not desired, a quicker, less expensive, and more sanitary way of disposing of a body is by *cremation*. The corpse, in a wooden coffin, is placed in the cremation chamber, which has been previously heated to 600 to 700°C. The body and coffin serve as additional fuel, and the temperature rises to about 1000°C; the temperature is maintained at this level until combustion is completed, usually about 1 to 2 hours. The cremation chamber should be so designed that no smoke issues, since rising particles might carry with them live bacteria. In the fire, all microbes are destroyed rapidly, and all that remains after cremation is the mineralized body skeleton, which is broken up into pieces and crushed to a fine ash with a mortar and pestle. The resulting ash, weighing about 5 lb, is placed in a suitable urn for burial or disposal.

We have learned in this chapter some of the principles relating to the spread of disease in populations. This knowledge is of value in both public health and preventive medicine, as well as in environmental sanitation in hospitals. The means of dispersal of a parasite from an infected to a susceptible host depends on the site of infection and on properties of the parasite that enable it to survive in the environment. Many parasites are dispersed through air, water, and food, some via inanimate objects, some by animals; other parasites are only dispersed by direct contact between people. Once knowledge is gained of the means of dispersal of a pathogen, control of the disease it causes is possible.

A disease is said to be epidemic when it occurs simultaneously in an unusual number of individuals within a community, whereas an endemic disease is one that is constantly present in a population. Epidemics arise either because of changes in the susceptibility of populations or because of changes in the virulence of pathogens. Control of epidemics can never be an individual matter but must involve participation of the entire community. Control begins with recognition of the presence of an epidemic, which requires appropriate diagnostic procedures. Specific control methods must be developed for each disease and depend on the manner of spread of the organism. With development of effective public health measures, many serious diseases have been eliminated from populations. Even diseases for which effective drugs are not available may still be eliminated by appropriate sanitation or immunization procedures.

Hospitals are especially hazardous because they house individuals with many kinds of infectious diseases and because many patients in weakened condition are unusually susceptible to infection. Because of these hazards, effective programs of hospital sanitation are essential, and hospital personnel must be well trained in proper preventive measures.

Summary

Study questions

1 Describe four ways in which microorganisms might leave the human body to be transferred to another person.
2 Trace the route of an infectious microorganism from the moment it is expelled from the mouth by a sneeze until it becomes lodged in the lungs. How might this transfer be prevented?
3 How may animals be involved in transmission of diseases to humans?
4 Distinguish between mortality and morbidity.
5 Examine the data in Figure 9.5 and compare with the data in Figure 1.1. Explain the differences in terms of the means available for controlling microbial infections.
6 Differentiate between endemic and epidemic diseases.
7 Describe three types of situation that may favor the development of an epidemic.
8 Why was measles considered a cyclic disease in the years before the measles vaccine was available? What effect has the vaccine had?
9 What is meant when we say an individual is a carrier? How are carriers important in public health?
10 What is quarantine, and how is it used in the control of infectious diseases?
11 Discuss ways in which epidemics may be controlled.
12 What means are available for the complete eradication of disease? Do you think it likely that *any* infectious disease will be completely eradicated?

THE SPREAD OF DISEASE

13 Why is the risk of infection especially great in hospitals? How can proper hospital sanitation procedures reduce the hazards that occur there?

14 How are the following organisms involved in hospital infections: *Staphylococcus aureus, Pseudomonas aeruginosa, Escherichia coli*?

15 Describe some dangers to the nurse during treatment of infectious patients. How can these dangers be prevented?

16 Why are the procedures for surgical skin disinfection so elaborate? What hazards must be overcome?

17 Describe how you would set up a sterile-supply department in a hospital. Indicate what types of items would be needed; discuss sterilization procedures and organization of the department.

18 What types of tests are used to ensure that hospital sterilizers are functioning properly?

19 Contrast steam and gas sterilization as they are used in hospital practice. List the kinds of objects sterilized by each. What are the advantages and disadvantages of each type of sterilization?

20 When a patient is placed in isolation, what sorts of precautions may be taken to prevent infection of other people?

21 Examine the quarantine control cards illustrated in Figure 9.12. Why are greater precautions taken for some kinds of infections than for others?

22 What microbiological reasons are there for disposing of the dead by cremation or embalming and burial?

Suggested readings

American Hospital Association, *Infection control in the hospital*, 3rd ed., American Hospital Association, 840 N. Lake Shore Drive, Chicago, 1974. An excellent brief textbook on the principles and control of hospital infection. Gives specific practical recommendations for sterilization of hospital goods and for contamination control.

American Public Health Association, *Control of communicable diseases in man*, 11th ed., American Public Health Association, 1790 Broadway, New York, 1971. Primarily in outline form, this standard work gives specific recommendations and data for the control of all infectious diseases. It also gives a brief summary of the characteristics of each disease.

Block, S. S., ed., *Disinfection, sterilization, and preservation*, 2nd ed., Lea and Febiger, Philadelphia, 1977. Good discussion of agents used to control microbial growth.

Cockburn, T. A., "Eradication of infectious diseases," *Science*, **133**, 1050–1058, 1961. Although somewhat old, this article enunciates clearly the principles involved in the complete eradication of infectious diseases.

Le Riche, W. H., and J. Milner, *Epidemiology as medical ecology*, Williams and Wilkins, Baltimore, 1971. An elementary textbook on epidemiology and disease control.

Sartwell, P. E., ed., *Preventive medicine and public health*, 10th ed., Appleton-Century-Crofts, New York, 1973. The standard textbook on epidemiology and public health, with excellent detailed treatment of the spread and prevention of disease.

Top, F. H., and P. F. Wehrle, eds., *Communicable and infectious diseases*, 8th ed., C. V. Mosby Co., St. Louis, 1976. A standard medical textbook on infectious diseases and their control. Good epidemiology chapters.

Wishnow, R. M., and J. L. Steinfeld, "The conquest of the major infectious diseases in the United States: a bicentennial retrospect," *Ann. Rev. Microbiol.*, **30**, 427–450, 1976.

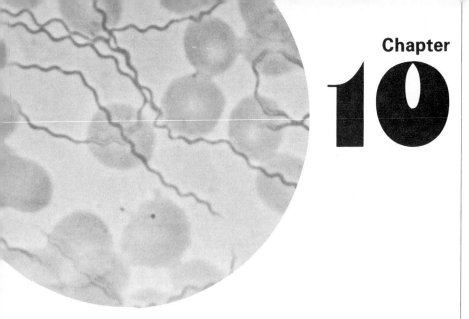

10

Bacterial diseases

Infectious diseases are caused by a wide variety of microorganisms, both procaryotic and eucaryotic. The most serious and numerous diseases of humans are caused by bacteria, yet some very important diseases are also caused by protozoa. The human diseases caused by fungi are less numerous and tend to be less serious, yet when well-established, fungal diseases are among the most difficult to eradicate. In this chapter we discuss the most important infectious diseases caused by bacteria; the diseases caused by fungi, protozoa, and viruses will be discussed in the next chapters. For each disease, we focus on the causal organism, its mode of transmission, manner of production of disease symptoms, immunization procedures for the disease, chemotherapy, and epidemiology. Procedures used for culturing and identifying the pathogen are also given where appropriate. This chapter can be read straight through to give some idea of the nature and extent of bacterial diseases, or it can be used as a reference source when information about a specific disease is sought.

10.1 Streptococcus

The genus *Streptococcus* contains a wide variety of members, with quite distinct habitats, whose activities are of considerable practical importance to human beings. Some members are pathogenic to humans and animals. In preantibiotic days, streptococcal infections constituted one of the leading causes of death. As producers of lactic acid, certain streptococci play important roles in the production of buttermilk, silage, and other fermented products.

Pathogenic species of the genus *Streptococcus* induce hemolysis on blood agar, owing to formation of a toxin (Figure 10.1). These streptococci are divided into immunological groups based on the presence of specific antigens. These antigenic groups are designated by letters: groups A through O are currently recognized, and those hemolytic streptococci found in human beings usually contain the group A antigen. These bacteria are called the *pyogenic* streptococci and are most frequently associated with disease of man. *Pyogenic* means "pus-forming" and refers to the characteristic symptoms induced by these organisms when infecting the skin or peripheral areas of the body. These symptoms result from the production of a variety of bacterial enzymes (Table 10.1) that cause destruction of phagocytic and other cells. These destroyed cells accumulate at the site of infection, leading to the formation of pus. Not all strains of pyogenic streptococci produce all the products listed in the table but every strain produces a good proportion of them. One product, erythrogenic toxin, is responsible for the characteristic rash of scarlet fever.

Group A, or pyogenic, streptococci are classified into a number of antigenic *types*, based on the immunological nature of a cell-surface protein called *M protein*. This M protein is associated with a resistance of the organism to phagocytosis. There are more than 40 M protein types, which are given numerical designations, that is, type 1, type 12, type 14, and so on. The study of M protein types is of use in following the spread of a specific strain of streptococcus through a population during an epidemic. Some types are often associated with a particular disease condition; for instance, type 12 is usually the causal agent of glomerular nephritis, a disease of the kidney.

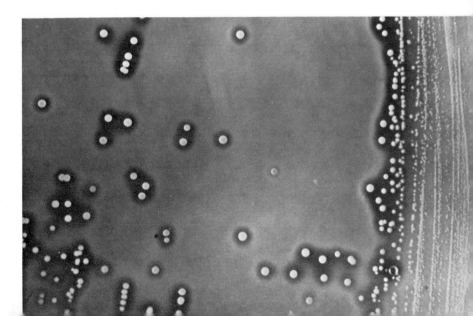

FIGURE 10.1
Demonstration of hemolysis on a blood agar plate. The zone of clearing around each colony is due to lysis of red blood cells. The organism is a *Streptococcus* species isolated from a patient with a sore throat.

Product	Distribution	Comment	
Erythrogenic toxin	Produced by most strains	Causes scarlet fever rash	**TABLE 10.1**
Streptolysin S	Produced by most strains	Hemolysin (affects phospholipid of membrane)	Extracellular products of group A streptococci
Streptolysin O	Produced by most strains	Hemolysin (affects cholesterol of membrane)	
Streptokinase	Produced by most strains	Causes dissolution of blood clots	
Deoxyribonuclease	Produced by all strains	Hydrolyses DNA	
Ribonuclease	Produced by most strains	Hydrolyses RNA	
Hyaluronidase	Produced by most strains	Hydrolyses hyaluronic acid in connective tissue	
Streptococcal proteinase	Produced by most strains	Nonspecific protease	
Amylase	Variable, but produced by many strains	Hydrolyses starch	

A wide variety of diseases are associated with streptococcal infection. These include mastitis, peritonitis, streptococcal sore throat, pneumonia, puerperal sepsis (a disease of the uterus following childbirth), erysipelas (a generalized infection of the body, sometimes called "blood poisoning"), glomerular nephritis, and rheumatic fever. In most streptococcal infections, the causal organism is transmitted through the air, and infection is initiated in the upper respiratory tract. If unchecked, the organism may spread to other parts of the body where more generalized infections can commence. Individuals vary greatly in their susceptibility to streptococcal infection. Many people are symptomless carriers of group A streptococci in their upper respiratory tracts and may serve as reservoirs for the infection of more susceptible individuals. Virtually all strains of group A streptococci are highly sensitive to penicillin and other antibiotics active against Gram-positive bacteria, and resistant strains rarely develop; therefore, most acute streptococcal infections may be readily treated. In preantibiotic days, however, group A streptococcus diseases were among the most frequent causes of death in humans. The one streptococcal disease that is still of major concern is rheumatic fever. The pathogenesis of this disease is still obscure, but the symptoms seem to be allergic in nature, and about 3 percent of the population is susceptible. Although treatment of the symptoms of rheumatic fever is not possible, control of the disease can be effected by continuous administration of penicillin to susceptible individuals, which hinders the initiation of minor respiratory infections and thus prevents the onset of the allergic reactions.

The bacterium *Streptococcus mutans* has been implicated as a prime causal agent of dental caries (tooth decay). The organism is a common component of dental plaque. Its cariogenicity is related to its ability to produce an extracellular slime or gum that enables the bacterium to adhere firmly to the tooth surface. The slime material is called *dextran* and is produced only from sucrose; this is one reason that sucrose is so cariogenic. The organisms adhere to the tooth surface, especially in crevices where they are difficult to remove by brushing. Sugars in the diet are converted to acids, which cause decalcification of the hard dental tissue of the tooth. Once the breakdown of the hard tissue has begun, proteolysis of the matrix of the tooth enamel occurs, through the action of proteolytic enzymes released by other bacteria. Microorganisms penetrate further into the decomposing matrix, but the later stages of the process may be exceedingly slow and are often highly complex.

Susceptibility to tooth decay varies greatly among individuals and is affected by inherent traits in the individual as well as by diet and other factors. The structure of the calcified tissue plays an important role. Incorporation of fluoride into the calcium phosphate crystal matrix makes the tooth more resistant to decalcification by acid; the use of fluorides in drinking water or toothpastes thus aids in controlling tooth decay. Careful attention to routine dental hygiene (tooth brushing, dental floss) is also necessary to reduce the amount of plaque that is present on the teeth.

The pneumococci (*Streptococcus pneumoniae*) are a group of streptococci closely related to the other streptococci. They differ from the other streptococci in that they tend to lyse readily, either spontaneously in the late logarithmic phase or after the addition of bile salts (hence, the pneumococci are said to be "bile soluble"). These bacteria are the most frequent causative agents for pneumonia, although pneumonia can also be caused by other bacteria, as well as by viruses.

10.2 Staphylococcus

The genus *Staphylococcus* contains common parasites of man and animals, including some that occasionally cause serious infections. Staphylococci are nonmotile, nonsporulating, Gram-positive cocci, which carry out cell division in several planes and hence can form irregular clumps. Despite the absence of spores, staphylococci are relatively resistant to drying and hence can be readily dispersed in dust particles through the air. They are able to grow on media containing a high concentration (7.5 percent) of sodium chloride, and this property is exploited in their selective isolation from natural materials, since very few other bacteria from nonsaline environments show such tolerance. In humans, two forms are recognized: *S. epidermidis*, a nonpigmented, nonpathogenic form usually found on the skin or mucous membranes; and *S. aureus*, a yellow-pigmented form commonly associated with pathological conditions, including boils, pimples, impetigo, pneumonia, osteomyelitis, carditis, meningitis, and arthritis. Those strains of *S. aureus* most frequently causing human disease produce a number of extracellular enzymes or toxins. At least four different hemolysins have been recognized, a single strain often being capable of producing more than one hemolysin, and the production of these is responsible for the hemolysis seen around colonies on blood agar plates. Another substance produced is *coagulase*, an enzymelike factor that causes fibrin to coagulate and form a clot (see Section 8.5). The production of coagulase is generally associated with pathogenicity. It seems likely that blood clotting induced by coagulase results in the accumulation around the cocci of fibrin and thus renders them resistant to phagocytosis. In addition, the formation of such fibrin clots results in the walling off of the bacteria, making them less likely to come into contact with host defense agents. Strains are tested for coagulase production by mixing 0.5 ml of a dense suspension of bacteria with 5 ml of human citrated plasma and incubating at 37°C. A coagulase-positive strain will usually cause clotting after 3 hours or after overnight incubation. Most *S. aureus* strains also produce *leukocidin*, a complex factor that causes the destruction of leukocytes; if a *S. aureus* cell is phagocytized, release of leukocidin can lead to the destruction of the leukocyte, and the *S. aureus* cell is enabled to escape unharmed. Production of leukocidin in skin lesions such as boils and pimples results in much cell destruction and is one of the factors responsible for pus formation (Figure 10.2). Other extracellular factors produced by some strains include proteolytic enzymes, hyaluronidase, fibrinolysin, lipase, ribonuclease, and deoxyribonuclease.

FIGURE 10.2
The structure of a boil. The bacteria causing boils are often of the genus *Staphylococcus*. *a* A ripe boil. The head of the boil is teeming with bacteria and pus. *b* A ruptured boil. The pus and bacteria have escaped.

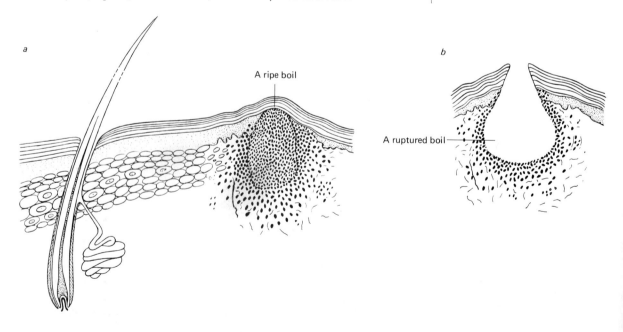

a

A ripe boil

b

A ruptured boil

The most common habitat of *S. aureus* is the upper respiratory tract, especially the nasopharyngeal passages, and many people are carriers throughout most of their lives; however, the strain being carried probably varies from time to time. Most infants become infected with *Staphylococcus* during the first week of life and usually acquire the strain associated with the mother or with another close human contact. In most cases, these strains do not cause pathological conditions, and serious staphylococcal infections occur only when the resistance of the host is low because of hormonal changes, debilitating illness, or treatment with steroids or other antiinflammatory drugs. Hospital epidemics have occurred in recent years, which have usually involved antibiotic-resistant strains. Extensive use of antibiotics has often resulted in the selection of resistant strains of *S. aureus*, and hospital epidemics with antibiotic-resistant staphylococci may occur when patients whose resistance to infection is lowered (owing to other diseases, surgical procedures, or drug therapy) become infected from hospital personnel who have become normal carriers of antibiotic-resistant staphylococci. Control of such hospital epidemics requires careful attention to the usual rules of aseptic technique (described in Section 9.9) and the elimination of personnel known to be carriers of drug-resistant strains.

The most common type of food poisoning is that caused by enterotoxin-producing strains of *S. aureus*, as discussed in Section 15.2.

10.3 Neisseria

The genus *Neisseria* consists of Gram-negative cocci that live in human beings. They are nutritionally highly specialized, grow well only near body temperature, and are extremely sensitive to inhibitory materials. Two species are pathogenic for man: *Neisseria gonorrhoeae*, the causal agent of gonorrhea; and *N. meningitidis*, the most frequent causal agent of spinal meningitis. These two organisms are often given the colloquial names of gonococcus and meningococcus, respectively.

Gonorrhea is a veneral disease that apparently occurs only in human beings; despite numerous attempts, experimental animals have not been successfully infected. It is one of the most widespread human diseases, and in spite of the availability of excellent drugs, it is still common even in countries where the cost of drugs is no economic problem. In contrast to syphilis (another important venereal disease of humans), gonococcus infection rarely results in serious complications or death. The disease symptoms are quite different in the male and female (Figure 10.3): In the female, the symptoms are usually a mild vaginitis that is difficult to distinguish from vaginal infections caused by other organisms; the infection may easily go unnoticed. In the male, however, the organism causes

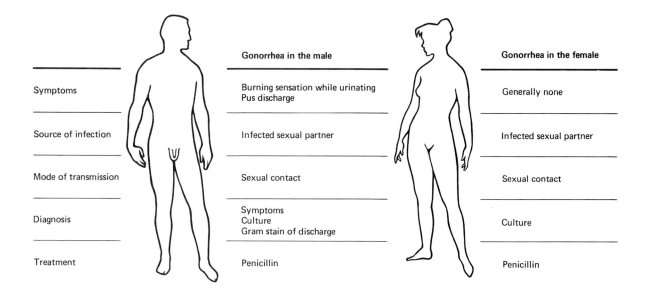

	Gonorrhea in the male	Gonorrhea in the female
Symptoms	Burning sensation while urinating Pus discharge	Generally none
Source of infection	Infected sexual partner	Infected sexual partner
Mode of transmission	Sexual contact	Sexual contact
Diagnosis	Symptoms Culture Gram stain of discharge	Culture
Treatment	Penicillin	Penicillin

FIGURE 10.3
Contrasts between gonorrhea in the male and female.

a painful infection of the urethral canal, and the disease is often given the colloquial name of "strain" or "clap." In addition to gonorrhea, the organism also causes eye infections in the newborn and adult.

The causal agent, *N. gonorrhoeae*, is an extremely fastidious organism and is quite difficult to culture on initial isolation from pathological material (Figure 10.4). Many components of ordinary culture media are inhibitory, but starch, serum, or heated whole blood, when added to culture media, adsorbs these toxic materials and makes growth possible. On initial isolation most strains require an atmosphere containing 2 to 10 percent CO_2, and their temperature limits are very narrow; growth does not occur below 30°C or above 38.5°C. The organism is killed quite rapidly by drying, sunlight, and ultraviolet light. This extreme sensitivity probably explains in part the venereal nature of the disease; that is, the organism is transmitted from person to person only by intimate direct contact. Infants may become infected during birth by mothers who carry the organism, showing the infection in the eyes. Prophylactic treatment of the eyes of all newborns with silver nitrate is mandatory in most states and has helped to control the disease in infants. The toxicity of the organism is probably due exclusively to an endotoxin and no extracellular products significant in pathogenicity seem to be produced.

The organism enters the body by way of the mucous membranes of the genitourinary tract. Treatment of the infection with penicillin is highly successful; a single injection usually results in elimination of the organism and complete cure. Strains moderately resistant to penicillin have been known since at least 1958 but have not presented any special problem since they can be controlled with higher doses and repeated injections of the antibiotic.

The recommended dosage today is 4.8 million units of penicillin, contrasted to the 200,000 units which was sufficient before resistant strains arose. Recently, a new strain of gonococcus has been identified which is resistant to penicillin in a different manner. No amount of penicillin will kill this new organism; instead, the organism destroys the drug by action of an enzyme, penicillinase. Another antibiotic, spectinomycin, can be used and is effective against this new strain, but the problems involved in identifying these special cases are great. Despite the ease with which most gonorrhea can be cured with penicillin, the incidence of gonococcus infection remains relatively high. The reasons for this are twofold: (1) Acquired immunity does not exist, hence repeated reinfection is possible; and (2) symptoms in mucosal tissues are such that even at the height of the disease it may go unrecognized. Thus, a promiscuous infected person can serve as a reservoir for the infection of many other people. The disease could be controlled if the sexual contacts of infected persons were quickly identified and treated, but it is often difficult to obtain the necessary information and even more difficult to arrange treatment.

Neisseria meningitidis, the causal agent of *meningitis*, is also a parasite only of human beings; its natural habitat is the nasopharynx. The organism possesses much the same fastidious nature as *N. gonorrhoeae*, and similar culture conditions are used in its isolation. Meningococcus is also sensitive to drying, heat, and other

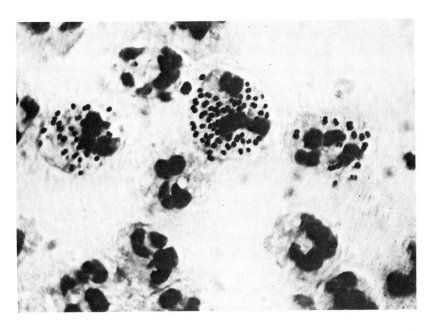

FIGURE 10.4
Photomicrograph of gonorrhea-causing bacteria (*Neisseria gonorrhoeae*) within a human phagocyte; magnification 1,800×. *Courtesy of Theodor Rosebury.*

adverse environmental conditions, but it is sufficiently resistant to provide the possibility for transmittal from person to person via the respiratory tract, although relatively close contact is required. The number of carriers of meningococcus in the upper respiratory tract is quite high; most carriers have infections with no noticeable symptoms. In certain instances, the organism invades the bloodstream from the nasopharynx and sets up a generalized infection of the body, and in some of these cases, the organism invades the central nervous system and becomes established in the meninges (the membranes surrounding the brain and spinal cord). The symptoms of meningitis are severe headache, muscular spasm, stiff neck, and exaggerated reflexes; if untreated, these are followed by convulsions, coma, and death. When these symptoms develop, the organism can usually be cultivated from samples of the cerebrospinal fluid. Pathogenicity results from the presence of an antiphagocytic capsule and the production of a characteristic endotoxin. The organism is extremely sensitive to sulfonamides, penicillin, and most other antibiotics, and drug-resistant strains have rarely developed. Successful treatment of meningitis requires the use of a drug that will penetrate the meningeal membrane, and the sulfonamides are often preferred because they possess this property. However, penicillin can also be used; although it does not normally penetrate to the spinal fluid in healthy persons, it does so when the meninges are acutely inflamed.

Because of the high incidence of inapparent infections of the nasopharynx, it is virtually impossible to control the spread of *N. meningitidis* through the human population. Fortunately, the clinical disease meningitis occurs only rarely, seeming to erupt in cycles of about 8 to 10 years. However, it is fairly common in military camps and barracks, especially among new recruits. It is thought that overcrowding in such situations frequently results in transmission of the organism from carriers to susceptible individuals, and the reduced resistance consequent to fatigue and other factors attendant on the training process may result in onset of systemic and nervous-system infections. The frequent presence of antibodies bactericidal to *N. meningitidis* in the serum of carriers suggests that these antibodies may play a role in preventing the spread of the organism from the nasopharynx to the bloodstream and nervous system.

10.4 Bacillus

The genus *Bacillus* consists of aerobic, spore-forming, Gram-positive rods. Members of the genus *Bacillus* are easy to isolate from soil or air and are among the most common organisms to appear when soil samples are streaked on agar plates containing various nutrient media. Spore formers can be selectively isolated from soil, food, or other material by exposing the sample to 80°C for 10 to 30 minutes,

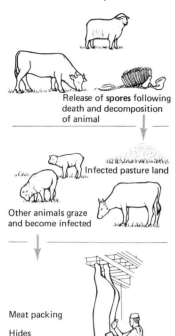

Infected cattle and sheep

Release of **spores** following death and decomposition of animal

Infected pasture land

Other animals graze and become infected

Meat packing

Hides

Wool

Humans handle carcass, hide, or wool and become **infected**

FIGURE 10.5
Transmission of the anthrax bacillus among animals and humans.

a treatment that effectively destroys vegetative cells while many of the endospores present remain viable. When such pasteurized samples are streaked on plates and incubated aerobically, the colonies that develop are almost exclusively of the genus *Bacillus*.

Anthrax is the only human disease caused by a *Bacillus* species: *B. anthracis*. Anthrax was the first disease shown conclusively to be caused by a bacterium, and its study by Koch provided one of the foundations for the development of microbiology (see Section 1.3). Primarily a disease of farm animals, it is only occasionally transmitted to human beings. *Bacillus anthracis* is a very large bacillus, 1 to 1.5 μm wide by 4 to 8 μm long, and the cells usually remain attached after division; long chains are formed on agar to create colonies of characteristic morphology. The anthrax bacillus forms a capsule that protects it from phagocytosis and, probably because of this capsule, is highly invasive, growing well throughout the body; in later stages of the disease, large numbers of bacilli are found in the blood. In the body the organism produces an exotoxin that is responsible for most of the disease symptoms. The toxin causes shock, electrolyte imbalance, swelling, hemoconcentration, and acute kidney failure.

The organism infects human beings only when they come in contact with diseased animals or their hides or animal products (Figure 10.5). Many kinds of wild and domestic animals are susceptible, including cattle, sheep, horses, goats, pigs, mink, dogs, deer, birds, and even frogs and fish, although not all animals are equally susceptible. An animal usually becomes infected by ingesting bacilli or endospores from an infected carcass. At death, large numbers of bacilli are present in the blood and tissues, and if the carcass is opened, sporulation can occur (in warm countries, high environmental temperatures favor this process). If sporulation has occurred, the organism can remain viable in bone for long periods of time and is transmitted to uninfected animals when they eat the bones. Animals are usually attracted to the carcass because of their craving for salt or their need for bone minerals; the latter is observed especially among animals living on forage in phosphate-deficient areas. In dry soil and in areas of pastureland, anthrax endospores can also remain alive for very long periods, but in most agricultural soils, they do not persist for more than a few years. The organism is not able to grow directly in soil.

Infection in man is primarily an occupational hazard in the meat-packing and tanning industries. Invasion most often occurs by way of the skin, usually through a small scratch or abrasion, and the primary lesion develops as an inflamed pustule or blister. In most cases, the disease in man is self-limited, but the organism may spread from the initial site of infection and initiate a fatal systemic infection. Another form of anthrax in man, called "woolsorter's disease," is a respiratory infection resulting from the inhalation of endospores.

Immunity to anthrax develops through production of antibodies against the toxin. The first attenuated vaccine was the anthrax vaccine of Pasteur and his colleagues (see Section 1.6). Despite Pasteur's early success, subsequent work has shown that proper attenuation is difficult, and even today, no anthrax vaccine is known that is considered safe for man. Immunization of animals has also been unsatisfactory. Control of the disease in animals is primarily by eradication measures aimed at eliminating diseased animals from herds and destroying carcasses. The infection of man by means of contaminated shaving brushes made from animal hair has been controlled by sterilizing the brushes. The human disease can be effectively treated with penicillin, the tetracyclines, or erythromycin, provided the symptoms are detected before bacteremia has developed.

10.5 Clostridium

The clostridia are spore-forming and Gram-positive; all are obligate anaerobes, not only being unable to grow in the presence of air but also usually being killed by O_2 unless they exist in the endospore form. The main habitat of clostridia is the soil, where they live primarily in anaerobic pockets. However, several clostridia that live primarily in soil are capable of causing disease in humans under specialized conditions. These are *C. botulinum*, causing botulism, *C. tetani*, causing tetanus, and *C. perfringens* and a number of other clostridia, causing gas gangrene. These pathogenic clostridia produce specific toxins that are responsible for the disease symptoms.

Botulism is a food poisoning rather than an infection, since the microorganism grows not in the host but in the food itself, producing a toxin, and it is this ingested toxin that causes the disease symptoms (see Sections 15.2).

Tetanus is a generally fatal disease, the symptoms of which are due to a potent toxin that acts on the nerves of the central nervous system, causing spastic paralysis (Figure 10.6). The causal agent, *C. tetani*, frequently infects dirty wounds, and if extensive tissue damage results in anaerobic conditions, it can grow. It remains localized at this initial site of infection, but the toxin spreads to distant parts of the body. Once the toxin combines with the nerves, its action cannot be reversed. Hence, treatment of tetanus is difficult, and control of the disease is by immunization with toxoid. General practice is to immunize infants and children and to maintain immunity through life either by routine periodic immunization or by immunization when a serious wound occurs.

In *gas gangrene* the causal agent, *C. perfringens*, also infects wounds, but it is more invasive than *C. tetani* because a series of toxins are produced that cause extensive destruction of tissue. The term *gangrene* refers to dead tissue, and the gangrenous condition

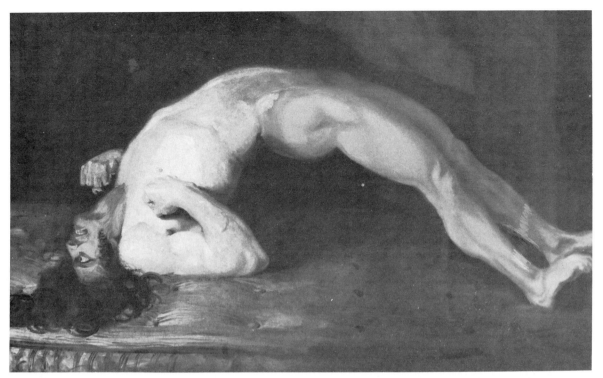

FIGURE 10.6
A soldier dying of tetanus.
*Photograph of a painting by
Charles Bell; in the Royal
College of Surgeons, Edinburgh.*

may result from either microbial or nonmicrobial causes. During *C. perfringens* infection, the dead tissue frequently contains gaseous products of the bacterium's activities, and the infected region may become liquid or hollow. In such dead tissue, anaerobic conditions develop, permitting the organism to grow more extensively. Some of the toxins also spread to distant parts of the body and cause destruction of heart tissue and other internal organs, leading to death. Immunization against all the toxins is not possible, so that control is mainly by preventing the spread of the pathogen through use of antibiotics and surgery. If the infection is in an extremity, the portion of the member that is diseased can be removed, thus eliminating the pathogen from the body. Gas gangrene develops most frequently as a result of infection of large and very dirty wounds. Infection can be prevented by immediate treatment of such wounds by surgical cleaning and antibiotics.

C. perfringens also can cause food poisoning, of a milder sort than that caused by *C. botulinum.* The organism produces a toxin while growing in food; when the food is ingested, the toxin acts on the gastrointestinal tract to cause vomiting and diarrhea. Food poisoning is considered in greater detail in Section 15.2.

The genus *Corynebacterium* comprises a group of aerobic, nonmotile, nonsporulating, Gram-positive rods. The *Corynebacterium* cell often has swollen ends so that the rod has a club-shaped appearance: hence the origin of the name (*koryne* is the Greek word for "club"). Members of the genus are widespread: some are common inhabitants of the soil, others are causal agents of plant diseases, and still others are organisms living with or pathogenic to man and higher animals. One species, *C. acnes*, is a common inhabitant of the human skin and has been implicated in the skin condition *acne*.

The best-known and most widely studied species is *C. diphtheriae*, the causal agent of *diphtheria*. At one time diphtheria was a major cause of death in children, but today it is quite rare. Diphtheria is of historical significance because it was the first infectious disease whose symptoms were shown to be due to an exotoxin, and it was also the first disease to be controlled by immunization procedures. The organism is strictly an inhabitant of the respiratory tract, being unable to invade other parts of the body. It is transmitted from person to person by the respiratory route. Upon its establishment in the upper respiratory tract, the organism multiplies on the mucous membranes and produces a potent exotoxin that causes damage to adjacent cells and creates a favorable environment for further growth. The inflammatory response of the throat tissues to infection results in the formation of a characteristic structure called a *pseudomembrane*, which consists of altered tissue cells and bacteria. This pseudomembrane can result in mechanical blockage of the throat, leading to death by suffocation. In most cases, however, death results from toxemia arising from the spread of the toxin throughout the body. Thus, even though *C. diphtheriae* is weakly invasive, its powerful toxin makes it highly virulent.

Administration of formalin-treated toxin (toxoid) results in the formation of antitoxin antibodies, and infants and small children are generally given routine immunizations. Antitoxin antibodies completely neutralize the action of the toxin but do not prevent the establishment of the organism in the upper respiratory tract. Inapparent infections are therefore common, and carriers can be the source of infection of unimmunized individuals. Occasional small epidemics of diphtheria can arise owing to the migration of unimmunized individuals into an area where the population is predominantly immunized but carries inapparent infections. To determine whether an individual is immune to diphtheria, the *Schick test* is used. This involves injecting under the skin a minute amount of diphtheria toxin; if a person is immune, antibodies present will neutralize the toxin, whereas in a nonimmune individual the toxin will induce formation of an area of redness and swelling. Diphtheria therapy is best effected by the use of antitoxin, since this will neutralize any circulating toxin. Antibiotic therapy is also used, but never alone, since the antibiotic will not affect toxin already circulating in the body.

10.7 Mycobacterium

The genus *Mycobacterium* contains rod-shaped nonsporulating aerobic organisms that exhibit acid-fast staining characteristics. Acid-fastness is a property by which cells stained with such dyes as basic fuchsin resist decolorization with dilute acid; cells of other bacteria are readily decolorized. Found only in the mycobacteria and a few related species, acid-fastness is conferred by the presence of large amounts of waxy substances on the cells, to which the dye molecules bind tightly. The acid-fast stain is thus a differential stain of great value for identifying mycobacteria in natural materials or in cultures. The genus *Mycobacterium* contains a variety of species that occur in soil, but the most important species are those pathogenic to man: *M. tuberculosis*, the causal agent of tuberculosis, and *M. leprae*, the causal agent of leprosy.

Tuberculosis has been one of the great scourges of mankind. Pulmonary tuberculosis, also called "consumption" or "phthisis," has been recognized as a disease entity for hundreds of years. Robert Koch first showed in 1882 that tuberculosis was caused by a bacterium when he successfully cultured *M. tuberculosis* in the laboratory and succeeded in establishing an experimental infection in guinea pigs. The organism, which is an obligate aerobe, has simple nutritional requirements and will grow on a synthetic medium containing acetate or glycerol as a sole carbon source and ammonium as a sole nitrogen source. Growth is stimulated by fats; egg yolk, which is high in fats, is often added to culture media to achieve more luxuriant growth, a glycerol-egg yolk medium often being used in the primary isolation of the organism from infected materials. Perhaps because of the high lipid content of its cell walls, the organism is able to resist such chemical agents as alkali or phenol for considerable periods of time. This resistance is used in the selective isolation of *M. tuberculosis* from sputum, this material first being treated with sodium hydroxide for 30 minutes before being neutralized and streaked on the isolation medium. Since avirulent and weakly virulent strains are widespread, any isolate must be tested to confirm virulence, and for this, the guinea pig is the animal of choice because it is highly susceptible to *M. tuberculosis* infection. Pathological material or a culture suspension of virulent organisms injected subcutaneously will cause the formation of characteristic tubercle nodules at the site of injection, and the spread of the organism from the initial site may result in secondary tuberculosis nodules in the spleen or peritoneal cavity within 4 to 5 weeks. Death of the animal usually occurs in 6 weeks to 3 months, and autopsy will reveal lesions in organs throughout the body.

The interaction of the human host and *M. tuberculosis* is an extremely complex phenomenon, being determined in part by the virulence of the strain but probably more importantly by the specific and nonspecific resistance of the host. It is convenient to distinguish between two kinds of human infection: primary and postprimary (or reinfection). Primary infection is the first infection that any individual receives and often results from inhalation of droplets containing viable bacteria derived from an individual with an active pulmonary infection. Dust particles that have become contaminated from sputum of tuberculous individuals are another source of primary infection. The bacteria settle in the lungs, grow, and become surrounded by macrophages that ingest the bacteria. In a few individuals with low resistance the bacteria are not effectively controlled, and an acute pulmonary infection is set up, which leads to the extensive destruction of lung tissue, the spread of the bacteria to other parts of the body, and death. Susceptibility to acute infection is more common in non-Caucasians, in infants and children, or in individuals whose nutritional status is poor.

In most cases, however, acute infection does not occur; rather, an immune response develops during the initial infection, and as a result, the macrophages become activated and effectively destroy the bacteria. The infection remains localized and is usually inapparent; later it subsides. But this initial infection makes the individual hypersensitive to the bacteria or their products and consequently alters the response of the individual to subsequent infections. If *tuberculin*, a protein fraction extracted from the bacteria, is injected intradermally into the hypersensitive individual, it elicits at the site of injection a localized immune reaction characterized by hardening and reddening of the site 1 to 2 days after injection. An individual exhibiting this reaction is said to be tuberculin positive, and many adults give positive reactions because of inapparent infections earlier in life. A positive tuberculin test does not indicate active disease but only that the individual has been exposed to the organism at some time. It is in tuberculin-positive individuals that the postprimary type of tuberculosis infection occurs. When renewed pulmonary infections occur in tuberculin-positive individuals, they are chronic types that involve destruction of lung tissue, followed by partial healing and a slow spread of the lesions within the lungs (Figure 10.7). Spots of destroyed tissue may be revealed by X-ray examination, but only in individuals with extensive tissue destruction are viable bacteria found in the sputum or stomach washings. In many cases infections in tuberculin-positive individuals are a result of reactivation and growth of bacteria that have remained alive and dormant in the lungs for long periods of time. Malnutrition, overcrowding, stress, and hormonal imbalance often are factors predisposing to secondary infection.

Drugs such as streptomycin, isonicotinic acid hydrazide (INH),

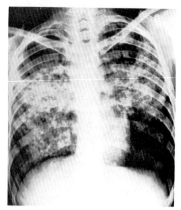

FIGURE 10.7
An X-ray photograph showing an advanced case of pulmonary tuberculosis. From Paul T. Chapman, "Tuberculosis," in F. H. Top and P. F. Wehrle, (eds.), *Communicable and infectious diseases,* 7th ed., The C. V. Mosby Co., St. Louis, 1972.

and *p*-aminosalicylic acid (PAS) are the most effective antituber-culosis agents, the latter two being almost specific in their action against mycobacteria. Treatment with these drugs either singly or in combination is of considerable benefit, although rather long-term therapy is necessary to eliminate the organism from all the sites within destroyed lung tissue where it has become established. Drug therapy does not of course lead to the direct healing of damaged tissue, but by destroying the bacteria it prevents the establishment of new sites of infection and new tissue damage.

The severity of tuberculosis disease in Europe and North America had begun to subside in the late nineteenth century, even in the absence of chemotherapeutic measures, owing to improved nutritional and socioeconomic conditions. Most people became tuberculin positive at an early age, and so the disease was of the postprimary or reinfection type. The introduction of chemo-therapeutic measures in the post-World War II era changed this situation considerably (Figure 10.8). The elimination of the bacteria from infected individuals has greatly reduced the infection rate in the population as a whole, and today, tuberculin-positive individuals

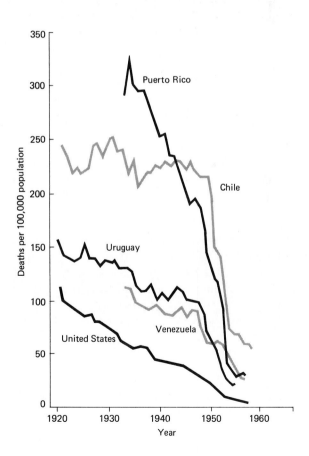

FIGURE 10.8
Efforts by public health authorities have resulted in a marked decline in deaths due to tuberculosis, as illustrated here for five countries.

comprise a smaller proportion of the population than they once did. At one time, extensive surveys were made using X rays and the tuberculin test to detect infected individuals, who could then be treated, but the incidence is now so low that the cost of such extensive surveys is hardly justified by the number of new cases detected.

A live vaccine, the BCG strain of *M. tuberculosis*, has been available for many years, but the desirability of its use has been the subject of much debate. Vaccination of tuberculin-negative individuals with BCG can convert them to tuberculin-positive status, with an increase in resistance to infection, but the vaccination confers the same hypersensitive state induced by natural infection and so does not eliminate the chance of the postprimary type of infection. Since vaccination renders the individual tuberculin-positive, it also eliminates the possibility of using the tuberculin test at a later date to detect a new infection. Vaccination in many countries is now recommended only for tuberculin-negative individuals who have a high probability of infection, such as children of tubercular parents.

Tuberculosis in cattle at one time was a serious disease. The bovine strains of *M. tuberculosis* are also highly virulent for human beings, and in the days before public health measures were widespread, many people became infected from contaminated dairy products. Because the organism entered the body by way of the gastrointestinal tract, the site of infection was usually not the lungs but the lymph nodes, and the organism subsequently localized in the bones and joints. Pasteurization of milk and other dairy products and elimination of diseased cattle have virtually eradicated this type of human tuberculosis in Europe and North America.

The disease *leprosy* (Hansen's disease) is caused by another mycobacterium, *M. leprae*, which has been difficult to culture although it grows profusely but slowly in intracellular sites in leprous lesions (Figure 10.9). Two types of infection are known: cutaneous, in which the organism primarily affects the skin, eventually causing extensive disfiguration; and neural, in which the organism infects the peripheral nerves, leading to loss of sensation. Both types of disease may be present in the same patient. Leprosy is not a highly contagious disease, and the incidence in most parts of the world is very low, although in ancient times it was apparently much more common, perhaps due to crowding and poor sanitation. Therapy with the drug diaminodiphenylsulfone (DDS) is effective, but treatment must be prolonged.

The enteric bacteria comprise a relatively homogeneous group: they are Gram-negative, nonsporulating rods, either nonmotile or motile by peritrichously occurring flagella; and they are facultative

a

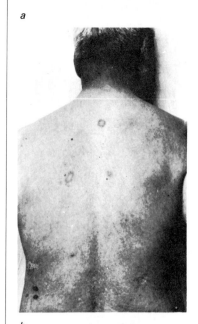

b

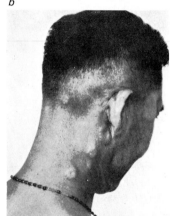

FIGURE 10.9
Symptoms of leprosy. *a* Macular skin lesions. *b* Plaque skin lesions. From C. H. Binford, "Leprosy," in F. H. Top and P. F. Wehrle, (eds.), *Communicable and infectious diseases*, 7th ed., The C. V. Mosby Co., St. Louis, 1972.

10.8
Enteric bacteria

anaerobes with relatively simple nutrition, fermenting sugars to acids or alcohols. Among the enteric bacteria are several genera pathogenic to man or animals as well as other genera of practical importance. These are listed in Table 10.2, along with the characteristics used in differentiating them. To distinguish among these characteristics, a variety of selective and differential media have been developed for isolating and characterizing specific genera of the enteric bacteria. Most selective media contain peptone or meat infusion, lactose, agents such as bile salts that inhibit the growth of bacteria of other genera, agar, and a pH indicator to measure the extent of production of acid from lactose. Once organisms are isolated, complete identification can be carried out by using various differential media: (1) An agar medium containing iron chloride will indicate the production of H_2S by the formation of black iron sulfide. (2) Motility can be indicated by making a stab in a soft agar of about 0.3 percent concentration. This agar concentration is sufficient to form a soft gel without hindering motility; during incubation motile cells swarm out from the line of inoculation and, through successive cell divisions, form a diffuse region of growth. If the culture is nonmotile, swarming will not occur, and growth will be confined to a compact band at the center of the tube. (3) Hydrolysis of gelatin can be detected when this substance is used in the medium as solidifying agent (instead of agar). (4) Production of urease can be shown with media containing urea and a pH indicator (the action of the enzyme on urea releases NH_3 with a concomitant rise in pH). In the diagnostic laboratory, a wide variety of even more specialized media and tests are used.

Members of the genus *Escherichia* are almost universal inhabitants of the intestinal tract of man and warm-blooded animals,

TABLE 10.2
Differentiation of the major subgroups of the enteric bacteria

	Lactose utilization	Indole production	Citrate utilization	H_2S production	Motility	Pigmentation	Gas from glucose	Urea decomposition
Predominantly acid producers								
Escherichia	+	+	−	−	+ (−)	−	+	−
Shigella	−	+	−	−	−	−	−	−
Salmonella	−	−	+	+	+	−	+ (−)	−
Erwinia	+ (−)	+ (−)	+ (−)	+ (−)	+	Yellow	+ (−)	−
Proteus	−	+ (−)	+ (−)	+ (−)	+	−	+	+
Predominantly alcohol producers								
Enterobacter (and *Klebsiella*)	+	−	+	−	+ (−)	−	+	−
Serratia	−	−	+	−	+	Red	+ (−)	−

KEY: +, characteristic present; −, characteristic absent; +(−), characteristic variable.

although they are by no means the dominant organisms in these habitats. Only rarely is *Escherichia* pathogenic, and then only when host resistance is low. Some strains have been implicated in diarrhea in infants, occasionally occurring in epidemic proportions in children's nurseries or obstetric wards, and *Escherichia* may also cause urinary tract infections in older persons or in those whose resistance has been lowered by surgical treatment or by exposure to ionizing radiation. Most strains are susceptible to streptomycin, the tetracyclines, and chloramphenicol.

Although *Salmonella* and *Escherichia* are very closely related, in contrast to *Escherichia*, members of the genus *Salmonella* are usually pathogenic, either to man or to other warm-blooded animals. In humans, the most common diseases caused by salmonellas are typhoid fever and gastroenteritis. The pathogenicity of salmonellas is due primarily to the action of endotoxins.

The important disease *typhoid fever* is caused by the species *Salmonella typhi*. This organism is able to resist digestion after phagocytosis and can live and reproduce intracellularly in macrophages and other phagocytic cells. *Salmonella typhi* is transmitted from person to person in food or water that has been contaminated from fecal sources. Initial replication of the organism occurs in the intestinal tract, and once it is established there some of the organisms enter the lymph system draining the intestine, travel to the bloodstream, and then become disseminated throughout the body (Figure 10.10). The organism grows especially well in the biliary tract, the spleen, and the lymph nodes but may reproduce in many other organs. After the organism is established in the tissues, the characteristic fever develops, probably as a result of the release of endotoxin (Figure 10.11). Diarrhea is not a common symptom of

FIGURE 10.10
Bacterial infection in typhoid fever.

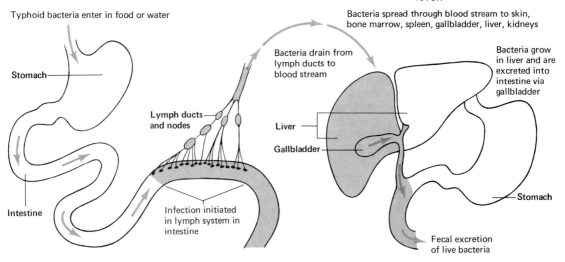

Typhoid bacteria enter in food or water

Stomach

Intestine

Lymph ducts and nodes

Infection initiated in lymph system in intestine

Bacteria drain from lymph ducts to blood stream

Bacteria spread through blood stream to skin, bone marrow, spleen, gallbladder, liver, kidneys

Bacteria grow in liver and are excreted into intestine via gallbladder

Liver

Gallbladder

Stomach

Fecal excretion of live bacteria

BACTERIAL DISEASES

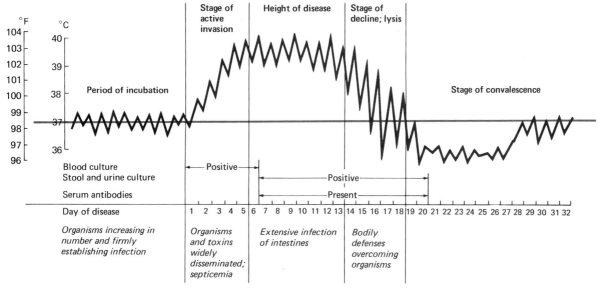

FIGURE 10.11
Typhoid fever: course of
disease.

typhoid fever as it is with salmonella gastroenteritis (see below). One
diagnostic sign of the disease is the transitory appearance of *rose spots*
on the trunk. The intestinal phase of the disease can be treated with
a variety of antibiotics active against Gram-negative bacteria. Even if
the organism is eliminated from the intestinal tract by antibiotic
therapy, however, organisms derived from the biliary tract may
continually appear in the feces; this is probably the situation during
the chronic carrier state. The chronic intracellular phase can be
treated effectively only with chloramphenicol, an antibiotic that
penetrates phagocytic cells readily. Since this antibiotic is only bacteriostatic, antibiotic therapy must be continued long enough so that
the host defenses can eliminate all viable bacteria, and indeed,
chloramphenicol is not always able to eliminate the chronic carrier
state. Typhoid fever is a disease that has been controlled most
dramatically by such public-health measures as pasteurization of
milk, treatment of sewage, purification of water, and elimination of
chronic carriers as food handlers (discussed in Chapters 14 and 15).
However, even today typhoid fever is not completely eradicated, and
a breakdown in water-supply treatment could easily lead to a return
of the disease. Sporadic cases are still reported around large cities,
especially during the summer months. A killed-bacteria vaccine for
S. typhi is available and is widely used for those parts of the world
where epidemics are still common.

Gastroenteritis caused by salmonella species other than *S. typhi* is
often called "food poisoning," since infection occurs most frequently
after eating contaminated food. It is not, however, true food poisoning but an infection derived from food; that is, the symptoms

develop as a result of multiplication of the organism in the intestinal tract. In contrast to *S. typhi*, the enteritis-inducing species rarely spread from the intestine. Between 8 and 48 hours after inoculation, onset of the symptoms occurs, and it is usually sudden with headache, chills, vomiting, and diarrhea, followed by a fever that lasts a few days. The disease is usually self-limited, but antibiotic therapy may be desirable. Diagnosis is from symptoms and by culture of the organism from the feces. The disease may be controlled by the use of public health measures, especially in monitoring food handlers and food preparation (see Sections 15.2 and 15.8).

The genus *Shigella* is closely related to *Escherichia* and *Salmonella*, although *Shigella* differs from *Escherichia* in being unable to utilize lactose or to produce gas during the fermentation of sugars, and shigellas are also nonflagellated and nonmotile.

Shigella is commonly pathogenic to man, causing a rather severe gastroenteritis usually called "bacterial dysentery" (to distinguish it from amoebic dysentery, discussed later). The pathogenicity of some strains of *Shigella* seems related to their ability to produce a powerful endotoxin. *Shigella* is found almost exclusively in man and primates, infection of other animals being rare. Communities with poor sanitation measures are often affected, and quite frequently, military troops in combat zones, where proper sanitation is unavailable, are stricken. Since the organism apparently is incapable of invading the bloodstream as *Salmonella typhi* often does, the disease is confined to the intestine and is usually self-limited; rarely is it fatal, spontaneous cure occurring within a few days.

The causal agent of *bubonic plague*, *Yersinia pestis*, has in recent years been found to be closely related to *Escherichia*, although traditionally it had not been classified in the same family. Until the twentieth century bubonic plague was one of the greatest scourges of humankind and was responsible for epidemics that killed millions of people. In the Middle Ages, a plague epidemic caused the death of one-quarter of the population of Europe; epidemics much less extensive have occurred in various parts of the world up to the present. There is good reason to believe that the organism was considerably more virulent in the extensive epidemics of the Middle Ages than it is today, but why this should be so is not known. The disease in the Middle Ages was often associated with severe hemolysis, producing dark skin in the dying person: hence the name "black death" for bubonic plague. This symptom is not seen today. The disease is still endemic in parts of Asia and Africa, and in a modified form, sylvatic plague, it is found in rodents in the western United States.

Plague is a disease more common in rodents than in humans.

10.9
Yersinia and plague

FIGURE 10.12

Transmission of plague bacteria by rat flea to rat and human.

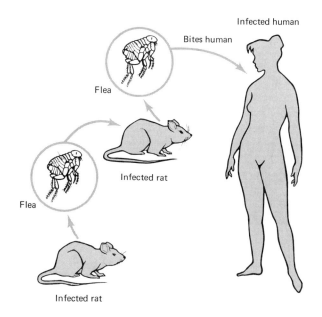

The causal organism is transmitted from one animal to another by the bite of the flea (Figure 10.12). In areas of the world where people live in close proximity to rodents, the flea may transmit the organism from an infected rodent to a human, but rarely is the organism passed directly from one person to another. As a result of fleabite, the bacteria are inoculated into the lymph system and move to the regional lymph nodes, where they multiply and cause the formation of enlarged lymph glands (called *buboes*; hence the name "bubonic" plague). In the later stages of the disease, the bacteria become disseminated throughout the body. The fatality rate of bubonic plague is high in untreated cases, often approaching 100 percent. Death results from production by the bacteria of a potent endotoxin, but other factors probably also are involved. Invasiveness is determined at least in part by a capsule that prevents phagocytosis. Interestingly, the optimum temperature for growth of *Y. pestis* is 28°C rather than the normal human body temperature of 37°C, but at 28°C the virulence factors are not produced. The bacteria also multiply in the flea; possibly the low temperature optimum encourages multiplication in the insect. In some cases in humans, the bacteria become established in the lungs, leading to the condition called *pneumonic plague*. This type is highly contagious, the bacteria being transmitted from person to person by droplets.

Recovery from plague involves the formation of opsonizing antibodies that seem to confer lifelong immunity. The disease is treated with antibiotics such as the tetracyclines, chloramphenicol, and streptomycin, and if antibiotic therapy is begun early enough, recovery may be assured. Control of the disease in humans is effected by elimination of rats, the chief reservoir of the organism in urban areas. Elimination of sylvatic plague in wild rodents is virtually impossible, and these animals thus constitute a reservoir from which the organism might move back into urban centers were rat control not carried out effectively.

10.10 Franciscella and tularemia

The causal agent of the disease *tularemia* is *Francisella tularensis*. Tularemia, a widespread disease of wild rodents such as squirrels and rabbits, was first discovered in 1911 in Tulare County, California, and it is from there that the disease receives its name. It acutely affects rodents, the organism being transmitted by blood-sucking insects. Only occasionally is the organism transmitted to humans, usually through the handling of an infected carcass; the disease is most common in hunters, who acquire the infection during the dressing of a rabbit or squirrel: hence the common name "rabbit fever." *F. tularensis* enters the body through the skin, and infection is concentrated in the lymph glands, producing symptoms of headache, body pain, and fever; it sometimes leads to death. The symptoms are probably due to the action of an endotoxin. Antibiotics active against Gram-negative bacteria can be used in therapy, streptomycin being the drug of choice because of its bactericidal action. The disease is best prevented by avoiding contact with the viscera of wild rodents and by using rubber gloves when cleaning susceptible animals.

10.11 Haemophilus and Bordetella

Organisms of the genus *Haemophilus* are characterized by a growth requirement for heme, the red pigment of blood: hence the origin of the genus name. A number of species of *Haemophilus* are known, all of which are found in association with animals or man. The most common of these, *H. influenzae*, received its name because it was first erroneously described as the causal agent of influenza, which in reality is a viral disease (see Section 12.5). *H. influenzae* is a capsulated organism that is invasive because the capsule prevents phagocytosis. It occurs as a pathogen primarily in children, causing pharyngitis, sinusitis, epiglottitis, meningitis, otitis media, and occasionally pneumonia.

An organism morphologically and physiologically similar to *Haemophilus* is *Bordetella pertussis*, the causal agent of *whooping cough*. It is also capsulated but does not require heme as a growth factor. The organism is only weakly invasive, becoming established in the upper respiratory tract but seldom penetrating to the bloodstream. Its growth in the respiratory tract induces violent coughing, leading to transmittal of the organism from person to person by droplets. A vaccine composed of virulent bacteria that have been killed by an organic mercurial is extremely effective in controlling the disease. This vaccine is usually given to infants in a combined injection with diphtheria and tetanus toxoids (see Section 8.15).

10.12
Brucella

The genus *Brucella* consists of small, Gram-negative, nonsporulating rods, which are usually pathogenic to animals or humans. Several species are closely related, being differentiated primarily by certain cultural characteristics and by the host in which they are most commonly found. *Brucella abortus* has its main reservoir in cattle, *B. melitensis* in goats and sheep, and *B. suis* in pigs, but any of the species may be found in all these animals or in humans. Any disease caused by a brucella is called "brucellosis." Various disease syndromes are induced, of which the most important are infectious abortion in cattle and undulant fever in humans. The organisms are aerobic and have a complex nutrition, and certain species require an atmosphere of 5 to 10 percent CO_2 for growth. Pathogenic isolates produce a surface antigen that enables them to grow intracellularly in phagocytes, and in the body an intracellular site is the natural habitat. In cattle, *B. abortus* shows a marked specificity for the reproductive tract, due to the presence in fetal tissues of erythritol, an alcohol that greatly stimulates the growth of *B. abortus*. Brucellosis is widespread in cattle, and these animals can be infected in a variety of ways, such as by the venereal route, ingestion, the skin, and inhalation. Once introduced into a herd, the organism spreads readily and may eventually infect most of the animals. Not all pregnant cows abort, but 30 to 50 percent abortion is not uncommon. Many animals recover completely, but a recovered cow may continue to harbor virulent organisms that will continue to infect other animals.

The organism is transmitted from cows to humans most commonly in contaminated milk. Multiplication occurs in the mucous membranes of the gastrointestinal tract, and the organism then invades the lymphatics and the bloodstream, becoming selectively localized in the spleen, liver, lymph nodes, bone marrow, and kidneys. Erythritol is not present in humans, and the organism shows no tendency to localize in the reproductive tract. Disease symptoms are due primarily to the presence of an endotoxin, the most characteristic signs being chills, fatigue, headache, and backache. Another symptom is a fever that increases at night and drops in the daytime, from which is derived the name *undulant fever* (Figure 10.13). Therapy with streptomycin and the tetracyclines is effective but must be prolonged because the intracellular localization of the organisms results in their being protected from the antibiotic. Antibiotic therapy often results in an aggravation of symptoms

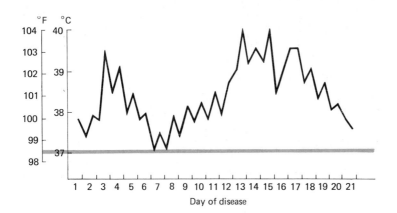

FIGURE 10.13
Undulating body temperature in a patient with brucellosis.

immediately following administration, apparently caused by induction by the antibiotic of bacterial lysis and sudden release of large amounts of endotoxin. Antibody develops readily in response to *Brucella* infection, and a rise in antibody titer is used in diagnosis, since the symptoms themselves are often too ill-defined to permit a precise identification of the disease. Recovery does not result in complete immunity, however, and reinfection may occur. The best diagnostic procedure is to culture the organism, usually from a blood sample. The brucellas are rarely transmitted from person to person, so that control of the disease can be effected through elimination of infected cattle. After immunological surveys of dairy herds to indicate the presence of *Brucella*, infected animals are segregated from the herd or slaughtered. Uninfected herds are vaccinated with a live attenuated vaccine. Pasteurization of milk is also an important procedure in preventing the spread of the organism to human beings.

The genus *Pseudomonas* comprises aerobic Gram-negative rods motile by means of polarly arranged flagella. Most species are harmless soil organisms, but a few are pathogenic to humans or animals. *Pseudomonas aeruginosa* is frequently associated with infections of the urinary tract in humans. This species is not an obligate parasite, however, since it is also common in the soil. As a pathogen, it appears to be primarily an opportunist, initiating infections in individuals whose resistance is low, such as hospital patients. In addition to urinary infections, it can also cause systemic infections, usually in individuals who have had extensive skin damage due to burns. The organism is naturally resistant to many of the widely used antibiotics, so that chemotherapy is often difficult. Polymyxin, an antibiotic that is not ordinarily used in human therapy because of

**10.13
Pseudomonas**

its toxicity, is effective against *P. aeruginosa* and can be used with caution.

The pathogen *P. pseudomallei*, the causal agent of *meloidosis*, is also primarily a soil organism, being isolated frequently in tropical areas, but it is able under special conditions to initiate infection in humans and in a variety of animals, including cows, pigs, sheep, goats, cats, dogs, rodents, and horses. The site of infection and the symptoms of meloidosis are highly variable, and diagnosis can be made only when the causative organism is isolated. On the other hand, *P. mallei*, the causal organism of *glanders*, is primarily a parasite; it is not isolated from soil or other nonanimal habitats. Glanders is a disease found almost solely in horses, donkeys, and mules, although the organism can be transmitted from one of these animals to humans and initiate infection. Two distinct diseases are known in horses: glanders, in which the primary focus of the organism is the lungs; and farcy, in which the organism usually enters by way of the skin and spreads to the lymphatics. Glanders has been effectively eliminated in North America and western Europe through the destruction of infected animals and through quarantine. The elimination of horse-drawn vehicles as a major mode of transportation has undoubtedly played a part in the successful control of the disease in humans.

10.14 Vibrio

The genus *Vibrio* consists of Gram-negative, polarly flagellated cells that are usually curved (comma-shaped) rods. Most are nonpathogenic, but some vibrios are causal agents of disease in animals and humans. The most important is *Vibrio cholerae*, which is the specific cause of the disease *cholera*. Cholera is one of the most common infectious human diseases and one that has had a long history. The organism is transmitted almost exclusively via water, and studies of its distribution in the nineteenth century played a major role in demonstrating the importance of water purification in urban areas (see Section 1.4). Today, the disease is virtually absent from the Western world, although it is still common in Asia. *V. cholerae* is an aerobe that has relatively simple nutritional needs. Also, it is quite insensitive to alkaline conditions, being able to grow at pH 9.0 to 9.6, at which most other intestinal bacteria are inhibited. This resistance to alkalinity makes possible the use of a high pH selective medium for primary isolation of *V. cholerae*. Although the bacteria grow readily in the intestinal tract, they do not invade the rest of the body. The organism produces two substances that affect the intestinal mucosa: an enzyme that attacks the substance holding the intestinal cells together in a tissue; and a toxin that affects the permeability of the intestinal wall and thus causes a profound electrolyte imbalance and water loss. The combined action of these two materials results in an enormous loss of water through the large intestine, up to 10 to 12 liters per day, causing extreme dehydration followed by shock and death. However, in many cases the disease can be mild and self-limited, running its course in a week or so. Antibiotic therapy (with streptomycin or the tetracyclines) is effective only in the very early stages of the disease. More important in treatment are measures taken to correct fluid imbalance, such as intravenous injection of isotonic fluids, which usually effect a dramatic recovery. Control of cholera primarily depends on adequate sanitation measures, such as water purification and sewage treatment. Vaccines of killed cells are available, but they are not always successful and confer immunity of only short duration. They are used only in areas where the danger of contracting the disease is great.

10.15 Spirochetes

The spirochetes are bacteria with a unique morphology and mechanism of motility. They are widespread in aquatic environments and in the bodies of warm-blooded animals. Many of them cause diseases of animals and humans, of which the most important is *syphilis*, caused by *Treponema pallidum*. The spirochete cell is typically a slender, flexuous body in the form of a spiral, often of considerable length. Motility probably occurs by means of the axial filament, a fiber or bundle of fibers attached at the cell poles and wrapped around the cell in a spiral fashion. These structures are somewhat analogous to flagella, but the pattern of motility is quite different from that of flagellated bacteria, which move through the water in a rapid whiplike pattern; the spirochetes move in the manner of a snake.

The spirochetes are classified primarily on the basis of cell size, number of spirals, structure of the axial filament, habitat, and pathogenicity (see Table 10.3).

Of *Treponema* species, *T. pallidum*, the causal agent of *syphilis*, is the best known. The cell of *T. pallidum* is extremely thin, and for this reason it is almost invisible with the ordinary light microscope, whether unstained or treated by conventional aniline dyes. It is usually stained by a silver impregnation method, in which the thickness of the cell is increased sufficiently by the silver so that the cell can be seen with the light microscope. Living cells are visible unstained by use of the dark-field microscope (Figure 10.14), and dark-field microscopy has been extensively used to examine exudates from suspected syphilitic lesions. In nature, *T. pallidum* is restricted to humans, although artificial infections have been established in rabbits and monkeys. Although the causal agent of syphilis has never been successfully cultured, nonpathogenic organisms morphologically resembling *T. pallidum* that were derived from syphilitic

FIGURE 10.14
Photomicrograph of the spirochete that causes syphilis, *Treponema pallidum*; magnification 1,350×, dark-field microscopy. From Theodor Rosebury, *Microbes and morals*, The Viking Press, Inc., New York, 1971.

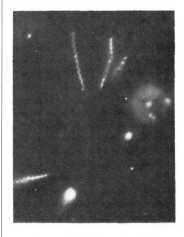

BACTERIAL DISEASES

patients have been cultured. Those treponemas that have been cultured are obligate anaerobes requiring complex media, and all grow slowly, with generation times of 4 to 18 hours. Some relevant information on the pathogenic *T. pallidum* can be obtained by studying the effects of various environmental conditions on the motility of organisms taken directly from syphilitic lesions. By such studies it has been shown that the organism is quite sensitive to increased temperature, being rapidly killed by exposure to 41.5 to 42.0°C. The heat sensitivity of *T. pallidum* is also reflected in the fact that the organism becomes most easily established in cooler sites of the body, such as the genital organs of the male, although once established in other areas of the body, it will multiply there. Experimental infection of rabbits is also most extensive in cooler sites such as the testicles or skin; indeed, artificial cooling of rabbit skin results in a dramatic increase in the number of lesions. The organism is rapidly immobilized by heavy metals, which were formerly used in therapy, and by drying. The extreme sensitivity to drying of the treponema at least partially explains why transmittal between persons is only by direct contact, usually sexual intercourse.

The disease in humans exhibits variable symptoms. The organism does not pass through unbroken skin, and initial infection most probably takes place through tiny breaks in the epidermal layer. In the male, initial infection is usually on the penis, whereas in the female, it is most often in the vagina, cervix, or perineal region. In about 10 percent of the cases, infection is extragenital, usually in the oral region. During pregnancy, the organism can be transmitted

TABLE 10.3

Major genera of spirochetes and their characteristics

Genus	Dimensions (μm)	General morphology	Number of fibers in axial filament	Habitat	Diseases
Cristispira	40–120 × 0.5–3.0	Coarse spirals; axial filament easily visible by phase-contrast microscopy	>100	Crystalline style (digestive organ of molluscs)	None
Spirochaeta	30–500 × 0.25–1.0	Large, tight spirals; axial filament not easily visible	2	Aquatic, free-living	None
Treponema	3–18 × 0.25–0.5	Coarse or tight spirals, often invisible with light microscope; cannot be stained with ordinary dyes	2–15	Man and other mammals; some aquatic	Syphilis, yaws
Borrelia	8–16 × 0.25–0.5	Coarse or tight spirals, often invisible with light microscope; can be stained with ordinary dyes	Not known	Man and other mammals; arthropods	Relapsing fever
Leptospira	5–10 × 0.1–0.2	Tightly coiled spirals, often invisible with light microscope; cannot be stained with ordinary dyes	2	Man and other mammals	Leptospirosis

from an infected woman to the fetus; the disease acquired in this way by an infant is called *congenital* syphilis. The organism multiplies at the initial site of entry and a characteristic primary lesion known as a *chancre* is formed within 2 weeks to 2 months (Figure 10.15). Dark-field microscopy of the exudate from syphilitic chancres often reveals the actively motile spirochetes. This is the *primary* stage of the disease process. In most cases, the chancre heals spontaneously, and the organisms disappear from the site. Some, however, spread from the initial site to various parts of the body, such as the mucous membranes, the eyes, joints, bones, or central nervous system, and extensive multiplication occurs. A hypersensitive reaction to the treponema takes place, which is revealed by the development of a generalized skin rash; this rash is the key symptom of the *secondary* stage of the disease. At this stage the patient's condition may be highly infectious, but eventually the organisms disappear from secondary lesions and infectiousness ceases. The subsequent course of the disease in the absence of treatment is highly variable. About one-fourth of the patients undergo a spontaneous cure, and another one-fourth do not exhibit any further symptoms, although the infection may persist. In about half the patients, the disease enters the *tertiary* stage, with symptoms ranging from relatively mild infections of the skin and bone to serious infections of the cardiovascular system or central nervous system. Involvement of the nervous system is the most serious phase of the illness, since generalized paralysis or other severe neurological damage may result. In the tertiary stage, only very few organisms are present, and most of the symptoms probably result from hypersensitivity reactions to the spirochetes.

The immunological phenomena involved in syphilis are quite complex. Several kinds of antibody have been identified in sera of patients infected with *T. pallidum*. One of these, the Wassermann antibody, reacts with a fatty substance extracted from beef heart. The reaction is detected by complement fixation and can also be measured by precipitation, which is more rapidly performed but may be less specific than is complement fixation. It is not known if the antibody is formed initially in response to a *T. pallidum* antigen or if it develops against a normal tissue component that is released as a result of hypersensitivity; however, its presence correlates highly with *T. pallidum* infection. On the other hand, Wassermann antibody has also been found in patients who have never been infected with *T. pallidum* but have had malaria, lupus erythematosus, leprosy, or one of several other diseases, thus suggesting that this antibody is not directed against a unique *T. pallidum* antigen. Despite the lack of complete specificity, serological tests involving the Wassermann antibody for the preliminary detection of syphilitic individuals are widely used because the inability to culture *T. pallidum* has made it difficult to develop serological tests directed

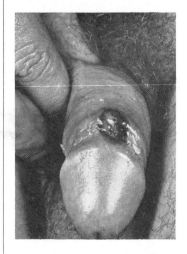

a
b

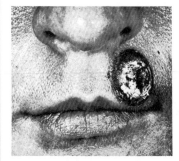

FIGURE 10.15
Primary lesions of syphilis. *a* On the penis. *b* On the lip. From S. Olansky and L. W. Shaffer, "Syphilis," in F. H. Top and P. F. Wehrle, (eds.), *Communicable and infectious diseases*, 7th, ed., The C. V. Mosby Co., St. Louis, 1972.

against antigens of the organism itself. It is possible to infect rabbits with *T. pallidum*, and the organisms extracted from lesions can be used to detect specific *T. pallidum* antibodies in human sera. Two procedures have been used: an immobilization test and a fluorescent-antibody test. The *T. pallidum immobilization* (TPI) test makes use of the fact that the spirochete is immobilized when it comes into contact with the specific antibody. Although this test is simple in principle, it is technically difficult to perform and therefore is used more as a research tool than as a diagnostic instrument. The *fluorescent treponemal antibody* (FTA) test is an indirect method, in which the adsorption of *T. pallidum* antibody from human serum onto *T. pallidum* cells is detected by use of fluorescent antibody directed against human gamma globulin. Serological tests are of great importance in diagnosing syphilis because of the variable symptoms exhibited by the disease and the lack of suitable culture methods.

Early methods of therapy involved administration of arsenic, mercury, or bismuth compounds, all of which are fairly toxic and relatively ineffective. Use of these drugs has now been completely superseded by highly effective penicillin therapy, and the early stages of the disease can usually be controlled by a series of injections over a period of 1 to 2 weeks. In the secondary and tertiary stages, treatment must extend for longer periods of time. Since penicillin kills only growing cells, the death rate is a function of how rapidly the pathogen is growing. Thus it is understandable that with such slow-growing organisms as *T. pallidum*, penicillin therapy must be prolonged.

Despite the relative effectiveness of penicillin in curing syphilis, the disease is still common. This is due mainly to the social problems of locating and treating sexual contacts of infected individuals, as we have already mentioned in relation to gonorrhea.

Spirochetes of the genus *Borrelia* cause *relapsing fever* in humans. The organisms are transmitted from their natural reservoir in rodents to humans via ticks and lice. The disease is characterized by a sudden onset of fever within 3 to 10 days of infection; the fever usually lasts for about 4 days and is accompanied by the appearance of large numbers of organisms in the blood and urine (Figure 10.16). As the fever declines, the number of organisms also decreases. As they disappear from the blood, the spirochetes become less motile and have pleomorphic shapes, often clumping in typical rosettes. For the next 3 to 10 days, there are no symptoms, but then the fever returns and the blood again has large numbers of organisms. Such attacks recur several times (hence the name relapsing fever), becoming progressively less severe until they finally subside.

The mechanism of the relapse is quite interesting, since the organisms in each successive attack are antigenically different. During each attack, circulating antibodies arise, specific for the particular strain present; these antibodies are responsible for the clumping

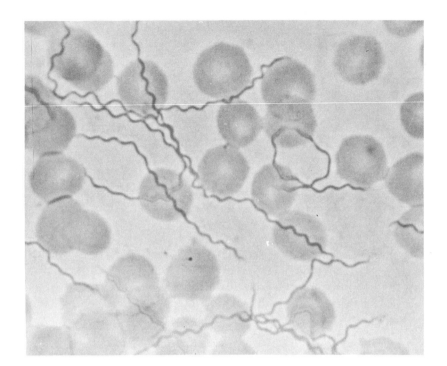

FIGURE 10.16
Large numbers of *Borrelia recurrentis* appear in the blood during an attack of relapsing fever. The round cells are red blood cells. Magnification 2,300×, stained smear.

and subsequent disappearance of the organisms. The next relapse and accompanying burst of organisms into the blood depends on the selection and growth of mutants that have different antigens, to which the host is not immune. These grow and cause a relapse, followed in turn by the elaboration of antibodies specific to the new strain, and subsequent recovery.

Diagnosis of the disease is made by microscopic examination of the blood during an attack. The organisms are observed either in stained smears or by darkfield microscopy. Tetracyclines are most effective in treating relapsing fever; penicillin and chloramphenicol can also be used. The louse-borne disease is confined primarily to eastern Africa, but the tick-borne forms are found in many areas of the world, including the western United States. The best prevention of the disease is by control of the vectors, ticks and lice.

The genus *Leptospira* contains a large number of diverse, widely distributed species. Some are harmless aquatic organisms, others are harmless parasites of animals, and some are pathogens, causing the disease *leptospirosis*. Although many of the pathogenic forms can infect man, the natural reservoirs are primarily domestic and wild animals (Figure 10.17). Rodents are the natural hosts of most leptospiras, although dogs and pigs are also important carriers of certain strains. Leptospiras can be readily cultured in complex media containing serum. Although they are obligate aerobes, the

BACTERIAL DISEASES

addition of a small amount of CO_2 to the atmosphere usually stimulates growth. Different strains are distinguished serologically by agglutination tests, and a large number of types have been recognized. Leptospiras ordinarily enter the body through the mucous membranes or through breaks in the skin. After a transient multiplication in various parts of the body, the organism localizes in the kidney and liver, causing nephritis and jaundice. The organism passes out of the body in the urine, and infection of another individual is most commonly by contact with water contaminated with the urine of infected animals. Sewer workers and those who swim in stagnant ponds and canals run the greatest risk of infection by this means. Therapy with penicillin, streptomycin, or the tetracyclines is possible but may require extended courses to eliminate the organism from the kidney; this is probably because of the slow growth and protected location of the leptospiras. Domestic animals are vaccinated against leptospirosis by means of a killed virulent strain; dogs are usually immunized routinely with a combined

FIGURE 10.17
Spread of *Leptospira* spirochetes.

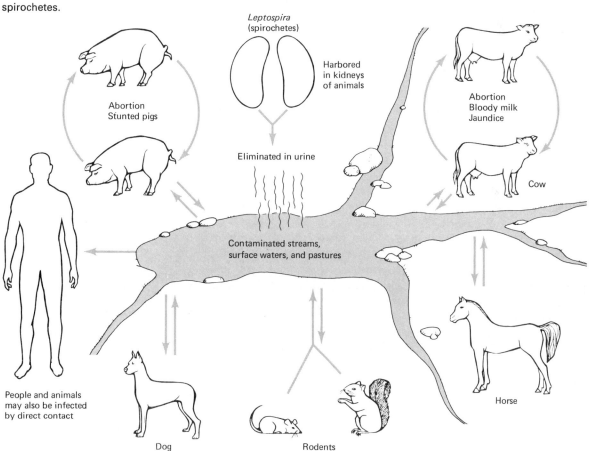

distemper-leptospira-hepatitis vaccine. In man, prevention is effected primarily by elimination of the disease from animals. The species that infects dogs, *L. canicola*, does not ordinarily infect man, but the species attacking rodents, *L. icterohemorrhagiae*, does; hence, elimination of rats from human habitations is of considerable importance in preventing the organism from reaching humans.

Organisms without cell walls that do not revert to walled organisms are currently classified in a group of organisms known as *mycoplasmas*. Members of this group are widespread in nature and have been frequently isolated from warm-blooded animals. Certain mycoplasmas are pathogenic, causing serious diseases in cattle and birds, and at least one species has been implicated in a human disease, primary atypical pneumonia. The first representative to be discovered and studied was the causal agent of pleuropneumonia in cattle, and for a long time all members of the group were called *pleuropneumonialike* organisms (abbreviated PPLO); however, PPLO has now been supplanted by the group designation of mycoplasma. Since many of the members of this group are not closely related, it is likely that various generic names will be designated as scientists become increasingly familiar wiith the group through further research. At the moment, however, it is simplest to refer to all organisms in the group as members of the mycoplasma group.

Mycoplasma cells are usually very small, often of dimensions near the resolving power of the light microscope, and they are highly irregular in size and shape, which is a consequence of their lack of rigidity. Growth is not inhibited by penicillin, cycloserine, or other antibiotics that inhibit cell-wall synthesis, but the organisms are as sensitive as other bacteria are to antibiotics that act on targets other than the cell wall. Use is made of the natural penicillin resistance of the mycoplasmas in preparing selective media for their isolation from natural materials. The culture media used for the growth of most mycoplasmas have usually been quite complex. Growth is poor or absent even in complex yeast extract-peptone-beef heart infusion media unless supplements such as fresh serum are added.

Using appropriate culture media and isolation techniques, it is possible to isolate mycoplasmas from normal and pathological material of warm-blooded animals. Although many isolates are probably harmless, a few are pathogenic. *Mycoplasma pneumoniae* causes a condition known as *primary atypical pneumonia*. The disease occurs primarily in children and young adults; it is usually marked only by upper respiratory symptoms, and very rarely by pneumonia, or involvement of the lungs. Tetracyclines have been shown effective in the treatment of mycoplasmal pneumonia. However, currently avail-

**10.16
Mycoplasma**

able diagnostic tests are slow, so treatment of patients remains difficult. Other diseases of humans in which the mycoplasmas have been implicated include urethritis, kidney infections, and infections of the oral cavity.

Bovine contagious pleuropneumonia, caused by *M. mycoides*, has been a serious problem to the cattle industry throughout the world. Another important mycoplasmal disease is agalactia of sheep and goats, a disease that primarily affects the mammary glands.

10.17 Rickettsia

The rickettsias are small bacteria that have a strictly intracellular existence in vertebrates, usually in mammals, and that are also associated at some point in their natural cycle with blood-sucking arthropods, such as fleas, lice, or ticks. Rickettsias cause a variety of diseases in humans and animals, of which the most important are typhus fever, Rocky Mountain spotted fever, scrub typhus (tsutsugamushi disease), and Q fever. Rickettsias take their name from Howard Ricketts, a scientist of the University of Chicago, who first provided evidence for their existence and who died from infection with the agent causing typhus fever, *Rickettsia prowazekii*. Rickettsias have not yet been unequivocally cultured in nonliving media and hence must be considered obligate intracellular parasites. Rickettsias have been readily cultivated in laboratory animals, lice, mammalian tissue cultures, and the yolk sac of chick embryos; this last is the host of choice for most laboratory work involving rickettsial species.

The rickettsias are Gram-negative, coccoid or rod-shaped, with cells in the size range of 0.3 to 0.7 μm wide by 1 to 2 μm long. Electron micrographs of thin sections show organisms with a normal bacterial morphology, and both cell wall and cell membrane are visible. The rickettsias have normal cell division, with doubling times of about 8 hours. The penetration of a host cell by a rickettsial cell is an active process, requiring both host and parasite to be alive and metabolically active. Once inside the phagocytic cell, the bacteria multiply primarily in the cytoplasm and continue replicating until the host cell is loaded with parasites, at which time the host cell bursts and liberates the bacteria into the surrounding fluid.

The rickettsias die quickly when out of their hosts; they must be transmitted from animal to animal by arthropod vectors. When the arthropod obtains a blood meal from an infected vertebrate, rickettsias present in the blood are inoculated directly into the arthropod, where they penetrate to the epithelial cells of the gastrointestinal tract, multiply, and appear later in the feces. When the arthropod feeds upon an uninfected individual, it then transmits the rickettsias either directly with its mouthparts or by contaminating the bite with its feces. However, the causal agent of Q fever, *Coxiella burnetii*, can also be transmitted to the respiratory system by aerosols.

The mode of transmission of the major rickettsial disease agents is variable (Table 10.4). Of the diseases they cause, the most important and widespread is *typhus fever*, also called "epidemic louse-borne typhus," caused by *R. prowazekii*. Typhus is an acute infectious disease that has a high fatality rate. It has probably afflicted human beings since ancient times. The symptoms of typhus frequently resemble those of typhoid, the disease caused by *Salmonella typhi*, and in earlier years, the two diseases were frequently confused; hence the origin of the name *typhoid*, which means "typhus-like." Typhus fever has occurred frequently in armies during military campaigns, often with disastrous consequences; indeed, this was one of the reasons for the downfall of Napoleon's army in 1812. Typhus is a frequent accompaniment of famine and other human misfortune. In contrast to most of the other rickettsial diseases, typhus occurs naturally only in humans. The causal agent is transmitted from person to person by two species of lice, the human body louse and the human head louse. The rickettsias are liberated from the louse in its feces. When the louse bites, it makes a small hole in the skin and defecates at the same time. The fecal matter containing the agent gets rubbed into the puncture and is carried to the bloodstream. The organism multiplies inside cells lining the small blood vessels, and clots occur in the vessels as a result of destruction of infected cells. A characteristic symptom of typhus is a rash, appearing first in the trunk and then spreading over the whole body

TABLE 10.4
Major rickettsial diseases, their causal agents, and modes of transmission

Disease in man	Causal agent	Mode of transmission
Epidemic typhus	*Rickettsia prowazekii*	. . . Man → Louse → Man → Louse . . .
Endemic murine typhus	*R. mooseri*	. . . Rat → Rat flea → Rat → Rat flea → Rat . . . ↘ Man
Rocky Mountain spotted fever, boutonneuse fever, other spotted fevers	*R. rickettsii*	. . . Tick → Tick → Tick → Tick . . . ↘ Dog → Tick → Man
Scrub typhus (tsutsugamushi fever)	*R. tsutsugamushi*	. . . Mite → Field mouse → Mite → Field mouse . . . ↘ Man
Rickettsial pox	*R. akari*	. . . Mite → House mouse → Mite → House mouse . . . ↘ Man
Q fever	*Coxiella burnetii*	. . . Tick → Small mammal → Tick → Cattle . . . Airborne ↘ Man

Reprinted from J. W. Moulder, *The Biochemistry of intracellular parasitism*, by permission of the University of Chicago Press, copyright 1962 by The University of Chicago.

except for the face, palms, and soles. Fever and general torpor of the patient result, with death on the ninth or tenth day of illness. Many of the symptoms may be due to the elaboration of a toxin, but no toxin has yet been positively identifed. The tetracyclines and chloramphenicol are highly effective against *R. prowazekii*, and the disease is treated with these drugs. The control of epidemics, however, is directed primarily at the louse. Insecticides such as DDT are applied to all persons known to have been in contact with infected individuals, and in widespread epidemics all persons in the community must be so treated. Two types of vaccine are available: a killed *R. prowazekii* preparation and a live avirulent strain. Vaccination of military troops sent into areas where typhus is endemic is recommended. At present typhus is not a problem in Western civilization, but presumably it could become so again if standards of personal cleanliness and public hygiene were considerably reduced.

Rocky Mountain spotted fever, caused by *Rickettsia rickettsii* was first recognized in Idaho and Montana in the 1900s, but has now been observed all over the United States and in parts of South America. Other rickettsial spotted fevers similar to Rocky Mountain spotted fever are found in other parts of the world. All the rickettsias that cause spotted fevers multiply in both the nucleus and the cytoplasm of host cells, whereas typhus and other rickettsias grow exclusively in the cytoplasm. The spotted fevers are all transmitted by ticks; the onset of the disease occurs 1 to 2 weeks after the tick bite and is marked by chills, fever, and a rash that begins peripherally on the ankles, wrists, and forehead, and spreads to the trunk. The disease responds well to chloramphenicol and tetracyclines. Ticks are the vectors and also the primary reservoirs for the organisms; the rickettsias are passed through generation after generation of tick without harming their arthropod host.

The agent which causes *Q fever* is *Coxiella burnetii* and differs from other rickettsias in several ways: (1) it is unusually stable outside host cells; (2) infection in humans occurs not by the bite of an infected arthropod but by inhalation of contaminated particles; (3) the symptoms of the disease include chills, fever, and pneumonitis, but no rash; (4) the organism differs antigenically from other rickettsias. The disease is acquired by inhaling contaminated dusts and aerosols, especially near cattle sheds. Infected cattle harbor the organisms in placenta and amniotic fluid, as well as in nasal and salivary secretions. The organisms remain viable in animal hair and in dried secretions and dust for a long time. Treatment with tetracyclines is very successful once the disease is correctly diagnosed.

**10.18
Chlamydia**

The genus *Chlamydia* consists of obligate parasites that are smaller than *Rickettsia* and even more dependent on their hosts for nutritional factors required for growth and reproduction. The life cycle of a typical member of the genus *Chlamydia* is complex. Two cellular types are seen in a typical life cycle: a small, dense cell, which is relatively resistant to drying and which is the means of dispersal of the agent; and a larger, less dense cell, which divides by binary fission and which is the vegetative form of these microorganisms. Unlike the rickettsias, the chlamydias are not transmitted by arthropods but are primarily airborne invaders of the respiratory system; hence the significance of resistance to drying of the small cells. When a chlamydia enters a cell, it begins to undergo binary fission, and after a number of divisions the vegetative cells become converted into small, dense cells that are released when the host cell disintegrates and can then infect other cells. Generation times of 2 to 3 hours have been reported, which are considerably shorter than those for the rickettsias.

Trachoma is caused by one of the most successful parasites of humans, *C. trachomatis*. The disease is limited to man and is the most common cause of blindness in the world. It is estimated that over 400 million people have the disease and that of these, 6 million are totally blind. Conditions of poor public sanitation and personal hygiene encourage the disease, which probably is transmitted primarily by eye-to-finger-to-eye contact, and by towels and clothing. Transmission by flies is also suspected but not proved. In many communities nearly all children are infected in their early years; the infection becomes chronic in a few of these cases, and blindness occurs primarily in this group. The disease affects only the epithelial cells of the eye, and its onset is marked by inflamed conjunctiva. Development of characteristic follicles and scars in the conjunctiva is followed by vascularization and infiltration of the cornea, which produces partial or complete blindness. Various complications may ensue, and as a result bacterial infections of trachomatous eyes are common.

Psittacosis is a chlamydial disease affecting a large number of different species of birds, including domestic fowl, turkeys, and various species raised by bird fanciers. The disease was first discovered in psittacine birds such as parrots: hence the name *psittacosis* ("parrot fever"); the disease was considered at one time to be restricted to them. It is now clear that the causative agent, *C. psittaci*, can be transferred from birds to humans, and indeed, birds are the primary source of human infection. Poultry and birds kept as pets or for show are probably the most important sources of infection for humans, for the disease is seen most frequently in bird handlers. It is a minor public-health problem; control in the United States is effected mainly by regulations on the importation of birds. The only major outbreak in the United States was in 1929 and 1930, when 170 cases of psittacosis with 33 deaths were reported; in this epidemic parrots were the principal vector. The disease in both birds and humans is characterized by long periods of latency, the organism remaining alive within the body but causing no overt symptoms. In humans the main foci of infection are the lungs, but other organs of the body are also affected. The organism produces a toxin that is responsible for at least some of the disease symptoms. Chlortetracycline is the drug of choice for both humans and birds, although the antibiotic does not always eliminate the organism in latent infections. Immunity in human beings is incomplete after infection, and active immunization is not recommended. Individuals in high-risk occupations should be checked immunologically from time to time for the presence of antibodies against the psittacosis agent, and they should be subjected to antibiotic therapy if they become infected.

Study questions

No specific questions are provided for this chapter. The student may find it useful to prepare an outline of the groups to be studied in this course, noting the properties of each, its habitat, the diseases caused, as well as what properties of the organism are important for its success as a disease agent. Methods of treatment or control of the diseases should also be listed. As an aid in organizing the study of the many groups of bacteria presented, it should be noted that they are presented in the text in an ordered sequence. The first groups are Gram-positive cocci, followed by Gram-negative cocci. Next come Gram-positive rods, then the many groups of Gram-negatives, and then four groups of bacteria of unique character.

Suggested readings

Davis, B. D., R. Dulbecco, H. N. Eisen, H. S. Ginsberg, and W. B. Wood, Jr. *Microbiology*, 2nd ed., Harper & Row, New York, 1973. Widely used textbook, with separate sections on various microbial diseases.

Lennette, E. H., E. H. Spaulding, and J. P. Truant (eds.), *Manual of clinical microbiology*, 2nd ed., American Society for Microbiology, Washington, D.C., 1974. Presents detailed methods for the isolation and characterization of pathogenic bacteria from clinical specimens. Also lists common pathogens, and gives properties used in identification.

Olds, R. J., *A colour atlas of microbiology*, Wolfe Medical Books, London, 1975. Many excellent photographs of bacterial colonies and stained cells.

Top, F. H., Sr., and P. F. Wehrle (eds.), *Communicable and infectious diseases*, 8th ed., C. V. Mosby Co., St. Louis, 1976. A standard medical textbook on diseases and their control.

Youmans, G. P., P. Y. Paterson, and H. M. Sommers, *The biologic and clinical basis of infectious diseases*, W. B. Saunders Co., Philadelphia, 1975. A text for medical students dealing primarily with bacterial diseases.

Zinsser, H., *Microbiology*, 16th ed., Wolfgang Joklik (ed.), Appleton-Century-Crofts, Inc., New York, 1976. Standard textbook on medical microbiology.

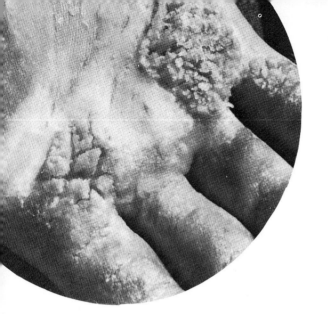

Chapter

11

Fungal and protozoal diseases

In contrast to the bacteria, fungi and protozoa are only rarely pathogenic for humans or animals; relatively few of the many different members of these groups are pathogenic. Most of the fungal diseases of humans are not especially serious, although some of the protozoal diseases are quite severe. One factor that greatly complicates treatment of the diseases in humans is the eucaryotic nature of the organisms. Most of the agents effective against fungi or protozoa are also toxic to humans. Only a few compounds have been found that are effective against the disease agents and yet harmless to the host. So, although the diseases often are not severe or fatal, they are frequently chronic, since they are so difficult to treat.

11.1 Fungal diseases

Most fungi are harmless to humans, occurring as saprophytes in nature. The molds we see most commonly are those growing on bread or fruits, or those decomposing organic matter in the soil. Only a few fungi are pathogenic; it has been estimated that only about 50 species cause human disease.

It should be noted, however, that many fungi cause plant diseases, and fungi have probably caused more suffering to humans in this role than they have as human pathogens. For example, more than 1 million people died from starvation and subsequent disease in

Ireland in the late 1840s because of a fungal potato blight. At that time a typical peasant family consumed about 8 pounds of potatoes per person per day, accounting for more than 80% of all calories consumed. The potato blight completely destroyed the potato harvest.

Most fungi that are pathogenic to humans are not restricted to growth in the host, and their primary habitat is the soil; they are merely opportunistic invaders of individuals with lowered resistance to infection. Transmission of the organism from person to person is rare, so that fungal diseases are generally not contagious. Fungal diseases of humans, called *mycoses* (Figure 11.1), can be divided into two main groups: (1) systemic, or deep-seated, mycoses, in which the pathogen is widely disseminated, growing in various organs and tissues; and (2) superficial mycoses, involving infections of the skin, hair, or nails (Table 11.1).

SYSTEMIC MYCOSES

Systemic mycoses are diseases caused by yeasts and also by fungi that can grow in a yeastlike manner in the body (or at 37°C in culture) or as filaments at lower temperatures. Usually, the infecting organisms reside in the soil and enter the body by being inhaled. They initiate a mild respiratory disease that has no characteristic symptoms and often resembles a cold. Subsequently, the organisms may become disseminated throughout the body and a chronic infection ensues, of which the symptoms are highly variable; general debilitation is common, but fatal infections are now rare.

Several fungi cause pulmonary infections that mimic tuberculosis, especially in their early stages. *Cryptococcus neoformans*, causing the disease *cryptococcosis*, is one such organism. Inhalation of its

TABLE 11.1 Fungi causing human disease	Disease	Causal organism	Main disease foci
	Systemic mycoses		
	Cryptococcosis	*Cryptococcus neoformans*	Lungs, meninges
	Coccidioidomycosis	*Coccidioides immitis*	Lungs
	Histoplasmosis	*Histoplasma capsulatum*	Lungs
	Blastomycosis	*Blastomyces dermatitidis*	Lungs, skin
	Candidiasis	*Candida albicans*	Oral cavity, intestinal tract
	Aspergillosis	*Aspergillus fumigatus*	Bronchi
	Superficial mycoses (dermatomycoses)		
	Ringworm	*Microsporum audouini*	Scalp of children
	Favus	*Trichophyton schoenleinii*	Scalp
	Athlete's foot	*Epidermophyton* and other genera	Between toes, skin

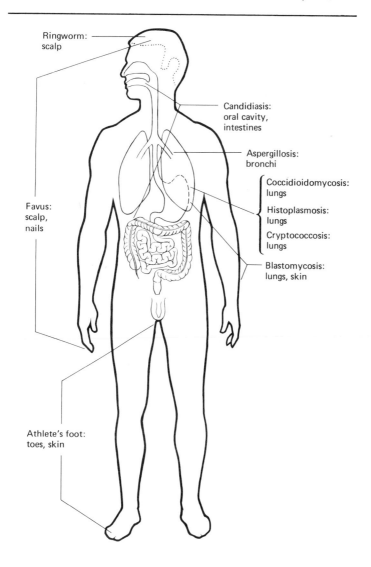

Superficial

Systemic

FIGURE 11.1
Important fungal diseases of humans.

Ringworm:
scalp

Candidiasis:
oral cavity,
intestines

Aspergillosis:
bronchi

Coccidioidomycosis:
lungs

Histoplasmosis:
lungs

Cryptococcosis:
lungs

Favus:
scalp,
nails

Blastomycosis:
lungs, skin

Athlete's foot:
toes, skin

yeast-form cells initiates pulmonary infection; in severe cases the organism disseminates widely, involving the meninges, brain, viscera, skin, or bones to varying extent. The organism appears in tissues or spinal fluid as masses of yeast cells. Cryptococcosis formerly was invariably fatal; since the introduction of polyene antibiotics, notably amphotericin B, the disease can now often be controlled.

Certain of the deep-seated mycoses are found only in restricted geographical areas. *Histoplasmosis* is found in high incidence in certain parts of the midwestern United States, especially in rural areas of the Ohio and Mississippi Valleys. The causal agent, *Histo-*

plasma capsulatum, is found in large numbers in soils receiving droppings of chickens and other birds, and infection occurs by inhalation of airborne conidia from such contaminated soils. Many human beings may become infected and develop pulmonary disease; most people have only mild symptoms which may go unnoticed or seem to be a normal cold. With recovery the pulmonary lesions become fibrotic and calcified and mimic tubercular lesions in X ray. In a small number of cases, the infection becomes progressive and is widely disseminated, with lesions in almost all organs and tissues.

Coccidioidomycosis (also called "San Joaquin Valley fever") is generally restricted to desert areas of the southwestern United States. The fungus *Coccidioides immitis* grows in desert soils, and the spores are disseminated on dry windblown particles and inhaled. In endemic areas, as many as 80 percent of the inhabitants may be infected, although only a small number of these develop serious symptoms. The precise factors responsible for pathogenesis are not understood, but many of the symptoms may be allergic in nature.

Some fungi are not normally pathogenic but may become so under certain conditions of lowered host resistance. These opportunistic fungi may become virulent pathogens for those suffering from diabetes, certain cancers, and in those treated with immunosuppressive agents, such as radiation. They may also appear in those treated with massive doses of orally administered broad-spectrum antibiotics as a consequence of the elimination of the normal intestinal bacteria. The most common of these opportunistic fungi is the yeast *Candida albicans. Candida* is often present in the normal flora of the mouth, vagina, and intestinal tract. When it becomes invasive, a variety of lesions may appear, depending on location. Perhaps the most common syndrome is *thrush,* in which the organism grows in white patches on the mucous membranes of the mouth and pharynx. Thrush is most common in the first days of life in newborns and in the terminal stages of a wasting disease.

SUPERFICIAL MYCOSES

Superficial mycoses are caused primarily by mycelial fungi. These organisms, frequently called *dermatophytes,* have an affinity for such regions as the skin, hair, and nails (Figure 11.2). Some of these fungi can grow within the superficial tissues, causing itching and reddening of the skin; some can grow on the surface of the hair; and others can grow within the hair fiber itself. The initiation of a dermatophytic infection generally depends on host factors such as hormonal balance and various other physiological influences. Age is a factor in susceptibility to two dermatomycoses: athlete's foot, which is common in adults and rare in children; and ringworm of the scalp, which is frequent in children but rare in adults. Adults may be resistant to ringworm because of increased secretions by the sebaceous glands, especially the secretion of fatty substances with

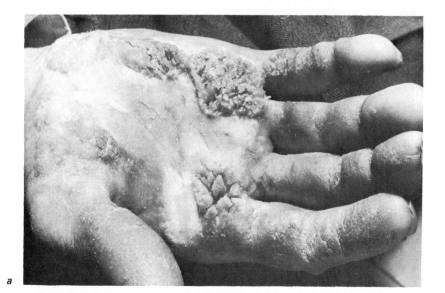

a

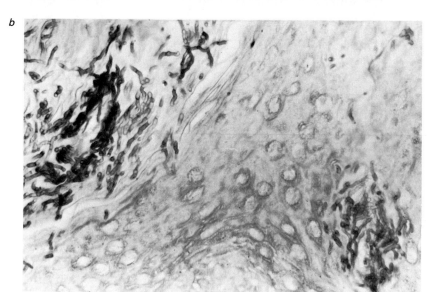

b

FIGURE 11.2
a Fungus infection of the hand caused by *Epidermophyton.*
b Photomicrograph of the fungus in infected tissue, magnification 350×.
Courtesy of P. M. D. Martin; from E. Moore-Landecker, *Fundamentals of the fungi,* Prentice-Hall, Inc., Englewood Cliffs, N.J., 1971.

antifungal activity. One of these antifungal substances, undecylenic acid, is now used as a component of many ointments marketed for the control of fungal infections of the skin. Moisture is an important factor in the development of many surface infections. Athlete's foot usually occurs between the toes because moisture accumulates in this region, promoting fungal growth.

There are only a few antibiotics effective for fungal infections, as discussed in Section 4.6. Griseofulvin has been used considerably for superficial mycoses. Although it is ineffective if applied directly to the skin, it is effective if given orally because the drug can accumulate in the newly synthesized keratin-containing tissues and render them resistant to fungal infection. Because such tissues gradually rejuvenate, griseofulvin therapy must be continued for prolonged periods, until the newly synthesized tissues have replaced those sloughed off. Deep-seated mycoses are not affected by griseofulvin but can be treated with nystatin. No immunization procedures for fungal infections are currently in use.

MYCOTOXINS

Some fungi elaborate chemical substances called mycotoxins. Some of these toxins are extremely active against humans, causing a wide variety of diseases: blood diseases, disturbances of nerve function, liver damage, kidney malfunction, limb deformation during embryonic growth, hormonal disturbances resulting in changed sexual behavior and decreased fertility, and cancer. Because of the extreme toxicity of some of the toxins, the diseases are often fatal. The most dramatic and widespread disease caused by a mycotoxin is liver cancer, which is caused by the *aflatoxins*, a group of fungal products of the organism *Aspergillus flavus*.

The fungi that produce mycotoxins (Table 11.2) can grow in foods, feeds, and grain, producing microscopic colonies invisible to the consumer. If the food is not obviously moldy, it may be eaten, the toxin ingested, and a disease elicited. In some areas of Africa where climatic conditions are very favorable for the growth of *A. flavus* on peanuts and grains, the incidence of liver cancer in the population is markedly higher than in areas where aflatoxin is not so commonly present in foods. Farm animals may also be affected by eating moldy feed, and serious economic effects on animal production can occur. Thus, an understanding of the factors influencing

TABLE 11.2

Some toxin-producing fungi, their toxins, and the disease syndromes they cause

Fungus	Toxin[a]	Disease
Aspergillus flavus	Aflatoxin	Liver damage, liver cancer
Aspergillus ochraceus	Ochratoxin	Fatty liver
Aspergillus versicolor	Sterigmatocystin	Liver damage, liver cancer
Aspergillus clavatus	Patulin	General cell damage, cancer
Fusarium graminearum	Zearalenone	Estrogenic hormone
Penicillium citrinum	Citrinin	Kidney damage
Stachybotrys species	Stachybotrytoxin	General cell damage, hemorrhage

[a] Over 20 distinct toxins have been identified; those in the table are the most widespread or best-known.
Adapted from H. K. Frank, "Zentralblatt Bakteriologie Hygiene: I," *Abt. Origin. B.*, **159**, 1974, 424–434, and C. J. Mirocha and C. M. Christensen, *Annual Review of Phytopathology*, **12**, 1974, 303–330.

fungal growth and mycotoxin production in foods and feeds is of great importance. The agricultural importance of mycotoxins is considered further in Section 16.9.

Most protozoa that are motile by means of flagella are harmless aquatic or soil organisms, but a few are parasitic on or pathogenic to animals and humans. The *trichomonads* are protozoal parasites of the vertebrate alimentary and genitourinary tracts. These flagellates have three to six flagella, one of which typically is trailed behind the cell during movement. Most of the parasitic forms produce cysts, which are the infective stages by which the species is transferred from host to host. There are three species parasitic in humans: *Trichomonas tenax* (or *T. buccalis*), found in the mouth; *T. hominis*, in the colon; and *T. vaginalis*, in the genitourinary tract. The last species, despite its name, is also found in the genitourinary tract of males. A surprisingly high proportion of the human population harbors one or more of these three species: *T. hominis* is found in 2 percent and *T. tenax* in 10 percent of the population; and *T. vaginalis*, in 25 percent of females and 4 percent of males. It has not been proved that these trichomonads are pathogenic to human beings, however. *Trichomonas vaginalis* has often been implicated in inflammation of the vagina, but whether the organism is a cause or a result of the vaginitis is not known. *Trichomonas tenax* probably is nonpathogenic and most likely is transmitted by kissing.

The most important pathogenic flagellates are the *trypanosomes*. These organisms cause a number of serious diseases of humans and vertebrate animals, including the feared disease *African sleeping sickness*. The trypanosomes are highly variable in morphology, both from species to species and within a single species at various stages of the life cycle. In *Trypanosoma*, the genus infecting humans, the protozoa are rather small, around 20 μm in length, and are thin, crescent-shaped organisms with a single flagellum that originates in a basal body and folds back laterally across the cell, where it is enclosed by a flap of surface membrane. Both the flagellum and the membrane aid in propelling the organism, making possible an effective movement even in blood, which is rather viscous. Organisms of several sizes and shapes are seen in the blood, with short and broad forms, long and slender forms, and intermediate forms all seen in the same blood smear (Figure 11.3). *Trypanosoma gambiense* is the species that causes the chronic but usually fatal African sleeping sickness. In humans, the parasite lives and grows primarily in the bloodstream, but in the later stages of the disease, invasion of the central nervous system may occur, causing an inflammation of the brain and spinal cord that is responsible for the characteristic

11.2
Protozoal diseases

FLAGELLATED PROTOZOA

FIGURE 11.3
Photomicrograph of the causal agent of African sleeping sickness, *Trypanosoma gambiense*, in blood.

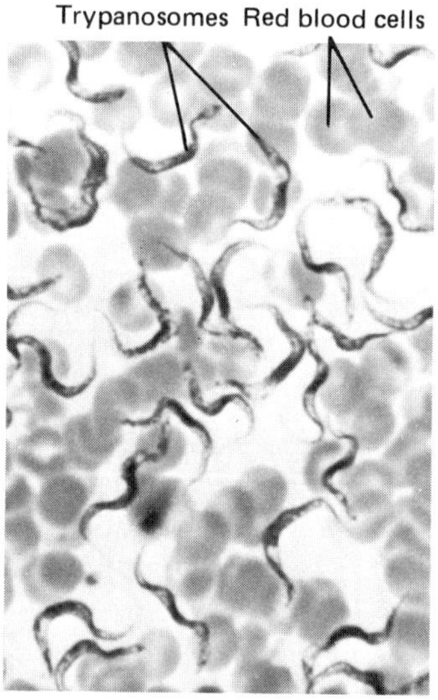

Trypanosomes   Red blood cells

FIGURE 11.4
Life cycle of *Trypanosoma gambiense*.

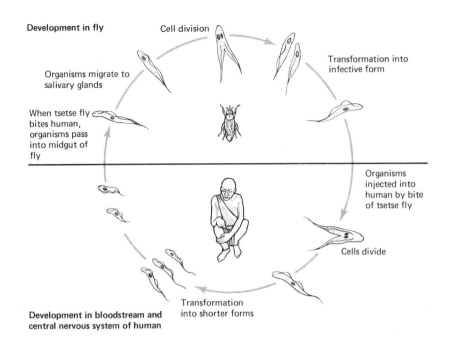

Development in fly

Cell division

Transformation into infective form

Organisms migrate to salivary glands

When tsetse fly bites human, organisms pass into midgut of fly

Organisms injected into human by bite of tsetse fly

Cells divide

Transformation into shorter forms

Development in bloodstream and central nervous system of human

neurological symptoms of the disease. The parasite is transmitted from person to person by the tsetse fly, a blood-sucking fly of the genus *Glossina*, found only in Africa (Figure 11.4). The parasite divides in the intestinal tract of the fly and then invades intestinal cells, where further multiplication is accompanied by an alteration in morphology. The parasite then invades the salivary glands and mouthparts of the fly, from which sites it may be transferred to a new human host. The spread of the disease can be prevented by destruction of the tsetse fly. Control of the fly proved relatively difficult until the development of rather potent insecticides, such as DDT; however, through their use and through the clearing of underbrush, in which the flies find shelter, the tsetse fly has been eliminated from many parts of West Africa that previously had been heavily infested. Several new drugs have also been developed for treating the disease in humans, and these have also reduced the incidence of the disease by reducing the number of sources from which the flies could become infected. Another trypanosome, *T. brucei*, infects domestic animals such as cattle and pigs, causing symptoms similar to African sleeping sickness and seriously restricting the ability to raise cattle in certain parts of Africa. This parasite, which is probably a variant of the form that attacks humans, is also transmitted by a species of tsetse fly, whose control is effected at the same time that the vector of *T. gambiense* is eliminated. Control of the tsetse fly has made possible the establishment of a cattle industry in many parts of Africa previously unsuitable for this purpose.

A wide variety of amoebas are parasites of humans and other vertebrates, their usual habitat being the oral cavity or the intestinal tract. *Entamoeba histolytica* can serve as an example of these parasitic forms. This organism is found in the intestinal tract of a high percentage of individuals living in regions where sanitation is poor, and it is found occasionally even in members of higher-income groups. In many cases, infection causes no obvious symptoms, but in some individuals it produces ulceration of the intestinal tract, which results in a diarrheal condition called *amoebic dysentery*. The organism is transmitted from person to person in the cyst form. Cyst germination occurs in the intestinal tract, and the amoebas divide by binary fission to establish a population, feeding directly on intestinal bacteria (Figure 11.5). Under certain conditions not yet defined, the amoebas form cysts, in which form they leave the intestinal tract in the feces. The cysts are resistant to drying and hence can survive for

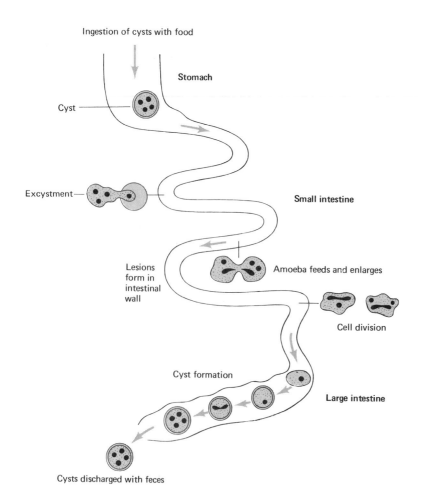

FIGURE 11.5
Life cycle of a parasitic amoeba (*Entamoeba histolytica*) in the human intestine.

long periods of time outside the body. *Entamoeba histolytica* is transmitted by fecal contamination of water supplies and by houseflies. The disease it causes can often be treated by the use of antibacterial antibiotics taken orally, but these probably act only indirectly on the entamoebas by reducing or eliminating the intestinal bacteria upon which they feed. Amoebic dysentery is diagnosed by microscopic examination of the stool for cysts, which can be separated and concentrated by centrifugation in a solution of zinc sulfate. The latter material is of high specific gravity, and the cysts float to the surface; a loopful of the surface film can be stained with iodine and examined under the microscope.

MALARIA PARASITE

The malaria parasite is a member of a large class of protozoa called *Sporozoa*, all of which are obligate parasites. They are characterized by a lack of motile adult stages and by a nutritional mode of life in which food is generally not ingested but is absorbed in soluble form through the outer wall, such as occurs in bacteria and fungi. Although the name "sporozoa" implies the formation of spores, this group does not form true resting spores, as do bacteria, algae, and fungi, but instead, produces analogous strucures called *sporozoites*, which are involved in the transmission of the species to a new host. Numerous kinds of vertebrates and invertebrates serve as hosts for Sporozoa, and in some cases, an alternation of hosts occurs, with some stages of the life cycle occurring in one host and some in another. The most important members of the class Sporozoa are the coccidia, which usually are parasites of birds, and the plasmodia (malaria parasites), which infect birds and mammals, including humans. Our discussion will be restricted to the plasmodia.

The malaria parasite is one of the most important human pathogens and has played an extremely significant role in the development and spread of human culture. Four species infect humans: *Plasmodium vivax*, *P. falciparum*, *P. malariae*, and *P. ovale*. These differ in the degree of severity of symptoms they cause and in certain aspects of their life cycles. We shall consider here only *P. vivax*, which is the most widespread species and the one about which the most information is available. This parasite carries out part of its life cycle in humans and part in the mosquito, by which vector it is transmitted from person to person. Only mosquitoes of the genus *Anopheles* are involved; and since these inhabit primarily the warmer parts of the world, malaria occurs predominantly in the tropics and subtropics.

The life cycle of *P. vivax* is complex. The human host is infected by plasmodial sporozoites, small elongated cells produced in the mosquito, which localize in the salivary gland of the insect. When biting, the female mosquito inserts her proboscis into a capillary, thereby inoculating the sporozoites directly into the bloodstream

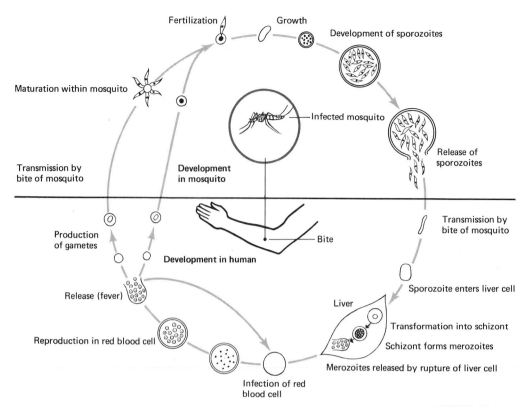

Fertilization

Growth

Development of sporozoites

Maturation within mosquito

Infected mosquito

Release of sporozoites

Transmission by bite of mosquito

Development in mosquito

Transmission by bite of mosquito

Production of gametes

Bite

Sporozoite enters liver cell

Development in human

Liver

Transformation into schizont

Release (fever)

Schizont forms merozoites

Reproduction in red blood cell

Merozoites released by rupture of liver cell

Infection of red blood cell

FIGURE 11.6
Life cycle of the malaria parasite, *Plasmodium vivax*.

(Figure 11.6). The sporozoites are carried throughout the body and are removed from the blood by the reticuloendothelial system (including lymph nodes, spleen, and liver). Replication occurs in the liver, and during this stage (termed the *exoerythrocytic* stage), the parasites are absent from the blood. The sporozoite becomes transformed to a *schizont*, which replicates by enlarging and segmenting into a number of small cells called *merozoites*; these cells are liberated from the liver into the blood stream. Some of the merozoites infect the red blood cells, initiating the *erythrocytic schizont stage* (Figure 11.7). The cycle in the red blood cells proceeds as in the liver and usually occupies a definite period of time, 48 hours in the case of *P. vivax*. It is during the erythrocytic stage that the characteristic symptoms of fever and chills occur, the chills occurring when a new brood of parasites is liberated from the erythrocytes. Not all the cells liberated from red cells are able to infect other erythrocytes; those which cannot, called *gametocytes*, are infective only for the mosquito. If these gametocytes happen to be ingested when another insect of the proper species of *Anopheles* bites the infected person, they mature within the mosquito into *gametes*. Two gametes fuse, and a zygote is formed; the zygote migrates by amoeboid motility to beneath the outer wall of the insect's intestine, where it enlarges and forms a number of sporozoites. These are liberated, some of them reaching the salivary gland of the mosquito, from where they can be inoculated into another person; and the cycle begins again.

This complex cycle can be broken by eradication of the *Anopheles* mosquito; destruction of the vector has essentially eliminated malaria from the United States and Europe (Figure 11.8). Even so, some infected persons may continue to serve as reservoirs for the plasmodium, and a renewed spread of the disease may result if the mosquito returns to these areas. The erythrocytic phase of the disease can be cured by a variety of drugs, of which the best known are quinine, Atabrine, and chloroquine. These drugs affect only the erythrocytic stage, however, and although they control the symptoms to a great degree, they do not completely cure the disease. The exoerythrocytic stages can be eliminated by use of the drug primaquine, and this drug in combination with one of the three others just mentioned can effect a complete cure. *Plasmodium vivax* has been very difficult to grow in culture, and because of its specificity for humans, experimental infection has also been difficult. For these reasons, most of the experimental work on malarial parasites has been done with species of *Plasmodium* that infect birds or rats, and it

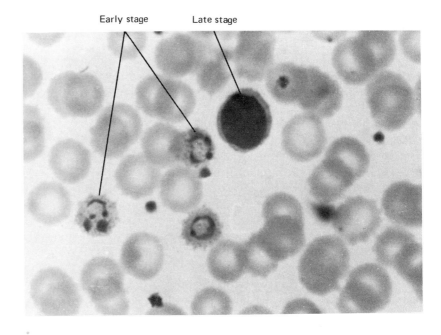

Early stage   Late stage

FIGURE 11.7
Photomicrograph of red blood cells infected with the malaria parasite.

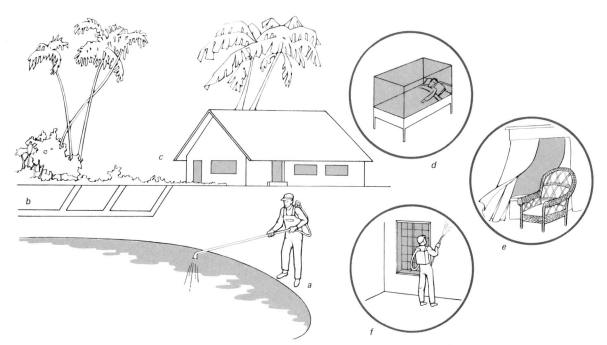

FIGURE 11.8
Preventing the spread of malaria
by mosquito control. *a* Oiling
surface of stagnant water.
b Concrete drains to draw off
excess water from soil. *c* Clean
yards. *d* Mosquito nets on beds.
e Clean and airy rooms with
screened windows. *f* Residual
spraying with oil on walls of
houses.

is with these forms that most of the studies on the development of new drugs have been carried out. Conclusive evidence for the diagnosis of malaria in humans is obtained by examining blood smears for the presence of infected cells; the various stages of replication in the erythrocytic phase are easily observed. Diagnosis of the disease in the tissues (exoerythrocytic stage) is much more difficult.

One of the most interesting discoveries about malaria concerns the mechanism by which people living in regions of the world where it is endemic acquire a genetic resistance to infection. As we noted earlier, malaria is a disease primarily localized in warmer parts of the world, where its mosquito vectors are most common, and has undoubtedly been endemic in Africa for thousands of years. In Africans, resistance to malaria caused by *P. falciparum* is associated with the presence in their red cells of hemoglobin S, which differs in structure from normal hemoglobin A. The malaria parasite cannot grow as well in red cells containing hemoglobin S as it can in normal ones. Individuals with hemoglobin S are less able to survive at high altitudes, where oxygen pressures are lower; but in tropic lowland Africa this disadvantage is not manifested, and obviously, the advantage of increased resistance to malaria has been the selective factor. Thus the malaria parasite has been a factor in the evolution of human beings. Other microbial parasites have also probably promoted evolutionary changes in their hosts; but in no other case do we have such clear evidence as in the case of malaria.

Study questions

As in Chapter 10, there are no specific questions for this chapter. It might be useful to prepare an outline similar to the one for the bacterial diseases. Include for each the properties of the organism, its habitat, a description of the disease caused, and methods of treatment or control.

Suggested readings

Beck, J. W., and E. Barrett-Connor, *Medical parasitology*, The C. V. Mosby Co., St. Louis, 1971. A good text in the field.

Brown, H. W., *Basic clinical parasitology*, 4th ed., Appleton-Century-Crofts, New York, 1975. Good standard text.

Conant, N. F., D. T. Smith, R. D. Baker, and J. L. Callaway, *Manual of clinical mycology*, 3rd ed., W. B. Saunders Co., Philadelphia, 1971. A large number of clinical photographs of diseases, micrographs of various fungi.

Emmons, C. W., C. H. Binford, J. P. Utz, and K. J. Kwon-Chung, *Medical mycology*, 3rd ed., Lea and Febiger, Philadelphia, 1977. An advanced and complete textbook.

Lennette, E. H., E. H. Spaulding, and J. P. Truant (eds.), *Manual of clinical microbiology*, 2nd ed., American Society for Microbiology, Washington, D.C., 1974. Detailed methods for the isolation and characterization of pathogenic fungi and parasites from clinical specimens. Listing of common pathogens and properties used in identification.

Markell, E. K., and M. Voge, *Medical parasitology*, 4th ed., W. B. Saunders Co., Philadelphia, 1977. A standard text in the field.

Moore-Landecker, E., *Fundamentals of the fungi*, Prentice-Hall, Inc., Englewood Cliffs, N.J., 1971. A good beginning text.

Olds, R. J., *A colour atlas of microbiology*, Wolfe Medical Books, London, 1975. Excellent photographs of mold colonies; stained hyphae, conidia, and other micrographs.

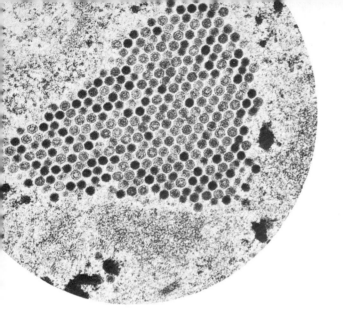

Viruses and viral diseases

12

A virus is a genetic element containing either deoxyribonucleic acid (DNA) or ribonucleic acid (RNA) that is able to alternate between two distinct states: intracellular and extracellular. In the extracellular, or infectious, state viruses are submicroscopic particles containing nucleic acid surrounded by protein; viruses may occasionally contain other components as well. The role of the virus particle is to carry the viral nucleic acid from the cell in which the particle has been produced to another cell where the viral nucleic acid can be introduced and the intracellular state initiated. It is in this phase that replication occurs, during which more nucleic acid and other components of the virus are produced. Cells that viruses can infect and in which they can replicate are called *hosts*. The host performs most of the functions necessary for virus replication.

Viruses can vary widely in size, shape, chemical composition, range of organisms attacked, and kinds of cell damage induced. Viruses are known that infect animals, plants, bacteria, blue-green algae, and fungi. Practically all virus particles are of submicroscopic size, which means that the individual particle can rarely be seen with the light microscope (Figure 12.1). Virus particles are fairly easy to visualize with the electron microscope, however, so that our knowledge of virus structure is now greatly advanced. In fact, it was in the field of virology that the electron microscope made its first great contributions to biology. Electron microscopists and virologists still work in close collaboration in most virus research.

The naming of viruses does not follow the rules and procedures used for naming cellular organisms. In many cases, the virus is named for the disease it causes (for example, rabies virus, polio virus), while in other cases it is named for the organ or tissue it infects (for example, adenovirus, which infects the adenoids). In the case of some viruses, code numbers are used to distinguish several viruses infecting the same host. From time to time there have been proposals to give viruses genus and species names, but as yet these proposals have not found extensive backing.

12.1
Virus structure and reproduction

The structures of viruses are exceedingly diverse. With the advent of the electron microscope, it was possible to study the overall sizes and shapes of virus particles, and in recent years new preparation methods have made it possible to study their internal structure as well (Figure 12.2). The core always consists of nucleic acid, either DNA or RNA, surrounded by a protein coat. Some viruses are roughly spherical in shape, others are rod-shaped, and some have

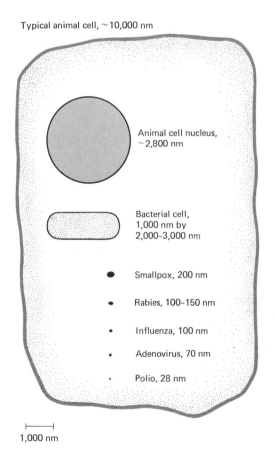

Typical animal cell, ~10,000 nm

Animal cell nucleus, ~2,800 nm

Bacterial cell, 1,000 nm by 2,000–3,000 nm

Smallpox, 200 nm

Rabies, 100–150 nm

Influenza, 100 nm

Adenovirus, 70 nm

Polio, 28 nm

1,000 nm

FIGURE 12.1
Relative sizes of various particles compared to the size of a bacterial and an animal cell. The thickness of a page of this book is about 100,000 nm (1,000 nm equals 1μm).

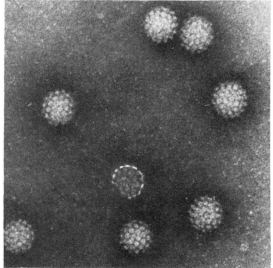

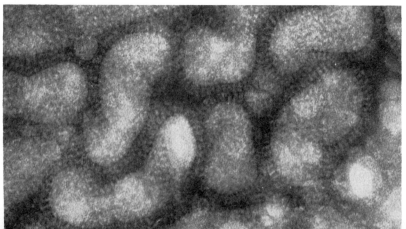

FIGURE 12.2
Electron micrographs of animal virus particles. *a* Polio virus particles; magnification 120,000×. Courtesy of the Virus Laboratory, University of California, Berkeley. *b* Human wart virus; magnification 154,400×. From W. F. Noyes, *Virology,* **23**, 65–72, 1964. *c* A crystal of adenovirus particles; magnification 32,500× *Courtesy of Councilman Morgan.* *d* Influenza virus; magnification 258,300×. From P. W. Choppin and W. Stoeckenius, *Virology,* **22**, 482, 1964.

more complex structures, with envelopes surrounding the central nucleic acid-protein core. The sizes of viruses are expressed in a unit called the *nanometer* (nm), which is $1/10^9$ meter (equivalent to $\frac{1}{1,000}$ of a micrometer). The range of sizes found in various viruses is from 30 nm, about the size of a ribosome, up to about 300 nm, the latter being resolvable with a light microscope.

In the reproduction of animal viruses, the whole virus particle usually penetrates the cell, being carried inside by phagocytic action of the cell. Once inside the cell, the viral nucleic acid becomes separated from the protein coat, and replication can then proceed. After the new nucleic acid and protein are produced, these are assembled into new virus particles that are released from the cell. Release of mature virus particles may occur with lysis of the infected cell or without lysis by a kind of budding process. In some viruses, release is not necessarily accompanied by cell death, and virus release may continue for many hours or days, with a single cell liberating thousands of virus particles. The replication of virus nucleic acid in animal cells can occur either in the cytoplasm or in the nucleus, depending on the virus. The viral nucleic acid itself may be infectious, although in all cases, the infectivity of the nucleic acid is less than that of the intact virus. However, although the protein coat is not absolutely required for virus infection to occur, it plays an important role in natural virus infections. In addition to its part in the adsorption and penetration processes, the protein coat protects the nucleic acid of the virus from attack by nucleases present in the extracellular environment. Infectious nucleic acid is rapidly inactivated by enzymes, whereas whole virus particles are usually completely unaffected by them.

Viruses vary enormously in the degree of damage that they inflict on the host (Figure 12.3). With some viruses no effect of virus infection is detectable; these are called *latent* viruses, and often their existence is not even suspected. Other viruses are highly virulent and kill cells very soon after infection. One effect of virus infection that is being increasingly well documented is the conversion of normal host cells into tumor or cancer cells. Some viruses in certain hosts act as moderate viruses that do not kill cells but induce alterations in them that lead to the cancerous state. Cancer cells are characterized by uncontrolled growth within the animal body, which apparently arises from changes in cell-surface properties. The manner in which a normal cell is converted into a tumor cell through virus infection is not yet understood.

12.2 Interference with viral activity

Since viruses depend on their host cells for many functions of virus replication, it is difficult to inhibit virus multiplication without at the same time affecting the host cell itself. Because of this, the spectacu-

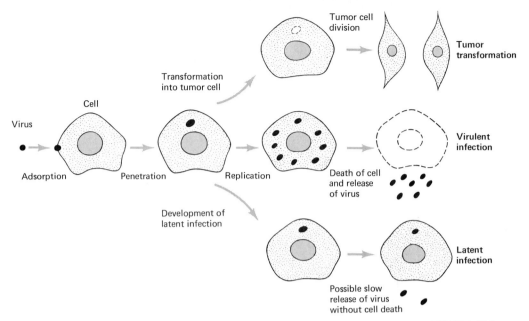

Tumor cell division

Tumor transformation

Transformation into tumor cell

Cell

Virus

Adsorption

Penetration

Replication

Death of cell and release of virus

Virulent infection

Development of latent infection

Latent infection

Possible slow release of virus without cell death

FIGURE 12.3
Various consequences of virus infection.

lar successes in the discovery of antibacterial and antifungal agents (see Section 4.6) have not been followed by similar success in the search for specific antiviral agents. A few antiviral compounds are successful in controlling virus infections in laboratory situations, and certain of these have been used in restricted clinical cases; but no substance has yet been found with more than limited practical use.

A natural type of antiviral agent called *interferon* has some promise for use in the control of virus infections. Interferons are antiviral substances, produced by many animal cells in response to virus infection, that interfere with viral multiplication. Interferons are formed in cells in response not only to live virus but also to virus inactivated by radiation or to viral nucleic acid. Interferon is produced in larger amounts by cells infected with viruses of low virulence, whereas little is produced against highly virulent viruses. Interferons are not virus specific but host specific; that is, an interferon produced by one type of animal (for example, a chicken) in response to influenza virus will also inhibit multiplication of other viruses in the same species but will have little or no effect on the multiplication of influenza virus in other animal species. Interferon has little or no effect on uninfected cells; thus, it seems to inhibit viral synthesis specifically. It is not known precisely how interferon acts. Because interferon is nontoxic, it would seem to be the ideal antiviral agent. However, interferon is difficult to purify, and hence large quantities are not available. Furthermore, interferon is preventive in its action rather than therapeutic; it has no effect on the

VIRUSES AND VIRAL DISEASES

course of the virus multiplication in cells already infected. Also, it is effective only during relatively short periods and thus would have to be administered periodically. Recent work has been concerned with a search for nonviral substances that might induce interferon production and thus confer immunity to virus infection.

12.3
Virus diseases of higher animals

A wide variety of viruses cause disease in higher animals, and our understanding of the nature and manner of action of animal viruses is advancing rapidly, spurred on by the development of two techniques for virus culture and assay: (1) Culturing of viruses in the chick embryo has become an important tool in animal-virus research, since it permits researchers to grow many viruses simply and reproducibly away from the animal body. There are a number of tissues and cavities in the embryo in which viruses can be cultured (see Figures 12.4 and 12.5). (2) Use of animal-cell cultures has led to enormous progress in the understanding of virus-cell interaction and has also made possible the development of new diagnostic methods and of new vaccines. Animal cells used to support virus growth may be either nongrowing primary tissues isolated from various organs of an animal or serially propagated cell lines that are capable of growing indefinitely away from the animal body. Cell cultures are maintained in complex media containing salts, glucose, amino acids, vitamins, other growth factors, usually some natural materials such as serum or embryo extract, and antibiotics (to reduce

FIGURE 12.4
An embryonated egg and sites at which various viruses can replicate.

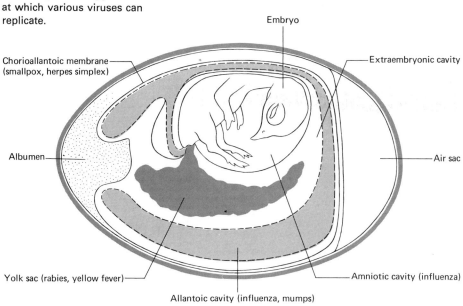

Embryo

Chorioallantoic membrane (smallpox, herpes simplex)

Extraembryonic cavity

Albumen

Air sac

Yolk sac (rabies, yellow fever)

Amniotic cavity (influenza)

Allantoic cavity (influenza, mumps)

bacterial contamination). In such media, animal cells may grow or subsist for considerable lengths of time. Cell cultures may be grown in suspension or in thin sheets or monolayers on glass, agar, or plasma or fibrin clots. Monolayers are especially useful for observing pathological changes that may be induced by viruses since the cells of the monolayer can be observed directly under the microscope. Cell monolayers are used to obtain plaques suitable for virus quantification (Figure 12.6). Dispersed cells may be useful for mass virus propagation and for biochemical studies of virus reproduction. Cell cultures offer advantages over laboratory animals in being cheaper, more reproducible, and simpler; most important, they make it possible to use human cells for propagating viruses that will not grow in nonhuman hosts.

In the following pages we consider in some detail a few representative viruses and indicate some of their characteristics and activities.

12.4 Polio

Poliovirus is the best studied and most important member of a group of viruses known as the *picornaviruses* or *enteroviruses*. Individual virus particles are very small, about 30 nm in diameter, and they are composed of about 75 percent protein and 25 percent RNA, with complete absence of carbohydrate and lipid. A single infected cell may produce as many as 10,000 virus particles, and cells that produce virus usually die and lyse.

Poliovirus replicates primarily in the cells of the intestinal tract of humans. It usually causes only mild symptoms that cannot be distinguished from many other trivial illnesses, and the disease may be called "summer cold" or "intestinal flu." The virus may occasionally pass into the blood or lymph system with no ill effects. In rare instances, it invades the central nervous system and causes the paralytic disease known as poliomyelitis or infantile paralysis (Figure 12.7). Thus, although infection with the virus may be very common, paralytic polio disease is relatively rare.

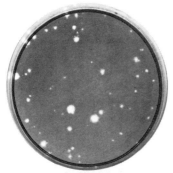

FIGURE 12.6
Plaques induced on a cell culture monolayer by the cancer-inducing Rous sarcoma virus. The cell culture monolayers have been stained with a dye in order to increase the contrast of the plaques.

FIGURE 12.7
Poliomyelitis has been a human affliction for a long time. In this Egyptian hieroglyph, the man has an atrophied right leg, probably as a result of paralytic poliomyelitis infection. From the Carlsberg, Glyptothek, Copenhagen. *Courtesy of The Bettmann Archive.*

Although the virus naturally infects only humans, it can be passed to the chimpanzee, the monkey, and other primates, in which animals it can also induce paralytic disease. It has also been possible to infect mice by injecting them intracerebrally, and certain strains of the virus have been adapted to chick embryos. The greatest experimental advances in our understanding of polio came with the development of methods for growing the virus in human and monkey tissue cultures.

The virus particles are excreted in the feces in large numbers, and fecal contamination of food and water results in spreading of the virus to other individuals. In areas of poor sanitation, most infants are infected early, when they still possess immunity transmitted to them by their mothers. The virus remains localized in the intestinal tract but induces an active immunity that may be lifelong. Paralysis is rare, and when it does occur it is found primarily in infants. In more developed countries with better sanitary conditions, infection of infants is rare. The growing infant loses its maternal immunity and may become a child or young adult without being infected. Infection that occurs later in life is more likely to lead to the paralytic disease, and hence infantile paralysis is more common in developed countries and in older children and young adults (Figure 12.8).

The development of cell-culture methods made it possible to grow the virus in large numbers, opening the way for the production of vaccines. Two kinds are used: killed-virus and live-virus vaccines. The *killed-virus vaccine*, introduced by Jonas Salk in 1953, is prepared by treating poliovirus with a dilute solution of formaldehyde until all virus particles are inactivated. This vaccine is administered by injection, and to induce proper immunity three separate injections must be made: two injections a month apart, followed by a booster 6 to 7 months later. In many parts of the world, it is difficult to ensure that individuals will report for the complete series of injections. Even when the full series has been given, long-lasting immunity is not always induced, especially in young children. Both of these disadvantages are overcome by use of the *live-virus vaccine*, first introduced by Albert Sabin during the period 1955 to 1957. This vaccine consists of virus that has been attenuated so that it no longer causes paralysis, although it still replicates in the intestine. It may be given orally rather than by injection, and a single dose is sufficient since the virus reproduces in the body and can continue to induce immunity for some period of time. The immunity induced is more long-lived than that of the killed-virus vaccine, and the live-virus vaccine is thus more suitable for use on populations where only a single dose can be administered. A possible disadvantage of the live-virus vaccine is that the virus might revert to a virulent form and cause paralytic infection, but this has not occurred even after vaccination of millions of individuals.

FIGURE 12.8
Poliomyelitis patient with paralyzed diaphragm muscles is kept alive by being placed in an iron lung. From F. H. Top and K. M. Johnson, "Enteroviruses: poliomyelitis," in F. H. Top, (ed.), *Communicable and infectious diseases,* 6th ed., The C. V. Mosby Co., St. Louis, 1968.

FIGURE 12.9

Annual poliomyelitis incidence
rates in the United States,
illustrating the consequences of
the introduction of polio
vaccine. From J. J. Witte, *Am. J.
Pub. Hlth.*, **64**, 939–944, 1974.

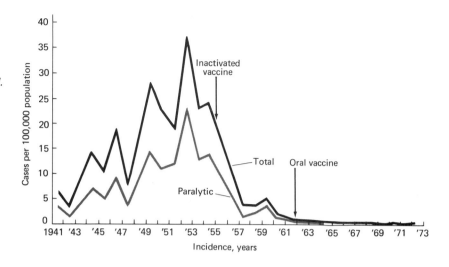

The benefits of vaccination are shown by the dramatic drop in the number of cases of polio after vaccine was introduced into the United States (Figure 12.9).

12.5
Influenza

Frequently in the winter people suffer from a short-lived fever associated with soreness and redness of the respiratory passages and a dry cough. Such a condition is commonly called "flu" or *influenza*, although not all respiratory infections of this type are caused by the same virus. The influenza virus is round or oval-shaped and 80 to 120 nm in diameter. It is an enveloped virus and fairly complex chemically, with less than 1 percent RNA, about 7 percent carbohydrate, 18 percent lipid, and 75 percent protein. The RNA is contained in an inner helical core surrounded by protein, and these two are held together by a lipid envelope.

Human influenza virus exists in nature only in humans. It is transmitted from person to person through the air, primarily in droplets expelled during coughing and sneezing. The virus infects the mucous membranes of the upper respiratory tract and occasionally invades the lungs. Localized symptoms occur at the site of infection. Systemic symptoms include an abrupt fever for 3 to 7 days, chills, fatigue, headache, and general aching. Recovery is usually spontaneous and rapid. Most of the serious consequences of influenza infection occur not because of the viral infection but because bacterial invaders may be able to set up severe infections in persons whose resistance has been lowered. Especially in infants and elderly people, influenza is often followed by bacterial pneumonia; death, if it occurs, is usually due to the bacterial infection.

A most significant aspect of influenza is its occurrence in

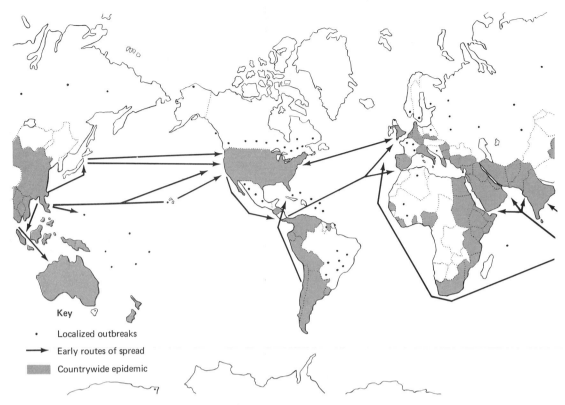

Key

- • Localized outbreaks
- → Early routes of spread
- ▨ Countrywide epidemic

epidemics, often of worldwide proportions. The most severe epidemic was that of 1918, but the one best studied was the 1957 epidemic of the so-called Asian flu, which provided an opportunity for a careful study of how a worldwide epidemic develops. The epidemic began with the development of a new form of the virus of marked virulence and differing from all previous strains in antigenicity. Since immunity to this strain was not present in the population, the virus was able to advance rapidly throughout the world (Figure 12.10). It first appeared in the interior of China in late February, 1957, and by early April had been brought to Hong Kong by refugees. It spread from Hong Kong along air and naval routes and was apparently transferred to San Diego, California, by naval ships. An outbreak occurred in Newport, Rhode Island, on a naval vessel in May. Other outbreaks occurred in various parts of the United States. Peak incidence occurred in the last two weeks of October, during which time 22 million new bed-cases developed. From this period on, there was a progressive decline. Other less serious epidemics of influenza have occurred in the 1960s and 1970s.

Control of influenza epidemics can be carried out by vaccination, although since there are a number of strains it is essential that

FIGURE 12.10
Route of spread of Asian influenza virus from its origin in China. This epidemic in 1957 and 1958 was followed by a similar one in 1968 and 1969. From A. D. Langmuir, *Am. Rev. Resp. Dis.*, **83**, 2, pt. 2, 1961.

VIRUSES AND VIRAL DISEASES

the vaccine be derived from the strain causing the epidemic. A killed-virus vaccine is used, the virus being grown in the chick embryo and inactivated with formaldehyde. Vaccines prepared from several different virus strains can be mixed, producing what is called a *polyvalent* vaccine. If a new strain has arisen, vaccine for it will of course not be available, but by careful worldwide surveillance, it is sometimes possible to obtain cultures of a new strain before it has reached epidemic proportions and thus to produce vaccine ahead of the epidemic. This is costly, however, and may result in production of large amounts of vaccine that are never needed. The more common procedure is to restrict vaccination to those most likely to succumb to severe or fatal illness, such as the aged, pregnant women, and those suffering from chronic debilitating diseases. The families of these persons should also be vaccinated. Duration of immunity is not long, usually only for a few years, so that revaccination is necessary if a new epidemic occurs.

12.6 Smallpox

The disease smallpox has been known for centuries. We discussed the history of this disease as well as the development of immunization procedures to protect against it in Section 1.6. Smallpox virus is much larger than polio or influenza viruses and is in fact just barely visible with the light microscope. The virus is chemically complex, containing about 85 percent protein, 5 percent DNA, 5 percent lipid, and 5 percent carbohydrate. Multiplication occurs in the cytoplasm and may proceed at four or five localities in the cell at the same time. Since DNA synthesis in eucaryotes normally occurs only in the nucleus, the synthesis of smallpox DNA in the cytoplasm is an unusual occurrence.

Smallpox disease has been known throughout recorded history and has occurred in epidemics many times in Western countries. The smallpox virus is found naturally only in humans and monkeys, even though it can be transferred artifically to other animals or to the chick embryo. It is present in large amounts in pustules on the skin of infected people and spreads through the air, entering a new host through the upper respiratory tract. The virus first grows in the mucous membranes of the upper respiratory tract and then multiplies in the lymphoid tissues that drain the respiratory tract. Just before the development of symptoms, the virus may be found in the bloodstream, and from there it passes to the skin, mucous membranes, and organs such as the heart, liver, spleen, and kidney. During the initial phase, fever, headache, and backache occur, followed by prostration. A rash appears on the skin, and soon after this the temperature falls to normal and the patient feels better. Vesicles teeming with virus then form crusts on the skin, which develop into the characteristic pockmarks that disfigure the skin.

In severe cases death occurs, usually due to hemorrhage and generalized toxemia.

Recovery from smallpox infection confers complete immunity. As described in Section 1.6, the earliest immunization procedure took advantage of this fact, using a small amount of virulent small-pox virus to cause a hopefully limited infection, which would in turn confer immunity against a normal full-blown case. The immunization procedure developed by Edward Jenner, using cowpox virus, conferred equally good immunity without the risk of acquiring the disease.

Until recently, smallpox was endemic in India, Southeast Asia, Africa, and South America and was essentially nonexistent in Europe and North America. Now, however, there is a growing tendency to consider the disease as the first to be eradicated worldwide. There have been few if any reports of smallpox cases from any country in recent time. The cases that have been reported most recently have been isolated and have thus not constituted a threat at the epidemic level. Smallpox vaccination has recently been discontinued in the United States, since incidence of the disease has dropped to such low levels. It seems likely that vaccination in other countries may one day be discontinued as well, if the disease actually does prove to be eradicated. The World Health Organization (WHO) will most certainly keep smallpox under close scrutiny to determine whether the eradication is complete.

Measles virus (sometimes also called rubeola virus) grows in nature only in humans, although monkeys can be infected experimentally and may also become infected naturally when kept in captivity. Only primates have been successfully infected, and research on measles has therefore progressed slowly; only since the introduction of tissue-culture methods has any significant progress been made.

Measles virus is related to mumps and parainfluenza viruses, canine distemper virus of dogs, and the rinderpest virus of cattle and sheep. The particles are roughly spherical, about 150 nm in diameter, and contain a core of RNA and protein. Surrounding this core is a lipid-containing membrane.

Measles is highly contagious and the virus is transmitted from person to person by the respiratory route. The virus infects cells of the nasopharynx and then spreads throughout the body. Virus can be found in the blood and in the nasopharyngeal secretions. First symptoms are chills, followed by sneezing, running nose, redness of eyes, cough, and fever. Characteristic spots, called *Koplik's spots*, appear on the mucous membranes and around the salivary glands and are an early diagnostic symptom of the disease. Fever and cough become worse and a rash appears, first on the forehead and behind

12.7 Measles

the ears, then on face, neck, limbs, and trunk. There may be inflammation of the eyes and sensitivity to light. Pustules are not formed; the rash may be an allergic response to presence of virus products in the body rather than a result of direct viral multiplication in the skin cells. A variety of complications may occur, including bronchopneumonia, otitis media, and encephalomyelitis. Recovery usually confers long-term immunity. Measles occurs in non-immunized populations in epidemics at about 3-year intervals, and in many countries nearly 90 percent of the people over 20 years of age have had the disease. Vaccines have been developed, using viruses grown in cell cultures, and have drastically reduced the incidence of measles (see Figure 9.7).

12.8 Rubella, or German measles

The disease caused by rubella virus is similar to measles but is milder, is of shorter duration, and has fewer complications; infection may even be inapparent. Because rubella seemed to be such a mild disease, it received little attention in the early days of virology. However, more recently it has become obvious that rubella infection of mothers during the first trimester (3 months) of pregnancy allows placental transmission of the virus and induces serious damage in the fetus. The infant may be stillborn; if it survives, severe deformities are common. The rubella epidemic in the United States in 1964 caused disabilities in nearly 20,000 infants, resulting from infection *in utero*. Immunization of all susceptible individuals is required, not so much to prevent what is a rather mild disease in children, but rather to prevent infection of pregnant women by eliminating the virus from its major transmitters, children between 1 and 12 years old.

12.9 Adenoviruses

The adenovirus group illustrates well the importance of tissue cultures in the discovery of new viruses. A large number of acute respiratory diseases exist that cannot be attributed to bacterial infection, and these were thought to be caused by viruses. When tissue-culture techniques were introduced (around 1953), workers discovered viruses in respiratory secretions that were able to cause destruction of cultures of human cells. A number of these viruses have been found to be closely related and were called adenoviruses since they were first detected in human adenoid tissues. Although adenoviruses are responsible for only a small percentage of human respiratory disease, they are of interest because they are so widespread.

There are over 50 different adenoviruses, which are related in their chemical, structural, and biological properties. These viruses have been found in nature primarily in humans, although some types are found in other primates. Attempts to transfer the virus to a

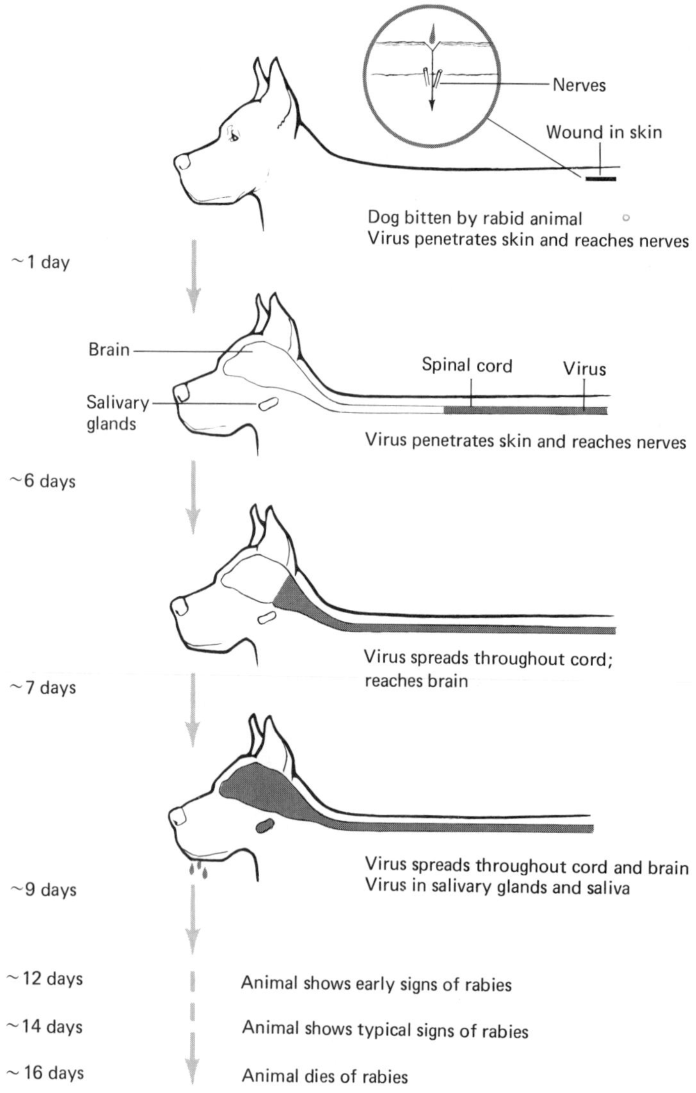

FIGURE 12.11
Progress of the rabies virus through an infected dog.

Nerves

Wound in skin

Dog bitten by rabid animal
Virus penetrates skin and reaches nerves

~1 day

Brain

Salivary glands

Spinal cord   Virus

Virus penetrates skin and reaches nerves

~6 days

Virus spreads throughout cord; reaches brain

~7 days

Virus spreads throughout cord and brain
Virus in salivary glands and saliva

~9 days

~12 days   Animal shows early signs of rabies

~14 days   Animal shows typical signs of rabies

~16 days   Animal dies of rabies

wide variety of experimental animals, including chick embryos, have been unsuccessful. The virus particles contain DNA and protein, are between 60 and 90 nm in size, and do not have outer envelopes. The virus nucleic acid replicates within the nucleus. Pathological effects on the cells are due to inhibition of cell protein and RNA synthesis.

The adenoviruses are transmitted by the respiratory route and infect the tissues of the upper respiratory tract and conjunctiva. They probably remain localized in this region and are not found in the blood or other parts of the body. A number of clinical conditions due to adenoviruses have been described, the most frequent being acute respiratory disease, pharyngitis and pharyngoconjunctival

VIRUSES AND VIRAL DISEASES

fever, and conjunctivitis and keratoconjunctivitis. In most cases the symptoms are a mild, flulike illness with fever, chills, headache, malaise, loss of appetite, mild sore throat, and mild hoarseness and cough. In the latter two conditions, inflammation of the conjunctiva of the eye occurs. Certain adenoviruses can cause tumors in experimental animals. This was discovered by inoculating the animals with a wide variety of adenoviruses that had been isolated from human beings and grown in tissue culture. There is no evidence as yet that adenoviruses can cause tumors in humans, however.

12.10 Rabies

Rabies virus attacks the nervous tissue of all warm-blooded animals. Rabies has been known since ancient times as a disease of dogs, but it is also common in wild animals such as rodents, foxes, skunks, and bats. The virus multiplies in the salivary gland and appears in the saliva; because of this it is readily transmitted by biting (Figure 12.11). The virus particles are about 75 nm wide and 180 nm long and contain RNA and protein inside an outer membrane. Rabies virus is highly neurotropic; after the virus has entered the bloodstream following the bite, it passes to the nervous tissue where multiplication takes place, and the nerve cells degenerate. Characteristic inclusion bodies in the cytoplasm of nerve cells, called *Negri bodies*, are formed during virus multiplication and are demonstrated by a special staining method. The length of time before symptoms appear is highly variable, ranging from 2 to 40 weeks, depending on the size, location, and depth of the wound and on the amount of virulent saliva introduced. When the virus has been passed for many generations in chick embryos, it becomes attenuated and is no longer capable of invading the central nervous system. This modified virus can then be used in the production of rabies vaccine for dogs. It can also be used to protect humans who have been bitten by rabid animals. Pasteur's development of rabies vaccine was recounted in Section 1.6.

Dogs may develop symptoms of two different types (Figure 12.12). In the aggressive (*rabid*) form the dog first exhibits a change in behavior showing restlessness, a tendency to roam, and an inclination to bite, all of which make the transmittal of the virus to susceptible individuals highly probable. Later the dog becomes partially paralyzed and has difficulty in drinking. It staggers around until complete paralysis sets in, and death follows. In the *dumb* form of the disease, the opposite symptoms develop: the animal is not vicious, has no tendency to bite or roam, and is not excitable. The most characteristic symptoms are paralysis of the lower jaw and limbs. In both forms the course of disease is short once symptoms set in, and death occurs in 3 to 7 days.

In human beings the most characteristic early symptom is an

FIGURE 12.12
Two forms of rabies in dogs.
a Furious, or rabid, form.
b Dumb form.

a

b

abnormal sensation around the site of infection. There may be a dull, constant pain in the nerve pathways leading from the wound. The patient becomes hypersensitive, complaining of drafts, loud noises, or bright lights, and becomes increasingly nervous, anxious, and apprehensive. Muscle spasms occur, the pulse becomes rapid, and there is dilation of the pupils, crying, increased salivation, and excessive perspiration. When fluids are consumed, they cause painful, spasmodic contraction of the muscles involved in swallowing, and the fluids are expelled violently. After this the sight, smell, or sound of liquids induces spasms in the muscles of the throat because these suggest the act of swallowing. For this reason the name *hydrophobia* (literally, "fear of water") has often been applied to rabies. Death usually occurs during convulsions, although in some cases the acute excitement phase subsides and paralysis develops, followed by stupor, apathy, coma, and death.

An animal may be suspected of having rabies from its characteristic behavior. Diagnosis of the disease in a dog thought to be infected is usually based on microscopic examination of brain tissue for Negri bodies. Intracerebral inoculation with saliva or brain tissue from the dog is performed on mice, and if the virus is present the mice become paralyzed within 6 to 8 days.

In humans, diagnosis is usually based on knowledge of exposure to the disease and on development of clinical symptoms. On autopsy, Negri bodies can usually be seen. The human mortality rate is considered close to 100 percent, and no therapeutic agent exists for the disease once it has taken hold. If it is known that a person has been bitten by a rabid animal, that individual must immediately receive treatment to inactivate the virus while it still remains localized. Because of the relatively long incubation period, vaccination of a person who has been bitten may be successful. If the incubation period is less than 30 days, however, the development of immunity may not be rapid enough to prevent onset of the disease. Vaccine for humans is prepared from brain tissue of infected rabbits, usually by killing the virus with phenol or ultraviolet light. Since an allergic reaction to rabbit-brain tissue may occur, vaccine is rarely used unless there is good evidence of exposure to rabies. More recently, the attenuated live-virus vaccine used for dogs has been used in humans since it does not cause sensitization.

The spread of rabies infection is quite complex, since the virus can maintain itself in nature only if infected animals can transmit the virus before they die. As we mentioned above, rabies virus is also found in nature in many wild animals, usually carnivores, including foxes, skunks, mink, weasels, wolves, and bats. The fact that bats carry rabies is important, since these animals may be symptomless and serve as a reservoir for the virus in nature. Insect-eating bats have attacked persons, and some of these bats have been shown to be infected. In Central and South America the vampire bat, a blood-sucking animal, frequently attacks cattle and transmits the virus to them. When infected through vampire bats, cattle develop the paralytic, or dumb, form of the disease rather than the rabid type. Rabies infection may be a major factor in controlling the population sizes of wolves, foxes, wild dogs, and other caninelike species: when the population of these animals builds up, an increase in incidence of rabies can bring about a sharp drop in their numbers.

12.11
Tumor viruses

A tumor is composed of living cells that are modified in some way from normal body cells, such as by (1) a decrease or total loss of some of the specialized functions; (2) an increase in a vegetative function, such as the ability to grow more vigorously; and (3) the acquisition of certain new functions, such as ability to invade surrounding tissues and to continue to grow in parts of the body distant from the point of origin. Tumors are quite diverse and may vary from small warts to extensive growths that invade and bring destruction to whole organs of the body. Although tumor cells may quite likely arise in various ways, some tumors clearly have been shown to be induced by viruses, and it is these that we shall consider briefly here. Both DNA and RNA viruses can cause tumors in animals.

A tumor of chickens, known as the *Rous sarcoma*, was the first tumor shown to be induced by a virus. Since then a number of other tumors of birds have been found to have viral origin. Shope papilloma and Shope fibroma are virus-induced tumors of rabbits, and a mammary-gland tumor of mice has been discovered to be induced by a virus transmitted in the mother's milk. The polyoma virus (discussed below) causes a wide variety of tumors in mice and other animals. With the electron microscope, it is possible to see viruses or viruslike particles within infected cells. It is difficult to obtain really convincing evidence of virus-induced tumors in humans, since Koch's postulates are difficult to apply; however, particles that resemble viruses can be seen in cells from some human tumors, providing considerable support to the theory that viruses may induce them. If this is so, new means of prevention and treatment might be possible.

Polyoma virus has been well studied and will illustrate some aspects of the recent search for tumor viruses. This virus was first isolated during research on tumors in mice and has been found to cause over 20 different kinds of tumors, including those of various glands and organs, the skin, connective tissue, bone, kidney, blood vessels, liver, and nervous tissue. Because of the large variety of tumor types induced, it was called the *polyoma* virus (from *poly*, "many," and *oma*, "tumor"). This virus also induces tumors of varying types when injected into other mammals, such as the hamster, rabbit, and rat; monkeys and humans, however, are not affected.

The polyoma virus is approximately spherical in shape and about 45 nm in diameter; it consists of only DNA and protein. When polyoma virus infects embryonic cells it causes a virulent infection, the cells are killed and lysed, and new virus particles are produced. In a virulent infection, extensive changes in the structure of the nucleus appear, and then large numbers of virus particles can be seen under the electron microscope; cells so infected lyse and release virus particles. When the virus infects adult cells, on the other hand, it does not cause a virulent infection; the cells instead become transformed into tumor cells, and no mature virus is produced.

The mechanism of conversion of a normal cell into a cancer cell is uncertain. However, it is clear that virus multiplication does not occur in a cell converted to a tumor cell. The viral DNA enters the nucleus, but no discernible structural changes occur. The new tumor cells do not produce virus and cannot be induced to do so, but it is known that the virus persists in these cells in some form.

A major difference between a tumor cell and a normal cell is in their responses to crowding. The normal cell stops growing when it comes into contact with other cells, a phenomenon called *contact inhibition*. The tumor cell, on the other hand, continues to grow even under crowded conditions. This ability may reflect the cancerous nature of the cell and suggests that some mechanism controlling the population density of normal cells is lacking in cancerous cells.

A number of viruses cause infections of the upper respiratory tract, such as influenza virus and the adenoviruses, but the symptoms of the common cold differ from those caused by these other agents. The first sign of the common cold is usually a sore throat; fever and other systemic manifestations are rare, and the predominant symptom is an increased flow of nasal mucus.

Cold viruses are not highly infectious: even when nasal secretions are inoculated directly into volunteers, the percentage of takes is only 30 to 40 percent. When a volunteer is placed in an isolated room and inoculated with nasal secretions from a person with active cold infection, the disease (when it appears) runs a quick course (Figure 12.13). No experimental animal except the chimpanzee has been found to be susceptible, and this animal is too expensive to keep and too difficult to work with for routine laboratory use. Attempts to culture the viruses in chick embryos have been unsuccessful, but techniques for growing the viruses in human cell cultures are now available, and many strains have been isolated in tissue culture from persons with typical colds.

12.12 Common cold

FIGURE 12.13

The course of a cold in a human volunteer inoculated intranasally with a cold virus. The severity of the symptoms can be roughly quantified by the number of handkerchiefs used each day. In the lower section, the gray squares indicate positive virus culture and the open squares, negative virus culture. From D. A. J. Tyrrell and M. L. Bynoe, *Brit. Med. J.*, **1**, 393, 1961.

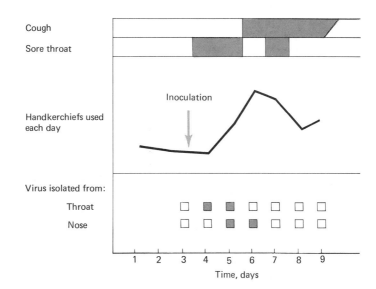

12.13 Hepatitis

Infectious hepatitis is a disease of the liver caused by a virus (hepatitis A virus) that is transmitted by the oral route and enters humans primarily through fecal contamination of water, food, or milk. The virus itself is not too well known, as it has proved difficult to carry into culture. The disease is endemic throughout the inhabited areas of the world, and outbreaks of epidemics occur in many regions. Explosive water-borne epidemics have occurred for a long time in areas where proper control of water supplies is not maintained, and food-borne epidemics have recently become of considerable concern.

Among foods transmitting the virus, the most significant recently have been shellfish (oysters and clams) taken from polluted waters. These animals live on the sea bottom in estuaries and shallow marine bays and feed by passing through their bodies large amounts of silt-laden seawater. The food-rich particles in the seawater are filtered out in the animal and serve as its source of food. Pathogenic microorganisms, including infectious hepatitis virus derived from domestic raw sewage in the water, will also be filtered out and concentrated by the animal. No problem with infectious hepatitis is experienced in eating such shellfish if they are cooked, since heating destroys the virus, but oysters in particular are often eaten raw, and many recent epidemics have occurred in areas where shellfish waters are subject to sewage pollution (Figure 12.14).

The best means of controlling infectious hepatitis is to prevent its spread by improving the general sanitation of the region through use of proper sewage-treatment systems, elimination of flies, and prevention of fecal contamination of food and milk supplies by infected food handlers. Unfortunately, identifying infected food handlers is not at all easy because suitable methods of detecting the virus are wanting. (General practices of food sanitation are further discussed in Section 15.8, and sewage-treatment systems are considered in Chapter 14.)

Diagnosis of the disease is made on the basis of the symptoms, especially the observation of liver damage. Symptoms include loss of appetite, fever, nausea, vomiting, and pain in the abdomen, followed generally by enlargement of the liver and jaundice. However, jaundice, which is a yellowing of the skin, tissues, and body fluids due to liver malfunction, is not a constant symptom of infectious hepatitis.

Because the virus has not been satisfactorily cultured, there is no vaccine available for inducing artificial immunity. Passive immunity can be induced by administering antibodies obtained from immune people. Such antibodies are generally obtained by collection of serum from individuals who have recovered from the disease and preparation of a gamma globulin fraction, since it is in this fraction that antibodies are present.

Serum hepatitis (caused by hepatitis B virus) resembles infectious hepatitis in many respects but differs in having a much longer incubation period and in the route of infection. As suggested by its name, serum hepatitis is contracted primarily through the use of infected blood and blood products in transfusions. Contaminated instruments and syringes and needles are also commonly implicated in transmission of the organism. No specific therapy is available for treatment of the disease. The best control is by constant vigilance in the use of all materials used for blood sampling and transfusions and by avoiding the use of blood donors having a history of hepatitis.

FIGURE 12.14
Warning sign in a shellfish area where infectious hepatitis is a problem.

Study questions

1 How does virus replication differ from cell division?
2 What are three types of host response to virus infection?
3 What role does interferon play in virus infection?
4 By what methods are viruses cultured in the laboratory? How have these techniques aided virus research?

For the rest of this chapter, there are no specific questions. Instead, the student may wish to prepare a study outline for each of the viral diseases included in this course in a manner similar to that followed for Chapters 10 and 11: type of virus; disease symptoms; modes of transmission; and methods of treatment and prophylaxis.

Suggested readings

Davis, B. D., R. Dulbecco, H. N. Eisen, H. S. Ginsberg, and W. B. Wood, Jr., *Microbiology*, 2nd ed., Harper & Row, New York, 1973. This excellent advanced textbook has extensive coverage of viruses and virus diseases of humans.

Fenner, F., B. R. McAuslan, C. A. Mims, J. Sambrook, and D. O. White, *The biology of animal viruses*, 2nd ed., Academic Press, New York, 1974. The best general text.

Lennette, E. H., E. H. Spaulding, and J. P. Truant (eds.), *Manual of clinical microbiology*, 2nd ed., American Society for Microbiology, Washington, D.C., 1974. Detailed methods for the isolation and characterization of pathogenic viruses from clinical specimens.

Top, F. H., Sr., and P. F. Wehrle (eds.), *Communicable and infectious diseases*, 8th ed., C. V. Mosby Co., St. Louis, 1976. Good chapters on viral diseases.

Youmans, G. P., P. Y. Paterson, and H. M. Sommers, *The biologic and clinical basis of infectious diseases*, W. B. Saunders Co., Philadelphia, 1975. A good general medical text, with some material on viruses.

The importance of microorganisms in the environment and the means of controlling microbes in food and water are the major themes of this part. Broadly defined, the environment is the surroundings of an organism: the chemical, physical, and biological forces that act on an organism and on populations of organisms. Organisms adapt to their environments, but they are also influenced by them, often in harmful ways. The human environment is more complex than the environment of other organisms, because humans to a great extent create their own environment, through construction of houses and cities. Most human beings no longer live close to the soil; food and water are not available in the immediate surroundings but are brought from distant parts. Ensuring adequate and safe supplies of food and water constitutes one of the major themes of environmental science, and because microbial influences on food and water are so important, environmental microbiology constitutes one of the major parts of environmental science.

In Chapter 13, we discuss briefly some of the general principles of environmental microbiology, with emphasis on how microorganisms are affected by and affect their environments. The role of microorganisms in the major nutrient and energy cycles in nature is

ENVIRONMENTAL MICROBIOLOGY

discussed. This chapter also includes some applied aspects of environmental microbiology dealing with the petroleum and mining industries.

The microbiology of water and wastewater are discussed in Chapter 14. Water is a major vehicle for the transmission of several important pathogens, and the means of ensuring safe drinking water are discussed here. Because the water supply used to produce drinking water in one area may have been influenced by pollution from other areas, adequate methods of sewage treatment are necessary to ensure that pathogens are not released into the environment. Most sewage-treatment procedures involve processes in which the major role is played by microorganisms, so that the microbiology of sewage treatment comprises a second major section of Chapter 14.

Foods are excellent environments for microbial growth, and microorganisms have both harmful and beneficial effects in food and dairy products. Chapter 15 describes both these areas in some detail. We consider food spoilage, how foods are processed and stored to prevent spoilage, and how food poisonings are prevented. The use of microorganisms to produce such foods as cheese, buttermilk, sausage, and pickles is also considered.

Environmental microbiology

Microorganisms play far more important roles in nature than their small sizes suggest. Their activities are so extensive and diverse that even when they are present in relatively low numbers they may have significant effects on the environment. We already discussed in Part 3 some of the most dramatic effects of microorganisms: their roles as causal agents of infectious disease. But pathogenic microorganisms constitute only a small fraction of the kinds of organisms present in the world. The vast majority of microorganisms do not cause disease, but that fact does not make them less important or less interesting; they carry out a wide variety of other processes of importance to the well-being of humans. Because microorganisms are usually invisible, their existence in an environment may go unsuspected; yet without them, higher organisms would quickly disappear from the earth.

Although the earth provides suitable environments for microbial growth, we do not find the same organisms everywhere. In fact, virtually every environment, no matter how slightly it differs from others, probably has its own particular complement of microorganisms that differ in major or minor ways from the organisms of other environments. Because microorganisms are small, their environments are also small. Within a single handful of soil, many microbial environments exist, each providing conditions suitable for the growth of certain microorganisms. When we think of microorganisms living in nature, we must learn to "think small."

The earth can be divided into three zones. The *lithosphere* is the solid portion of the earth, composed of rocks and soil. Microbes constitute an important part of the soil. The *hydrosphere* represents the aquatic environments of the earth—the oceans, lakes, rivers—and in these, microorganisms are common. The *atmosphere* is the gaseous region surrounding the earth, relatively dense near the surface but thinning to nothing in its upper reaches. Microorganisms are present in the atmosphere but they do not grow there; they are merely being carried from one place to another.

The *biosphere* represents the mass of all living organisms present on earth. Higher organisms, plants and animals, provide excellent environments for the growth of microorganisms, and as a result we find large populations of microbes associated with living organisms, often causing no harm and perhaps even being beneficial. We discussed the normal flora of the human body in Section 8.1. Dead animals and plants also are well colonized by microorganisms, and the decomposition of their remains constitutes one of the major functions of microorganisms in the world. Without microbial decomposition, think of the vast piles of dead animals and plants that would have accumulated on earth!

13.1 Symbiosis

When two organisms live together, the relationship may sometimes be beneficial for both organisms. Such a mutually beneficial relationship is called *symbiosis* or *mutualism*. We discuss two agriculturally important symbioses in Chapter 16: the rumen symbiosis and the rhizobium-legume relationship. Here we discuss several nonagricultural examples of microbial symbioses.

LICHENS

Lichens are leafy or encrusting growths of a rather insignificant appearance that are widespread in nature and are often found growing on bare rocks, tree trunks, house roofs, and on the surfaces of bare soils (Figure 13.1). The lichen plant, usually called a *thallus*, consists of two organisms, a fungus and an alga, living together to their mutual benefit. Lichens are extremely interesting associations because they demonstrate so clearly the value and importance of symbiosis. They are found in environments where other organisms do not grow, and it is almost certain that their success in colonizing such extreme environments is due to the mutual interrelationships between the alga and fungus partners.

The lichen thallus usually consists of a tight association of many fungus hyphae, within which the algal cells are embedded (Figure 13.2). The shape of the lichen thallus is determined primarily by the fungal partner, and a wide variety of fungi are able to form lichen associations. The diversity of algal types is much smaller, and many

a

b

c

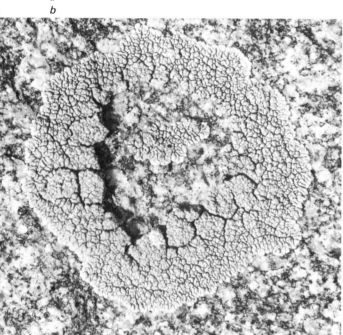

FIGURE 13.1
Lichens in their natural
environments. *a* Foliose lichen
growing on a dead branch.
b Crustose lichen colony growing
on the surface of a granite rock.
The colony is about 3 cm in
diameter. *c* Crustose lichens
growing on a large pot in a
garden.

different kinds of lichens may have the same algal component. The algae are usually present in defined layers or clumps within the lichen thallus.

Although both the fungus and the alga can be cultivated separately in the laboratory, they are almost always found living together in nature. The alga is photosynthetic and is able to produce organic matter from carbon dioxide of the air. Some of the organic matter produced by the alga is then used as nutrient by the fungus. In many cases, the fungus produces special structures called *haustoria* that penetrate into the algal cells and provide an avenue for the transport of materials from the alga to the fungus. Since the fungus is unable to photosynthesize, its ability to live in nature is dependent on the activity of its algal symbiont.

ENVIRONMENTAL BIOLOGY

The fungus clearly benefits from the association, but how does the alga benefit? The fungus provides a firm anchor within which the alga can grow protected from erosion by rain or wind. In addition, the fungus absorbs from the rock or other substrate upon which the lichen is living the inorganic nutrients essential for the growth of the alga. Another role for the fungus is to protect the alga from desiccation; most of the habitats in which lichens live are dry (rock, bare soil, roof tops), and fungi are in general much better able to tolerate dry conditions than algae.

If a lichen is removed from its natural habitat and placed under certain conditions in the laboratory, the symbiosis can be induced to break down. Either the fungus proliferates and destroys the alga, or the alga overgrows the fungus, depending on conditions. It is possible to cultivate the alga and fungus separately in the laboratory, but it is difficult to bring the cultures back together and have them reform a lichen thallus. Lichen resynthesis from cultures never occurs in laboratory media where the isolated components are able to grow well, but only where starvation conditions are induced. It appears that starvation "forces" the alga and fungus to reunite.

Although lichens live in nature under rather harsh conditions, they are extremely sensitive to air pollution and quickly disappear

FIGURE 13.2
Photomicrograph of a vertical section through a lichen thallus, showing the location of the algal layer. Note also the tangle of fungal hyphae. Magnification 220×.

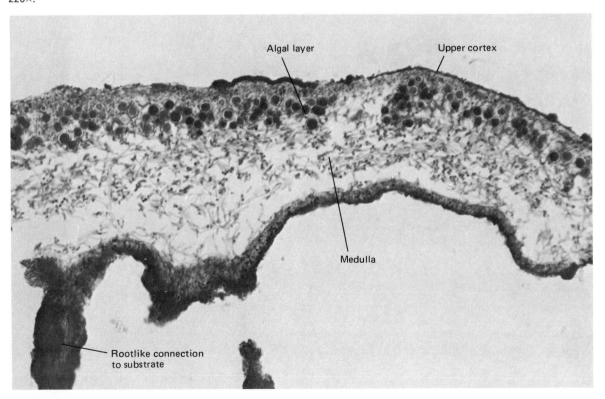

Algal layer

Upper cortex

Medulla

Rootlike connection to substrate

from large cities where heavy air pollution occurs. One reason for the great sensitivity of lichens to air pollution is that they absorb and concentrate elements from rainwater and air and have no means for excreting them, so that lethal concentrations are reached. Lichens have been extensively used as indicators of air pollution and may provide some early warnings of impending problems. Studies made on the richness of the lichen flora in cities and their surroundings show that the number of species of lichens found on various habitats decreases as one moves from the countryside to the center of the city (Figure 13.3).

Many lichens produce characteristic pigments of striking color. The chemistry of these pigments is used in classification. More interestingly, many of these pigments are useful dyestuffs and have been used by primitive and pioneer societies for dyeing cloth for clothes.

INSECT SYMBIOSES

Insects and microorganisms have developed a wide variety of symbiotic relationships. These symbioses are especially important to insects that obtain their food from plant sources. Plants are low in protein, and the microbial symbionts may enable the insects to live on these protein-poor diets. There are two types of microbial symbioses with insects: *endosymbioses*, in which the microorganisms live inside the cells of the animal, and *ectosymbioses*, in which the organisms live extracellularly, either in the intestine or in the environment of the animal.

One of the most interesting examples of an ectosymbiosis is the partnership that leaf-cutting ants have entered into with filamentous fungi (Figure 13.4). The ants actually cultivate the fungus as their principal source of food. Primarily tropical, the ants live in great colonies in which they construct chambers within which the fungi grow. Leaf fragments are carried by the ant to the nest, and the fungus grows on these fragments and ramifies extensively throughout the chambers. As the fungus is grazed by the ants, it is replaced by further growth. The ant encourages growth of the fungus by applying small droplets of fecal exudate to the cut pieces of leaf. This fecal exudate contains enzymes that digest leaf proteins and also contains amino acids that the fungus can use as a nitrogen source.

Other examples of ectosymbioses are ambrosia beetles, which bore tunnels in trees within which they cultivate fungi as food, and wood-eating termites, which possess protozoa in their intestines that digest the cellulose from the wood, making the nutrients available to the insect. These termites sometimes also have bacteria in their intestines that can fix N_2 from the air, thus making it possible for the animal to live on nitrogen-deficient substances such as wood.

A wide variety of insects possess *endosymbiotic intracellular bac-*

FIGURE 13.3
The decrease in richness of lichens as the city center is approached. Data for Newcastle-upon-Tyne, an industrial city in northern England. Lichens can be used as indicators of long-term trends in air pollution. From O. L. Gilbert, "Lichens as indicators of air pollution in the Tyne Valley," in G. T. Goodman, R. W. Edwards, and J. M. Lambert (eds.), *Ecology and the industrial society*, John Wiley & Sons, Inc., New York, 1965.

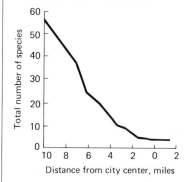

ENVIRONMENTAL BIOLOGY

FIGURE 13.4
Leaf-cutting ants in a fungus garden. The large ant is the queen; the smaller are workers. The white material is fungus mycelium. *Courtesy of Neal A. Weber.*

teria that are probably essential for the survival of their hosts. These bacteria resemble the rickettsia, a group of intracellular parasitic bacteria (see Section 10.17). Over 10 percent of all species of insects possess intracellular bacteria, and they are especially common in three large insect groups: the cockroaches, the leafhopper-aphid group, and the weevils. Also, many blood-sucking bugs have symbionts. The symbiotic bacteria are housed in special cells called *mycetocytes*, which may occur in special organs called *mycetomes*. Mycetocytes are often near the ovaries or are embedded in the fat body of the insect. Transmission of the symbionts from parent to offspring has been shown to be via the maternal line, and it is thought that the symbionts may become incorporated into the egg and then multiply as the egg becomes converted into an insect. These endosymbiotic bacteria are probably not involved in digestion, since they do not live in digestive organs, and their intracellular location does not make them accessible to undigested food. At least one postulated function is in the synthesis of vitamins required by the insect, although this is not the whole answer, as vitamins will not permit normal growth of symbiont-free insects. That the symbionts are essential for the life of the insect is clear, since if the symbiont is eliminated by chemical treatments or other means, the insect develops only very poorly (Figure 13.5).

A wide variety of invertebrate animals, including protozoa, sponges, corals, worms, molluscs, echinoderms, and tunicates, harbor algal symbionts. Although most animals harboring algal symbionts are marine, freshwater types are also known. The algae are usually unicellular species that live within the cells of the animal, where they are able in some manner to resist digestion by the animal cells. The algae are often found in well-defined regions or layers of the animal (Figure 13.6). The algae are not absolutely essential for the life of the animal, but their importance is shown by the fact that animals containing symbionts grow more efficiently in the light than do symbiont-free animals, and they survive for much longer periods of time in nutrient-poor environments. The algae probably play several roles: during photosynthesis they synthesize oxygen that the animal can use, and they also convert carbon dioxide into organic compounds that are transferred to the animal tissue. The algae may also contribute to the health of the animal by removing waste products. The algae benefit from the association because their nutrients pass through the animal tissues; the animal functions as an absorptive organ for the algae, concentrating nutrients from the surrounding water. Also, the animal by its movements or growth habit remains in shallow water, where light intensities are high, thus permitting better photosynthesis.

Mycorrhiza means "root fungus" and refers to a symbiotic association which occurs between plant roots and fungi. The roots of the majority of terrestrial plants are probably mycorrhizal. Most forest trees, especially conifers, beeches, and oaks, are known to be mycorrhizal, and in a dense forest, almost every root of every tree will be mycorrhizal. The fungi that participate in these symbiotic associations also form typical mushroom fruiting bodies (see Section 2.7), and there is a physical connection between the fungus in the root and the fruiting body above the ground. Most mycorrhizal fungi probably aid the plant by absorbing inorganic nutrients from the soil, the filamentous network of fungal material providing an extensive surface through which absorption can occur. In return, the fungus obtains organic nutrients excreted from the root.

The beneficial effect of the mycorrhizal fungus is best observed in poor soils, where trees that are mycorrhizal will thrive but non-mycorrhizal ones will not. If trees are planted in prairie soils, which

ALGAE AND INVERTEBRATES

FIGURE 13.5
Size contrast between cockroaches with and without bacterial symbionts. Both sets are the same age. Symbiont-free cockroaches (upper row) are produced by injecting the enzyme lysozyme, which digests bacteria, into young roaches. Photographed 11 months after injection. From H. Malke and W. Schwartz, *Z. Allgem. Mikrobiol.*, **6**, 34, 1966.

MYCORRHIZAL FUNGI

a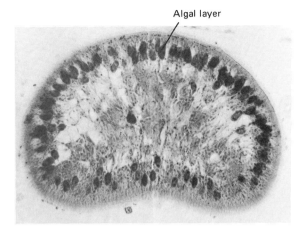

b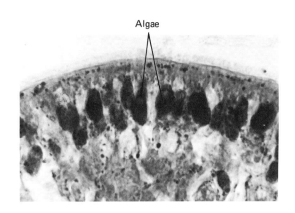

FIGURE 13.6
Symbiotic relationship between a marine snail and an alga. *a* Cross section through the snail, showing how the algae are oriented in a peripheral layer. The dorsal surface of the snail is at the top. Magnification 75×. *b* Photomicrograph at higher magnification (200×) showing the single layer of algae. From D. L. Taylor, *J. Mar. Biol. Assoc. U.K.*, **51**, 301–313, 1971.

ordinarily lack a suitable fungal inoculum, they do poorly, whereas if they are artificially inoculated at the time of planting, they grow much more rapidly (Figure 13.7). However, in nutrient-rich conditions, the mycorrhizal plant does not grow any better than an uninoculated control, and growth may even be slightly retarded.

FIGURE 13.7
Six-month-old pine seedlings, showing the beneficial effects of mycorrhizae. Left, nonmycorrhizal; right, mycorrhizal. The trees are growing in a prairie soil lacking native fungal inoculum, and the mycorrhizal tree was inoculated. *Courtesy of S. A. Wilde.*

An ecosystem is usually defined as the total community of organisms living together in a particular habitat, together with the physical and chemical environment in which they live. Each organism interacts with its physical and chemical environment and with the other organisms in the system; so the ecosystem can be viewed as a kind of superorganism with the ability to respond to and modify its environment. A good example of an ecosystem is a lake. The sides of the lake define the boundaries of the ecosystem, and within these boundaries the organisms live and carry on their activities, greatly modifying the characteristics of the lake as well as each other (Figure 13.8).

Energy enters an ecosystem mainly in the form of sunlight and is used by photosynthetic organisms in the synthesis of organic matter. Some of the energy contained in this organic matter is dissipated by the photosynthesizers themselves during respiration, and the rest is available to herbivores, which are animals that consume the photosynthesizers. Of the energy entering the herbivores, one portion is dissipated by them during respiration, and the rest is used in synthesizing the organic matter of the herbivore bodies. Herbivores are themselves consumed by carnivorous animals, and these carnivores are eaten by other carnivores, and so on. At each step in this chain of events, a portion of the energy is dissipated as heat. Any plants or animals that die, whether from

13.2
Microorganisms and ecosystems

Lake surface

Invertebrates

FIGURE 13.8
A lake is a good example of an ecosystem. The changes in environmental factors with depth, during summer stagnation, is shown. *Courtesy of A. E. Walsby.*

ENVIRONMENTAL BIOLOGY

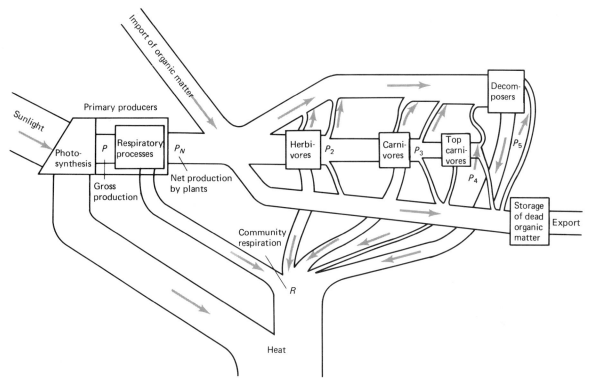

Import of organic matter

Sunlight

Primary producers

Photo-synthesis

P Respiratory processes

Gross production

P_N

Net production by plants

Herbi-vores P_2

Carni-vores P_3

Top carni-vores

P_4

Decom-posers

P_5

Storage of dead organic matter Export

Community respiration

R

Heat

FIGURE 13.9
Energy-flow diagram for an ecosystem. The sizes of the boxes and channels are proportional to the relative amounts of energy at each stage. All energy that is not stored or exported as organic carbon is degraded as heat and lost. Adapted from E. P. Odum, *Fundamentals of ecology*, 3rd ed., W. B. Saunders Co., Philadelphia, 1971.

natural causes, injury, or disease, are attacked by microorganisms and small animals, collectively called *decomposers*. The decomposers also utilize energy released by plants or animals in the form of excretory products. All these reactions constitute a *food chain* or *food web*. Because there is a loss of energy at each stage of the food chain, ultimately practically all of the biologically useful energy that is used to convert materials to organic matter by the photosynthesizers is dissipated; usually only very small amounts are stored. Because of this, energy is said to flow through the ecosystem. The quantitative relations can be expressed by an energy-flow diagram, such as the one shown in Figure 13.9.

Although the energy fixed by photosynthesizers is ultimately dissipated as heat, the chemical elements that serve as nutrients usually are not lost from the ecosystem. For instance, carbon from CO_2 fixed by plants in photosynthesis is released during respiration by various organisms of the food chain and becomes available for further utilization by the plants. Nitrogen, sulfur, phosphorus, iron, and other elements taken up by plants are also released through the activity of the decomposers and are thus made available for reassimilation by other plants. Therefore, although energy flows through the ecosystem, chemical elements are carried through cycles within the system. In some parts of the cycle the element is oxidized, whereas in other parts it is reduced; for many elements a biogeochemical cycle can thus be defined, in which the element undergoes changes in oxidation state as it is acted upon by one organism after another. In addition to this redox (oxidation–reduction) cycle, it is also possible to define a transport cycle, which describes the movement of an element from one place to another on earth, as for instance, from land to air or from air to water. Such a transport cycle may or may not also involve a redox cycle. For instance, when oxidation or reduction leads to conversion of a nonvolatile substance to one that is volatile, the latter can then be transported to the air, so that the transport cycle is coupled to the redox cycle. In other cases, oxidation or reduction does not lead to a change in state and has no influence on transport.

In the next few sections, we describe several of the important biogeochemical cycles in which microorganisms are involved.

The biogeochemical cycle of carbon is shown in Figure 13.10. Energy from the sun is used by higher plants and algae to convert carbon dioxide into organic matter. When the plants die, this organic matter becomes decomposed and eventually carbon dioxide is formed again. The organisms involved in the decomposition

13.3
The carbon cycle

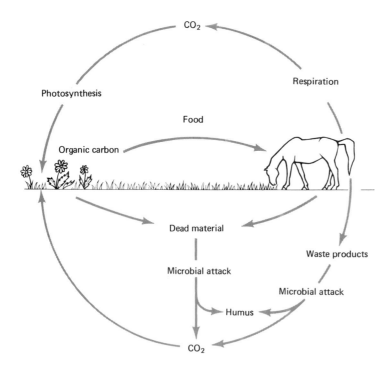

FIGURE 13.10
The carbon cycle.

process and the biochemical steps that occur depend to a considerable extent upon whether decomposition occurs aerobically or anaerobically. In soil decomposition is generally aerobic, whereas in lakes it is aerobic in the top part but anaerobic in the mud and in the deeper waters.

Organic matter decomposed anaerobically is acted upon almost exclusively by bacteria. The initial steps involve fermentation to organic acids, followed by the conversion of these acids to methane, hydrogen, and CO_2. If organic acids accumulate, this often leads to the inhibition of further decomposition processes, and the organic material may then slowly accumulate. Coal is thought to have originated when enormous accumulations of organic matter in the anaerobic zones of shallow marshes and lakes of the past were subsequently modified by purely geological processes. In many coals the remains of plants are still easily visible in fossil form. Petroleum is thought to have been created in a similar way from organic deposits in marine environments. Organic matter deposited as coal or oil represents carbon and energy lost from the ecosystem. Much of this material is converted back to CO_2 when these fuels are combusted.

In aerobic situations, decomposition is a combined effort of animals and microorganisms. The animals are involved to some extent directly in decomposition, by eating and digesting plant remains, but probably their most important role is to fragment large plant parts into smaller pieces, making them much more susceptible to microbial attack. The shredding and chewing action of animals produces a rich array of small particles that are quickly colonized by bacteria and fungi. In aerobic environments, decomposition usually goes essentially to completion, with almost all the carbon returned to CO_2. A small amount of carbon remains in an undegradable or very slowly degradable fraction, humus.

13.4
Petroleum microbiology and the carbon cycle

Petroleum contains a large array of organic carbon compounds that are potential nutrients for microorganisms. As long as petroleum remains buried in the earth, away from oxygen, it is resistant to microbial attack because there are no microorganisms which can attack it anaerobically. However, as soon as it is brought into contact with oxygen during petroleum-recovery activities, it is rapidly colonized by a variety of aerobic microorganisms. One of the main groups of organic compounds present in petroleum are *hydrocarbons*, containing only carbon and hydrogen. During refining operations, hydrocarbons are separated from the crude petroleum and are converted into the major fractions used in gasoline and oil. There are large numbers of microorganisms, both bacteria and fungi, that can attack and degrade hydrocarbons. Because of the wide variety of microbial activities possible with crude petroleum and its derivatives, petroleum microbiology is a large and important field.

Prospecting is one of the most interesting ways in which microorganisms have been of aid to the petroleum industry. Associated with liquid and solid petroleum is a gaseous fraction consisting of methane, ethane, and propane. In petroleum-producing regions these gases may seep to the surface and provide nutrients for the growth of specific hydrocarbon-utilizing bacteria. Where one finds bacteria capable of oxidizing these gases, there is a strong suggestion that a petroleum deposit is nearby. Looking for methane-utilizing organisms in searching for petroleum is not practical, since methane is produced biologically in many systems that are not related to petroleum. Ethane, however, is not produced biologically in significant amounts and is almost always associated only with petroleum, so that detection of ethane-utilizing organisms can be used quite reliably to discover petroleum reserves. Since geological methods of locating petroleum deposits have thus far been adequate, microbiological petroleum prospecting has not found wide use. However, it may become more important in the future, as petroleum reserves become depleted.

The decomposition by microorganisms of petroleum and petroleum products is of considerable importance. It is virtually impossible to keep moisture from bulk storage tanks; it accumulates as a layer of water beneath the petroleum. At the petroleum-water interface, bacteria develop in large numbers (Figure 13.11), and fungi, yeasts, and actinomycetes may also grow. Microbial growth has become an especially serious problem in the kerosene-based fuels used in jet airplanes. When such microbe-containing fuels are burned, fuel strainers rapidly become clogged, leading to power loss or stalling. In addition, microbial growth on the inside surfaces of the fuel tanks of aircraft can lead to corrosion of the tanks. Several control methods are used: (1) The fuel is filtered through membrane filters, which remove microorganisms. (2) Inhibitors of microbial growth are added. (3) Corrosion can be minimized by coating the inside surfaces of the fuel tanks with more resistant substances, such as polyurethane. (4) Aircraft fuel tanks are washed out at regular intervals with a 2-percent solution of potassium dichromate, an antimicrobial agent. (5) Some hydrocarbon fractions of the fuel are more readily attacked than are others; if these fractions can be removed from the fuel, its storage life may be lengthened.

    Another detrimental effect of microorganisms is observed during the drilling of an oil well. Large amounts of water are used at the site of the drill bit, the function of the water being to cool and

FIGURE 13.11
Hydrocarbon-oxidizing bacteria in association with oil droplets. The bacteria are concentrated in large numbers at the oil-water interface. Magnification 600×.

Bacteria

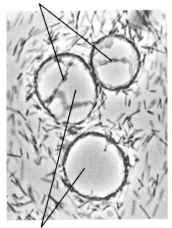

Oil droplets

ENVIRONMENTAL BIOLOGY

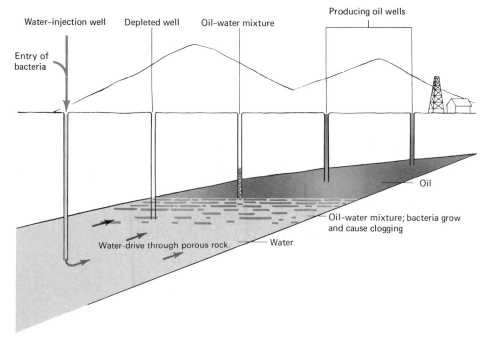

FIGURE 13.12
Injection of water into deep wells that are becoming depleted will aid in forcing remaining oil out, but bacterial growth in the injection-water systems may cause clogging.

lubricate the drill bit and to bring the rock cuttings to the surface. When the water is brought back to the surface, it is pumped into a pit where the rock particles settle out, and the liquid is then recycled. The water used, called a "drilling fluid" or "drilling mud," usually contains organic substances that are added to thicken it and make it flow better. Some of the additives used are starch, modified cellulose, lignin, and natural gums, most of which are excellent substrates for the growth of bacteria. Bacterial growth in the drilling fluid can so greatly change its properties that it no longer functions properly, so that inhibitors are nearly always added. The chemical used must be inexpensive, effective, and noncorrosive. Representative types of inhibitors include quaternary ammonium compounds, organic amines, phenols, and formaldehyde.

Microbial activity in the oil well itself can have detrimental effects that may seriously affect the recovery of petroleum. The most important is the growth of microorganisms in the pores of the rock within which the petroleum is trapped, thus plugging the pores and preventing the movement of the petroleum out of the rock reservoir and into the well.

Microbial growth in petroleum will occur whenever water and air are present. In the original (undrilled) petroleum reservoir, neither of these substances is present, but when the well is opened they are admitted to the reservoir. Water is often injected into the reservoir intentionally to force out the petroleum and thus increase recovery. This injection water naturally carries air and bacteria with it, and with the rich source of organic matter present in the petroleum deposit, conditions are good for bacterial growth (Figure 13.12). Many of the bacteria that grow produce capsules, and the slimy bacterial masses effectively block the pores.

Since it is virtually impossible to maintain a sterile petroleum reservoir, inhibitors such as those used in drilling are added to the injection waters to prevent plugging caused by bacteria. The inhibitor used must of course be active against the bacteria involved, but other important properties are low cost, ease of handling, solubility in both water and oil, lack of corrosion of pipes and pumps, and lack of toxicity to humans. If there is any danger of the injection water entering the ground-water supply used for humans, the inhibitor must also be nonpolluting to natural waters and should be biodegradable.

The biogeochemical cycle of nitrogen is shown in Figure 13.13. Nitrogen gas (N_2) is the most abundant gas in the atmosphere, making up 78 percent by weight. Much of the nitrogen on earth is in the atmosphere, and from this source, most of the nitrogen of living organisms is ultimately derived.

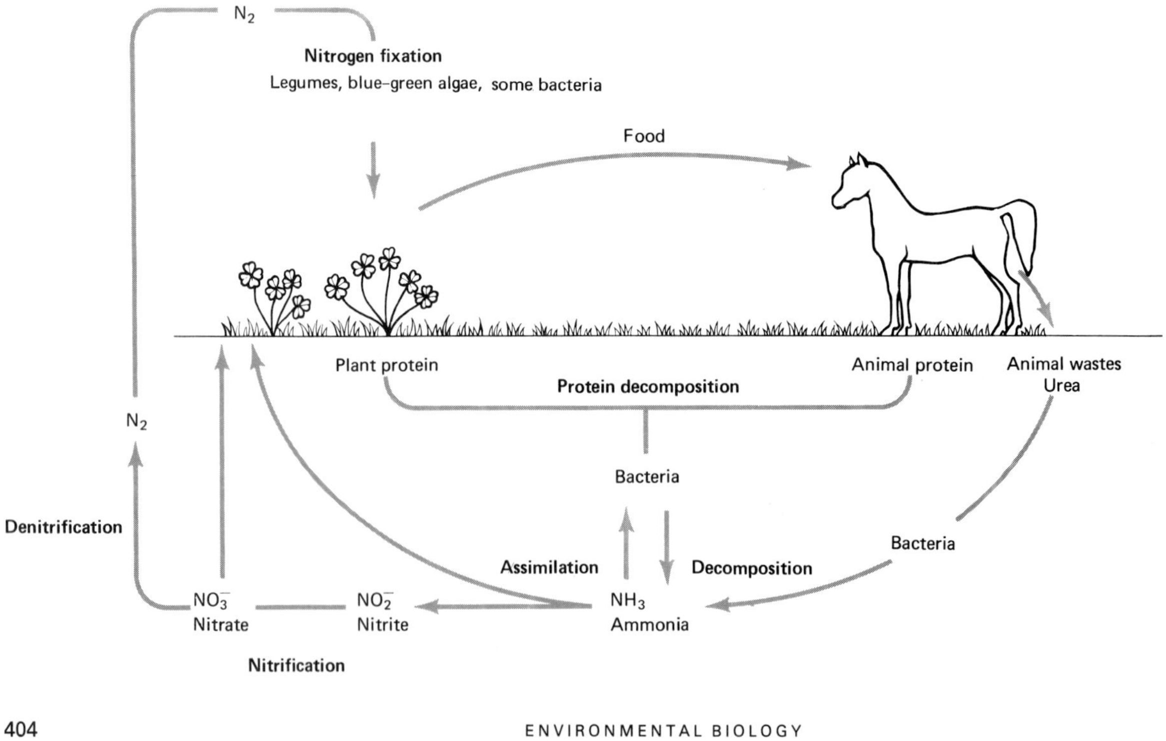

13.5
The nitrogen cycle

FIGURE 13.13
The nitrogen cycle.

The conversion of N_2 into combined forms is called *nitrogen fixation* and is one of the most important microbiological processes on earth. All the microorganisms that carry out nitrogen fixation are procaryotes, and nitrogen-fixing organisms include a variety of different bacteria and blue-green algae. We can distinguish two kinds of nitrogen-fixing organisms: free-living and symbiotic nitrogen fixers. A symbiotic nitrogen-fixing organism carries out this process only when living in association with a plant (in symbiosis), whereas a free-living organism is able to fix nitrogen when living directly in soil or water with no other organism involved.

Free-living nitrogen-fixing organisms include two genera of bacteria that are widespread in soil, *Azotobacter* and *Clostridium*, as well as several genera of blue-green algae. Of the bacteria, *Azotobacter* species are aerobic, whereas *Clostridium* species are anaerobic. Studies of nitrogen fixation in soils of various types have shown that fixation occurs most readily under anaerobic conditions, suggesting that *Clostridium* is most important. Nitrogen fixation by these free-living bacteria requires the presence of considerable amounts of organic matter that serves as energy source for the bacteria. Nitrogen fixation is therefore more common in soils rich in organic matter than in organically poor soils.

The other major group of free-living nitrogen-fixing organisms is blue-green algae. Not all blue-green algae fix nitrogen, but those that do so may be responsible for considerable addition of nitrogen to the habitats in which they live. Since blue-green algae grow using light as sole energy source and CO_2 as sole carbon source, nitrogen fixation by these organisms does not require the presence of organic matter, as it does with bacteria. These procaryotic algae are widespread in lakes, streams, and the ocean, and are also found frequently in the surface crusts of desert and other arid soils. Another important habitat of blue-green algae is in rice paddies (Figure 13.14). When the paddy is flooded, the blue-green algae develop rapidly in the warm shallow water and are probably responsible for a significant amount of combined nitrogen added to the system.

When nitrogen is fixed, it is reduced to ammonia (NH_3), and the ammonia produced is then converted into organic nitrogen, primarily in the form of amino acids in the proteins of the cells. Thus initially the nitrogen that is fixed is held within the protoplasm of the nitrogen-fixing organism and is not available to plants. Later, when the nitrogen-fixing organisms die, their cells are decomposed by other bacteria, and the organic nitrogen is converted back to NH_3, a process called *mineralization*. The ammonia is then available directly to plants as a nutrient, or as described later, it can be converted further to nitrate, which is also available as a plant nutrient.

The most important *symbiotic nitrogen-fixing organisms* are members of the genus *Rhizobium*, which live in association with

FIGURE 13.14
A rice paddy in southern India. Nitrogen-fixing blue-green algae often grow extensively in this partially aquatic system.

plants called legumes. This type of nitrogen fixation is discussed in Section 16.4.

The conversion of ammonia to nitrate, called nitrification, is an important microbial process that occurs commonly in soils where ammonia is present. Ammonia is often added to soil as a fertilizer in the form of anhydrous ammonia and is also produced by the microbial decomposition of organic nitrogen. It is also the first product of nitrogen fixation, just discussed. Nitrification is brought about by a special group of bacteria called the *nitrifying bacteria*, which obtain their energy for growth from the oxidation of ammonia or nitrite. These bacteria are autotrophs, being able to grow in completely inorganic media, using CO_2 as sole source of carbon.

Nitrification occurs in two stages: (1) the oxidation of ammonia (NH_3) to nitrite (NO_2^-), brought about by one group of bacteria; and (2) the oxidation of nitrite to nitrate (NO_3^-), brought about by another group of bacteria. The combined action of both these groups leads to the conversion of ammonia to nitrate. Nitrification requires the presence of oxygen and, hence, occurs only in aerated, well-drained soils. Under anaerobic conditions, such as in water-logged soils, it is inhibited. Nitrate is taken up well by plants, and in fact, it is taken up more efficiently than is ammonia. But nitrification may not necessarily be beneficial in agricultural practice; since nitrate is very soluble in water, it is rapidly leached from the soil and is then unavailable in any form to the plants.

Nitrification also occurs commonly in aerobic sewage-treatment plants and is responsible for the formation of nitrate from the organic nitrogen of sewage, since ammonia released by mineralization of organic nitrogen is nitrified to nitrate. As we shall discuss in Section 14.15, nitrification under these conditions can lead to considerable nutrient enrichment of lakes and rivers, allowing growth of many algae and other organisms and leading to a deterioration of water quality.

Denitrification is an anaerobic process that results in the conversion of nitrate to gaseous products, usually N_2 and N_2O, which then escape into the air. Denitrification is thus a harmful process in that it leads to a loss of combined nitrogen. A number of bacteria are able to denitrify. Denitrifying bacteria are facultative anaerobes, and denitrification occurs only under anaerobic conditions, being inhibited by the presence of O_2. Denitrification is a type of anaerobic respiration (see Section 6.2), and organic compounds are necessary for it to occur. The process occurs commonly in the anaerobic sediments of lakes and estuaries and in soils that become flooded.

13.6
The sulfur cycle

The biogeochemical cycle of sulfur is shown in Figure 13.15. Sulfur is an important plant nutrient, and several forms of this element exist in nature. There are three forms that are of practical significance: *sulfate* (SO_4^{2-}), which is the most oxidized form; *sulfide*, in the form of either hydrogen sulfide (H_2S) or sulfides such as iron sulfide (FeS), which is the most reduced form; and *elemental sulfur* (S^0), which is intermediate in oxidation level. The conversion of sulfur from one state to another is carried out in nature most commonly by microorganisms.

SULFATE REDUCTION

Under anaerobic conditions, sulfate may be reduced to hydrogen sulfide, a toxic product. This conversion is carried out by a special group of bacteria called the *sulfate-reducing bacteria*; the most widespread genus is *Desulfovibrio*. If a soil becomes waterlogged and hence anaerobic, sulfate reduction can occur; the toxic hydrogen sulfide that accumulates may cause damage to plants and animals. Sulfate reduction and the resultant hydrogen sulfide production may also cause serious corrosion of iron and steel pipes buried in soils, the hydrogen sulfide attacking the metal, making holes and pits in the pipe.

SULFUR OXIDATION

The oxidation of elemental sulfur to sulfate is a microbial process of considerable significance. When elemental sulfur is oxidized; sulfuric acid is produced, reducing the pH of the soil. Soils too alkaline for

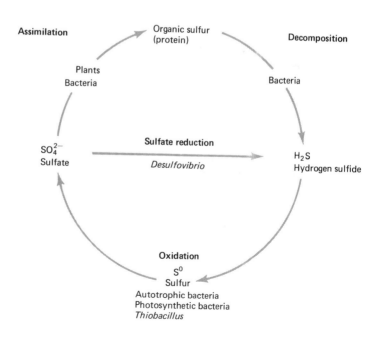

FIGURE 13.15
The sulfur cycle.

normal plant growth can thus be acidified by adding elemental sulfur to them. Elemental sulfur, a common form of sulfur fertilizer, is not utilizable directly by plants; however, upon its conversion by microbes to sulfate, it becomes available as a plant nutrient. The most common sulfur-oxidizing bacteria are members of the genus *Thiobacillus*, and representatives of this genus are widespread in soils.

A third sulfur transformation, the oxidation of sulfides to sulfuric acid, is usually harmful because it leads to development of excessive acidity in soil and water. Sulfides are fairly common in low-lying soils along seacoast areas, especially in warmer parts of the world. As long as the soils are flooded they remain anaerobic, and the sulfides remain reduced. If the soils are drained for agricultural purposes, however, the aerobic conditions that develop lead to rapid oxidation of sulfide and the formation of excessive sulfuric acid, resulting in soil pH too low for the growth of plants. Another kind of soil showing high acidity is that which forms in the spoil banks of many surface-mining areas where sulfides are present in large amounts. The acidity makes the growth of plants and the successful revegetation of such areas impossible. This problem is discussed in more detail in the next section.

13.7 Mining microbiology and the sulfur cycle

In the mining of coal and minerals (copper, lead, and zinc), microbial activities in the sulfur cycle can lead to effects that are either beneficial or harmful. A serious form of water pollution, acid mine drainage, occurs frequently in coal mining operations and occasionally in mineral mining. In addition, the same bacteria that cause acid mine drainage can also play a beneficial role in the extraction of minerals from some kinds of low-grade copper, lead, and zinc ores. In both cases, the bacteria act by attacking sulfide minerals present in the coal or ore deposit. In the case of coal, a mineral frequently present in association with it is an iron sulfide known as *pyrite*; this substance has a goldlike glitter, a factor responsible for its sometimes being called "fool's gold." In the case of copper, lead, and zinc, the minerals themselves are in the form of sulfides, usually associated with iron sulfide. The attack of bacteria on these sulfides results in both formation of sulfuric acid and solubilization of the metal.

Not all coal seams contain iron sulfides, so that acid mine drainage does not occur in all coal-mining regions; where acid mine drainage does occur, however, it is often a very serious problem. Mixing of acid mine drainage with natural waters in rivers and lakes causes a serious degradation in the quality of the natural water, since both the acid and the dissolved metals are toxic to aquatic life. In addition, such polluted waters are unsuitable for human consumption and industrial use. An understanding of the factors involved in acid mine drainage may help us control it. Since bacteria are a prime factor in its production, an understanding of their properties and activities may aid in preventing its occurrence.

Attack on the sulfide minerals involves the breakdown of iron sulfide into sulfuric acid and ferrous iron and is done by the autotrophic bacterium *Thiobacillus ferrooxidans.* The sulfuric acid is not changed further, but the ferrous iron liberated can be oxidized by this bacterium, producing ferric iron. This forms an insoluble yellow precipitate, called "yellow boy" by miners, which coats polluted streams and rivers, making these waters unsightly. *Thiobacillus ferrooxidans* is a very acid-tolerant bacterium and in fact prefers acid to neutral conditions; therefore it is able to remain active under the acidic conditions that it produces. Oxygen is required for growth and for the oxidation of sulfides, and the organism develops only where aerobic conditions prevail. It is autotrophic and is therefore able to grow on a completely inorganic medium; in the coal and ore-bearing materials that are uncovered, there is almost always a plentiful supply of the nitrogen, phosphorus, potassium, sulfur, magnesium, and other nutrients needed for growth.

The properties of this organism help to explain how acid mine drainage develops. As long as the coal is unmined, oxidation of its sulfide minerals cannot occur, since neither air nor the bacteria can reach it. When the coal seam is exposed, it quickly becomes contaminated with the bacteria, and the presence of oxygen makes oxidation of the sulfide minerals possible. If conditions are appropriate, the acidity produced can leach out of the mine into a nearby stream.

There are essentially three kinds of coal mines, and they vary in the degree to which they can be a source of acid mine drainage. A *shaft* mine is one in which a shaft is sunk below ground to reach a coal seam. In such a mine, the coal seam is below the water table, and acid water formed at the mine face does not naturally leach into the surrounding streams. However, shaft mines are often subject to flooding, and water must be continually pumped out to keep them dry. If this water is acidic, it will cause acid pollution unless steps are taken to neutralize it.

A *drift* mine is one in which a shaft is dug into the side of a hill where a coal seam is present. Drift mines are found primarily in hilly or mountainous regions and are very common in the leading coal-mine regions of the eastern United States. Since the mine is above the water table, acid water drains out or is leached out into a nearby stream, and because of this, drift mines are a serious source of acid mine drainage (Figure 13.16). It is virtually impossible to prevent leakage of acid water from an active drift mine, so that the water

Acid + iron

FIGURE 13.16
Formation of acid mine drainage in an underground (drift) mine.

must be removed and treated if acid pollution is to be avoided. A still more serious source of acid mine drainage is abandoned drift mines, which can be continuous sources of acid water for many years after they are no longer being mined. The best way to control acid mine drainage from such abandoned sites is by sealing them. The usual method for sealing is to fill the hole with concrete blocks and mortar. Sealing does two things: it prevents the acid water from draining out, and it restricts the entrance of oxygen into the mine. Unfortunately, sealing is an expensive operation, and with abandoned mines it is not clear who should bear the expense, since the company that operated the mine may no longer be in existence.

A *strip mine*, also called an *open cast*, or *surface*, *mine*, is one in which the coal is so near the surface that it can be mined by removing (stripping) the overlying rock and soil (called *overburden*). Coal seams within 30 to 50 meters (100 to 150 feet) of the surface can be mined by this method. Stripping of overburden is done by large electrically operated shovels called *draglines* (Figure 13.17*a*) since they sit on top of the overburden and drag the shovel across it. The coal is removed from the seam, and then the overburden is replaced. To strip a wide area, the dragline moves in a straight line to the end of the property, then turns and goes back along an adjacent line. The overburden from one pit is deposited in the pit formed by the previous stripping. Because the rock and soil is considerably loosened by the dragline, it can never be completely redeposited in the pit, and a series of low hills, called *spoil banks*, are formed (Figure 13.17*b*). When the limits of the property are reached, the stripping ceases, and the final hole formed remains unfilled. This hole, called the *final cut*, is bordered on one side by spoil and on the other by a steep wall of unmined rock and coal called the *high wall*. Acid mine drainage in strip mines may be much less than in shaft or drift mines, or it may be similar, depending both on the geological conditions of the mine and on the way in which the dragline operator handles the overburden (Figure 13.17*c*). Most of it is usually low in sulfide minerals and thus is not potentially acid-producing, but the material immediately over the coal seam may be high in sulfide minerals. Since the overburden immediately over the

a

b

FIGURE 13.17
Coal-mining operations. *a* The large shovel, called a *dragline*, is used to remove overburden during strip-mining operations. *b* Spoil bank formed as a result of strip mining. Note the absence of vegetation. *c* Origin of acid mine drainage in strip mine operations.

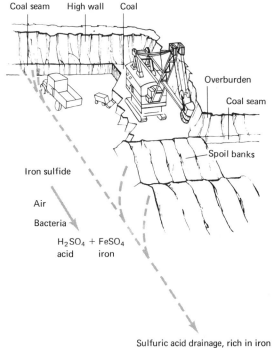

coal seam is the last to be removed from the pit, it will be on top of the spoil bank. If it is left on the top of the spoil bank, where it is exposed to air and to contamination by *T. ferrooxidans*, it may become a source of acid mine drainage. Modern practice requires that any acid-producing spoil be buried deep in the pit so that it cannot be a source of acidity. Unfortunately, in many older strip-mine areas this practice was not followed, and such older areas may be sources of acid drainage for many years. Further, the soil of such spoil banks becomes acid, making it difficult or impossible to establish vegetation. Such acid spoil banks are veritable deserts (see Figure 13.17*b*), even if they are in regions where plentiful rainfall occurs.

Another source of acidity in strip-mine operations is the final cut. The final cut usually fills with water to the level of the water table and thus becomes a small lake. Such a lake might be a recreational or industrial resource but often cannot be because of the acidity that develops. The high wall contains coal and sulfide minerals that are exposed, and the acid produced drains naturally into the lake. Such a lake is often very acid, pH 2 to 3, and is completely unsuitable for fish and most other animals.

When the coal is removed from the mine, it is always intermixed with rocks and other nonburnable material which must be removed before the coal can be marketed. This material, called *coal refuse*, is removed in the coal preparation plant and is usually discarded in nearby coal refuse piles. This refuse is often very high in iron sulfides and is a major source of acid pollution. Within the coal refuse pile, *T. ferrooxidans* quickly develops, and extremely acidic conditions result. Rain falling on the pile leaches acid out, and streams draining such piles often become very acidic. Since many refuse piles are abandoned, the responsibility for controlling acid mine drainage from them rests on no one legally and they are serious sources of pollution. Eventually, all the sulfide becomes oxidized and all the acid leaches out, but this takes many years.

Since acid mine drainage would not develop in the absence of bacterial activity, it might seem logical to prevent its occurrence by using chemicals or other agents that kill or inhibit the growth of bacteria. A number of such agents are available, including antibiotics, antiseptics, and organic acids. In addition, viruses that attack *T. ferrooxidans* have been found, as have other microorganisms that are predatory and kill and consume *Thiobacillus*. None of these agents have yet proved of practical value in controlling acid mine drainage. They may be too expensive to use in the enormous quantities necessary to have significant impact, or it may be difficult or impossible to deliver them at appropriate concentrations to the active sites of acid production deep within the ore materials, or they may not be sufficiently active in inhibiting or killing the bacteria. To date, the most effective means of controlling acid mine drainage is the use of mining practices that keep air and bacteria from the acid-producing materials.

The acid produced from coal-mining operations can be neutralized by use of lime. Neutralization is the most widely employed method of eliminating this water-pollution nuisance. However, one problem with use of lime as a neutralizing agent is that the lime particles become coated with a layer of ferric hydroxide, so that fresh lime must be added periodically.

In addition to the acidophilic *T. ferrooxidans*, there are a number of other microorganisms that not only tolerate but actually prefer acid conditions. Some of these microorganisms are found in quite high numbers in waters influenced by acid mine drainage. Among these are the algae *Euglena*, *Ulothrix*, *Chlorella*, *Chlamydomonas*, and *Eunotia*, several bacteria related to the pseudomonads, and several acidophilic protozoa. The algae often form bright green mats on the bottoms of streams and are usually especially abundant close to the source of acid water. They are less common further downstream from the source, because here the heavy deposit of yellow boy covers the stream bed and blocks out the light that algae need for growth.

ROLES OF THIOBACILLUS FERROOXIDANS IN MINERAL-MINING OPERATIONS

The ores from which copper, lead, and zinc are obtained are often sulfide ores, and *T. ferrooxidans* is able to oxidize these sulfide minerals in a manner similar to the way it oxidizes iron sulfide. In addition, iron sulfide is usually present in association with these other sulfide minerals in the ore body and is also oxidized. As a result of the oxidation of the sulfide minerals, not only is sulfuric acid produced, but also the metal of the sulfide mineral is solubilized. Acid mine drainage can thus develop from mineral mines, although in practice it rarely does; since the acid water is high in the mineral being mined, the mining company usually controls the acid mine drainage in the recovery of the mineral.

The most important role of *T. ferrooxidans* in mineral mining is the extraction of the mineral from low-grade ores, a process called *microbial leaching* (Figure 13.18). Leaching is used most extensively in copper mining, and as copper becomes scarcer and low-grade ores therefore become more valuable as copper sources, leaching will become even more important. In the leaching process the ore, usually containing 0.5 percent or less of copper, is placed in huge piles, and acid water is sprayed over the top of the pile (Figure 13.19). This water slowly percolates down through the pile and creates the appropriate conditions for the growth of *T. ferrooxidans* within the pile. This organism oxidizes the sulfide mineral, creating more sulfuric acid and liberating copper. At the bottom of the leach dump, the copper-laden acid water exits and is carried to a processing plant where the copper is removed. After the copper is removed, the water is still very acidic and is high in ferrous iron.

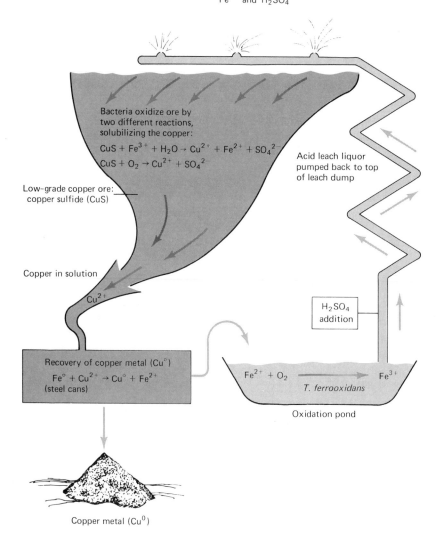

Sprinkling of acid leach liquor on copper ore:
Fe^{3+} and H_2SO_4

Bacteria oxidize ore by
two different reactions,
solubilizing the copper:

$CuS + Fe^{3+} + H_2O \rightarrow Cu^{2+} + Fe^{2+} + SO_4^{2-}$

$CuS + O_2 \rightarrow Cu^{2+} + SO_4^{2-}$

Acid leach liquor
pumped back to top
of leach dump

Low-grade copper ore:
copper sulfide (CuS)

Copper in solution

Cu^{2+}

H_2SO_4
addition

Recovery of copper metal (Cu^c)

$Fe^\circ + Cu^{2+} \rightarrow Cu^\circ + Fe^{2+}$
(steel cans)

$Fe^{2+} + O_2 \xrightarrow{\quad\quad} Fe^{3+}$

T. ferrooxidans

Oxidation pond

Copper metal (Cu^0)

FIGURE 13.18
Microbial activity in leach
dumps is responsible for
recovery of large amounts of
metals from low-grade ores.

This water is usually carried to an oxidation pond where it is further
acted upon by *T. ferrooxidans*, which oxidizes the ferrous iron to the
ferric form. The water, now laden with ferric iron, is recycled
through the leach dump, and the ferric iron itself reacts with further
sulfide mineral to solubilize more copper. The bacteria thus play two
roles: in the leach dump itself, and in the oxidation pond.

The actual operation of the leach dump is carried out without
any careful attempt to promote bacterial action. The process of
leaching was actually developed by copper companies without any

ENVIRONMENTAL BIOLOGY

real knowledge that bacteria were involved and is being continued today mostly by using empirical methods. Although less widely used, leaching is also carried out in some uranium-mining operations. The importance of microbial leaching is shown by the fact that in the United States, about 25 percent of the copper recovered from ores comes from leaching operations.

A recent development that shows some promise of improving copper-mining operations and at the same time protecting the environment is the procedure called *in situ mining*. In this process, the copper ore is solubilized through bacterial action while still underground, by percolating acid water through the formation. In beginning an *in situ* mining operation, the ore body is fractured underground by use of explosives, to make it permeable to the leach water and air. During the initial stages of leaching, the bacteria carried into the ore body with the leach solution must colonize the ore particles; once this has occurred, solubilization of the mineral occurs. *In situ* leaching must be carried out on an ore body that is on a hillside or above the general water table, so that the leach solution pumped in at the top can be collected by gravity flow. It is also essential that the rock underlying the ore body be impermeable to water, so that the leach solution does not percolate into the underlying ground water, where it might cause water pollution and where

FIGURE 13.19
A typical leach dump. The leach water leaving the bottom of the dump is collected in a small pond and pumped to the copper-recovery site.

the solubilized copper would be lost. *In situ* leaching presents some exciting possibilities for the development of mining operations relatively free of environmental impact.

Summary

In this chapter, we have learned something about the activities of microorganisms in the environment. In the first part of the chapter, we discussed a variety of ways in which microorganisms form beneficial associations with animals and plants, forming symbiotic associations. The existence of symbiosis should emphasize that disease is not the only way in which microorganisms can interact with higher organisms. Although there is a tendency to think of microorganisms affecting higher organisms only in detrimental ways, this obviously is not true, and it may even be the case that more microorganisms enter into beneficial associations than harmful ones.

In the second part of this chapter, we learned about some of the roles of microorganisms in ecosystems. An ecosystem is defined as the total community of organisms living in a particular habitat, together with the physical and chemical environments in which they live. Microorganisms play many important roles in ecosystems. They function as decomposers, converting dead plants and animals back into their basic constituents. They carry out oxidations and reductions on various elements, such as carbon, nitrogen, and sulfur, and thus are the prime biological agents in the biogeochemical cycles. A major form of organic carbon on earth is petroleum, present in vast underground deposits. Microorganisms can carry out a variety of processes related to petroleum, both harmful and beneficial, and petroleum microbiology is a large and important field of study.

The ability of microorganisms to convert nitrogen gas (N_2) from the air into combined nitrogen, a process called *nitrogen fixation*, is one of the most important contributions that microorganisms make to ecosystem function. Nitrogen fixation makes available to higher organisms nitrogen from from the air, an inexhaustible nitrogen source.

The oxidation and reduction of sulfur compounds in nature is carried out primarily by microorganisms. Such transformations are important in the sulfur cycle, but there are also significant practical implications of microbial activities on sulfur. Coal and metal ores often contain or are associated with sulfide minerals, and the oxidation of these sulfide minerals by microorganisms can have either harmful or beneficial effects. In coal-mining operations, the oxidation of sulfide minerals results in the formation of sulfuric acid, an important pollutant, leading to the formation of acid mine drainage. In mineral mining, especially of copper, the ability of *Thiobacillus ferrooxidans* to oxidize copper sulfide minerals is used to recover useful ore from low-grade deposits in a process called *leaching*. The ore-containing rock is placed in large piles, and dilute acid is percolated through it. The acid-loving bacteria develop on the sulfide minerals, oxidize them, and release the mineral into solution, where it can be recovered from the leach water that exits from the bottom of the leach pile. This process is of major importance in mineral recovery, especially as high-grade deposits become exhausted and more attention must be directed to low-grade deposits.

1 What is the biosphere?
2 What is symbiosis? Contrast with antibiosis and pathogenesis.
3 Describe the manner in which the fungus and alga interact in the formation of a lichen thallus.
4 What benefit does the fungus derive from the lichen symbiosis? the alga?
5 How can lichens be used as sensitive indicators of air pollution?
6 Contrast ecto- and endosymbioses of microorganisms with insects.
7 Describe the relationship that the leaf-cutting ants have entered into with fungi. What is the principal source of food for the ants?
8 What evidence shows that cockroaches must have symbiotic bacteria for normal development? Where are the symbiotic bacteria of the cockroach located?
9 Compare the function of the alga in marine invertebrates with the function of the alga in the lichen symbiosis.
10 What is the likely function of mycorrhizal fungi? How can the beneficial importance of mycorrhizal fungi be demonstrated?
11 Define *ecosystem*. Give two examples of ecosystems.
12 What is a food chain?
13 Discuss the energy relations of ecosystems. How do decomposing microorganisms affect the energy relations of ecosystems?
14 Discuss the roles that microorganisms play in the carbon cycle. Contrast the processes of the carbon cycle under aerobic and anaerobic conditions.
15 What is petroleum microbiology? Describe two detrimental effects that microorganisms can have in petroleum production. In what way can microorganisms be of benefit in petroleum prospecting?
16 Why is the process of nitrogen fixation so important?
17 Discuss some of the organisms involved in nitrogen fixation.
18 What is nitrification? Contrast with denitrification.
19 Diagram the sulfur cycle, and indicate the kinds of microorganisms involved in different steps.
20 How do sulfur bacteria cause the formation of acid mine drainage? How can acid mine drainage be prevented?
21 Describe the strip-mining process, and indicate the places where sulfur bacteria can develop.
22 How do the same sulfur bacteria that cause acid mine drainage carry out a beneficial effect in the recovery of minerals from low-grade ores?
23 What is *in situ* mining? What role do bacteria play in this type of mining?

Suggested readings

Bland, John, *Forests of Lilliput*, Prentice-Hall, Inc., Englewood Cliffs, N.J., 1971. Good elementary treatment of lichens and mosses. Excellent photographs.

Brock, T. D., *Principles of microbial ecology*, Prentice-Hall, Inc., Englewood Cliffs, N.J., 1966. A fairly elementary textbook dealing with the fundamentals of microbial ecology.

Brock, T. D., *Biology of microorganisms*, 2nd ed., Prentice-Hall, Inc., Englewood Cliffs, N. J., 1974. General textbook of microbiology with extensive chapters on microbial symbioses and on biogeochemical cycles.

Buchner, P., *Endosymbiosis of animals with plant microorganisms*, John Wiley & Sons, New York, 1965. Very detailed textbook on the symbiotic microorganisms of insects.

Davis, J. B., *Petroleum microbiology*, American Elsevier Publishing Co.,

New York, 1967. A detailed and very useful textbook of petroleum microbiology, with extensive discussion of practical aspects.

Delwiche, C. C., "The nitrogen cycle," *Sci. Am.*, **223**, 136–146, 1970. An excellent review with emphasis on the microbial processes.

Harley, J. L., *The biology of mycorrhiza*, 2nd ed., Leonard Hill, London, 1969. An excellent brief textbook on mycorrhizae.

Henry, S. M. (ed.), *Symbiosis*, Vols. 1 and 2, Academic Press, New York, 1966 and 1967. Separate chapters in these two volumes deal with the various microbial symbioses.

Kellogg, W. W., R. D. Cadle, E. R. Allen, A. L. Lazarus, and E. A. Martell, "The sulfur cycle," *Science*, **175**, 587–596, 1972. Although this article does not cover microbial aspects directly, it provides an excellent view of the global sulfur cycle.

Odum, E. P., *Fundamentals of ecology*, 3rd ed., W. B. Saunders Co., Philadelphia, 1971. The standard textbook on ecology, with excellent treatment of energy relations of ecosystems.

Scientific American, "The biosphere," *Sci. Am.*, **223**, 1970. A whole issue of this magazine dealing with the biogeochemical cycles.

Trager, W., *Symbiosis*, Van Nostrand Reinhold Co., New York, 1970. Brief paperback on symbioses with emphasis on animal-microbe relationships.

Weber, N. A., *Gardening ants: The attines*, Memoir 92, The American Philosophical Society, Philadelphia, 1972. A detailed and well-illustrated monograph.

Zajic, J., *Microbial biogeochemistry*, Academic Press, New York, 1969. Extensive discussion of the role of microorganisms in the leaching of ores.

Microbiology of water and wastewater

Chapter 14

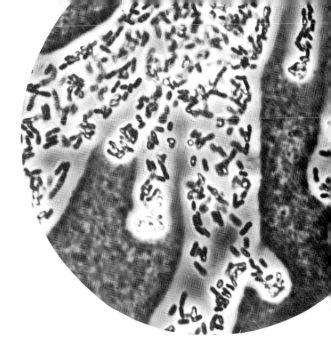

Water is essential for life, but it can also be a hazard, as it can carry pathogenic microorganisms and toxic chemicals. Probably the greatest contribution that urbanization has made to the advance of human civilization has been the provision of pure and reliable water for its inhabitants. The amount of water used by a single person varies greatly throughout the world. In advanced, urbanized countries, water use per capita is highest, although wide variations in use occur within a single country, depending on local conditions. In the United States, average water use in private residences and apartment houses is about 100 gallons per person per day. Of this water, about 20 percent is used in the kitchen, 25 percent in the toilet, 25 percent in the shower or bath tub, 10 percent in the wash basin, and 20 percent in the laundry.

14.1
Water supply

Water arises initially as precipitation (rain or snow), which falls to the earth, percolates through the soil, and moves into rivers and streams. These water courses carry the water to the lakes and the sea, from which it is evaporated and returned to the atmosphere. It can again fall to the earth as precipitation, completing what is called the *hydrologic cycle* (Figure 14.1).

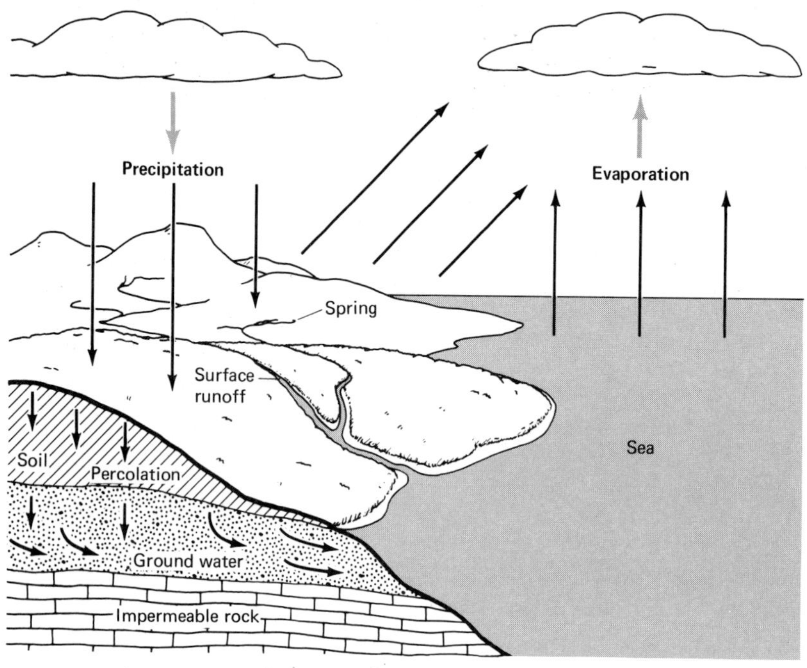

FIGURE 14.1
The hydrologic cycle.

Not all water is fit for human consumption. Drinking water must not be too salty, and it must be free of contaminants harmful to the human body. The average individual drinks between 1 and 2 liters of water a day, depending on the climate and season. More water is drunk in summer than in winter. Not all of this drinking water comes out of the tap in the home. The water in beverages comes from the bottling or canning plant, water in milk comes from the dairy farm, and so forth. Thus, even if one is certain of the purity of one's own drinking-water supply, there is no guarantee that all of the water consumed is free of contaminants.

DRINKING WATER

The water system of a city is the essential ingredient that makes the city possible. It is virtually impossible to have a high-density population without some sort of central water-supply facility. Sources of water vary widely (Table 14.1). The most widely used sources are surface waters, such as lakes and rivers. If the river is not very large, it may be necessary to make an impoundment with a dam, creating a reservoir within which water can be stored. Ground waters provide excellent water sources, but are less widely available than surface waters. Surface waters are often subject to pollution so that they must be treated before use. After treatment, the water may be pumped to a reservoir for storage and then distributed to individual homes and businesses via an extensive network of underground

URBAN WATER SUPPLY

TABLE 14.1 Sources of water for domestic uses	Source	Characteristics
	Ground water	Reliable, constant temperature, low pollution, limited supply, hardness often high
	Dug well	Shallow, undependable, often contaminated, inexpensive
	Drilled well	Deeper, more dependable, less contaminated, moderately expensive
	Spring	Inexpensive but rarely available, often hard water
	Surface water	Often unreliable, varies in temperature, often polluted
	Rivers	Often turbid, often polluted
	Lakes	Lower turbidity than rivers, not widely available
	Reservoirs	Costly, variable water quality, algal growth may be a problem
	Rainwater	Soft water, uncontaminated, unreliable, difficult to store
	Cistern	Stores rainwater, may become polluted from surface runoff

pipes (Figure 14.2). Great care must be taken to ensure that water pipelines remain intact, since sewer pipes are usually also present underground, and a real possibility exists of leakage of sewage into the water supply (called a *cross-connection*), resulting in a serious health hazard.

KINDS OF CONTAMINANTS IN WATER AND THEIR SOURCES

A vast array of potentially harmful materials may be transported by water. Even if the source water is free of harmful materials, many possibilities exist for subsequent contamination of the water by the vast array of harmful materials present in the urban environment. Contaminants can be divided roughly into three categories: physical, chemical, and biological. The principal *physical contaminant* is turbidity, caused by the presence in the water of particulate materials that scatter light. These particulates may be harmless materials, such as clays derived from soil, but they render the water unsightly. Also, particulate materials may interfere with the disinfection process used to render drinking water free of pathogens (see Section 14.4). A particulate of recent concern is asbestos, which is present in some supplies as a result of either natural causes or pollution; asbestos is a carcinogen when breathed and conceivably can cause cancer if present in drinking water. Research is being undertaken to assess the significance of asbestos in drinking water. If particulate materials are present in drinking water, they can be removed by filtration (discussed in Section 14.4).

Chemical contaminants in drinking water can be divided roughly into two major types: organic and inorganic. Inorganic contaminants include *metals*, such as lead, copper, zinc, iron, manganese. Iron and manganese are often present in undesirable amounts in the source

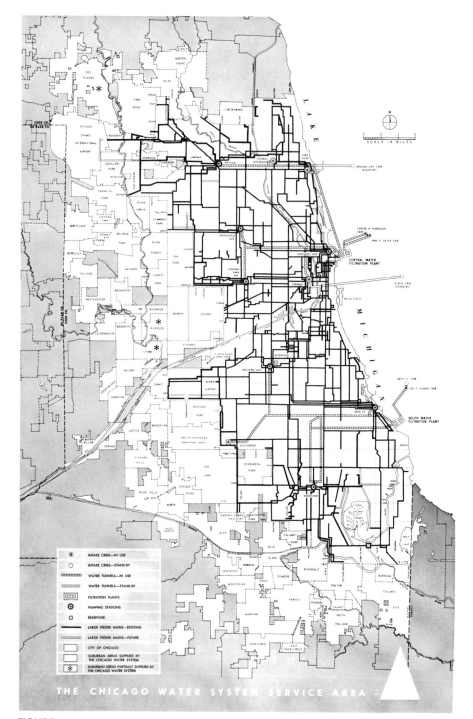

FIGURE 14.2

The overall water-distribution system for the city of Chicago, typical of the water-supply systems of large urban areas. The Chicago system services not only the city proper but many surburban areas as well.

water, especially if ground water is used, and may have to be removed by chemical and physical treatments. Lead, copper, and zinc reach the water most commonly through the pipes themselves. Lead pipes are no longer used, although they were extensively used in Roman times; indeed, lead contamination of the Roman water supply may have contributed to the downfall of the Roman Empire. Copper and zinc are used today in water pipes, and significant amounts may leach out of the pipes and enter the water. The problem is most severe with soft waters, since the metals are more soluble in soft than in hard water.

Sodium may be present in undesirably large amounts in some water supplies or in softened waters, making the water unsuitable for people with certain kinds of heart condition.

Fluoride is an interesting element, since it can be a harmful contaminant of drinking water, yet it is often intentionally added to drinking water in small amounts. In certain parts of the world, the water supplies are naturally high in fluoride. Since fluoride can become incorporated into the chemical structure of bone and teeth, it can modify the structure of these materials. People who drink water that is unusually high in fluoride may develop brown mottling on their teeth, but they are also completely free of tooth decay (dental caries, see Section 10.1). It has been found that if the fluoride concentration of drinking water is around 1 part per million (ppm), harmful mottling does not occur, but resistance to tooth decay still develops. For this reason, many water-supply agencies intentionally add fluoride to achieve this concentration.

Sulfate is a natural constituent of water, but the concentration varies widely depending on the source of the water. Sulfate can also enter water in large amounts through pollution resulting from coal, gypsum, and mineral mining operations. When concentrations of sulfate greater than 250 ppm are present, the drinking water may cause diarrhea in some people.

Nitrate is a natural constituent of water but may be a contaminant resulting from use of nitrogen fertilizers in agricultural practice. Nitrate in excessive amounts can be harmful, especially to infants, since it is converted in the intestinal tract into *nitrite*, which can enter the blood, bind to hemoglobin, and cause a condition known as *methemoglobinemia*. Recently, the maximum permissible level of nitrate in drinking water has been lowered from 45 ppm to 10 ppm, reflecting recent knowledge and concern about its potential harmful effects.

A wide variety of *organic* chemicals can be found in drinking water in trace amounts. Some of these chemicals are natural constituents, but most of them are derived from pollution sources. There has been concern for some time about the presence of pesticides in drinking water, and maximum permissible levels for certain common pesticides have been set. There has been recent concern about the presence of chloroform (CHCl₃) and related compounds (CCl_4, $CHCl_2Br$, etc.) in drinking water, since chloroform has been shown to be carcinogenic when fed in large amounts to animals. It has been established that chloroform is actually formed in water during disinfection processes using chlorination (see Section 14.4), and it is natural to wonder whether the very small amounts of chloroform present in drinking water might be capable of causing some cancers in humans. However, analysis of the data suggest that, even in cities that have fairly high chloroform levels in their finished drinking water (up to 300 parts per billion have been found), it is unlikely that sufficient chloroform is ingested from this source to cause cancer.

Biological contaminants of most concern are the pathogenic or potentially pathogenic bacteria, viruses, and protozoa of intestinal origin that may cause human disease. We discuss these in some detail in Section 14.2. In addition to these biological agents, algae may also be present in drinking waters, especially if the waters are derived from lakes or reservoirs. These algae can be responsible, at least in part, for undesirable odors and flavors in waters. We discuss the problem of algal growth in lakes and reservoirs in Section 14.15.

Organisms pathogenic to humans that are transmitted by water include bacteria, viruses, and protozoa (Table 14.2). Organisms transmitted by water usually grow in the intestinal tract and leave the body in the feces. Fecal pollution of water supplies may then occur, and if the water is not properly treated, the pathogens enter the new host when the water is consumed. Because water is consumed in large quantities, it may be infectious even if it contains only a small number of pathogenic organisms. The pathogens lodge in the intestine, grow, and cause infection and disease.

14.2 Pathogenic organisms transmitted by water

Probably the most important pathogenic bacteria transmitted by the water route are *Salmonella typhi*, the organism causing typhoid fever, and *Vibrio cholerae*, the organism causing cholera. Although the causal agent of typhoid fever may also be transmitted by contaminated food and by direct contact from infected people, the most common and serious means of transmission is the water route. Typhoid fever has been virtually eliminated in many parts of the world, primarily as a result of the development of effective water treatment methods (see Section 1.4). However, typhoid fever does still occur occasionally, usually in the summer months, when swimmers are active in polluted water supplies. A breakdown in water-purification methods, contamination of water during floods, or cross-contamination of water pipes from leaking sewer lines occa-

BACTERIA

TABLE 14.2
Water-borne disease outbreaks in the United States, 1946–1974

Disease[a]	Causal agent	Outbreaks	Cases
Bacterial			
Typhoid fever	*Salmonella typhi*	58	836
Shigellosis	*Shigella* species	48	10,813
Salmonellosis	*Salmonella paratyphi*, etc.	16	16,801
Gastroenteritis	*Escherichia coli* (pathogenic strains)	4	188
Leptospirosis	*Leptospira* species	1	9
Tularemia	*Francisella tularensis*	2	6
Viral			
Infectious hepatitis	Hepatitis A virus	66	2,184
Poliomyelitis	Polio virus	1	16
Protozoal			
Dysentery	*Entamoeba histolytica*	5	75
Giardiasis	*Giardia lamblia*	21	5,464
Unknown etiology			
Gastroenteritis		237	55,626

[a] One water-borne disease still important in some parts of the world but not in the United States is cholera, caused by the bacterium *Vibrio cholerae*.

Data from G. F. Craun and L. J. McCabe, "Review of the causes of waterborne-disease outbreaks," *J. Am. Water Works Association*, **65**, 74–84, 1973, and Center for Disease Control, *Foodborne and waterborne outbreaks*, 1976.

sionally results in epidemics of typhoid fever. Water that has been contaminated can be rendered safe for drinking by boiling for 5 to 10 minutes or by adding chlorine (see Section 14.4). Usually, contaminated water is safe to use for laundry or other domestic cleaning purposes, provided care is taken that it is not swallowed.

The causal agent of cholera is transmitted only by the water route. At one time cholera was common in Europe and North America, but the disease has virtually been eliminated from these areas by effective water purification. The disease is still common in Asia, however, and travelers from the West are advised to be vaccinated for cholera. Both *V. cholerae* and *S. typhi* are eliminated from sewage during sewage treatment and hence do not enter water courses receiving treated sewage effluent. More frequent than typhoid, but less serious a disease, is salmonellosis caused by species of *Salmonella* other than *S. typhi*. As shown in Table 14.2, the largest number of cases of water-borne bacterial disease in the United States during the last several decades have been due to salmonellosis.

Bacteria are effectively eliminated from water during the water-purification process (see Section 14.4), so that they should never be present in properly treated drinking water. Most outbreaks of water-borne disease in the United States are caused by breakdowns in the treatment system or by post-contamination in the pipelines. The latter problem can be controlled by maintaining a concentration of free chlorine in the pipelines.

It may seem surprising that drinking water can be a vehicle for bacterial disease agents even when the water looks clear and uncontaminated. However, the number of bacteria necessary to start an infection is not especially large. Feeding trials using human volunteers have shown that with some bacteria only a few cells are necessary to cause infection. Since people drink fairly large amounts of water, between 1 and 2 liters a day, even if bacterial numbers are low in the drinking water, sufficient bacteria may be received to initiate an infection.

Viruses transmitted by the water route include polio virus and other viruses of the enterovirus group, as well as the virus causing infectious hepatitis. Polio virus has several modes of transmittal, and transmission by water may be of serious concern in some areas. Before introduction of the polio vaccines, it was commonly encountered during the summer months in polluted swimming areas.

> VIRUSES

Infectious hepatitis (Section 12.13) is caused by a virus (hepatitis A virus) that resembles the enteroviruses, but because it has not been successfully cultured its characteristics are poorly known. Hepatitis is the most serious water-borne viral disease agent at present (see Table 14.2); however, it is also transmitted in foods, and most of the infectious heptatitis cases probably arise by food-borne rather than water-borne means.

It should be recalled that viruses are not cells and lack cellular structure. Because of this, they are more stable in the environment and are not so easily killed as cells. However, both polio and infectious hepatitis virus are eliminated from water by purification, and the maintenance of 0.6 ppm free chlorine in a water supply will generally ensure its safety.

A pathogenic protozoan transmitted via the water route is *Entamoeba histolytica*, the causal agent of amoebic dysentery (see Section 11.2). This amoeba lives in the intestine and forms resistant structures called *cysts*, which are excreted with the feces and are able to survive in the environment for long periods of time. If they contaminate the water, it can then serve as a source of infection. Amoebic dysentery is most common in those parts of the world where sanitation standards are low. Water containing cysts can be made safe by boiling, filtration, or chlorination. Usually, chlorination or rapid sand filtration is not 100 percent effective, and the recommended procedure is to filter the water through diatomite (diatomaceous earth) or use slow sand filtration, followed in both cases by chlorination.

> PROTOZOA

Another protozoal agent that has been recently implicated in a number of outbreaks of water-borne disease is *Giardia lamblia*, the causal agent of giardiasis. *G. lamblia* does not cause as severe a disease as *E. histolytica*, the symptoms of giardiasis being diarrhea, flatulence, and discomfort. There has been a noticeable increase in incidence of giardiasis in the United States over the past 5 years. The cysts of *G. lamblia* are resistant to conventional chlorination treatment and can remain viable in the water supply so that filtration is necessary to ensure their removal.

14.3 Ensuring the safety of drinking water

Even water that looks clear and pure may be sufficiently contaminated with pathogenic microorganisms to be a health hazard. Some means are necessary to ensure that drinking water is safe. One of the main tasks in water microbiology is the development of laboratory methods that can be used to detect the microbiological contaminants which might be present in drinking water. It usually is not practical to examine drinking water directly for the various pathogenic organisms that might be present. As discussed in Section 14.2, a wide variety of organisms may be present, including bacteria, viruses, and protozoa. To check each drinking-water supply for each of these agents would be a difficult and time-consuming job. In practice, *indicator organisms* are used instead. These are organisms associated with the intestinal tract, whose presence in water indicates that the water has received contamination of an intestinal origin. The most widely used indicator is the coliform group of organisms.

COLIFORMS AS INDICATORS

The coliform group is defined in water bacteriology as all the aerobic and facultatively anaerobic, Gram-negative, nonspore-forming, rod-shaped bacteria that ferment lactose with gas formation within 48 hours at 35°C. This is an operational rather than a taxonomic definition and includes a variety of organisms, mostly of intestinal origin. In practice, the coliform organisms are almost always members of the enteric bacteria (see Section 10.8). The coliform group includes the organism *Escherichia coli*, a common intestinal organism, plus the organism *Klebsiella pneumoniae*, a less common intestinal organism. The definition also currently includes organisms of the species *Enterobacter aerogenes*.

The coliform group of organisms are suitable as indicators because they are common inhabitants of the intestinal tract, both of humans and warm-blooded animals, and are generally present in the intestinal tract in large numbers. When excreted into the water environment, the coliform organisms eventually die, but they do not die at a faster rate than the pathogenic bacteria *Salmonella* and *Shigella*, and both the coliforms and the pathogens behave similarly during water-purification processes. Thus, it is likely that if coliforms are found in drinking water, the water has received fecal contamination and may be unsafe. There are very few organisms in nature meeting the definition of the coliform group that are not associated with the intestinal tract. (One exception is an organism called *Aeromonas*, which is common in some source waters and may appear in a coliform analysis. *Aeromonas* can be distinguished from the intestinal coliforms by means of a special diagnostic test, the oxidase test.) It should be emphasized that the coliform group includes organisms derived not only from humans but from other warm-blooded animals as well. Since many of the pathogens (e.g., *Salmonella, Leptospira*) found in warm-blooded animals also infect humans, an indicator of both human and animal pollution is desirable.

All sampling and testing should be done using standard methods, following carefully controlled procedures. Such procedures are described in considerable detail in a publication of the American Public Health Association, Washington, D.C., entitled *Standard Methods for the Examination of Water and Wastewater*. New editions of this book are printed every few years; the latest is the 14th edition, published in 1976. By use of these methods, analysis of water in each water-supply area can be standardized so that comparisons from one district to another may be meaningfully made. Additionally, the Drinking Water Standards established by the U.S. Environmental Protection Agency specify that the methods prescribed by *Standard Methods* must be used.

Samples of water are taken directly from the source under study using sterile glass bottles with tightly fitting stoppers (Figure 14.3). When sampling from a tap, it is important to allow the water to flow for about 10 minutes before sampling so that microbe-laden deposits are washed out of the pipe. The hands or faucet must not touch the lip or the stopper of the bottle. When sampling from a lake or stream, the bottle is placed below the surface and moved forward while being filled so that water coming in contact with the hands does not enter the bottle. If the water is chlorinated, the sample bottle should contain sodium thiosulfate to dechlorinate the water, thus eliminating this growth-inhibiting substance and making it possible for any live organisms held in check by the chlorine to grow. The sample should be returned to the laboratory as soon as possible for processing, preferably within 6 to 12 hours, since there are changes in microbial number during storage. If the sample must be stored for any length of time before assay, it should be kept cool. No sample should be stored longer than 30 hours before analysis. Assays performed on samples shipped through the mail should be accepted cautiously. It should also be remembered that the water supply can change after sampling, so that results of an analysis apply only to the time of sampling. If any unusual results are obtained, a

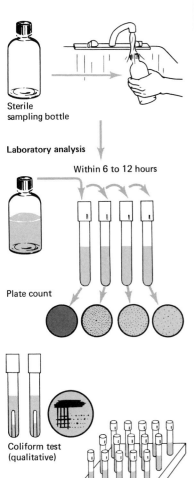

Taking sample

Sterile
sampling bottle

Laboratory analysis

Within 6 to 12 hours

Plate count

Coliform test
(qualitative)

Coliform MPN test
(quantitative)

FIGURE 14.3
Procedure for sampling and
bacteriological analysis of
drinking water.

fresh sample should be obtained and processed. The most satisfactory procedure is to carry out sampling and assay on a routine prescribed basis. Depending on the local situation, samples and assays may be performed daily, weekly, or monthly. It is important that samples be taken throughout the distribution system. The Drinking Water Regulations (see Section 14.5) prescribe precisely the sampling frequency for cities of various sizes.

PROCEDURES FOR THE STANDARD COLIFORM TEST FOR DRINKING WATER   Two types of procedures are used for the coliform test, the most-probable-number (MPN) procedure, and the membrane filter (MF) procedure. The MPN procedure employs liquid culture medium in test tubes, the samples of drinking water being added to the tubes of medium. In the MF procedure, the sample of drinking water is passed through a sterile membrane filter that removes the bacteria; the filter is placed on a culture medium for incubation.

There are three stages in the MPN procedure: presumptive, confirmed, and completed (Figure 14.4). In the *presumptive* test, a sample of water and dilutions from it are inoculated into tubes of a lactose broth medium, and the tubes are incubated for 24 to 48 hours. If gas is produced, the test is considered a positive presumptive test, and the further stages are followed. For the *confirmed* test, samples are streaked from the positive tubes at the highest dilutions onto plates containing a special indicator agar, usually eosine–methylene blue (EMB)–lactose agar. Because coliforms produce acid from lactose, they form colonies of very dark color and with a metallic sheen, caused by accumulation under acid conditions of the eosine and methylene blue dyes in the colonies. The presence of such colonies is considered a positive confirmed test. For the *completed* test, typical colonies are picked and inoculated into lactose broth. If gas is produced upon incubation, further tests are run to ensure that the organism so isolated is a typical Gram-negative nonsporulating rod and has certain biochemical properties characteristic of coliforms. If these observations are satisfactory, it is considered a positive completed test.

In well-regulated water systems, the presumptive coliform test is always negative, so the confirmed and completed tests do not have to be run. The determination of coliform numbers in the MPN procedure follows the principles enumerated in Section 4.5. For drinking water, five tubes are inoculated, each with 10 ml of water. The number of positive tubes is counted, and from a statistical table (Table 14.3), the MPN is determined. Confirmed and completed tests are then performed on positive tubes. For grossly polluted waters, all tubes may be positive if 10-ml samples are used. To obtain a count on such water, tubes should also be set up using 1 ml and 0.1 ml samples. The MPN is then determined using the inoculum size in which there are some negative and some positive tubes.

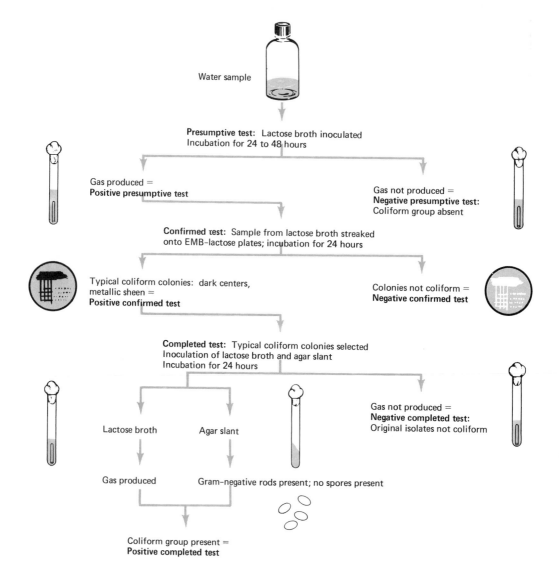

Water sample

Presumptive test: Lactose broth inoculated
Incubation for 24 to 48 hours

Gas produced =
Positive presumptive test

Gas not produced =
Negative presumptive test:
Coliform group absent

Confirmed test: Sample from lactose broth streaked
onto EMB-lactose plates; incubation for 24 hours

Typical coliform colonies: dark centers,
metallic sheen =
Positive confirmed test

Colonies not coliform =
Negative confirmed test

Completed test: Typical coliform colonies selected
Inoculation of lactose broth and agar slant
Incubation for 24 hours

Gas not produced =
Negative completed test:
Original isolates not coliform

Lactose broth   Agar slant

Gas produced   Gram–negative rods present; no spores present

Coliform group present =
Positive completed test

The general procedure of membrane filter methods is discussed in Section 4.2. For drinking water, at least 100 ml of water should be filtered, although in clean water systems, even larger volumes could be filtered. After filtration of a known volume of water, the filter is placed on one of several special culture media which are selective and which will indicate the presence of coliforms. The coliform colonies (Figure 14.5) are counted, and from this value the number of coliforms in the original water sample can be determined. The membrane filter method permits the determination of coliform numbers in 1 day instead of the 3 to 4 days otherwise required, since presumptive, confirmed, and completed tests are combined into one. For some applications, this newer membrane

FIGURE 14.4
Standard method for the analysis of drinking water for coliforms by the most-probable-number (MPN) procedure. For each sample, 5 tubes of lactose broth are inoculated, 10 ml/tube (50 ml total sample). In a well-regulated water system, all tubes should be negative for coliforms. Positive tubes should be confirmed, as indicated. For coliform standards for drinking water, see Section 14.5.

TABLE 14.3
MPN of organisms when various combinations of positive and negative results are obtained using five 10-ml portions of drinking water

Number of positive reactions out of 5 tubes of 10 ml each	Index MPN per 100 ml	95% confidence limits[a]	
		Lower limit	Upper limit
0	Less than 2.2	0	6.0
1	2.2	0.1	12.6
2	5.1	0.5	19.2
3	9.2	1.6	29.4
4	16.0	3.3	52.9
5	Greater than 16.0	8.0	Infinite

[a] The 95% confidence limits express the accuracy of the determination; i.e., 95% of the time, the actual value will be within the lower and upper limits given. The MPN index is the average value.
From *Standard methods for the examination of water and wastewater*, 14th ed., American Public Health Association, Washington, D.C., 1976.

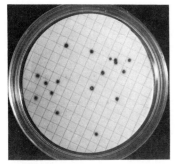

FIGURE 14.5
Coliform colonies growing on a membrane filter. Coliforms are indicated by the characteristic color of the colonies, which develop as a result of lactose fermentation, and by the change in color of the indicator dye present in the medium.

filter technique is replacing the older MPN method. However, it does not work well on water samples with large amounts of suspended matter, silt, algae, and bacteria, since these materials interfere with both the filtration and the development of colonies.

STANDARD PLATE COUNT The standard plate count involves preparing pour plates in which 1-ml samples of water or a dilution of the water are mixed with melted agar in a petri dish. After solidifying, the plates are incubated at 35°C for 24 hours, and the number of colonies is counted. The number of bacteria per milliliter of water is reported. Values can range from less than one per milliliter to many thousands, depending on the water supply. It should be emphasized that the values obtained refer only to the bacteria able to grow under the conditions of temperature and medium used; if other conditions are used, higher or lower numbers may be obtained, and different kinds of bacteria may be detected. The value of a single assay is, therefore, not very great, and the main use of total counts is in periodic sampling programs to detect unusual changes which might occur, thereby indicating a change in the quality of the water supply.

The standard plate count is a useful adjunct in any quality-control program, as noncoliform organisms in high numbers may cause significant health effects. In addition, high numbers of non-coliform organisms in a water system may interfere with the coliform determination, leading to false negatives. Most frequently, standard plate counts are high in drinking waters obtained from wells in which disinfection is not practiced. Such waters may be perfectly safe, but it is important to be aware of the high bacterial count.

Standard plate counts are also useful in evaluating the efficiency of water-treatment processes. By performing standard plate counts on raw water and on water at various stages through the purification system, the efficiency of the system in reducing bacterial contamination can be readily measured. For such purposes, it may be desirable to use an incubation temperature of 20°C rather than 35°C, since the lower temperature permits growth of more of the normal bacteria present in source waters, permitting a more accurate measure of the efficacy of the treatment process.

OTHER INDICATOR BACTERIA

A variety of other bacteria have been used as indicators of water quality, but none has been as generally useful as the coliform group. *Fecal streptococci*, sometimes called the *enterococci*, include a group of streptococci commonly found in the intestinal flora of man and animals. They can be counted by either MPN or MF techniques using media containing sodium azide, which inhibits virtually all bacteria other than the enterococci. Fecal streptococci are more useful in studies of pollution of recreational waters than for drinking-water evaluation. Other organisms found in the intestinal tract that have been studied as possible indicators include *Clostridium perfringens*, *Bifidobacterium*, *Staphylococcus*, and the yeast *Candida albicans*. The latter two organisms are not common in the normal flora but are associated with certain types of gastrointestinal disease. Although methods are available for quantifying all these organisms, none of them is likely to supplant the coliform group as the best indicator of drinking-water quality, although they may find some uses in pollution work related to sewage.

A subgroup of the coliform bacteria called the *fecal coliform group* is a useful indicator of sewage pollution but does not find direct use in drinking-water analysis. The fecal coliform group is discussed in Section 14.8.

It is a rare instance when available water is of such clarity and purity that no treatment is necessary before use. Water treatment is carried out both to make the water safe microbiologically and to improve its utility for domestic and industrial purposes. Treatments are performed to remove pathogenic and potentially pathogenic microorganisms and also to decrease turbidity, eliminate taste and odor, reduce or eliminate nuisance chemicals such as iron or manganese, and soften the water to make it more useful for the laundry.

14.4
Water purification

DOMESTIC WATER PURIFICATION

The kind of treatment that water is given before use depends on the quality of water supply. Rarely are large cities fortunate enough to have clear, cool, uncontaminated water for their water supplies; usually, it must be treated (Figure 14.6).

Water is first pumped to *sedimentation basins*, where sand, gravel, and other large particles settle out. A sedimentation basin should be used only if the water supply is highly turbid, since it has the disadvantages that algal growth may occur in the basin, adding odors and flavors, and that pollution of the water by surface runoff may occur. Bacteria may grow in the bottom mud and add further problems.

Most water supplies are subjected to *coagulation*. Chemicals containing aluminum and iron are added, which under proper control of pH form a flocculent, insoluble precipitate that traps organisms, adsorbs organic matter and sediment, and carries them out of the water. After the chemicals are added in a mixing basin,

FIGURE 14.6
Aerial view of water-treatment plant of Louisville, Kentucky. The arrows indicate the direction of flow of water through the plant. *Courtesy of Billy Davis and the Courier-Journal and Louisville Times.*

the water containing the coagulated material is transferred to a settling basin where it remains for about 6 hours, during which time the coagulum separates out. About 80 percent of the turbid material, color, and bacteria are removed by this treatment.

After coagulation, the clarified water is usually *filtered* to remove the remaining suspended particles and microbes. Filters can be of the slow sand, rapid sand, or pressure type. *Slow sand filters* are suitable for small installations such as resorts or rural places. The water is simply allowed to pass through a layer of sand 2 to 4 feet deep. Eventually, the top of the sand filter will become clogged, and the top layer must be removed and replaced with fresh sand. *Rapid sand filters* are used in large installations. The rate of water flow is kept high by maintaining a controlled height of water over the filter. When the filter becomes clogged, it is clarified by backwashing, which involves pumping water up through the filter from the bottom. *Pressure filtration* is similar to rapid sand filtration except that the water is pumped through under pressure. Pressure filters are used mainly in small installations such as swimming pools and industrial plants. From 98 to 99.5 percent of the total bacteria in raw water can be removed by proper settling and filtration.

Chlorination is the most common method of ensuring microbiological safety in a water supply. Chlorine is an oxidizing agent that reacts with any organic matter present in the water. In sufficient doses, it causes the death of most microorganisms within 30 minutes. In addition, since most taste- and odor-producing compounds are organic, chlorine treatment reduces or eliminates them. It also oxidizes soluble iron and manganese compounds, forming precipitates which can be removed. Chlorine can be added to water either from a concentrated solution of sodium or calcium hypochlorite or as a gas from pressure tanks. The latter method is used most commonly in large water-treatment plants, as it is most amenable to automatic control.

When chlorine reacts with organic materials, it is used up. Therefore, if a water supply is high in organic materials, sufficient chlorine must be added so that there is a residual amount left to react with the microorganisms after all reactions with organic materials have occurred. The water-plant operator must perform chlorine analyses on the treated water to determine the residual level of chlorine. A chlorine residual of about 0.2 to 0.6 ppm is an average level suitable for most water supplies. In some water systems, a combined treatment with chlorine and ammonia is used. Ammonia and chlorine react in the water to produce chloramines, which are less effective agents but do not impart the chlorine taste to the water that occurs when chlorine is used alone.

Water-softening treatment is often used with hard waters to make them more suitable for laundry and other cleaning purposes.

Softening treatment removes calcium and magnesium by precipitation or by ion-exchange processes. There is no microbiological benefit from water softening. Another treatment given to many waters is *fluoridation*, in which about 1 ppm of fluoride is added to the final water. Individuals who drink fluoridated water are less subject to tooth decay than are those drinking unfluoridated water.

After final treatment, the water is usually pumped to *storage tanks*, from which it flows by gravity to the consumer. Covered storage tanks are essential to ensure that algal growth does not take place (no light is available) and that dirt, insects, birds, and other objects do not enter. The final product is available in most parts of the world at amazingly inexpensive rates. Although most users take their water supply for granted, it should be clear from the foregoing discussion that provision of a safe, reliable water supply is a complicated matter, requiring much technical knowledge and skill.

EMERGENCY WATER-SUPPLY TREATMENT

We are rarely called upon to treat our own water, but in emergencies and in remote parts of the world, such treatment may be required. The most reliable and simplest method of purifying small quantities of water is *boiling*, a 5-minute treatment usually being sufficient. Water from a hot water tank, which has usually been kept at a temperature of 55 to 60°C (about 130 to 140°F) for many hours, is also microbiologically safe, since virtually all pathogens are killed at such temperatures. *Filtration* can be accomplished on a provisional basis by passing the water through a barrel or drum filled with sand. The rate of filtration should not exceed 50 gallons per day per square foot of filter surface area. The filtered water should then be chlorinated.

Chlorination alone is satisfactory for water that is not grossly polluted. Suitable preparations, both powders and liquids, can be purchased at grocery and drug stores. The active ingredient in powders is calcium hypochlorite and in liquids, sodium hypochlorite. The strength is given on the label as percent of available chlorine. Since these materials deteriorate with age, fresh supplies should be obtained from time to time. Chlorination tablets suitable for use on hiking and camping trips can also be obtained. The actual dose needed will depend on the available chlorine provided by the preparation, and this should be indicated on the label. As an example, common household bleach provides 5.25 percent available chlorine; used at full strength, six drops should be used to treat 1 gallon of water. It is important to allow the water to stand for 30 minutes after the chlorine is added before it is used. When large volumes of water are treated, continuous chlorination treatment may be arranged by allowing for a slow drip of chlorine solution into the water as it is used. Commercial chlorination units suitable for small installations can be purchased.

Iodine may also be used to disinfect water. A dosage of 5 to 10 ppm free iodine is effective within 20 minutes against most pathogens. Using a 2-percent tincture of iodine solution, eight drops will disinfect 1 quart of water, if the water is allowed to stand at least 20 minutes before use. Iodine tablets can be purchased at drug stores and outdoor equipment stores.

In some parts of the world, *bottled water* may be purchased. This is usually used because of bad odors or flavors of tap water, but it may also serve as a source of microbiologically pure water. The water may come either from a pure, free-flowing spring or from a suitable domestic water supply. In areas where water-supply lines have not been constructed, water from tank trucks may also be purchased. It is essential that the tanks be thoroughly cleaned and disinfected, and each tank of water should be dosed with chlorine at a rate of 1 to 2 ppm.

14.5 Drinking water standards

Drinking water standards in the United States go back to 1914, when standards were prescribed by the federal government for the drinking water provided by common carriers (railroads, etc.). Revised standards for common carriers were prescribed in 1925, 1943, and 1962. Some states established drinking water standards that were similar to the federal standard for common carriers, but until 1975 there were no requirements that these standards apply to drinking waters provided in cities and towns throughout the land. In 1974 the U.S. Congress passed the Safe Drinking Water Act, which provided a framework for the development by the Environmental Protection Agency of drinking water standards for the whole country. Although these standards are still under development, there seems to be general agreement that the measurement of coliforms provides the greatest assurance of the microbiological safety of drinking water. It should be noted that the drinking water standards are a legal rather than a scientific document and are established in such a way that they can be enforced. Enforcement of the standards is the responsibility of the states, although if a state refuses to accept primacy in this matter, the U.S. Environmental Protection Agency must step in and enforce the standards in that state.

As far as microbiology is concerned, several aspects of the standards are of special interest. These include the kinds of water systems to which the standards are applicable, the frequency of sampling for water systems of various sizes, the maximum allowable coliform number, and the procedures to be followed if a water supply does not meet the standard. The standards apply to both community and noncommunity water systems. A *community water system* is defined as one that supplies at least 15 service connections used by year-round residents or regularly supplies at least 25 year-round residents. *Noncommunity systems* serve primarily transient-population areas, such as camp grounds, gasoline service stations, restaurants, and so forth, which may have their own water systems.

SAMPLING FREQUENCY

The sampling frequency in the drinking water regulations is related to the size of the population served by the water system. Noncommunity water systems are required to sample only once each calendar quarter during the period when the system is in operation. Community water systems are required to sample for coliforms at regular time intervals, the frequency of sampling being related to the size of the population being served (Table 14.4). Small communities are required to sample less frequently than large communities. For the smallest community water systems, serving populations of 25 to 1,000

	Population served[a]	Minimum number of samples per month
TABLE 14.4 Required sampling frequencies for coliform determinations in community water systems of various sizes, as specified by the U.S. Drinking Water Regulations	25 to 1,000	1
	1,001 to 2,500	2
	2,501 to 3,300	3
	3,301 to 4,100	4
	4,101 to 4,900	5
	4,901 to 5,800	6
	. . . .	
	10,301 to 11,100	12
	. . . .	
	25,001 to 28,000	30
	. . . .	
	50,001 to 54,000	60
	. . . .	
	111,001 to 130,000	110
	. . . .	
	410,001 to 450,000	200
	. . . .	
	970,001 to 1,050,000	300
	. . . .	
	1,970,001 to 2,060,000	400
	. . . .	
	4,690,001 or more	500

[a] Only selected values are given. The complete table and further details can be found on page 59,571 of the *Federal register*, Vol. 40, no. 248, December 24, 1975.

people, only one sample per month is required. At the other extreme, large cities are required to sample several hundred times a month. Based on a history of no coliform bacterial contamination and on a sanitary survey of the system to show that the water is supplied solely from a protected ground-water source free of sanitary defects, a community water system serving fewer than 1,000 persons may reduce sampling frequency to only one sample per quarter.

Samples are taken at points throughout the distribution system that are representative of the conditions within the system. This is an extremely important point, since water that leaves the water-purification plant free of contamination may become contaminated within the distribution system. Pipes may develop leaks and become contaminated from outside; of most concern is the problem of cross-connections, which may result if there are breaks or defects in both the water pipes and the sewer pipes serving an area. Under these conditions, there is a real possibility of movement of intestinal pathogens from the sewer system to the water system, leading to significant contamination. By requiring that samples be taken throughout the distribution system, it is hoped that health hazards resulting from cross-connections will be avoided.

COLIFORM STANDARDS

The Drinking Water Regulations specify that laboratory procedures for coliform counts be performed in accordance with the recommendations outlined in the book, *Standard Methods for the Examination of Water and Wastewater*. The maximum coliform levels permitted by the regulations are given in reference to the type of method used, membrane filter (MF) or the most probable number (MPN).

When the MF technique is used, 100-ml samples must be filtered, and the number of coliform bacteria shall not exceed any of the following:

1 One per 100 ml as the arithmetic mean of all samples examined per month.
2 Four per 100 ml in more than one sample when less than 20 are examined per month.
3 Four per 100 ml in more than 5 percent of the samples when 20 or more are examined per month.

When the MPN technique is used, the sample size used can vary, but the intent of the regulations is that the coliform density should not exceed that specified by the MF procedure. In most water-supply practice, the MPN procedure is carried out using five tubes, each of which is inoculated with a 10-ml sample. After incubation, the number of tubes giving positive reactions is noted, and from Table 14.3, the coliform density can be calculated. In good water practice, all of the tubes inoculated should be negative, indicating a coliform density less than 2.2 per 100 ml.[1]

In the MPN procedure, it is of course necessary to confirm presumptive positives, using the procedures outlined in Section 14.3. With the MF, confirmation is not necessary unless there is some reason to believe that positive colonies may not be coliforms.

When the coliform bacteria in a single sample are greater in number than the regulations permit, the supplier must obtain at least two additional check samples collected at the same point. If these check samples are negative, no further action is required. If these check samples are positive, additional check samples must be analysed until at least two consecutive check samples show no positive tubes.

REPORTING AND PUBLIC NOTIFICATION

The water supplier is required to report to the state within 48 hours any failure to comply with the regulations, including failure to comply with the monitoring requirements. If a community water system fails to comply with the coliform standards, water users must be notified in writing in their next water bill and by announcement in communications media, including publication within 14 days in a newspaper in general circulation in the area served by the system, and by furnishing within 7 days a copy of the notice to radio and television stations serving the area. If no communications media are available in the area served, notices must be posted in post offices within the area served by the system. The notices must be written in nontechnical language, must not be printed in unduly small type, and must not be communicated by other means that would frustrate the purpose of the notice. The notice must disclose all facts regarding the problem and must state clearly that the drinking water regulations have been violated. The notices must also indicate any preventive measures that should be taken by the public, such as boiling water before drinking or complete avoidance of the water.

The intent of these requirements is to ensure that the public is totally aware of any problems that the water system might have and the measures that are being taken to rectify them. It is implicit in the regulations that if the public is informed, then not only can it act to

---

[1] However, the regulations do not refer directly to this calculated number but to the actual results of the MPN tabulation of the positive and negative tubes in the MPN procedure: (1) When 10-ml standard portions are used, the coliform bacteria shall not be present in more than 10 percent of the portions planted in any month. (2) The coliform bacteria shall not be present in three or more portions (that is, the 10-ml portions in individual tubes) in more than 1 sample when less than 20 samples are examined per month. (3) The coliform bacteria shall not be present in three or more portions in more than 5 percent of the samples when 20 or more samples are examined per month.

protect itself against hazardous water, but pressure will be brought to bear on the water supplier to rectify problems in the system.

Many water systems use chlorination as a means of disinfection. It is highly desirable to maintain a chlorine concentration (called a *chlorine residual*) in the water-distribution system, to help protect against cross-connections, breaks, and other events that might cause sewage pollution of drinking water. Recognizing the value of chlorination, and especially of maintaining a chlorine residual in the system, the regulations permit the substitution of chlorine-residual measurements on not more than 75 percent of the samples required for coliform determinations, with the additional requirement that there shall be at least daily determinations of chlorine residual.

CHLORINE SUBSTITUTION FOR COLIFORM DETERMINATIONS

A chlorine residual of at least 0.2 mg/liter free chlorine must be maintained throughout the distribution system. If water at a sampling site is found to be free of residual chlorine, then coliform analyses must be run.

Substitution of chlorine-residual measurements for coliform analyses can only be done with the approval of the state. In addition, a sanitary survey of the water system must be made, to show that the source water used is free of contamination and that there are no obvious hazards in the system which might lead to problems if frequency of coliform testing were reduced as permitted.

Currently, there is no requirement in the Drinking Water Regulations for the standard plate count. A standard was proposed in the preliminary regulations, requiring that there be no greater than 500 organisms/ml, but this standard was deleted in the final regulations because of objections by water suppliers and others interested in water-supply problems. The objections concerned the questionable health significance of the standard plate count and the considerable cost that many water suppliers would have to bear in order to make their systems meet the standard of 500 organisms/ml. Because the health significance of the standard plate count did not seem to be of overriding importance, the requirement was deleted from the regulations.

STANDARD PLATE COUNT REGULATIONS

The arguments in favor of a standard plate count are several. First, there are noncoliform bacteria, such as *Pseudomonas*, that are not of intestinal origin but are pathogenic. Such organisms may find their way into water, especially if well water is used. Second, a high total bacterial count may render the water unfit to drink, even if all pathogens are missing. Third, if a high number of noncoliform organisms are present, the coliform analysis itself is interfered with, since the other bacteria may inhibit growth of the coliforms. Thus, a negative coliform test in itself does not mean that coliforms are absent. Many water bacteriologists thus hope that the standard plate count will be restored to the regulations, at least for those types of systems where it would provide some measure of public-health protection.

14.6 Microbiology of water pipelines

The water pipeline, usually hidden from sight, is a crucial link between the individual and the source of water. Water lines are subject to a number of microbiological onslaughts, and a knowledge of these can help prevent any serious problems from developing. Water pipes may be constructed of steel, galvanized iron, brass, copper, wood, reinforced concrete, asbestos cement, or plastic. Galvanized iron, brass, copper, and plastic are used for small water systems and within buildings, and asbestos cement, iron, and steel are used in cities and other places where large amounts of water are distributed. Corrosion and clogging of pipelines are always potential problems (Figure 14.7).

CORROSION

Pipe corrosion by hydrogen sulfide is brought about by *sulfate-reducing bacteria*, which convert sulfate (SO_4^{2-}) to hydrogen sulfide (H_2S) when adequate organic matter is available. The hydrogen sulfide attacks the iron of pipes and converts it into iron sulfide, causing pitting and deterioration of the pipe. These bacteria are obligate anaerobes and are only of significance in the corrosion of pipes buried in areas in which oxygen is absent, such as waterlogged soils and marine and freshwater muds. Such corrosion is especially serious in areas where seawater has infiltrated the soil, since seawater is very high in sulfate, and an unprotected pipe buried under anaerobic conditions in such an area can be destroyed in a few years. Pipes can be protected by coating them with asphalt or other

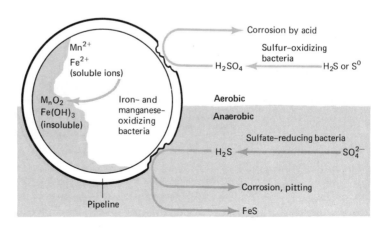

FIGURE 14.7
Corrosion and clogging of pipelines as a result of microbial action.

impervious material, or it may be preferable to use pipe made of noncorrodible material, such as plastic.

Corrosion by sulfuric acid is brought about by *sulfur-oxidizing bacteria,* which convert hydrogen sulfide or sulfur to sulfuric acid in the presence of oxygen. This strong acid attacks copper, zinc, and concrete, in addition to iron and steel, and the activities of these bacteria can be responsible for the corrosion of pipes made of any of these materials.

Iron- and *manganese-oxidizing bacteria* can cause clogging of water-supply pipelines. In many water supplies, especially wells, iron and manganese are present in soluble form. When this water is brought to the surface and aerated, iron and manganese bacteria oxidize these substances and large amounts of insoluble precipitate can form. Difficulties arise both because of the clogged pipeline and because water containing these precipitates is unsightly, has a metallic flavor, and causes the staining of laundry. Iron causes a rusty red stain, and manganese, a black stain. Iron and manganese can be removed from a water supply before distribution by aeration, followed by settling or filtration.

PRECIPITATE FORMATION

Pathogenic microorganisms that have not been removed in the water-purification process may be retained in viable form within the water pipe. Subsequent release of these organisms in a *slug,* or clump, may cause infection in individuals receiving that portion of water. Retention of microorganisms within the pipe may be within precipitates of manganese or iron but more likely within organic slimes, which build up in certain water supplies. Drinking water derived from surface water supplies often has small amounts of organic matter present, derived from the lake or stream providing the water. This organic matter, if not removed during flocculation and filtration, can serve as a food source for slime-producing bacteria, which then form large deposits within the pipes. Pathogenic organisms may be retained within these slimes, and certain of them may even grow. Slime build-up and growth of pathogenic bacteria within the pipes can be prevented by maintenance of a residual-chlorine concentration in the water. It is recommended that a chlorine residual of at least 0.2 to 0.3 ppm be maintained throughout the distribution system (see Section 14.5).

PATHOGENIC ORGANISMS

Without maintenance of a chlorine residual, slimes and precipitates harboring pathogenic organisms may remain firmly fixed upon the water pipes for long periods of time. However, they can be rapidly liberated if there is a large change in water pressure caused by unusual water demand, such as for fighting a large fire or heavy watering of lawns during dry periods. Such heavy usage can result in a marked pressure drop in the pipe; then, when demand returns to normal, the lowered pressure within the pipe acts as a vacuum, pulling large amounts of water through the pipe very rapidly. This sudden force of water rushing through the pipe can dislodge slimes or precipitates and carry them and their associated pathogens into residences. The problem can be remedied by periodic flushing of pipes, preferably annually, to eliminate build-up of slimes and precipitates. Maintenance of the recommended residual chlorine concentration will prevent slime growth in the first place.

14.7
Water pollution

Water pollution occurs when undesirable effluents enter water courses and so change water quality that the water is unfit for human use without treatment. Water pollution can be divided into three main sources: natural, domestic, and industrial.

Natural sources of water pollution include thermal and acid effluents from volcanic areas and are not common on earth. *Domestic sources* are primarily sewage and laundry wastes and are generated in houses, apartments, and other dwellings. In rural and some suburban areas, domestic wastes are handled at the individual residence and enter the environment through the soil either in partially treated or untreated fashion. In urban areas, domestic wastes are collected in sewage pipes and transmitted to a central location either for treatment or for discharge into a water course without treatment. Urban sewage wastes are a major potential source of water pollution, but since they are handled by established government agencies (city, county, or regional), they can usually be effectively controlled. Rural and suburban wastes, although quantitatively less significant, are often of major public health concern because they are not as easily controlled by regulatory agencies.

Industrial wastes are those wastes discharged by industries. In contrast to domestic sewage, which is usually of similar composition in all locations and hence can be treated by universally suitable methods, industrial wastes vary greatly from industry to industry and from location to location. Some industries generate wastes high in organic matter, and these wastes can usually be handled by methods similar to those used for domestic wastes. Such industries include dairy and food-processing plants, meat-packing houses, petroleum refineries, and pulp and paper mills. Other industries, however, generate wastes that are low in organic matter but high in toxic chemicals such as metals, acids, or alkalies. These include chemical plants, mining facilities, textile mills, and metal-refining and -processing plants. Each industry has its unique problem.

TYPES OF POLLUTANTS

A number of general categories of pollutants are recognized. These include:

1 *Oxygen-demanding wastes*, such as sewage, as well as industrial wastes of plant and animal origin. In the latter category are wastes generated by food plants, paper mills, and tanneries, among many others.
2 *Disease-causing agents*, such as pathogenic bacteria and viruses. These arise mostly from sewage but can also arise from tanneries and meat-packing plants.
3 *Plant nutrients*, substances that support growth of aquatic algae and weeds. The most important nutrients in this category are nitrogen and phosphorus compounds. These arise from sewage, from certain industries, and from drainage from fertilized agricultural lands. Conventional waste-treatment methods do not remove these materials.
4 *Organic chemicals*, such as soaps, detergents, industrial chemicals, and petroleum wastes. Many of these substances are toxic to aquatic life and possibly to humans. Some of these pass through conventional treatment systems unchanged.
5 *Inorganic chemicals*, such as metal salts and acids, arising primarily from mining and manufacturing processes.
6 *Sediments*, the particles of sand, soil, and minerals that wash from land into streams and give an unsightly appearance to waters. They may hinder operation of shipping and hydroelectric power facilities, owing to siltation, and may reduce fish and shellfish populations. Vast amounts of sediment are generated in construction projects, when soil washes easily from the disturbed land surface into waterways.
7 *Radioactive substances*, arising either from natural sources or, more seriously, from nuclear power plants, industrial operations, and medical and research procedures.
8 *Heat*, added to waters mainly from power plants and industrial cooling operations. Warmer water can absorb less oxygen than cooler water and hence may become anaerobic more quickly.

In most situations, the waste to be treated is a mixture of the foregoing types of pollutants, thus greatly complicating treatment and control procedures.

When sewage is added to a water course, one of the most dramatic events is the depletion of the dissolved oxygen of the water. This is due to the fact that the sewage contains organic matter, and bacteria in the water oxidize organic matter, in the process consuming oxygen. The decrease in oxygen content has several important consequences: (1) It makes the water course partially or completely anaerobic, and this leads to the development of odors, flavors, and toxic materials in the water. (2) When the water becomes anaerobic, many animals such as fish die, and their remains putrefy and add further foul odors and organic matter to the water (Figure 14.8). (3) Even if odors do not develop, water depleted of oxygen has a flat taste. (4) Decomposition of organic materials takes place much more

BIOCHEMICAL OXYGEN DEMAND (B.O.D.)

slowly in the absence of oxygen; the purification processes in the water course are therefore slowed, and a thick, unsightly, organic-rich sediment may accumulate on the bottom of the water course. (5) Certain undesirable animals, such as the red bloodworms (chironomid larvae), develop to very large numbers in waters depleted of oxygen.

The consumption of oxygen by bacteria is called the *biochemical oxygen demand*, usually abbreviated B.O.D., and the extent of oxygen consumption is determined by the amount of oxidizable organic matter present in the water (Figure 14.9). The B.O.D. is commonly used as a measure of the degree of organic pollution of waters. The

FIGURE 14.8
A fish kill resulting from the development of anaerobic conditions in a river, caused by excessive pollution. *Courtesy of the Wisconsin Natural Resources Department, Madison, Wisconsin.*

B.O.D. is determined by taking a sample of water, aerating it well, placing it in a sealed bottle, incubating for a standard period of time (usually 5 days at 20°C), and then determining the residual oxygen in the water at the end of incubation. During the 5-day incubation period, microorganisms present in the water grow, oxidize the organic matter, and consume oxygen. The amount of oxygen consumed is roughly proportional to the amount of biodegradable organic matter present.

Another test frequently used measures the *chemical oxygen demand* (C.O.D.). In this test, the organic matter is oxidized not with microorganisms but with a chemical oxidizing agent, usually potassium dichromate. The C.O.D. is usually higher than the B.O.D. because some organic materials in the water that are resistant to microbial oxidation and hence not involved in B.O.D. will be oxidized chemically and contribute to C.O.D. Although it is really B.O.D. that is of concern in water pollution, C.O.D. is often determined because it can be measured more quickly, since a 5-day incubation is not required.

When sewage is added to a river or other water course, pollution occurs. Pollution is followed by purification, the processes in which the quality of the water is returned toward normal. When purification occurs without human intervention, it is called *self-purification* (Figure 14.10) and occurs as a result of microbiological, chemical, and physical changes. Microbiological changes include death of many intestinal microorganisms present in the sewage and growth of normal aquatic microorganisms able to oxidize organic matter entering the system. Chemical changes include oxidation of organic matter, release of nitrate and phosphate, and reoxygenation of the water by solution of oxygen into the water from the air. The most important physical changes involve sedimentation, in which particulate matter settles out of the water onto the bottom of the water course.

It is the purpose of sewage-treatment systems (see Section 14.9) to replace these relatively slow, uncontrolled self-purification processes with rapid, well-controlled purification processes. The sewage-treatment process, carried out under careful control before the material enters the watercourse, attempts to ensure that the quality of the water into which sewage is to be discharged is not degraded.

SELF-PURIFICATION OF WATER

We discussed indicator organisms in Section 14.3, during our consideration of drinking-water quality. Microbial indicators also provide an excellent means of monitoring natural water for sewage pollution, because they can be readily detected even in relatively

14.8
Microbial indicators of sewage pollution

FIGURE 14.9
Determination of biochemical
oxygen demand (B.O.D).

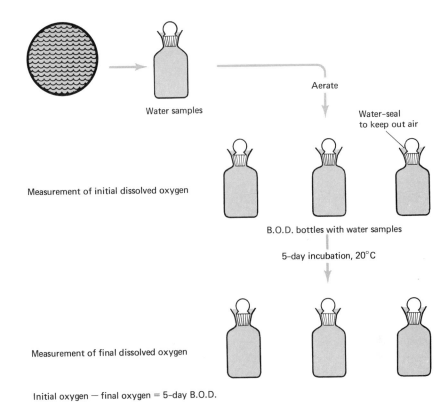

Water samples

Aerate

Water-seal
to keep out air

Measurement of initial dissolved oxygen

B.O.D. bottles with water samples

5-day incubation, 20°C

Measurement of final dissolved oxygen

Initial oxygen — final oxygen = 5-day B.O.D.

FIGURE 14.10
Self-purification of bodies of
water following sewage
discharge will occur if the
amount of organic matter is
small in relation to the volume
of water.

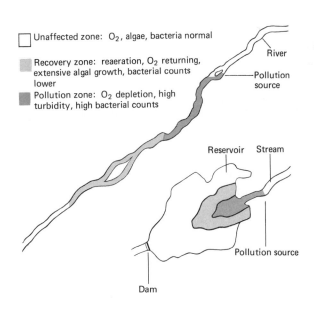

☐ Unaffected zone: O_2, algae, bacteria normal

▨ Recovery zone: reaeration, O_2 returning,
extensive algal growth, bacterial counts
lower

■ Pollution zone: O_2 depletion, high
turbidity, high bacterial counts

River

Pollution
source

Reservoir   Stream

Pollution source

Dam

small numbers using membrane filter or most-probable-number techniques. Any of the organisms used in evaluating drinking water for microbial purity can also be used as indicators of sewage pollution. The two indicators most frequently used are the coliform group and a subgroup of the coliforms called the *fecal coliforms*. We defined the coliform group in Section 14.3 and emphasized that this was an operational rather than a taxonomic definition and included a variety of organisms other than *Escherichia coli*. Some of the organisms in the coliform group may not even be derived from the intestine and therefore may not be of direct sanitary significance. For evaluating drinking-water quality, this deficiency is not a problem, since we are concerned with having a certain margin of safety in our drinking-water analyses and can accept a test that counts additional organisms. For studies of sewage pollution, however, it is desirable to have an indicator that is associated directly with the intestinal tract. For this reason, the fecal coliform test has been devised.

The fecal coliform group is defined as containing those coliforms able to grow at an elevated temperature of 44.5°C. This elevated temperature is selective for organisms associated with the intestinal tract and eliminates many of the nonintestinal coliforms able to grow at the standard temperature of 35°C. The fecal coliforms are thus better indicators of recent sewage pollution.

    The fecal coliforms can be used as indicators in two ways: (1) there are MF and MPN methods that permit the direct quantification of fecal coliforms from water samples; and (2) the elevated temperature test can be used as a confirmatory procedure for samples showing positive presumptive tests by the standard coliform procedure. In practice, the standard coliform test will most likely be run in any water-pollution study, and the proportion of positive samples that contain fecal coliforms will be determined. In situations where there has been recent sewage pollution, the proportion of fecal coliforms will be high. More distant from pollution sources or in nonpolluted waters, fecal coliform counts will be low or absent.

FECAL COLIFORMS

Sewage is collected into *sanitary sewers* from homes, businesses, and many industries and is carried to the sewage plant for treatment before discharge into streams (Figure 14.11). Water from rain and snow is also often collected in sewers called *storm sewers*; this water is not usually polluted and can be discharged directly into streams without treatment. In many cities, however, separate storm sewers and sanitary sewers are not maintained, and both effluents are carried through the same pipes. This is undesirable, since in times of

**14.9
Sewage-treatment
systems**

FIGURE 14.11
Collection of domestic sewage
for treatment at a municipal
plant.

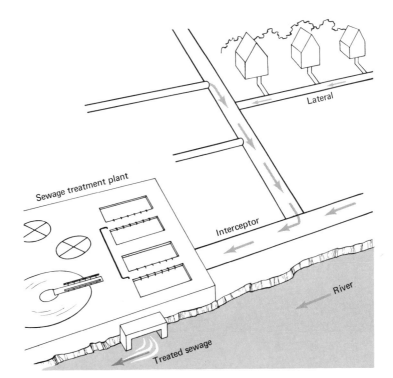

heavy rains or flood, the pipes carry much more water than can be processed by the sewage plant, so that the excess simply bypasses the plant, and water is dumped untreated into the stream. It is preferable to have separate storm and sanitary sewers, although this is costly.

Sewage treatment consists of a series of processes in which undesirable materials in the water are removed or rendered harmless. Oxygen-consuming organic matter (B.O.D.) is destroyed, silt, clay, and other debris are removed, pathogenic microorganisms are killed, and the total number of microorganisms is reduced. There are many designs for sewage-treatment systems; the best design to be used for a specific system generally depends upon local factors. Figure 14.12 shows the sequence of steps in a typical urban sewage-treatment system. Both biological and nonbiological treatments are used. Nonbiological treatments include coarse and fine screening, sedimentation, sand filtration, chemical treatment to induce flocculation, and incineration. Biological treatments include various digestion and oxidation processes in which organic matter and inorganic materials are removed from sewage through the action of living organisms. The key organisms in biological treatment are bacteria, although fungi, algae, protozoa, and even higher animals are involved in many treatment processes. Biological treatment processes can be divided into two groups: digestion processes, which are

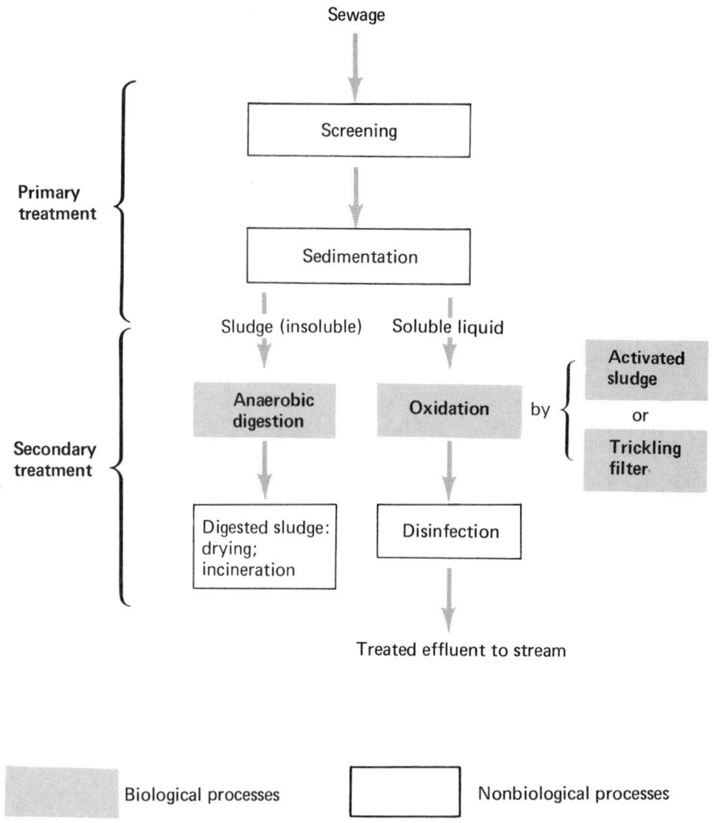

FIGURE 14.12
Steps in sewage treatment.

Sewage

Primary treatment

Screening

Sedimentation

Sludge (insoluble)

Soluble liquid

Secondary treatment

Anaerobic digestion

Oxidation

by

Activated sludge

or

Trickling filter

Digested sludge: drying; incineration

Disinfection

Treated effluent to stream

Biological processes

Nonbiological processes

anaerobic; and oxidation processes, which are aerobic. The principal digestion process is *anaerobic sludge digestion*, which functions in the treatment of insoluble organic-rich wastes; these are high in fiber and cellulose. Installations for oxidation processes include *trickling filters, activated-sludge systems*, and *lagoons* (oxidation ponds). A knowledge of the principles of operation of these processes is essential if they are to be managed in an efficient and effective manner.

14.10
Anaerobic digestion processes in sewage treatment

The insoluble material that settles in any part of a sewage-treatment system is called *sludge*. The major components of sludge are proteins and organic fibers rich in cellulose, and anaerobic sludge digestors are used to decompose these substances (Figure 14.13). The process is carried out in large tanks from which oxygen is completely excluded. Under the anaerobic conditions that prevail, fermentation occurs, and the sugars released in the digestion process are converted to methane and carbon dioxide. The methane is a burnable gas, and in many sewage plants, the methane produced in the digestion tank is used as a source of heat or power for the whole sewage-plant complex.

STEPS IN SLUDGE DIGESTION

The whole process occurs in four stages:

1 *The digestion phase*: initial digestion of fiber and other insoluble materials to sugars, amino acids, and other soluble products.
2 *The acid-production stage*: fermentation of the soluble materials to organic acids and alcohols by acid-producing bacteria; during this stage, acid accumulates, and the pH drops.
3 *The acid-regression stage*: conversion of organic acids and alcohols to hydrogen (H_2) and carbon dioxide (CO_2); during this stage, the pH rises again.
4 *The methane-production stage*: conversion of H_2 and CO_2 to methane (CH_4) by methane-producing bacteria.

The time taken for sludge to become digested is from 2 weeks to a month, and in a well-operated plant, 90 percent or more of the B.O.D. of the sludge can be removed; that is, 90 percent of the oxidizable organic matter is converted to H_2, CO_2, H_2O, and methane (which may itself be burned to CO_2 and H_2O). The undigestible materials remaining in the sludge are removed at intervals and dried, after which they are buried, burned, or used as an agricultural soil builder.

A serious problem in the management of many sludge digestors is the development of acid conditions. If stage 3, the acid-regression stage, does not immediately follow stage 2, the acid-production stage, the pH may drop to a value so low that the bacteria are killed or inhibited. When this happens, the whole process stops, gas production ceases, and B.O.D. removal does not occur. This condition may be corrected by addition of lime to neutralize the acidity, but it is prefer-

FIGURE 14.13
a Anaerobic sludge digestor. Only the top of the tank is shown; the rest is underground. *b* Inner workings of an anaerobic sludge digestor.

a

b

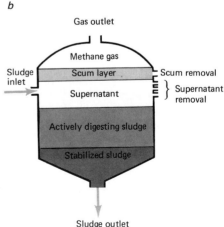

able to manage the system properly in the first place. The management of a sludge digestor is not easy, and a knowledge of the microbiology of the process is essential, although many plant operators unfortunately lack such knowledge.

Another serious problem is the disposal of the undigested sludge. This material must first be dewatered, to reduce its bulk. Dewatering is often difficult with certain kinds of sludges, and the material may have to be pumped to a large field for settling and drying. If this is the case, considerable care must be taken to ensure that the drying sludge does not pollute nearby streams or lands. The dried sludge may be useful as a soil builder, but it is important that the sludge does not contain any harmful materials, such as live viruses or heavy metals. Many sludges contain undesirably large amounts of heavy metals such as copper, chromium, lead, and zinc derived from the water pipes or from industrial processing. Because of the potential toxicity of these heavy metals, it must be ascertained that they remain firmly tied up in the sludge particles and do not leach into the soil solution so that they are taken up by the plants being cultivated. If it is predominantly organic in nature, the dried sludge can be burned in an incinerator, care being taken to ensure that no air pollution results from the incineration process. After incineration, the ash residue is buried. If the sludge is mainly inorganic and hence not burnable, it is usually buried. The preferred way to bury solid waste is in a *sanitary landfill*; this procedure is discussed in detail in Section 14.16.

14.11 Aerobic sewage-treatment systems

Aerobic systems are used to treat the soluble components of sewage that remain after the sludge has been removed. Tremendous volumes of liquid that must be treated in this way are generated by urban populations. The liquids contain a large variety of organic compounds, none present in very large amounts. Before these organic compounds can be acted upon by the living organisms of the treatment system, they must be concentrated in some way. Concentration involves the adsorption of the organic materials to some type of surface upon which the treatment organisms can grow. In the trickling filter process, a bed of rocks is used for this purpose, whereas in the activated-sludge process, the organisms themselves form *flocs*, or clumps of cells, that serve as the concentrating agent. Once the organic materials of the sewage have adsorbed onto the concentrating surface, they are then broken down by the organisms of the treatment system and converted into inorganic materials.

FIGURE 14.14
A trickling-filter installation.

TRICKLING FILTERS

A trickling filter is a bed of crushed rock about 6 feet thick, on top of which the sewage liquid clarified by sedimentation is sprayed (Figure 14.14). As the liquid slowly trickles through the bed, the organic matter adsorbs to the rocks, and microbial growth occurs. A film of microorganisms develops on the rocks, and within this film, oxidation of organic matter takes place.

The organisms in the film are of several types:

1 *Slime-forming bacteria*, which provide the basis of the film and carry out the initial oxidation. It is the slime produced by these bacteria that keeps the film stuck together and adherent to the rocks.
2 *Filamentous bacteria* and *fungi*, which live in association with the slime producers and help maintain the structure of the film. These organisms also oxidize organic matter.
3 *Protozoa*, which eat the bacteria and fungi, keeping the film from becoming so thick that the flow of sewage is blocked.
4 *Invertebrate animals*, especially rotifers and nematodes, which eat the protozoa.
5 *Autotrophic bacteria*, which oxidize the NH_3 and H_2S liberated during the decomposition of organic matter, converting these materials into NO_3^- and SO_4^{2-}.

The sum result of the activities of all of these organisms is *mineralization*, the conversion of organic matter to inorganic materials:

Organic carbon $\rightarrow$ carbon dioxide (CO_2)

Organic nitrogen $\rightarrow$ ammonia (NH_3) $\rightarrow$ nitrate (NO_3^-)

Organic sulfur $\rightarrow$ sulfide (H_2S) $\rightarrow$ sulfate (SO_4^{2-})

Organic phosphorus $\rightarrow$ phosphate (PO_4^{3-})

a

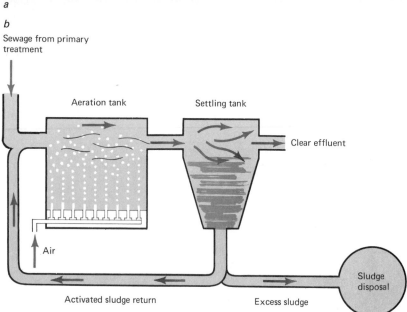

b

Sewage from primary
treatment

Aeration tank

Settling tank

Clear effluent

Air

Sludge
disposal

Activated sludge return

Excess sludge

FIGURE 14.15
Activated-sludge process.
a Aeration tank of an activated-
sludge installation. *b* Inner
workings of an activated-sludge
installation.

The effluent of a trickling filter is thus low in organic matter, the B.O.D. being reduced about 75 to 90 percent, but is high in inorganic nutrients that promote the growth of algae. The water body receiving treated sewage effluent thus often shows extensive growth of algae as a result of this enrichment (see Section 14.15).

The activated-sludge process (Figure 14.15) is another oxidation process used in many installations instead of the trickling filter. The wastewater to be treated is mixed and aerated in a large tank, and slime-forming bacteria similar to those present in trickling filters

ACTIVATED SLUDGE

grow and form flocs. These clumps constitute what is called the activated sludge; they are composed of several kinds of bacteria and in addition contain protozoa and other animals that live by feeding on the bacteria. One of the principal bacteria in the floc is the organism *Zoogloea ramigera*, a small, Gram-negative rod that forms an extensive extracellular slime. This organism forms complex, fingerlike colonies that provide the basis of the floc (Figure 14.16). The other bacteria, protozoa, and so forth, attach to the floc and build up the complex, active particle.

Oxidation of organic matter by the activated-sludge flocs occurs in a manner similar to that described for trickling filters. The treatment operates continuously, the wastewater being pumped into the tank and the effluent contain flocs passing out at another location. The holding time in the activated-sludge tank is generally only a few hours. The effluent containing the flocs is pumped into a holding tank or clarifier, where the flocs settle. Some of the floc material is then returned to the aerator as inoculum for the continuation of the oxidation process, and the rest is either dried and discarded or is pumped to the anaerobic sludge digestor for treatment.

A serious problem in the operation of some activated-sludge systems is *bulking*; the flocs do not settle sufficiently rapidly, and the clarification process is inhibited. Bulking usually occurs because of unusual growth of filamentous bacteria, which form a loose floc rather than the tight floc that is formed by *Zoogloea ramigera*. It is not clear what conditions promote the excessive growth of filamentous bacteria, but it is probably the presence in the wastewater of certain

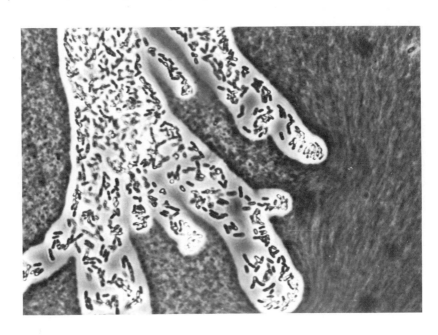

FIGURE 14.16
Photomicrograph of a floc formed by *Zoogloea ramigera*, the characteristic organism of the activated-sludge process. Note the large number of small, rod-shaped bacteria and the characteristic fingerlike projections of the floc. Negative stain preparation.

kinds of organic compounds that favor their growth over the *Z. ramigera* population. If bulking occurs, the sewage-plant operator is faced with an unclarified effluent that is high in organic materials. No simple remedy for bulking is available, but if it can be shown that it is caused by the presence in the wastewater of some unusual effluents, perhaps from industrial or food-processing plants, it may be possible to remove these effluents from the main treatment process and handle them separately.

About 75 to 90 percent decrease in B.O.D. occurs in the activated-sludge process. The clarified effluent is rich in inorganic materials such as nitrate, phosphate, and sulfate, as was that from the trickling filter, and can likewise promote extensive algal growth.

In recent years, it has been discovered that oxidation in the activated-sludge process can be made more efficient if the process is carried out with pure oxygen instead of air. Despite the greater cost of pure oxygen as opposed to air, the increased efficiency of the process makes it economically feasible. The efficiencies of various sewage-treatment processes are compared in Table 14.5.

Most organic substances in domestic sewage decompose readily during sewage treatment. In contrast, many chemicals present in industrial wastes are resistant to decomposition by the microorganisms present in normal sewage-treatment systems. Some of these compounds are toxic; if so, they can sometimes be effectively handled merely by diluting them to a point where they are no longer toxic. Other compounds are not toxic but are merely resistant to decomposition. Sometimes these materials cannot be treated biologically at all, and chemical means of treatment must be used. In some cases, it may be possible to develop a microbial population that will degrade the compound. If a compound can be decomposed by microorganisms, it is *biodegradable*; if it cannot be decomposed, even after attempts to obtain a population capable of degrading it, it is considered *nonbiodegradable*. It is often observed that when a new substance is first introduced into a sewage-treatment system, it decomposes slowly, but after a period of time, decomposition is rapid as a result of development of a suitable microbial population. Therefore, when beginning the treatment of a new material, it is important to ensure that the proper population is present before introducing material into the treatment plant.

If a material is not decomposed, it will pass through the system unchanged and will be discarded into the water body receiving the treated effluent. For instance, nonbiodegradable detergents will pass through a system unchanged and enter receiving waters where they can cause foaming. Since microbial populations able to degrade these compounds do not exist, the problem of foaming has been

SUBSTANCES RESISTANT TO DECOMPOSITION

TABLE 14.5
Efficiency of sewage-
treatment processes

Treatment	Reduction (%)		
	Suspended solids	Biochemical oxygen demand	Bacteria
Sedimentation	40–70	25–40	25–75
Sedimentation plus sand filter	90–98	85–95	Greater than 90
Sedimentation plus trickling filter	75–90	80–95	90–95
Activated sludge	85–95	85–95	90–98

reduced by legislation banning the sale of nonbiodegradable detergents.

14.12
Discharge of treated sewage

Although the treatment processes discussed in Section 14.11 bring about marked reduction in the harmful materials present in raw sewage (see Table 14.5), a potential for the deterioration of the body of water receiving the effluent still exists because of the large volumes of treated sewage that are discharged, even when the concentrations of harmful materials are quite low. Also, since many sewage-treatment systems do not operate properly 100 percent of the time, there are situations in which essentially untreated wastes might be discharged. It is vital that the sewage-plant operator monitor the effluent by performing analytical tests and procedures. Total and fecal coliform counts, B.O.D. determinations, and chemical analyses for harmful materials such as ammonia are essential. There should be no detectable fecal coliforms in the effluent. The level of B.O.D. permitted depends on state and federal regulations but should be maintained at a low, constant level. The turbidity of the effluent should also be low, and the total solids present, representing materials that do not settle out in the clarification processes, should be kept below levels specified by the regulations. Some variations are permitted, depending on the size and type of the receiving waters. Discharge of harmful materials into a large river or lake presents fewer problems than discharge into a small stream. Each sewage-treatment system presents a unique set of problems.

CHLORINATION

The potentially pathogenic microorganisms present in raw sewage are virtually all eliminated during sewage treatment, either by sedimentation and removal or because they die during the treatment process. However, to ensure that the final effluent is free of harmful microorganisms, it is often chlorinated before it is discharged into the receiving water body. Although chlorination is used primarily to destroy any intestinal microorganisms that might have survived passage through the plant and would contaminate the receiving water, it also leads to the decomposition of materials in the effluent that would otherwise cause unpleasant odors. Chlorine is injected in gaseous form into the effluent 15 to 30 minutes before the treated water is discharged into the water course.

Viruses present special problems in sewage treatment. Because of their small size and noncellular structure, they do not behave like bacteria and other pathogenic organisms during the treatment process. They are not killed by either anaerobic or aerobic treatment processes. Viruses attached to particles settle out in the sedimentation processes, and some adsorb to the flocs in activated sludge, but this does not lead to their destruction, and they may remain viable in the final dried sludge. Virus particles that remain free in the soluble liquid may pass through the whole treatment process and end up in the effluent, presenting a possible health danger if they are discharged into a stream. Viruses in the effluent can be eliminated by chlorination, although the doses required may be higher than those required to eliminate bacteria. Thus, an effluent that shows a negative coliform count is not necessarily free of viruses. The greatest hazard arises if the stream that receives the virus-laden effluent is used by another community further down as its water supply, since viruses may also survive many water-purification processes.

VIRUSES IN SEWAGE EFFLUENTS

In both trickling-filter and activated-sludge processes, mists that contain fecal bacteria are formed in the air. These mists, called *aerosols,* can be the source of pathogenic bacteria for nearby populated areas, since under windy conditions the bacteria, among which may be a few pathogens, are blown away from the plant. Fortunately, such conditions have not yet proved to be a public-health hazard, although aerosols are recognized as a potential problem by treatment-plant managers. Most intestinal pathogens are relatively sensitive to drying and are probably killed when the aerosol dries, although some viruses are not killed by drying.

HEALTH HAZARDS OF TREATMENT PLANTS

The conventional biological treatment methods described in this chapter are used in virtually all treatment systems. However, as the burden of pollutants increases and unusual pollutants become more common, the conventional methods become less and less satisfactory.

**14.13
Advanced treatment methods**

A 95 percent reduction of B.O.D. and of microbial load is not enough if the initial load is very large. Newer procedures are being developed that are nonbiological in nature and hence avoid some of the difficulties and problems that arise from biological treatment systems. These processes are being used either instead of conventional systems or as a final polishing stage in which the effluent from the conventional process is further treated.

COAGULATION-SEDIMENTATION

In this process, chemicals such as alum and lime are added to the effluent, causing the formation of flocs. This treated water is held in a sedimentation tank until the flocs settle, and the clarified effluent is then discharged. Coagulation-sedimentation has been used for a long time by certain industries but is just being introduced into domestic sewage treatment. Inorganic nutrients, especially phosphate, are adsorbed to and trapped in the floc material; thus, the effluent water is much lower in phosphate, reducing possible algal growth in the receiving stream.

ADSORPTION

This process involves passing the effluent over a bed of material that will adsorb chemicals dissolved in the water. The adsorbent in the bed is usually activated carbon granules, which will remove over 98 percent of the dissolved organic chemicals. After the carbon has become saturated, the organic matter is driven off by steam, and the carbon is reused. Adsorption is especially useful in removing those chemicals that are not biodegradable and hence pass through conventional plants unchanged.

ELECTRODIALYSIS

In this process, the effluent passes along one side of a plastic membrane that is permeable to salts. A current generated across the membrane causes the salts to pass across the membrane, and the water left behind is thus depleted. Since treated sewage effluent is higher in salts than are lakes and rivers, electrodialysis can be used to restore the water to a more desirable (low) level of salts.

REVERSE OSMOSIS

The effluent is forced under pressure down one side of a plastic membrane that is more permeable to water than to dissolved substances. The pressure causes the water to pass through the membrane, and water is collected on the other side in a relatively pure state, the dissolved substances being left behind. Also left behind are sediment, bacteria, and other particulate substances. Practical application of this method awaits development of strong, long-lasting membranes that are not too costly.

14.13 ADVANCED TREATMENT METHODS   459

Pure oxygen or ozone is used to oxidize organic matter. This is strictly a chemical process, as opposed to the conventional biological oxidation process and, although more expensive, is more complete.

After sewage has been given conventional and advanced treatments, the effluent obtained may be pure enough to be consumed again as drinking water. In several cities throughout the world where fresh water is in short supply, such recycling of sewage water into drinking water is carried out.

14.14
Lagoons, septic tanks, and privies

All the sewage-treatment systems discussed in Sections 14.9 to 14.13 involve large installations, complicated engineering, and skilled operation by large staffs of people. They are suitable for urban and industrial use but are beyond the means of resources of individual homeowners or small communities. It is vital that all sewage be treated before it is discharged in order to avoid public-health problems and environmental degradation. In situations where conventional systems are not feasible, certain types of treatment processes are available that can provide at least some measure of public protection. None of these methods is desirable in areas with dense population distribution; rather, they can be used in rural areas and in small, remote communities.

The sewage lagoon, sometimes called an *oxidation pond* or *stabilization pond*, is simply a large shallow pond into which sewage clarified by sedimentation is pumped. In the lagoon, bacteria oxidize organic substances to inorganic matter, such as carbon dioxide, phosphate, nitrate, and sulfate (Figure 14.17). These inorganic materials serve as nutrients for the growth of algae, which use sunlight and carry out photosynthesis. In a well-balanced lagoon, 80 to 95 percent of the B.O.D. may be removed. The oxygen produced by the algae during photosynthesis keeps the surface of the lagoon aerobic, thus preventing the development of anaerobic conditions, which generally lead to foul odors. A lagoon is usually odor-free. Pathogenic bacteria are usually destroyed in the lagoon.

In general, lagoons are used in small towns; they are not suitable for efficient treatment of large volumes of waste. Lagoons function best in warm, sunny climates where plenty of light is available to permit algal growth and where warm temperatures speed up the oxidation process. Lagoons are also used in cold climates but operate rather inefficiently in winter, since if a surface layer of ice forms for any length of time, anaerobic conditions develop, leading to inefficient sewage treatment.

FIGURE 14.17
Layout and processes in a
sewage lagoon.

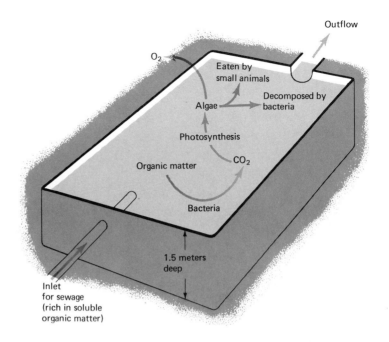

SEPTIC TANKS

A septic tank is a water-tight tank used at rural and some suburban residences to provide minimal treatment of sewage where municipal sewer lines are not available. The tank is a concrete, metal or masonry structure through which the raw sewage passes (Figure 14.18). During the time the sewage is in the tank, settling and minimal sludge digestion occur; the effluent liquid is then distributed into the soil through a tile field or leaching pit. Eventually the sludge builds up in the bottom of the tank to a point where operation is impaired, and the tank must then be pumped out or otherwise cleaned. Some microbiological digestion of sludge does occur in septic tanks, although under less efficient and less well-controlled conditions than in modern treatment plants. An evaluation of treatment efficiency that is commonly obtained in a septic tank is shown in Table 14.6. These values should be compared to those already given for standard treatment processes in Table 14.5; the low efficiency of the septic tank is clearly indicated.

The extent to which a given soil can accept septic-tank effluent will depend on its permeability. In a porous, highly permeable soil, the effluent spreads readily and no difficulties arise; in a hard clay soil of low permeability, the effluent cannot spread well but will only follow cracks and crevices in the soil, resulting in poor purification. The soil serves as a crude sand filter for septic-tank effluent, and much of the purification process occurs in the soil rather than in the tank. To determine the capacity of soil to accept septic-tank effluent,

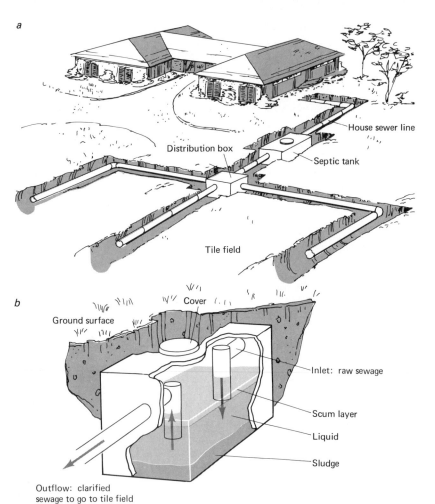

FIGURE 14.18
Septic tank, as used in small-scale sewage treatment in rural areas. *a* Distribution system. *b* Inner workings of the septic tank.

a

House sewer line

Distribution box

Septic tank

Tile field

b

Cover

Ground surface

Inlet: raw sewage

Scum layer

Liquid

Sludge

Outflow: clarified sewage to go to tile field

	Raw sewage	Septic tank effluent[a]
Biochemical oxygen demand (B.O.D.)	350 mg/l	150 mg/l
Percent reduction		57%
Total suspended solids	400 mg/l	140 mg/l
Percent reduction		65%
Total nitrogen	80 mg/l	50 mg/l
Percent reduction		37%
Fecal coliform count		1,200,000/100 ml

TABLE 14.6
Efficiency of sewage treatment in septic tanks

[a] Values are average concentrations for typical septic tank systems.
From *On-site wastewater disposal for homes in unsewered areas*, Bulletin of the Small-Scale Waste Management Project, September, 1973, University of Wisconsin, Madison. Available from the Agricultural Bulletin Building, Madison, WI 53706.

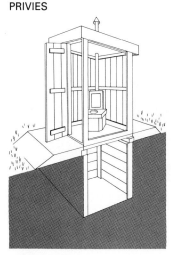

FIGURE 14.19
A privy, or outhouse.

14.15
Algae and water pollution

a soil percolation test is run. Such a test is required by law in many areas before a septic tank may be installed.

A privy (outhouse) is merely a hole or pit in the ground above which a toilet seat and some kind of shelter are mounted (Figure 14.19). Privies are used mainly in very rural areas, such as farms or primitive campgrounds, where population density is quite low, but they are also used in temporary installations, such as military outposts or construction sites. Decomposition of waste matter occurs within the pit, both by digestion and oxidation, but the process is slow and inefficient. Odor is frequently a problem, and disease transmission may be common and serious. Transmission of pathogens occurs often by flies, which may pick up pathogens on their feet and transport them to nearby dwellings. Screened shelters and covered toilet seats hinder the entry of flies and hence reduce transmission of pathogens. Pathogens can also be carried out of the pit into the surrounding soil by leaching and may be carried through crevices and cracks in the soil to the water table. In general privies are highly unsatisfactory, and the installation of more effective sewage-treatment systems should be encouraged.

A serious problem in many lakes and reservoirs used as sources of water is the growth of algae. Algae are undesirable because they cause bad odors and flavors in water and may produce toxic materials of potential danger to humans. Algal growth is favored by warm water temperatures, high sunlight, and adequate sources of nutrients, especially nitrates, phosphates, and carbon dioxide. Therefore, algal growth in temperate lakes and reservoirs is most common in summer and is rare in winter. Occasionally, in late summer and early fall, algal growth may be so heavy that the water resembles pea soup. This condition, called an *algal bloom* (Figure 14.20), occurs when algae float to the surface and drift into backwaters where they become concentrated. Bacteria attack and decompose the algae, causing a reduction in oxygen, which in turn leads to the death of fish and other animals and the development of foul and putrefying odors. In tropical areas, algal growth occurs at about the same rate all year, and the massive blooms seen in temperate climates are less common.

Reservoirs for domestic water supplies are often good habitats for algal growth, because they are generally relatively shallow and receive large amounts of algal nutrients from the watershed. In addition to the odor and flavor problems that may develop, heavy algal growth in a reservoir makes filtration and disinfection of the water difficult, thus markedly increasing the cost of water purification.

Algal growth in reservoirs is generally controlled by application of copper sulfate. In many water supplies, copper sulfate is applied routinely at 2- to 4-week intervals throughout the period from April to October. The exact amount of copper sulfate required depends on the alkalinity or acidity of the reservoir; in an average reservoir, about 2.5 to 2.75 pounds of copper sulfate per million gallons of water in the reservoir is recommended. Because copper sulfate is toxic to fish (and in high doses to humans also), its use in reservoirs must be carefully controlled. The recommended amounts are well below the known toxic levels. It is most desirable to carry out routine microscopic examinations of the reservoir for algae throughout the warm season. When microscopy reveals an increase in algal numbers, copper sulfate should then be applied, before the algal bloom begins. In this way, uncontrolled increases in algal numbers can be eliminated completely.

FIGURE 14.20
An algal bloom results from eutrophication. The thick algal scum decomposes, causing oxygen depletion and foul odors.

SWIMMING POOLS

Algal growth in swimming pools causes an unsightly slime on the walls of the pool and reduces water clarity. The best method of control is by chlorination, but if algal growth is heavy, copper sulfate treatment may be used. A dosage of 5 lb per million gallons will control most algae, but in difficult cases higher doses may be needed. The copper sulfate is distributed in these small water volumes by dragging a burlap bag containing the required amount of copper sulfate through the water.

EUTROPHICATION AND DOMESTIC SEWAGE

Sewage not only contributes organic matter for the growth of bacteria but nutrients for the growth of algae. The enrichment of water courses with algal nutrients is called *eutrophication* and is a serious economic problem because algal growth adds organic matter back into the water, thus increasing the B.O.D. and causing a deterioration of water quality. The algal nutrients are inorganic and are released by mineralization during the decomposition of the organic matter of the sewage by bacteria (Figure 14.21). Organic nitrogen is converted to ammonia; the ammonia is then oxidized to nitrate. Organic sulfur is converted to hydrogen sulfide, which is then oxidized to sulfate. Organic phosphorus is converted to inorganic phosphate. Nitrate and phosphate are especially important in water pollution because they are effective nutrient sources for algae.

Since conventional sewage treatment does not eliminate algal nutrients, eutrophication can be prevented only if advanced sewage-treatment methods are used. Some of the advanced methods discussed in Section 14.13 eliminate phosphate from treated sewage. The

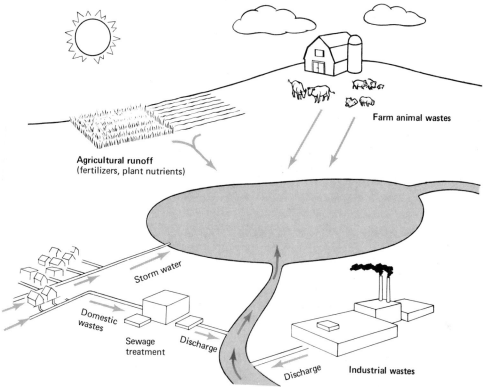

FIGURE 14.21
Sources of algal nutrients,
mainly nitrogen and phosphorus
compounds. Excesses of these
nutrients lead to eutrophication.

elimination of nitrate is more of a problem, as this anion is not precipitable by any of the agents used in advanced treatment systems. One method for elimination of nitrate is *denitrification*, a bacterial process discussed in Section 13.5. Bacteria that denitrify convert nitrate into N_2 gas, which is readily removed from the system. For denitrification to occur, some source of organic carbon is necessary to provide the energy source. In experimental advanced-treatment denitrification systems, a *Pseudomonas* that is able to denitrify using methanol has been utilized. Just enough methanol is added to completely consume the nitrate present; the *Pseudomonas* then oxidizes the methanol to CO_2. As yet, such a system has not been used in large-scale operations, but it shows promise.

14.16
Water pollution from the pulp and paper industry

A number of microbiological problems arise during the manufacture of paper, some within the plant itself and others within the bodies of water into which paper mill effluents are discharged. Paper production is one of the largest consumers of water; a single paper mill may consume 30,000,000 gallons of water per day, and a few use as much as

100,000,000 gallons per day. Much of this water is reused a number of times within the mill, but it is eventually discarded and can cause water pollution.

Paper consists of a matted or felted sheet of cellulose fibers. The source of these fibers is most commonly wood, but rags, flax, hemp, cotton, or straw may also be used. The raw material is first converted to *pulp* by cooking it, usually in the presence of certain chemicals; during the pulping process lignin, resins, oils, and other noncellulosic materials are extracted, leaving behind cellulose suspended in water. The two main chemical pulping processes are sulfite pulping and soda ash pulping.

In the sulfite process, wood chips are cooked at high temperature in the presence of calcium or sodium bisulfite. Lignin and other ingredients solubilized from the wood are separated from the cellulose pulp. The latter is washed and processed further; the former, called *spent sulfite liquor* or *sulfite waste liquor*, must be discarded. This liquor contains about 12 percent total solids, of which at least 20 percent is solubilized lignin and 15 to 20 percent is sugars. The biochemical oxygen demand (B.O.D.) of the spent sulfite liquor is very high (30,000 to 50,000 mg/liter), and about 2,000 liters of spent liquor are produced per 1,000 kg of pulp. The sugars present in the spent liquor are the main contributor to the B.O.D. One way to eliminate these is to use them as substrate to produce food or feed yeast (see Section 17.2), or to produce alcohol. Alternatively, a paper company can treat spent sulfite liquor by traditional sewage treatment systems, such as the activated-sludge or trickling-filter methods. The waste water resulting from either a yeast process or a sewage treatment process is evaporated and the lignin recovered and used as a fuel or for industrial purposes.

SULFITE PROCESS

In the soda ash or kraft pulping process, the wood chips are cooked in an alkaline solution containing sodium sulfide. Chemical reactions between the sulfide and organic products of the wood result in the production of odoriferous volatile compounds. Air pollution problems from the kraft process are therefore considerable, but water pollution problems are much less serious than in the sulfite process since it is desirable to recover the chemicals in the waste liquor for reuse, and in doing so the organic materials are removed. The waste liquor is evaporated to dryness and then burned; the organic materials (principally lignin and sugars) are thus oxidized. The heat generated by this burning is used to raise the temperature in the kraft digestor. Inorganic chemicals that remain unburned are dissolved in water, additional chemicals are added, and the material is reused to begin the digestion process again. The effluent water from a kraft pulping mill has a B.O.D. less than 10 percent that of a sulfite pulping mill.

SODA ASH PROCESS

In a modern manufacturing plant, the process of paper making from pulp is continuous (Figure 14.22). From the reservoir at one end, called the *headbox*, the stock is allowed to flow out continuously in a thin sheet onto a long wire mesh screen, called the *Fourdrinier wire*. The water and some fibers fall through the wire into a catch basin below; the bulk of the fibers aggregate and bond together on top of the wire to form paper. The residual water is called *white water*, its white color caused by the pulp fibers remaining in it. This white water is recycled by being brought back to the headbox. At the end of the Fourdrinier wire the wet paper is transferred to a felt-covered suction roll, where water is removed under vacuum, and then to felted press rolls, where more water is squeezed out and the paper is compacted. It still contains 60 to 70 percent of its weight as water, and it is passed now to the dryer to be heated until its moisture is greatly reduced. The paper is then formed into large rolls. The two ends of the paper machine are often called the wet end and the dry end; it is at the wet end that important microbiological problems arise.

FIGURE 14.22
The manufacture of paper is complicated by the growth of bacteria in the nutrient-rich processing waters. Discharge of used water causes enormous pollution problems.

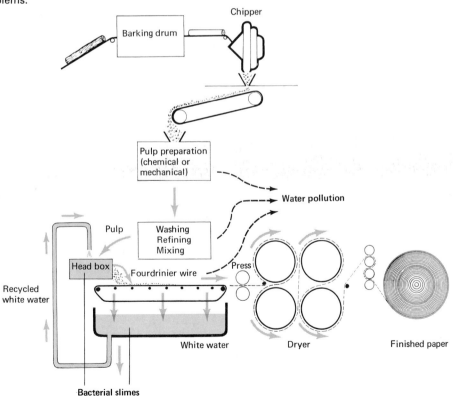

The most critical microbiological problems arise in the white water system. The pulp fibers provide an excellent medium for the growth of many microorganisms, and such coatings as starch or gelatin that are added to some stocks are excellent microbial foods. Since white water is recycled many times, there is ample opportunity for microbial buildup. Such contamination is usually seen as the accumulation on surfaces of a layer of slime, some of which gets incorporated into the paper. The slime itself derives from capsules or slime layers produced by bacteria. These most commonly are *Enterobacter aerogenes* and members of the genera *Pseudomonas, Flavobacterium, Achromobacter,* and *Bacillus.* Fungi and yeasts are also often present in the microbial slimes, including *Candida krusei, Pullularia pullulans, Oospora lactis, Cladosporium herbarium,* and *Aspergillus fumigatus.* The formation of slime can cause breaks in the paper sheet as it is being produced, or diminished overall paper quality.

In most paper-making systems it is essential to add bacterial agents at strategic points to control slime buildup. The most common of these agents, called *slimicides,* are organic mercury compounds, such as phenyl mercuric acetate, and chlorinated phenols, such as tri-chlorophenol. These compounds are usually added periodically at doses calculated to give an inhibitory concentration when fully mixed with the white water system.

Two serious problems in the use of slimicides should be mentioned. First, a small amount of the slimicide becomes incorporated into the fabric of the paper, and since some papers are used as food wrappers and others come into contact with humans in other ways, the long-term low-level toxicity of these slimicides should be known. Second, it is impossible to save all the white water; some does enter sewage systems and can carry slimicides into natural waters. The concentration of mercury downstream from mills using organic mercurials can build up with time, and levels toxic for humans or animals may be reached. Therefore, care must be taken to ensure that organic mercurials do not escape from the mill.

Another microbial problem in the mill is the deterioration of the woolen felts used in some stages of the process. Wool is an animal protein that is utilized by a variety of bacteria and fungi. At the wet end of the machine, microbial attack on felts often occurs, thus degrading the quality of the felts. No chemical treatment is completely satisfactory for controlling the microbial degradation of felts. The best solution is to replace wool with a synthetic fiber, provided one is available that has the proper felting characteristics for the paper type desired. Otherwise, the only solution is to replace the felt when it begins to deteriorate.

Paper that is used for wrapping food should be free of harmful bacteria. The heat used in drying the paper generally reduces the residual bacteria in the paper to an insignificant level. Usually the only bacteria found in paper are spore-forming organisms; coliform and pathogenic bacteria have never been found. For papers and paper-board to be used in the food industry, bacterial counts can be made by homogenizing a sample in sterile water with a high-speed blender, then performing a standard plate count. The results are expressed as colony count per gram of sample counted. Although no absolute bacterial standard for paper can be devised, it has been suggested that counts should not exceed 250 per gram.

14.17
Solid-waste disposal

Disposal of solid waste is one of the most complicated of urban problems. Solid waste from urban areas is a complex heterogeneous mixture consisting of garbage, paper, glass, ashes, metal, and miscellaneous products, and the proportions of ingredients vary considerably in different parts of the world. Another kind of solid waste is the undigested residue (sludge) from sewage-treatment operations. The most desirable means of handling solid waste is to recover the useful components from it, but recovery is not always possible, and in any event, some means of sanitary disposal must be sought for the remaining material. The aims of solid-waste disposal processes are to reduce the volume and weight of the refuse and to convert it into a less offensive form. Since a large fraction of municipal refuse is organic material, one means by which it is eliminated is microbial action. Microorganisms oxidize the organic components and convert them to carbon dioxide, water, and other simple substances that either go off into the air or drain relatively harmlessly into the soil or water. When organic refuse is disposed of in the environment in any way, its decomposition by microorganisms inevitably occurs. The goal of solid-waste disposal methods is to encourage decomposition and to arrange for it to occur in such ways that it will not be a public nuisance.

The traditional disposal method is dumping, or *tipping*, in which material is discarded loosely and at random into a convenient gully or ravine. Not only is dumping wasteful of space, since compaction does not occur, but it constitutes a health hazard, since rats, flies, and other vermin have access to the dumped materials. Further, spontaneous combustion may occur in the refuse, resulting in a burning dump, which is a source of air pollution. Finally, water percolating through the dump leaches out minerals, which can be carried to nearby streams and may be a source of water pollution.

The modern dump is called a *sanitary landfill*, or *controlled tip*. In a sanitary landfill (Figure 14.23), the refuse is spread in thin layers, and each layer is compacted by machine before the next is spread. When about 6 to 10 feet (2 to 3 meters) of refuse has been laid down, it is covered by a thin layer of clean earth that is also compacted. At the end of each day, the landfill is sealed with a thin layer of earth, thus preventing leaching into surrounding streams.

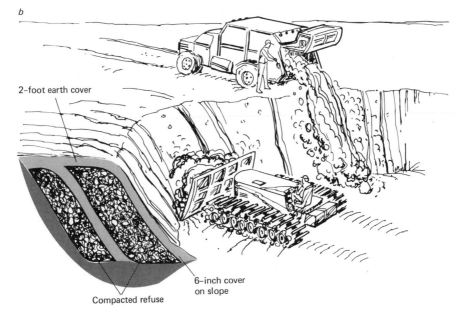

FIGURE 14.23
Sanitary landfill methods: *a* in level areas; *b* on slopes.

When no more space is available at the site, the landfill is sealed with 2 or 3 feet of compacted earth. A properly constructed sanitary landfill should have no burning and should present no problems from odors, flies, rats, or water pollution.

Microbiological activities in sanitary landfill consist of both aerobic and anaerobic processes. Initially, aerobic bacteria and fungi develop, but because of the compacted nature of the landfill, which inhibits free movement of air into the pile, the landfill soon becomes depleted of oxygen, and anaerobic microorganisms develop. The action of anaerobes accounts for the bulk of the decomposition of organic matter in a landfill. The processes are basically similar to those occurring in two other anaerobic systems: the anaerobic sludge digestor and the rumen. Cellulose and other insoluble organic compounds are digested, and the digestion products are fermented with the production of methane and carbon dioxide. Methane slowly escapes from the fill, but the rate of escape is usually not rapid; however, if a building is constructed on an old fill site (as is sometimes done), the building may trap methane, allowing buildup of an explosive concentration. Carbon dioxide produced in the landfill will dissolve in the water percolating through the fill, making it weakly acidic and able to dissolve rocks, thus increasing the mineral content of the water, although usually not enough to make the water a potential source of pollution.

The rate of decomposition is not fast. Within several months, substantial changes in the organic materials have occurred, but even after as much as 2 years, the decomposition process is not complete. As the organic materials slowly decompose, the landfill shrinks in volume, and this indicates one difficulty in attempting to use landfills as building sites. Landfills do make excellent sites for parks, however, once they are planted with desirable vegetation.

The heat produced as a result of microbial action (*thermogenesis*) causes an increase in the temperature of the landfill. After several weeks, temperatures in excess of 65°C develop deep within the fill, so that during the main stage of decomposition, thermophilic microorganisms make the greatest contribution to the process. One consequence of this heating process is that pathogenic microorganisms that might be present in the refuse are killed, thus sanitizing the landfill. However, it is important that temperatures do not rise too high, since spontaneous combustion might then occur. Temperature rise is determined by the amount of organic matter and by the degree of compaction. If the organic content of the landfill is controlled by proper layering and if the landfill is well-compacted, the temperature rise will remain in the desired range.

There are several disadvantages of the sanitary landfill method. Suitable sites for landfills may be scarce or difficult to purchase, and long-distance hauling of refuse to landfill sites may be prohibitively expensive. Also, winter conditions of ice and snow may make year-round operation of the landfill difficult. Finally, the needed soil cover material may not be available nearby. An alternative method of disposing of municipal refuse is *composting*, which may eliminate some of these disadvantages. Composting is a microbiological process in which organic materials are converted into *humus*, an organic-rich material that has good water-holding capacity and can serve as a soil builder. The main difference between a compost heap and a landfill is that the former is operated as an aerobic process, while as we have seen the latter is primarily anaerobic. Further, in the compost heap, the product formed is used as a soil conditioner; in the landfill, the organic materials are not used again.

Although the operation of a small residential compost heap is relatively simple, operation of a large composting system for handling municipal refuse is expensive and complicated. Perhaps because of this, widespread use of composting for stabilization of municipal refuse has not been done, although a number of reasonably large installations have been operated as pilot projects.

Summary

In this chapter, we learned some of the microbiological principles involved in the handling and purification of water and waste. Vast amounts of pure fresh water are needed for domestic purposes, and care must be taken to ensure that this water is free of harmful microorganisms. Water can be the vehicle for the transmittal of a number of pathogenic organisms, including bacteria, viruses, and protozoa. The most serious pathogens transmitted by water are *Salmonella typhi*, the causal agent of typhoid fever, and *Vibrio cholera*, the causal agent of cholera. Other agents include *Shigella*, *Salmonella* species other than *S. typhi*, the viruses causing polio and infectious hepatitis, and the protozoa causing amoebic dysentery and giardiasis.

Purification of water is carried out by sedimentation, filtration through sand, and coagulation. Chlorination is the most common method of ensuring the microbiological safety of a water supply. Microorganisms are good indicators of water pollution, and the quality of water is usually defined in terms of its microbial load. The coliform count, the most commonly used indicator of water pollution, measures the concentration of fecal bacteria in the water supply. The federal Drinking Water Regulations specify the coliform count for drinking water. The regulations also specify the sampling protocol that cities of various sizes must follow in order to ensure that the water supplied throughout the city is safe.

Pollution of water by domestic sewage and industrial wastes can result in serious degradation of water quality. One measure of organic pollution of water is the biochemical oxygen demand (B.O.D.), which measures the rate at which oxygen is consumed by the aquatic system. To remove B.O.D. and harmful microorganisms from sewage, a variety of treatment methods are employed, both biological and nonbiological. One biological method is the anaerobic digestion process, in which cellulose and other fibrous organic materials are digested and fermented by anaerobic bacteria with the production of methane. Other processes involve oxidation by aerobic bacteria of soluble organic materials present in sewage. Two aerobic systems are the trickling filter, in which the liquid is allowed to trickle over a bed of rocks upon which the microorganisms grow, and the activated-sludge process, in which the liquid is aerated in large tanks. The end product of sewage treatment systems is a liquid low in B.O.D. and lacking pathogenic microorganisms; it can be discharged into a river or lake without causing harm. An insoluble undigestible residue, the sludge, is also produced, which is usually disposed of by burial. Another form of waste generated in large amounts by society is solid waste, which consists of a mixture of garbage, paper, glass, ashes, metal, and miscellaneous products. Solid waste is best disposed of in sanitary landfills, in which the refuse is spread in compacted layers that are covered by clean earth. Microorganisms develop within the landfill and degrade much of the organic matter. When landfill sites are filled, they can often be converted into parks or other useful installations. We thus see that in all stages of the handling and disposal of sewage and solid waste, microorganisms play important roles.

Study questions

1 Both surface and ground waters are used as sources of drinking water. Contrast these two types of sources from the point of view of availability, likelihood of pollution, and degree and type of treatment necessary to make them potable.
2 What are the most important water-borne diseases? Identify the pathogen causing each. Group these as bacterial, viral, and protozoal.
3 Define coliforms. How does the coliform group differ from the organism *Escherichia coli*?
4 Describe briefly how the standard coliform test for drinking water is performed by the MPN procedure.
5 Describe briefly the MF procedure for coliforms in drinking water.
6 Compare the MPN and MF procedures, and give the advantages and disadvantages of each for coliform tests for drinking water.
7 What is the standard plate count? How is it performed? What value does it have in assessing the quality of drinking water?
8 Outline briefly the steps in the purification of water intended for domestic use. Discuss briefly the function of each step.
9 As an aftermath of floods, earthquakes, or other disasters, a city's water supply may be interrupted. What means of purification would you use to ensure a supply of safe water for drinking?
10 The presence of coliforms is generally taken as evidence of fecal contamination. Why? If coliforms are present, does this necessarily indicate *human* fecal contamination?
11 Discuss the federal Drinking Water Regulations from the viewpoints of sampling, laboratory evaluation, limits of contamination, and notification procedures.
12 The federal Drinking Water Regulations permit substitution of residual-chlorine determinations for coliform counts. Under what conditions are substitutions permitted, and what is the rationale for this alternative?
13 How may microorganisms be involved in the corrosion or clogging of pipelines?
14 Pathogenic organisms may be harbored in pipelines under certain conditions. Why? Is this of any health significance?
15 Sources of water pollution may be natural, domestic, or industrial. Compare the types of pollutants to be expected from each source. Which source would generate pollutants the most difficult to treat, and why?
16 Discuss the nature, measurement, and significance of B.O.D.
17 Describe the events that occur downstream from a raw-sewage outfall.

18 What is meant by self-purification of a stream? Why is this not a reliable way of handling domestic or industrial wastes?

19 Define *fecal coliform*. Contrast with total coliform. How are fecal coliform tests used in water pollution investigations?

20 Domestic sewage in populated areas is usually treated by a multistep process. Outline a typical treatment process, describing at each step what is occurring microbiologically and in what ways the entering sewage is changed.

21 What is sludge digestion? Is it aerobic or anaerobic? Describe the steps in sludge digestion.

22 Compare and contrast the trickling filter and the activated-sludge plant for the aerobic treatment of sewage.

23 What types of problems may arise if nonbiodegradable compounds are present in sewage?

24 Compare the efficiency of sewage treatment by activated-sludge and septic-tank systems. Based on this comparison, what is your opinion of the value of septic tanks in built-up (suburban) areas?

25 Try to find out what sewage-treatment methods are used in the area where you live. Does your area use any advanced treatment methods? Where is the treated sewage in your area discharged?

26 Many sewage-treatment plants are grossly overloaded and do not operate properly, thus discharging poorly treated sewage. Try to find out how well the sewage-treatment plant in your area is operating. Is it meeting the standards of your state?

27 What causes an algal bloom? How can algal growth be controlled in swimming pools? in reservoirs? in large bodies of water?

28 What is eutrophication? How can it be prevented?

29 It has been said that solid waste is our only growing resource. What is meant by this? The most widely used disposal method for solid waste today is the sanitary landfill. Describe the construction of such a landfill, and indicate the types of microbiological processes that are important within the fill.

Suggested readings

American Public Health Association, *Standard methods for the examination of water and wastewater*, 14th ed., American Public Health Association, Washington, D.C., 1976. This volume provides detailed procedures for the examination of drinking water and sewage for coliforms and other microbial parameters of water. It also has brief sections describing the principles behind the methods.

Craun, G. F., and L. J. McCabe, "Review of the causes of waterborne-disease outbreaks," *J. Am. Water Works Association*, **65**, 74–84, 1973. A survey of the water-borne disease outbreaks that have occurred in the United States in recent years, with brief discussion of the causes.

Geldreich, E. E., *Sanitary significance of fecal coliforms in the environment*, U.S. Department of Interior, Federal Water Pollution Control Administration, Publication WP-20-3, 1966. Useful guide for sewage plant operators and others on the definition, assay, and significance of fecal coliforms in the environment.

Mitchell, R., (ed.), *Water pollution microbiology*, John Wiley & Sons, New York, 1972. Detailed discussions of various aspects of water-pollution microbiology. Excellent chapters on the coliform test and on bacterial and viral pathogens in water.

National Research Council/National Academy of Sciences, *Drinking water and health*, Washington, D.C., 1977. An extensive and authoritative discussion of drinking water contaminants and standards. Chapter 3 deals with microbiology.

"U.S. Drinking Water Regulations," *Federal register*, Vol. 40, no. 248, December 24, 1975. Available in the document room of most libraries. The first nationally applied drinking water standards in the United States. In addition to a legal document promulgating the standards, this publication has appendices presenting some of the rationales for the regulations.

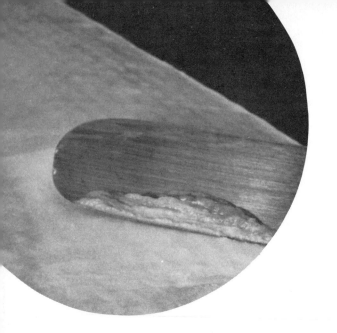

Microorganisms have both beneficial and harmful effects on foods, and the study of microbiology encompasses both these aspects. Beneficial roles of microorganisms include their use in preparing foods such as sausage, cheese, and pickles. On the other hand, microorganisms are responsible for some of the most serious kinds of food poisonings and toxicities and also cause spoilage of a wide variety of food and dairy products. It is convenient to divide our discussion into separate sections dealing with the general principles of food spoilage, including what pathogens are likely to be found and how they are detected, and the various methods of food preservation. There are also special sections on dairy and meat products because of the unique problems relating to preservation and microbial control in these highly perishable foods.

Food spoilage is one of the most serious economic problems of the food-processing industry. Foods are subject to attack by microorganisms in a variety of ways, and such attack is usually harmful to the quality of the food. The physical and chemical characteristics of the food determine its degree of susceptibility to microbial attack, the kinds of microorganisms that will infect it, and the kinds of spoilage that will result. Food-processing techniques have been developed that make it possible to reduce or prevent microbial

15.1
Food spoilage

spoilage and thus ensure long shelf life for the food product.

A food is considered spoiled when it is in such a state that a discriminating consumer will not eat it. Spoilage is not an absolute characteristic of a food, since custom plays an important role in considerations of whether a food is desirable to eat. For instance, although soured milk may be considered spoiled, buttermilk is not considered spoiled even though it is sour. An important distinction must be made, therefore, between foods that are not eaten because they taste bad and foods that are not eaten because they are poisonous or harmful. Bad taste is often a matter of custom and experience, and something considered bad tasting in one country or culture may be considered a delicacy elsewhere. Toxicity, on the other hand, is a reflection of the presence in the food of some poison or toxin that is harmful to humans.

Microbiologists are involved in food-spoilage problems in the following ways: (1) They identify types of microorganisms causing food spoilage, assess the conditions under which they develop, evaluate the potential harm to humans of different food spoilage organisms, and predict the circumstances under which spoilage will develop. (2) They develop and perfect methods for processing foods so that they can be stored safely. (3) They check stored foods for microbial spoilage and make recommendations for discarding unsuitable foods. (4) They evaluate food-processing plants from a microbiological viewpoint and recommend modifications to reduce microbial contamination and spoilage.

Spoilage may arise either through microbial action or through nonmicrobiological processes. In the latter category are chemical or biochemical changes that can occur in food during storage, such as browning, drying, or development of rancidity, some of which may be mistaken for microbial processes. Foods vary in the ease with which they undergo microbial spoilage:

1 *Highly perishable foods* are those that must be processed and stored carefully if they are to be preserved. These include meats, fish, poultry, eggs, milk, and most fruits and vegetables; in short, the basic daily foods in many parts of the world.
2 *Semiperishable foods* can be kept for fairly long periods of time if they are stored well. Examples include potatoes, some apples, and nuts.
3 *Stable* or *nonperishable foods* can be stored for long periods of time without difficulty. These foods, which are stable because they are low in moisture content, include sugar, flour, rice, and dry beans. In addition to these naturally stable foods, many perishable foods can be converted to stable ones by drying. Thus, dried fruits, meats, and vegetables are stable, although their flavor and texture may be quite different from that of the fresh products.

At any step along the route from farmer to consumer, foods can become infected with microorgnisms. At each step, the kinds of infecting organisms may be different, and the consequences of such infection may also be quite different. At the farm, foods can become infected by soil organisms, many of which are spore formers and hence resistant to heat, or by organisms from other sources in the farm environment. Microbial growth may actually occur in the product during storage or handling, so that the product is altered even before it reaches the processing stage. In the food-processing plant, infection may come from the water used in processing, from contaminated equipment, from the packages in which the final product is placed, or from the employees handling the food. After processing, the food may become infected during storage in warehouse, store, or home, especially if the product has been improperly packaged and develops leaks. Each link in the chain from farmer to consumer must be carefully controlled if the consumer is to be provided with a product that is safe, clean, and tasty.

The factor of foremost significance is the *moisture content* of the food. If the food is sufficiently dry, microbial growth is impossible. However, different microorganisms are able to grow at different moisture levels. Although most microorganisms grow only in moist foods, certain fungi and yeasts, called *xerophilic*, are able to grow at very low moisture levels. Thus, in drier foods, spoilage would be expected to be caused primarily by the xerophilic forms. Different foods vary in the minimum moisture content above which spoilage will occur. The minimum level in dry milk is 8 percent water; in dried whole eggs 10 to 11 percent; in flour 13 to 15 percent; in nonfat dry milk 15 percent; in dehydrated fat-free meat 15 percent; in dry beans 15 percent; in other dehydrated vegetables 15 to 20 percent; and in dehydrated fruits 18 to 25 percent. At moisture levels below those values, growth of even xerophilic organisms is not generally possible. A knowledge of these moisture levels permits the food processor to select the appropriate moisture conditions for storage (Figure 15.1). Interestingly, a liquid food high in sugar or salt (say, a syrup or brine) resembles a dry food even though it is high in water, since the sugar or salt causes water to pass out of the microbial cell, which then becomes dehydrated. Thus spoilage of syrups is caused by the same kinds of xerophilic fungi that spoil partially dry foods.

A second major factor affecting the stability of a food is its *physical structure*. A food that consists of small pieces is much more susceptible to spoilage than a food that consists of large pieces, since in the former much more surface is exposed and available for microbial growth. Further, some foods have protective coverings such as shells, waxy coatings, or scales that prevent or decrease the extent of microbial invasion. Additionally, when foods are cut, diced,

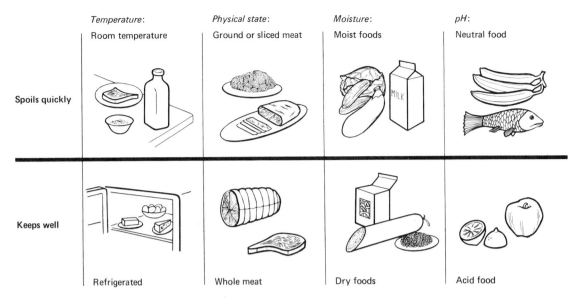

	Temperature:	Physical state:	Moisture:	pH:
Spoils quickly	Room temperature	Ground or sliced meat	Moist foods	Neutral food
Keeps well	Refrigerated	Whole meat	Dry foods	Acid food

FIGURE 15.1
Factors affecting the keeping qualities of foods.

chopped, or ground, moisture is released and microorganisms are spread more widely in this moisture, greatly increasing the susceptibility of the food to spoilage. For instance, ground beef spoils much more rapidly than the whole beef chunks from which it is prepared.

A third major factor affecting food stability is the *acidity* of the food, its pH. We discussed acidity and pH in Section 3.2. Foods vary widely in pH, although most are neutral or acidic. Microorganisms vary in their ability to grow under acidic conditions: many bacteria are unable to grow at pH values below 5, whereas a few, such as some of the bacilli and the lactic acid bacteria, can grow as low as pH 4. Below pH 4, only a few rare bacteria are able to grow, and these are not often found in foods. However, at these acid values, many molds and yeasts grow quite well, and it is these organisms that are usually responsible for spoilage of acidic foods. From a knowledge of a food's pH, a food processor can assess the extent and kind of spoilage and proceed accordingly.

A number of *chemical properties* of foods affect microbial growth. Foods, of course, provide the nutrients for microbial growth, just as they provide the nutrients for the human beings for which they are intended. However, foods vary widely in nutritional qualities for both man and microbe. Meats, for instance, are high in protein and fat, while many vegetables are high in starch and cellulose, and fruits may be high in sugars. In meat, microorganisms that produce enzymes capable of breaking down proteins predominate; such organisms are called *proteolytic*. In the softening or rotting of vegetables and fruits, microorganisms able to break down the pectin materials that hold the plant cells together predominate.

In addition to providing energy for microbial growth, foods

must provide the other essential nutrients: nitrogen, sulfur, phosphorus, and the rest. Nitrogen and sulfur come mainly from amino acids of the food protein, and phosphorus comes from the food nucleic acids. Foods are also often rich in vitamins, so that vitamin-requiring microorganisms usually have no trouble growing, although many food-spoilage microorganisms can synthesize their own vitamins as well, if necessary.

The conditions of storage greatly influence the tendency of foods to become spoiled. Of these conditions, the most crucial is *temperature* (see Figure 15.2). In general, the lower the temperature of storage, the less rapid the spoilage rate. Much depends upon climate: in cold climates, natural temperatures may be low enough so that foods can be safely stored without refrigeration, at least during the cold part of the year; but in tropical areas, unrefrigerated food may spoil within hours or days, and only if artificial refrigeration is available can long storage be ensured. In foods stored nonfrozen at refrigeration temperatures, microbial growth can occur, although slowly. The organisms in such foods are psychrophiles and include various genera of bacteria and fungi. Some of these organisms cause changes in flavor and odor of foods, and are therefore considered spoilage organisms. Foods most susceptible to such spoilage are the

CONDITIONS OF STORAGE

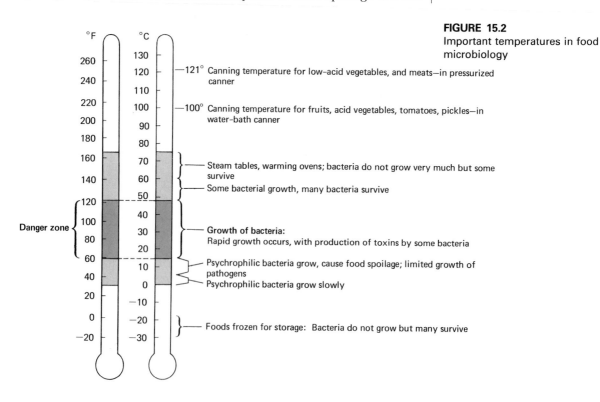

FIGURE 15.2
Important temperatures in food microbiology

—121° Canning temperature for low-acid vegetables, and meats—in pressurized canner

—100° Canning temperature for fruits, acid vegetables, tomatoes, pickles—in water-bath canner

— Steam tables, warming ovens; bacteria do not grow very much but some survive
— Some bacterial growth, many bacteria survive

— Growth of bacteria:
Rapid growth occurs, with production of toxins by some bacteria

— Psychrophilic bacteria grow, cause food spoilage; limited growth of pathogens
— Psychrophilic bacteria grow slowly

— Foods frozen for storage: Bacteria do not grow but many survive

Danger zone

highly perishable ones, such as meat and milk. Because of the presence and activity of psychrophiles, meat and milk cannot usually be stored at refrigeration temperatures for longer than a week or so.

The presence or absence of *oxygen* in the stored food affects the extent and type of microbial growth. All filamentous fungi require oxygen for growth, as do most yeasts and many bacteria. On the other hand, some bacteria are obligate anaerobes, growing only when oxygen is absent. Canned foods are usually completely anaerobic and hence are usually spoiled only by anaerobes. Such foods as jams and jellies are kept from developing mold growth by covering them with a layer of paraffin, which keeps out oxygen, or by sealing them tightly to avoid entrance of oxygen.

CONSEQUENCES OF MICROBIAL GROWTH IN FOODS

Extensive microbial growth is usually not required to render a food spoiled and worthless. In fact, the microbial growth need not even be visible. The kind of spoilage that develops depends to a large degree on the food itself, since the nature, texture, and chemical composition of the food will determine the kinds of organisms that will develop and therefore the kinds of alterations that will take place in the food. Bread becomes moldy as a result of visible growth of filamentous fungi, which are unsightly and bad tasting, or it becomes ropy because of growth of slime-forming *Bacillus* species, which break down the starch of the bread and form ropy strands that can be seen when the bread is broken. Vegetables are subject to soft rots, caused by *Erwinia* species that decompose the pectic substances holding the plant cells together; they are also subject to mold spots and rots, caused by growth of fungi on the surface of the food. Meat develops surface shines, owing to growth of slime-forming bacteria and yeasts on the surface of the meat; color changes, from red to shades of green, brown, or gray, due to microbial production of peroxides or hydrogen sulfide that react with the red pigment in meat; rancidity, through attack by microbes on the fats; taint, due to production of off-flavor in the meat; and putrefaction, caused by breakdown of the protein of meat and the release of bad-smelling nitrogen compounds such as ammonia, amines, indole, and skatole. Bacterial growth in the white or yolk of whole fresh eggs can lead to the production of green, black, or pink rots and bad odors, while in other cases, the appearance of the egg may be unchanged, but the odor will indicate spoilage. Milk may become sour through production of lactic acid by lactic acid bacteria; or it may become ropy, through growth of slime-forming bacteria. It may develop flavor changes, caused by growth of a wide variety of bacteria, or it may undergo proteolysis, caused by the breakdown by bacteria of the protein of milk, generally accompanied by an unpleasant smell (rather than the appetizing sour odor of most fermented milk products).

The growth of microbes in foods, although rendering the food unsightly and possibly bad tasting, does not necessarily render it unfit for human consumption. However, several microorganisms produce powerful toxins that can cause serious illness or death, and when one of these toxins is present, the food is highly dangerous. Because of the severity of the effects of these microbial toxins, any food showing obvious microbial growth should be discarded, even if there is no direct reason to believe that the food is toxic. Unfortunately, not all foods that contain microbial toxins show obvious spoilage, so that food poisoning can develop even from seemingly normal foods.

Botulism is the most severe type of food poisoning; it is often fatal and occurs following the consumption of food containing the toxin produced by the anaerobic bacterium *Clostridium botulinum*. Although it is considered a rare disease (only 0.03 percent of all outbreaks of food-borne illness reported to the Center for Disease Control in Atlanta, Georgia), botulism remains a threat in the United States. During the period 1899 to 1969, 659 outbreaks comprising 1,696 cases were reported and resulted in 959 fatalities. The fatality ratio was very high during 1899 to 1949 (above 60 percent); it decreased to less than 40 percent for the period 1950 to 1959 and to less than 30 percent for 1960 to 1969. This decline is undoubtedly a result of improvements in diagnosis, availability of *Clostridium botulinum* antitoxins, and increased skill in caring for patients.

C. botulinum normally lives in soil or water, but its spores may contaminate raw foods before harvest or slaughter. If the foods are properly processed so that the spores are killed, no problem arises; but if viable spores are present, they may initiate growth, and even a small amount of the resultant neurotoxin can render the food poisonous. Outbreaks of botulism have most frequently been related to home-processed foods, since the processing is sometimes not well controlled. From 1899 to 1969, 72 percent of the cases were caused by home-processed food, 19 percent were of unknown origin, and only 9 percent were caused by commercially processed food. The majority of outbreaks (60 percent) were caused by contaminated vegetables, 13 percent were traced to preserved fruits, and 12 percent to contaminated fish or fish products.

The toxin itself is destroyed by heat, so that properly cooked food should be harmless, even if it did originally contain the toxin. Most cases of botulism occur as a result of eating foods that are not cooked after processing. Canned vegetables and beans are often used without cooking in making cold salads. Similarly, smoked fish and meat and most of the vacuum-packed sliced meats are often eaten directly, without heating. If these products contain the botulinum toxin, then ingestion of even a small amount results in this severe and highly dangerous type of food poisoning.

15.2
Pathogens in foods

FOOD POISONINGS AND TOXINS

Another *Clostridium* species, *C. perfringens*, causes a milder form of food poisoning, which is not usually fatal. This organism produces an enterotoxin during sporulation in the intestine; the enterotoxin acts on the gastrointestinal tract to cause abdominal cramps and diarrhea. The organism normally exists in soil and sewage and in the intestinal tracts of humans and animals. There are some strains that have more heat-resistant spores, and it is these that may cause food poisoning. The sequence of events that can lead to *C. perfringens* food poisoning is shown in Figure 15.3. The food may

FIGURE 15.3

Sequence of events that may lead to *Clostridium perfringens* food poisoning.

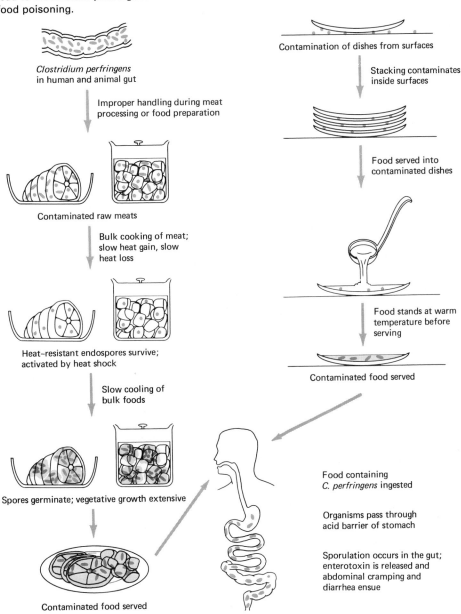

Clostridium perfringens in human and animal gut

Improper handling during meat processing or food preparation

Contaminated raw meats

Bulk cooking of meat; slow heat gain, slow heat loss

Heat–resistant endospores survive; activated by heat shock

Slow cooling of bulk foods

Spores germinate; vegetative growth extensive

Contaminated food served

Contamination of dishes from surfaces

Stacking contaminates inside surfaces

Food served into contaminated dishes

Food stands at warm temperature before serving

Contaminated food served

Food containing *C. perfringens* ingested

Organisms pass through acid barrier of stomach

Sporulation occurs in the gut; enterotoxin is released and abdominal cramping and diarrhea ensue be infected before it is processed, with some spores surviving to vegetate and grow later. Or noncontaminated food may come into contact with contaminated dishes, utensils, or food workers; the organisms thus introduced multiply before the food is eaten. The organism is quite commonly found in our cooked and uncooked foods, especially raw meat, poultry, and fish, in most dust samples, and in many water supplies. However, ingestion of a large dose of *C. perfringens* (10^8 to 10^9 organisms) is required for food poisoning to occur. Thus, although small numbers of the organism may frequently be consumed, with proper food processing and handling the organisms do not multiply to the point where they constitute an infective dose. A transient carrier state follows recovery and lasts for some weeks.

Staphylococcal food poisoning is the most common type of food poisoning and is caused by varieties of *Staphylococcus aureus*. This organism produces an enterotoxin that is released into the surrounding medium or food; if food containing the toxin is ingested, severe reactions are observed within a few hours, including nausea with vomiting and diarrhea. The kinds of foods most commonly involved in this type of food poisoning are custard- and cream-filled baked goods, poultry, meat and meat products, gravies, egg and meat salads, puddings, and creamy salad dressings. If such foods are kept refrigerated after preparation, they remain relatively safe, as the *Staphylococcus* is unable to grow at low temperatures. In many cases, however, foods of this type are kept warm for a period of hours after preparation, such as in warm kitchens or outdoors at

FIGURE 15.4
Development of staphylococcal food poisoning.

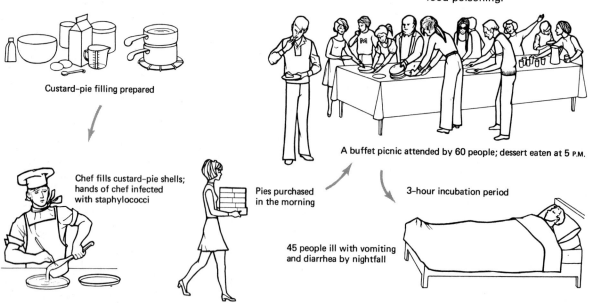

Custard-pie filling prepared

Chef fills custard-pie shells; hands of chef infected with staphylococci

Pies purchased in the morning

A buffet picnic attended by 60 people; dessert eaten at 5 P.M.

3-hour incubation period

45 people ill with vomiting and diarrhea by nightfall

FOOD MICROBIOLOGY

summer picnics. Under these conditions, the *Staphylococcus*, which might have entered the food from a food handler during preparation, grows and produces enterotoxin (Figure 15.4). Many of the foods involved in staphylococcal food poisoning are not cooked again before eating, but even if they are, this toxin is relatively stable to heat and may remain active.

Staphylococcal food poisoning can be prevented by careful sanitation methods so that the food does not become inoculated, by storage of the food at low temperatures to prevent staphylococcal growth, and by the discarding of foods stored for any period of hours at warm temperatures. Staphylococcal food poisoning often occurs in small epidemics among people eating at a common location, such as a banquet, dormitory, or large picnic. The reason for this is often that when foods are prepared in large quantities, sufficient refrigeration is usually not available to keep them all cold, and a portion must therefore be left at warm indoor or outdoor temperatures.

SALMONELLA FOOD INFECTION

Although sometimes called a food poisoning, gastrointestinal disease caused by food-borne *Salmonella* is in reality a food infection because symptoms arise only after the pathogen grows in the intestine. Virtually all species of *Salmonella* are pathogenic for humans: one, *S. typhi*, causes the serious human disease typhoid fever, and a small number of other species cause food-borne gastroenteritis. The ultimate source of the food-borne salmonellas is humans or warm-blooded animals. The organism reaches food by contamination from food handlers; or in the case of such foods as eggs or meat, the animal that produced the food may be the source of contamination (Figure 15.5). The foods most commonly involved are meats and meat products (such as meat pies, sausage, cured meats), poultry, eggs, and milk and milk products. If the food is properly cooked, the organism is killed and no problem arises, but many of these products are eaten uncooked or partially cooked. *Salmonella* causing food-borne gastroenteritis is often traced to products made with uncooked or slightly-cooked eggs, such as custards, cream cakes, meringues, pies, and eggnog. Previously cooked foods that have been warmed and held without refrigeration or canned foods held for a while after opening often support the growth of *Salmonella* if they have become contaminated by an infected food handler.

The organism must be present in the food in fairly large numbers if infection is to occur; this usually means that the organism must multiply in the food before it is eaten. Growth occurs most rapidly if the food has been stored under warm conditions, and hence infection most commonly occurs in tropical climates or in summer in temperate climates. The length of time after ingestion

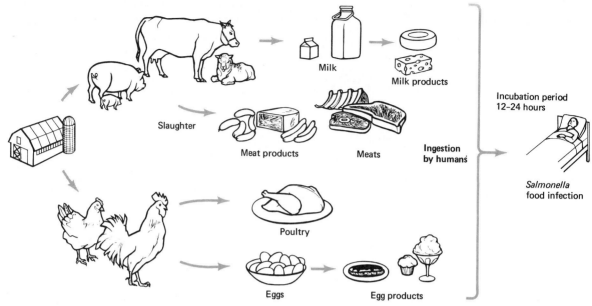

FIGURE 15.5
Spread of *Salmonella* from food animals to humans.

before symptoms appear is usually 12 to 24 hours, which is longer than the time required for *Staphylococcus* food poisoning to appear. This longer incubation time is due to the fact that *Salmonella* must multiply in the intestine before symptoms develop. The principal symptoms are nausea, vomiting, abdominal pain, and diarrhea, usually preceded by headache or chills. The disease is rarely fatal, and recovery usually occurs, even without treatment, after 2 to 3 days. Proper food sanitation practices minimize the incidence of this food-borne disease.

INVESTIGATING AN OUTBREAK

Once a diagnosis or tentative diagnosis of food-borne disease has been made, the cause of the outbreak must be determined. Although some outbreaks are relatively easy to figure out, others are extremely difficult. The purpose of an investigation is to determine what conditions led to the outbreak and, more importantly, to gather information that will help prevent a recurrence. All facts must be collected and studied as soon as possible after the outbreak occurs. The methods followed are those of the classic epidemic, and if the outbreak is severe, a trained epidemiologist from the state or local health department should head the investigation. Case histories should be obtained. The suspect food may have been widely distributed in homes and restaurants via a common supplier, or it may have been quite localized, as at a banquet or picnic. If a particular meal is suspected, case histories should be taken from both those

TABLE 15.1
Food-specific attack rate
table.

Foods	No. of persons who ate specific foods				No. of persons who did not eat specific foods				Difference in percent
	Ill	Not ill	Total	Percent ill	Ill	Not ill	Total	Percent ill	
Braised beef	74	17	91	81	2	9	11	18	+63
Peas	48	20	68	71	28	6	34	82	−11
Cabbage salad	36	12	48	75	40	14	54	74	+1
Buttered biscuits	46	12	58	79	30	14	44	68	+11
Peaches	62	22	84	73	14	4	18	78	−5
Milk	60	16	76	79	16	10	26	62	+17

From *Procedures to investigate foodborne illness*, 3rd ed., International Association of Milk, Food, and Environmental Sanitarians, Inc., Ames, Iowa., 1976.

who became ill and those who did not. Careful inspection of all food preparation methods should be made, and full information as to food storage, health of food workers, menu, and so on should be obtained. Samples of leftover foods should be taken for microbiological analysis to corroborate the findings. If an outbreak occurs in a group of people attending a particular meal, case histories should be collected from a large number of people, recording what each ate or did not eat. Then an attack rate table can be prepared (Table 15.1). From such a table, the food responsible for the outbreak may be determined. The food that caused the problem shows a higher attack rate in persons who ate the food than in those who did not. In the case shown in Table 15.1, the attack rate for those eating braised beef was 81 percent whereas for those not eating it, the rate was 18 percent. The difference in attack rates between eaters and noneaters is the important point. For other noncontaminated foods, eaters and noneaters had about the same rates, since it was only from eating the braised beef that they became ill.

15.3 Assessing microbial content of foods

All fresh foods have some viable microbes present. The purpose of assay methods is to detect evidence of abnormal microbial growth in foods or to detect the presence of specific organisms of public-health concern, such as *Salmonella*, *Staphylococcus*, or *Clostridium botulinum*. With nonliquid foods, preliminary treatment is usually required to suspend in a fluid medium those microorganisms embedded or entrapped within the food. The most suitable method for treatment is high-speed blending (Figure 15.6). (If a blender is not available, homogenization in a sterile mortar and pestle can be done.) The blender should operate at not less than 8,000 rpm (revolutions per minute) and not more than 45,000 rpm. The blender cup should be autoclavable. Examination of the food should be

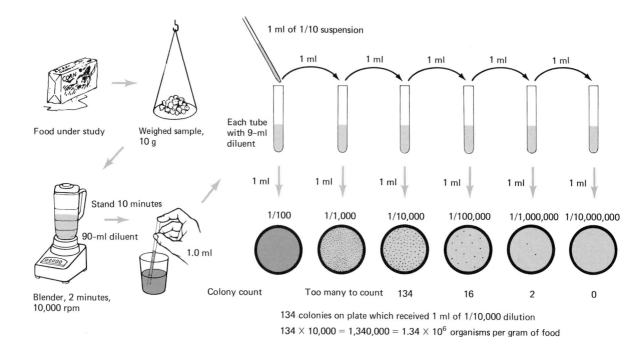

1 ml of 1/10 suspension

Food under study

Weighed sample, 10 g

Each tube with 9-ml diluent

1 ml   1 ml   1 ml   1 ml   1 ml

Stand 10 minutes

90-ml diluent

1.0 ml

Blender, 2 minutes, 10,000 rpm

1 ml   1 ml   1 ml   1 ml   1 ml   1 ml

1/100   1/1,000   1/10,000   1/100,000   1/1,000,000   1/10,000,000

Colony count   Too many to count   134   16   2   0

134 colonies on plate which received 1 ml of 1/10,000 dilution

$134 \times 10,000 = 1,340,000 = 1.34 \times 10^6$ organisms per gram of food

FIGURE 15.6
Procedure for performing viable count of a food.

done as soon after sampling as possible, and if examination cannot begin within 1 hour of sampling, the food should be refrigerated. A frozen food should be thawed in its original container in a refrigerator and examined as soon as possible after thawing is complete.

Into the preweighed blender jar is placed at least 10 g (grams) of sample; precautions should be taken to be sure that this sample is representative of the food specimen. A volume of diluent nine times as large as the sample is added, thus giving an initial dilution of 1/10. The diluent used should be a fluid that does not cause harm to the microorganisms present, and a suitable diluent is peptone–water (1 g peptone per liter of water, pH adjusted to 7.0, and autoclaved). The blender should be operated at a speed sufficient to homogenize the sample and for a sufficient time to give a total of 15,000 to 20,000 revolutions. Thus, a blender operating at 10,000 rpm should be run for at least 1.5 to 2 minutes. The homogenate is then allowed to stand for 15 minutes at room temperature to permit rehydration of dry organisms. The contents of the blender jar are then mixed, and serial dilutions are prepared. From these dilutions, platings are made using a standard culture medium that will permit the growth of the common bacteria found in foods. The plates are incubated at appropriate temperatures: 0 to 5°C for psychrophiles, 30 to 35°C for mesophiles, and 55°C for thermophiles. For mesophiles and thermophiles, incubation should be carried out for 48 hours, whereas for psychrophiles, it should be longer: 1 to 2 weeks. For colony counting, plates are selected from those dilutions yielding 30 to 300 colonies per plate, and the viable count per gram of food is then calculated. For *Escherichia coli* enumeration, a MPN count, such as described in Section 14.3, should be performed using a medium selective for this bacterium. For *Salmonella*, several selective media are available, and tests for its presence are most commonly done on animal food products, such as raw meat, poultry, eggs, and powdered milk, since *Salmonella* from the animal may contaminate the food. For staphylococcal counts, a medium high in salt (either sodium chloride or lithium chloride) is used, since of the organisms present in foods, staphylococci are the only common ones resistant to salt. Since *Staphylococcus aureus* is responsible for one of the most common types of food poisoning, staphylococcal counts are of considerable importance. Special procedures used in the enumeration of microorganisms in milk and meat are discussed in Sections 15.5 and 15.7. Table 15.2 presents some suggested standards in effect in various states. It should be noted that these figures are recommended only; compliance is voluntary.

15.4
Food preservation

From a knowledge of the factors involved in food spoilage, as outlined in Section 15.3, effective measures for food preservation can be devised. It has been through the development and perfection of effective methods for food preservation that human civilization has been able to develop into the highly industrialized and urbanized system found today in most parts of the world. In such an industrialized system, very few people live close enough to the land

TABLE 15.2
Suggested microbiological guidelines and standards for selected foods

Food	Plate count no./gram	Coliforms no./gram	Comments[a]
Precooked frozen	100,000	10 (100)[b]	Some states require absence of *Staphylococcus, Salmonella, Shigella.*
Frozen pies	50,000	50	
Gelatin	3,000	10	
Raw meat	5,000,000	50	In some states, *E. coli* is tested
Cooked meat	1,000,000	10	for instead of coliforms in general.
Fresh frozen meat	5,000,000	50	
Ice cream and frozen dessert	50,000	20	Federal purchases
Ice cream and frozen dessert	100,000 to 500,000	no test	Most states
Ice cream and frozen dessert	no test	10	20 states

[a] The guidelines vary widely from one state to another; there are no federal standards.
[b] One state allows 100 coliforms.

FIGURE 15.7
Sun-drying of fruit. *Courtesy of the Dried Fruit Association of California.*

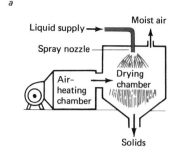

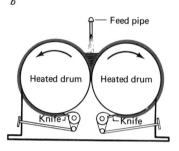

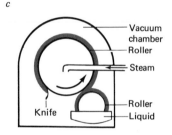

FIGURE 15.8
Drying methods for liquid food products: *a* spray-drying; *b* drum-drying; *c* vacuum-drying.

to be able to grow their own foods. Foods preserved in one way or another are thus essential. Fortunately, our knowledge of effective means of food preservation is highly advanced, and long-term storage of even highly perishable foods is possible. The major methods of preserving foods are drying, refrigeration and freezing, pickling, preserving with sugars, and canning. Each of these methods has certain advantages and certain disadvantages when used with various types of foods.

The most economical method of drying is *sun-drying*, used commonly in hot, dry areas (Figure 15.7). The product is merely spread out in a thin layer on a cloth or plastic sheet in the sun. The process is slow and uncertain, and unexpected rains may lengthen it. Usually, several days are required for reducing the moisture content to a level low enough to prevent spoilage.

Drum-drying is the most common commercial way of drying foods. The product is tumbled in a large drum, and warm dry air passes over the product and carries the moisture away (Figure 15.8). Drum-drying is more expensive than sun-drying but, of course, can be done anywhere, and the moisture content can be reduced to a lower level.

DRYING

FOOD MICROBIOLOGY

Freeze-drying, or *lyophilization,* is a process in which the product is frozen and the moisture removed under vacuum while the product remains in the frozen state. Its use in preserving microbial cultures was described in Section 3.7. In recent years, the process has become more frequently used for preservation of foods. The advantage of freeze-drying is that changes in flavor and texture are kept to a minimum. The disadvantage is that the process is expensive, as high vacuums are required, and these are difficult to achieve on a large scale.

Once a product is dried, it must be wrapped and sealed in such a way that it remains dry. Plastic and cellophane film are the cheapest wrapping materials, but they are easily broken and hence must be packed in a protective cover. A dried product kept dry will remain free of microbial growth indefinitely. With time, its flavor will probably deteriorate because of chemical changes, usually oxidation by air. A dried product stored in the complete absence of air may remain unchanged for an indefinite period.

LOW-TEMPERATURE STORAGE

Storage at temperatures just above freezing (2 to 5°C) greatly prolongs the shelf life of foods. Such temperatures are easily achieved with mechanical refrigeration powered by electricity or gas. The availability of mechanical refrigeration has considerably changed eating habits, as it has made possible the storage of many foods (Figure 15.9). Also, refrigerated transport by truck, rail, and air has made possible the distribution of fresh foods to areas quite distant from those where they are raised. Storage at these temperatures for extended periods is not possible, as spoilage will eventually occur, usually as a result of growth of psychrophilic microorganisms.

Long-term storage can be carried out, however, if temperatures below freezing are used. Foods freeze at temperatures below

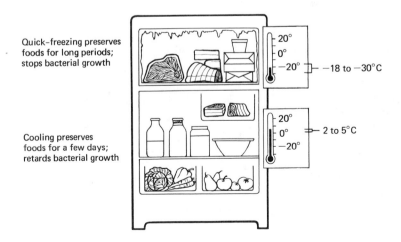

Quick-freezing preserves foods for long periods; stops bacterial growth

20°
0°
−20°   −18 to −30°C

Cooling preserves foods for a few days; retards bacterial growth

20°
0°
−20°   2 to 5°C

FIGURE 15.9
Low-temperature storage prolongs the shelf life of foods.

0°C, but complete and solid freezing occurs only at temperatures below about −20°C. It should be emphasized that even if a food looks solidly frozen, it may have pockets of liquid where microbial growth can occur. Storage of foods in the frozen state is one of the most effective means of ensuring long-term storage, although freezing changes to varying degrees the physical and chemical characteristics of foods. Quick freezing is usually more effective than slow freezing in preserving the quality of the food. It is important to remember that, in general, freezing does not kill all the microorganisms present in the food but may actually preserve their viability. Thus, foods should be frozen soon after harvest, before microbes have had a chance to grow. During long-term frozen storage, care should be taken to ensure that the food does not thaw and refreeze, since during the thaw cycle microbial growth may occur.

During frozen storage, changes in quality of foods can occur that are unrelated to microbial activity. Foods lose water slowly during frozen storage, resulting in a phenomenon called "freezer burn." To prevent drying, the product should be wrapped tightly before freezing. Oxidation of fats during frozen storage can lead to the development of rancidity. This can be prevented by keeping air away from the product. The quality of the product is best when it is frozen rapidly, since this minimizes modification of the consistency and structure of the product. Rapid freezing can best be assured when the product is frozen in relatively small volumes. A thawed product may actually spoil more rapidly than one that was never frozen. This is because freezing and thawing change the physical structure of many foods, breaking down cells and tissues and making them more accessible to microorganisms. Thus, thawed foods should be eaten soon after thawing and should never be stored unfrozen, even in the refrigerator.

PICKLED OR FERMENTED FOODS

Acid is often used to prevent the growth of microorganisms in perishable foods; this process is usually called *pickling* (Figure 15.10). Foods commonly pickled are cucumbers (sweet, sour, and dill pickles), cabbage (sauerkraut), milk and milk products (buttermilk, yogurt), and some meats and fruits. The food can be made acid either by addition of vinegar or by allowing acidity to develop directly in the food through microbial action, in which case the product is called a *fermented* food. In addition to the increased keeping qualities of pickled foods, their flavors are often distinctive, and they are delicacies in many cuisines.

The microorganisms most commonly involved in food fermentations are the lactic acid bacteria, the acetic acid bacteria, and the propionic acid bacteria. In most cases, the acids formed are derived

FOOD MICROBIOLOGY

Figure 15.10
All these foods are prepared by the use of microorganisms. Back row, left to right: vinegar, soy sauce, pickles, sauerkraut, buttermilk. Front row: sour cream, cheese, pickle, sausage, bread.

from sugars occurring in the foods. The microorganisms involved in the fermentation may be present in the raw food when it is harvested from the field, or they may be added as cultures by the food-plant operator.

When green vegetables such as cabbage, lettuce, and spinach are shredded or chopped and allowed to stand, a lactic acid fermentation usually develops as the naturally occurring lactic acid bacteria act upon the sugars in the liberated plant juices. Salt is usually added to such plant materials at the start of the process to prevent the growth of other bacteria, the lactic acid bacteria being more resistant to inhibition by salt than are most spoilage organisms. The amount of acid that develops depends on the amount of sugar in the food. If the natural sugar content is low, sugar can be added to increase the acid output. The most commonly fermented green vegetable is cabbage, and the product obtained is sauerkraut (Figure 15.11). Another common food fermentation is that in which cucumbers are converted to pickles. The cucumbers are placed in a salt solution in large vats and allowed to ferment for several weeks or longer, during which time the lactic acid and flavor produced by the microorganisms becomes soaked into the cucumbers, greatly altering their flavor, color, and texture.

Pickling of foods can also be carried out by mixing the product directly with vinegar, in which case the process is not really a fermentation. However, vinegar itself is a microbial product, made by the action of acetic acid bacteria on alcoholic juices such as wine or alcoholic apple juice (cider). The industrial manufacture of vinegar will be discussed in Section 17.8.

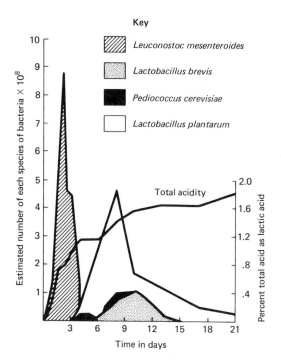

FIGURE 15.11
Changes in the bacterial flora of sauerkraut and development of acid during fermentation at 23° C with 2.25% salt. From C. S. Pederson and M. N. Albury, *N.Y. State Agr. Expt. Sta. Bull.*, 824, 1969.

PRESERVES, JELLIES, AND SALTED PRODUCTS

Because of the dehydrating effect of syrups and brines (discussed earlier in this chapter), foods soaked in these solutions can be preserved for long periods of time. Salting is used for meat products, especially fish. However, certain bacteria, called *halophiles*, are able to grow even in concentrated brines and cause spoilage of salted fish. Because of this and because of the greater availability of refrigeration, salted fish is no longer a major item of commerce in most parts of the world. Foods preserved in syrups are mainly fruits, although preservation of the product is less the goal than is the production of a tasty condiment. Jellies are made from fruit juices by adding sugar and a gelling agent such as pectin. Preserves are made by cooking fruit pieces in a sugar syrup, during which process the natural pectins of the fruit are extracted and cause gelling when the product is cooled. Both preserves and jellies are subject to spoilage by the growth of molds tolerant of high sugar concentrations. Since these molds are all aerobes, they grow only on the surface of the product, and such surface growth can be prevented by sealing the container either with a tightly fitting lid or with a layer of paraffin.

CANNING

Canning is a process in which foods are sterilized by heating in cans that are tightly sealed so that organisms cannot enter. When prop-

Hollow-center lid

Disc with rubber seal

FIGURE 15.12
Glass jar frequently used in home canning of food, with self-sealing lid.

FIGURE 15.13
Manner of construction and sealing of a metal can. Can is made of steel plated with tin and sometimes enameled on the inside.

erly sealed and heated, the food should remain sterile and unspoiled indefinitely and can be stored safely even without refrigeration. Home-canned foods are usually prepared in glass containers, whereas commercial products are most often in tin-coated steel cans. In any canning process, it is important to select containers that do not leak and to seal them properly. Glass containers for home-canned foods are usually sealed with lids containing rubber gaskets (Figure 15.12). The filled jars are heated for the proper length of time at the proper temperature, then cooled and sealed. After sealing, the jars should be checked for leaks before being stored in a cool, dry place. Commercial cans must be constructed of tin plate sufficiently heavy so that leaks do not occur, and the cans must be carefully inspected for holes, pits, or signs of corrosion. During canning operations, frequent inspection of can seams must be made to be sure that the can-sealing equipment is working properly (Figure 15.13).

For sterilization, all nonacid foods must be heated under pressure, whereas acid foods can be sterilized in a boiling water bath. Time and temperature for heating vary with the ease with which the food can be sterilized (Table 15.3). It is not desirable to heat foods much longer than necessary to sterilize, since prolonged heating affects the nutritional and eating qualities of the food.

Commercial canning operations (Figure 15.14) consist of the following stages: (1) *Blanching*, a preliminary treatment of the product with hot water or live steam. It is done with all vegetables, the purpose being to soften the tissues so that the can may be filled properly, to eliminate air from the product so that there is less air in the can, and to destroy enzymes of the vegetable that might cause harmful changes in the flavor, texture, or color of the product

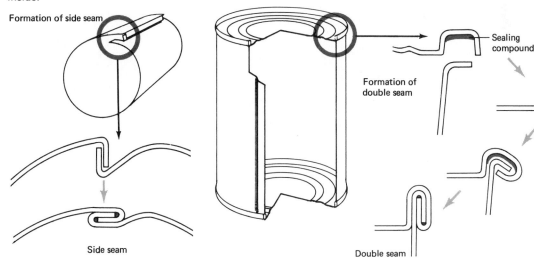

Formation of side seam

Side seam

Formation of double seam

Sealing compound

Double seam

TABLE 15.3
Home canners processing schedule

	Acid foods, canned in boiling water bath (100°C)	Type of pack	Time (min)
Fruits	Applesauce	Hot	20
	Berries	Raw	15
	Cherries	Raw	20
		Hot	10
	Peaches	Raw	25
		Hot	20
	Plums	Hot	20
Acid vegetables	Rhubarb	Hot	10
	Sauerkraut	Raw	30
	Tomatoes	Raw	35
		Hot	10

	Low-acid foods, canned in pressure cooker (10 lbs, 116°C)	Type of pack	Time (min)
Low-acid vegetables	Beans, green	Raw or Hot	20
	Beets	Hot	30
	Corn, whole kernel	Raw or hot	55
	Corn, cream style	Hot	85
	Peas	Raw or hot	40
	Squash	Hot	65
Meats	Beef, corned	Hot	75
	Beef stew	Hot	75
	Poultry, boned	Hot	75
Soups	Broth, clear	Hot	25
	Split pea	Hot	50

FIGURE 15.14
Commercial canning operations.

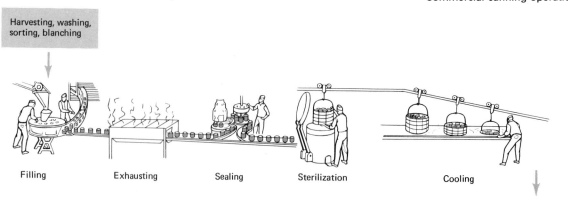

Harvesting, washing, sorting, blanching

Filling   Exhausting   Sealing   Sterilization   Cooling

Labeling, storage, delivery to consumer

before canning. (2) *Cleaning* the cans. (3) *Filling* the cans. It is important that each can be as full as practicable, to eliminate the dead space at the top. (4) *Exhausting* the can to remove air. This can be done either by heating the filled can for a few minutes or sucking out the air with a vacuum pump. It is important to have a vacuum in the final sealed can, since this ensures absence of oxygen and, hence, a reduction in corrosion and spoilage. (5) *Closing* the can. This is usually done by machine. It requires construction of a good seal on the seams, and the proper operation of the sealing machine must be checked from time to time. (6) *Marking* the can with a code number to indicate the product, date of packing, and the grade. This ensures identification of the can even in the absence of a label and, in the case of spoilage, permits determination of when and where the product was canned so that other cans of the same processing batch can be identified and eliminated from market. (7) *Sterilization* of the cans by heating in a steam bath or autoclave. The time and temperature for treatment are determined by the pH of the food; low-acid foods are heated at higher temperatures or for longer times, or both, than high-acid foods. The autoclave must have a temperature recorder that will provide a graph of the sterilization cycle, so that the time at sterilization temperature can be accurately determined. (8) *Cooling* the cans, preferably under controlled decrease of pressure to prevent buckling. Cooling by immersion in water can be a serious source of contamination, since during cooling, some water may be sucked into the can through any tiny holes present, and if this water has a microbial load, some organisms will of course enter the can. All cooling waters should be chlorinated to maintain at least 1 ppm residual chlorine at the discharge end of the can cooler. (9) *Handling and storage* of the cans under such conditions that contamination does not occur. Dropping, bumping, or rattling of cans should be avoided, since any impact to the can may lead to the development of leaks. Cans should be dried immediately after leaving the cooling system and stored dry, since bacteria then cannot develop on the outside of the can. If completely dry conditions cannot be maintained, water containing 3 to 5 ppm residual chlorine should be continuously sprayed.

Bacterial contamination can be greatly reduced or virtually eliminated by proper design and operation of a canning factory. Cleanliness is the prime factor, but elimination of equipment that could permit bacterial growth and hence be a source of contamination is also important. Spoilage may be due either to improper heating or to leakage. By determining the kinds of bacteria responsible for the spoilage, these two causes can be distinguished, since spoilage due to improper heating will virtually always be caused by a single heat-resistant, spore-forming species; whereas if leakage occurs, a mixture of nonspore-forming bacteria will usually be found. A major role of the microbiologist in the canning industry is in the

a   b   c   d

FIGURE 15.15

Changes in cans as a result of microbial spoilage. *a* Normal can; note that the top is indented due to negative pressure (vacuum) inside. *b* Slight swell resulting from minimal gas production. Note that the lid is slightly raised. *c* Severe swell due to extensive gas production. Note the great deformation of the can. This can is potentially dangerous, and it could explode if dropped or hit. *d* Exploded can. A can such as that in *c* was dropped, and the gas pressure resulted in a violent explosion. Note that the lid has been torn apart.

investigation of spoilage problems so that recommendations can be made to rectify deficiencies in the canning process.

When spoilage problems are being investigated, the can should first be examined externally for abnormal conditions. The canning industry has evolved terms for different abnormalities (Figure 15.15). A *flat* is a can that appears normal, showing no swelling. In *flat sour spoilage* no gas is produced but there is sufficient acid to make the product distinctly acidic. Flat sours develop because of underprocessing and are usually caused by the growth of thermophilic spore-forming bacteria, mainly *Bacillus stearothermophilus*, the spores of which are unusually heat resistant. *Flippers* are cans that appear flat but have one end that becomes forced out when the can is knocked against a flat surface. The flipper represents an early stage of infection by a gas-producing microorganism, enough gas having been produced to relieve the vacuum in the can. The *springer* represents a later stage of gas accumulation in which one end of the can is permanently bulged, and if pressure is applied to this end, it will flip in but the other end will flip out. A *soft swell* is a can with both ends bulged, but not so tightly that the ends cannot be pushed in with the thumbs, and a *hard swell* is a can bulged at both ends so tightly that no indentation can be made with thumb pressure. If sufficient gas pressure builds up, *bursting* may occur at one of the seams. Although gas production and the formation of swells are commonly caused by microbial spoilage, gas can also be produced by the chemical action of acids on the metal of the can. In this case, the gas consists only of hydrogen (H_2), whereas with microbial spoilage, CO_2 is the main gas; chemical and microbial causes of swells can thus be differentiated by an analysis of the kind of gas present in the head space of the can. Flat cans showing darkening of the contents accompanied by the odor of hydrogen sulfide exhibit a condition called *sulfide spoilage*. This is due to the growth of the sulfate-reducing bacterium *Desulfotomaculum* (formerly called *Clostridium nigrificans*), a spore-forming thermophile. The presence of sulfide spoilage indicates underprocessing.

After external examination, the can should be opened and subjected to bacteriological examination. The end of the can should be cleaned and then rubbed with a rag soaked in alcohol, after which the end of the can should be flamed, thus burning the alcohol off. A sterilized can opener of the puncture type should be used to open the can, and the opening in the can should be made just large enough to remove the sample. If the can is a swell, it should be covered with a cloth or inverted over a sink and punctured first to release pressure, then opened the rest of the way in an upright position. A sterile widemouth pipette should be used to remove liquid or semiliquid products, and a sterile forceps or spatula used to remove solid foods. About 15 to 20 ml of sample should be transferred to a sterile screw-capped test tube and stored for use in various tests. A microscopic examination of a sample should be made first by making a smear and staining with dilute crystal violet solution. The morphology of organisms seen and the presence or absence of spores should be noted. Since some bacteria, particularly thermophiles, die quickly following the period of growth at favorable temperature, careful examination should be made for the presence of empty microbial cells or rows of dark granules without cell walls that remain after the cells have disintegrated.

Cultures of the contaminants should be made by inoculating 2 to 4 g of material into tubes containing 10 ml medium, and both aerobic and anaerobic cultures should be prepared. Incubations should be done at both 30 to 35°C (for mesophiles) and 50 to 57°C (for thermophiles). To detect the presence of spores, several inoculated tubes should be treated by heating to 80 to 82°C for 10 to 15 minutes and then incubated. If growth is obtained in any culture, it should be examined microscopically and studied further.

If facilities for bacteriological work are not available to the canner, a simple alternative for controlling problems is to remove cans at random from the output and incubate them at 30 to 35°C and 50 to 57°C and examine them periodically for evidence of spoilage. The disadvantage of this procedure is that one must wait as long as 14 days for results, although swells may begin to show within as little as 2 days.

A wide variety of spoilage problems can develop, and the canner should be alert to all of them. If spoilage does occur, a careful analysis of the canning operation is necessary to identify the source of difficulty. Even if spoilage is not occurring, routine bacteriological checks on cans removed at random from the line should be carried out, especially for those items that are difficult to sterilize. Many canners produce very large numbers of cans in central plants for shipping long distances. Quality control is essential if consumers are to be confident that they are buying a reliable, entirely safe product. If it should happen that the deadly toxigenic bacterium *Clostridium botulinum* develops in the product, the whole output from the same processing operation must be destroyed, a quantity that could amount to many thousands of cans. In addition to the cost of finding and destroying cans containing a spoiled product, the bad public image created may adversely affect future sales by the canner. Alert bacteriological control of the canning process is a proper and inexpensive way to avoid such difficulties.

Milk and milk products are excellent, high-quality foods, providing both nutritional and culinary values. However, milk is extremely susceptible to spoilage by microorganisms, and the microbiologist plays a major role in the dairy industry in the control of milk quality. Further, microorganisms are used in a beneficial way in the manufacture of many dairy products, including buttermilk, cheese, and sour cream; but proper management of microorganisms in the production of these items is required to ensure quality and dependability of the product. Finally, milk may be a major agent in the transfer of disease-causing microorganisms from cows to humans, and great care must be taken to ensure that milk and milk products are free of such pathogens.

    Most milk is produced on specialized dairy farms; the technology of milk production is often highly advanced, and high milk yields are obtained. From 2 to 3 gallons of milk can be obtained from a single cow in a day (Figure 15.16). In modern dairy operations, cows are milked by machine, and the milk is transported to holding tanks where it is immediately cooled and stored until it can be transported to the dairy plant. At the plant, the milk is pasteurized, bottled, and held for distribution to the consumer. Some dairy plants also process milk for use in making various products such as cheese, dry milk, or butter. At each stage on the way from cow to consumer, the milk must be carefully handled and checked to be sure that its microbiological quality is maintained.

Cow's milk consists of a complex mixture of constituents including fat, protein, and carbohydrate; the most plentiful component is water, which comprises over 85 percent of the weight of milk (Table 15.4).

15.5
Dairy microbiology

THE NATURE OF MILK

TABLE 15.4	Constituent	Percent by weight
Average composition of cow's milk	Water	87
	Fat	3.8
	Lactose (milk sugar)	4.7
	Protein	
	Casein	3.0
	Lactalbumin	0.4
	Ash (minerals)	0.75
	Others (vitamins, flavor ingredients, etc.)	0.15

Milk is an ideal medium for the growth of microorganisms because it is rich in microbial nutrients, and even nutritionally exacting microorganisms are able to grow successfully in milk. However, there are some peculiarities about the composition of milk that should be noted. The sugar of milk is lactose, and many microorganisms are unable to utilize it as a source of carbon and energy. Those organisms that can utilize lactose possess a special enzyme called *lactase*, or (more commonly) β-galactosidase, which catalyzes the splitting of lactose into its constituent sugars, glucose and galactose, which then are utilized further. Another peculiarity is that the protein of milk is present primarily in the form of casein; if microorganisms are to use casein as a source of nitrogen, they must possess a protease enzyme that will break the casein down into its constituent amino acids, although there are usually traces of free amino acids in milk that allow slight growth of bacteria not able to digest protein. Sometimes milk also contains substances inhibitory to microbial growth, either substances present in the hay or other feed provided to the cow or, more commonly, antibiotics such as penicillin used to treat infections in the cow, which become transferred to the milk.

MICROORGANISMS IN MILK

Milk from the udder of a healthy cow contains very few microorganisms. Almost immediately after removal from the cow, however, milk can become contaminated, and from the time of milking to the time of consumption, contamination continues to be possible. Most microorganisms that enter milk have the potential for growth if the temperature is appropriate. In clean milk stored at low temperature very little microbial growth will occur, since the only organisms present come from the udder of the cow and these have temperature optima of 35 to 40°C (Figure 15.17). However, if the milk has become contaminated with organisms from the environment, microbial growth may occur even if the milk is kept cool, since at least some of the contaminants will be psychrophiles able to grow at low temperatures.

Microorganisms in milk can be classified in two groups: pathogenic and nonpathogenic. The pathogenic organisms are derived either from the animal or from persons handling the animal and its milk. A number of microorganisms that cause diseases of cattle also infect man, of which the most important are *Mycobacterium tuberculosis*, causing bovine and human tuberculosis, and *Brucella abortus*, causing abortion in cows and undulant fever (brucellosis) in humans. Microorganisms that are derived from milk handlers include the causal agents of typhoid fever (*Salmonella typhi*), bacterial dysentery (*Shigella dysenteriae*), and gastroenteritis (various *Salmonella* species). Even if milk is produced under the most careful conditions and has a low bacterial count, it may still be dangerous if it is infected with pathogenic organisms.

The nonpathogenic organisms in milk are usually derived from the environment and, although of little or no danger to humans, may still be responsible for undesirable effects such as souring, curdling, or production of off-flavors (Figure 15.18). The organisms entering milk after it leaves the cow are derived from the utensils, dust, manure, and water present around the dairy farm or from persons handling milk or equipment. These organisms include the lactic acid bacteria (*Streptococcus lactis*, *Lactobacillus casei*, and *L.*

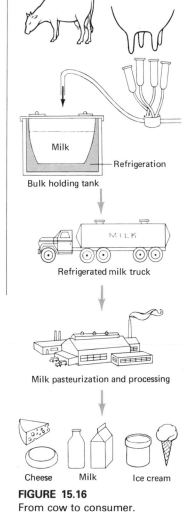

FIGURE 15.16
From cow to consumer.

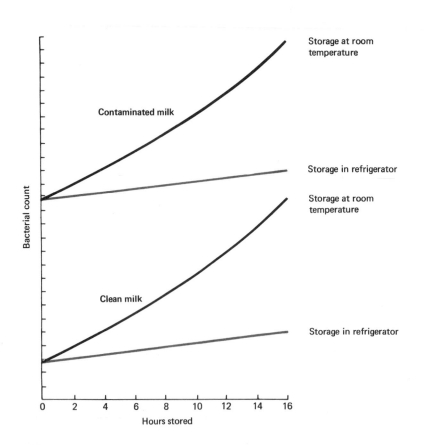

FIGURE 15.17
Effect of initial count and of storage on the number of bacteria in milk. Note that after storage in the refrigerator, the contaminated milk has an even higher bacterial count than the clean milk stored at room temperature.

FOOD MICROBIOLOGY

TABLE 15.5
Sensitivity of bacterial
pathogens to the
pasteurization process

Organism	Disease	Time (min) required for destruction of pathogens at 60°C
Mycobacterium tuberculosis	Tuberculosis	20
Corynebacterium diphtheriae	Diphtheria	1
Salmonella typhi	Typhoid fever	2
Shigella dysenteriae	Bacterial dysentery	10
Brucella abortus	Undulant fever (brucellosis)	10 to 15
Streptococcus agalactiae	Mastitis (in cows)	Under 30
Streptococcus pyogenes	Streptococcal infections (in humans)	Under 30

acidophilus), the coli-aerogenes group (*Escherichia coli*, *Enterobacter aerogenes*), spore formers (*Bacillus* and *Clostridium*), and the pseudomonads (*Pseudomonas* species). Another bacterium, *Streptococcus agalactiae*, is the cause of bovine mastitis, a disease of the udder of the cow, and this bacterium may be present in milk of untreated infected cows. This organism is not generally pathogenic to humans, although it is closely related to the human pathogen *Streptococcus pyogenes*, discussed in Section 10.1

PASTEURIZATION

The pasteurization of milk probably more than anything else has been responsible for the production of milk that is safe and reliable for human consumption (Table 15.5). In addition to being pathogen-free, pasteurized milk has much better keeping qualities than does unpasteurized milk, making it possible to distribute milk satisfactorily from dairy farms to urban centers. It should be emphasized that pasteurization is not the same as sterilization, since in pasteurization, the heat treatment is not sufficiently long or high to kill all organisms. Although pasteurization alters the flavor and reduces the nutritional value of milk slightly, these disadvantages are minor when compared to the advantages. Indeed, it is virtually impossible in many parts of the world today to purchase nonpasteurized (so-called raw) milk.

The times and temperatures chosen for pasteurization are those needed to completely kill any bacteria of the species *Mycobacterium tuberculosis* that may be present in milk. Many other bacteria are of course also killed, and at least 99 percent of the non-pathogenic bacteria are destroyed; however, it is the pathogen *M. tuberculosis* that is used as the indicator organism to show when pasteurization is successful. Those bacteria surviving pasteurization are either *thermoduric* species, which are able to survive temperatures used for pasteurization but not able to grow at such temperatures, or

FIGURE 15.18
The milk in the bottle on the left has soured and formed a curd.

thermophilic species, which are able to grow at pasteurization temperatures and may actually increase in numbers during the pasteurization process. Both of these kinds of microorganisms grow poorly at the low temperatures at which milk is stored and hence do not cause serious spoilage problems. The bacterial count of pasteurized milk generally ranges from 2,000 to 20,000 bacteria/ml, whereas the bacterial count of raw milk will generally range from 500,000 to 3,000,000. In terms of spoilage during low-temperature storage, most of the bacteria remaining will present little problem; it is the psychrophilic bacteria (able to grow at low temperature) that enter the milk after pasteurization that cause the main problems. By close attention to sanitary conditions in the dairy plant, contamination with psychrophilic bacteria can be minimized or eliminated, thus greatly extending the keeping qualities of the milk.

There are three major pasteurization methods: the low-temperature method, the high-temperature short-time process (so-called flash pasteurization), and the ultra-high-temperature method (Figure 15.19). The *low-temperature method*, or "batch-holding process," involves heating the milk in bulk at 63 to 66 °C for 30 minutes (Figure 15.20). The heating can be done after the milk is bottled (in-the-bottle pasteurization) but is more commonly done in bulk in tanks of 500 to 2,000 liters. The walls of the tank within which the milk is placed are heated with either hot water or steam to the desired temperature, and the milk is agitated with a motor-driven stirrer to ensure that the heat is transferred throughout the milk. The milk must be uniformly heated to the desired temperature, and care must be taken to ensure that no pockets or dead spaces in the tank escape the heat treatment. After heating, the milk is cooled quickly to avoid growth of thermophilic bacteria and the development of bad flavors.

High-temperature short-time pasteurization, or "flash pasteurization," (Figure 15.21) is a continuous-flow method, in which the milk is passed through a tube where it is heated to 72°C for at least 15 seconds and then quickly cooled. This method requires more elaborate equipment than does the batch pasteurization method; however, it is far better suited to large dairy operations, where milk is bottled continuously. Much more milk can be pasteurized in the same space because there is no necessity for a large area for bulky holding tanks. The equipment for flash pasteurization is more complicated to construct and operate, however. Further, although pathogens are destroyed just as effectively by both methods, thermoduric bacteria are less readily destroyed by flash pasteurization, so that if raw milk is high in thermoduric bacteria, the bacterial count of the pasteurized product may be high. Since the milk is heated only very briefly, undesirable flavors are lower in flash pasteurized milk, and on balance, the advantages of this method exceed the disadvantages in most dairy operations.

a

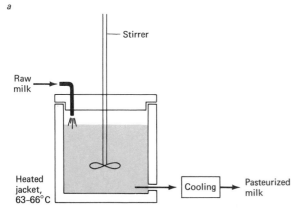

Stirrer

Raw milk

Heated jacket, 63–66°C

Cooling → Pasteurized milk

b

Milk flows through heating unit in 15 seconds

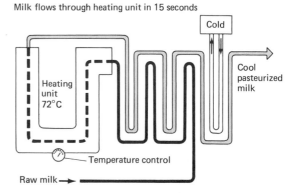

Cold

Cool pasteurized milk

Heating unit 72°C

Temperature control

Raw milk →

FIGURE 15.19
Comparison of a batch pasteurizer and a flash pasteurizer. *a* In the batch-holding process, milk is held in the tank for 30 minutes after pasteurization temperature is reached. *b* In flash pasteurization, hot milk passes out through pipes close to those through which raw milk enters. This proximity serves to give the raw milk a preliminary heating while the hot milk becomes cooler.

FIGURE 15.20
A batch pasteurizer.

Even higher temperatures and shorter treatment times can be used to pasteurize milk with the *ultra-high-temperature treatment.* Steam is used to heat the milk, either by direct injection or indirectly through a heat-exchange system, the product reaching temperatures over 130°C for less than 1 second. The milk is virtually sterile and has long keeping qualities, although flavor changes occur, and the

FIGURE 15.21
A flash pasteurizer.

milk may take on a boiled taste or a flavor sometimes called "cabbagey." Some reduction in vitamin content also occurs.

Pasteurization by the normal methods gives milk and its other fluid products an average shelf life of 7 to 10 days. Since the turnover time of such a rapidly consumed product as milk is fast, this shelf life usually is adequate, and the more extreme heating treatments to increase shelf life further are not worth the decrease in flavor and quality. For such products as cream, however, that have a relatively slow turnover time on the shelf, a longer shelf life is desirable. Storage times of 30 or more days can be attained by more vigorous heating followed by aseptic packaging, to give a product that may not be absolutely sterile but has so few organisms that it can be termed "essentially sterile."

In some cases, it is desirable to have sterile milk, and this can be easily accomplished if the milk is treated at sufficiently high temperatures. Sterilization is done directly in the bottle, which is closed in such a way that a contaminant-free seal is obtained. The filled bottles are then subjected to sterilizing temperatures, 121°C for 10 minutes, usually in large autoclaves. Sterile milk has almost indefinite keeping qualities and can be stored without refrigeration, but it has a distinctively cooked flavor, a faint brownish color (due to caramelized sugar), and reduced vitamin content. Sterilized milk is especially useful when no refrigeration is available, such as in military encampments or in underdeveloped countries.

Quality control of pasteurized milk begins with appropriate supervision of the pasteurizing apparatus and the recording of the duration of time that the milk is at pasteurization temperature. The single most useful test for ensuring that pasteurization has occurred properly is the *phosphatase test.* The enzyme phosphatase, present in raw milk, is destroyed by pasteurizing temperatures; by testing for the absence of this enzyme in pasteurized milk, one can be sure that the milk has indeed been properly treated. The test depends on the ability of the phosphatase of milk to decompose a phosphate-phenol compound and liberate the phenol part of the molecule. The substrate used is *p*-nitrophenol phosphate, which is colorless but yields the yellow-colored *p*-nitrophenol when the phosphate is removed by the action of the enzyme. Of course, marketed milk must not reveal the presence of any phosphatase.

Although good sanitary practice in the production of milk and milk products begins on the farm, the microbiologist usually does not become involved in the quality control of milk until it reaches the dairy plant. Milk testing begins on the platform where the milk is received. There is need for a rapid test for the microbiological quality of milk that might be run immediately on the platform by relatively untrained workers. Unfortunately, such tests are not avail-

TESTING MILK

FOOD MICROBIOLOGY

able, and only relatively crude, indirect tests are now carried out. Such tests are suitable for detecting seriously deteriorated milk but will not detect more subtle changes.

The *odor* of milk can be an indication of its quality. A trained person with a keen nose may be able to identify a product of poor quality by smelling the can lid as it is removed. If the milk has been cooled, detection of off-flavors is difficult, and it is desirable to heat a sample before smelling. *Acidity* is a poor test for quality but may be useful in evaluating milk for industrial purposes. An even simpler test is the *clot-on-boiling test*, in which a 2-ml sample of milk is boiled in a test tube for 5 minutes. Clotting will occur if there is sufficient acid in the milk to coagulate the casein at this higher temperature.

For rapid routine assessment of microbial activity in milk, *dye-reduction* tests are used. These tests are based on the principle that the respiratory activities of microorganisms will cause a change in the color of oxidation-reduction indicator dyes, and the color change can be easily and quickly determined. Two dyes used are methylene blue and resazurin. Because the reduction in the dye caused by microorganisms is reversed by oxidation from oxygen in the air, the tests are carried out in closed test tubes to prevent the entry of air. In the *methylene blue test*, 10 ml of sample is placed in a test tube, and 1 ml of a standard solution of methylene blue is added. The tube is stoppered and incubated at 37.5°C in a water bath and examined every half hour. When the dye is reduced the blue color disappears, and the time at which reduction is complete is noted. The reduction time is roughly proportionate to the bacterial count of the milk, as indicated in Table 15.6.

The *resazurin test* is performed in a manner similar to the methylene blue test, but the color changes are different. Resazurin is blue when oxidized, and as it becomes reduced the color changes through lilac, mauve, and pink to colorless. By detecting the intermediate stages of reduction, the test can be made more sensitive, and the time to complete it can be decreased. A modified resazurin test for quick checking on the platform is the 10-minute test. If color change to pink or pink-mauve has occurred within the 10-minute incubation time, the milk is rejected. This test is only suitable for detecting milk of the poorest quality but is considered superior to the odor, clot-on-boiling, or acidity tests.

TABLE 15.6 Methylene blue test	Quality of milk	Reduction time (hr)	Approximate number of organisms/ml
	Good	Over 4.5	200,000 or less
	Average	2.5–4.5	200,000–2,000,000
	Poor	Less than 2.5	2,000,000–10,000,000

The determination of the number of bacteria present in milk is part of routine examination procedures. Two types of determinations are used: the direct microscopic count and the viable count. A number of methods are available for performing these tests, each with its advantages and disadvantages.

The *direct microscopic count* is sometimes called the Breed test (Figure 15.22). With a special pipette, 0.01 ml of milk is deposited on a clean glass slide and spread uniformly over an area of 1 sq cm. The area may be marked on the slide, or a template marked with the area may be placed under the slide. It is important that the sample be spread uniformly. The slide is dried, treated to remove fat, and stained. The slide is then examined with an oil-immersion lens, and the number of organisms in 30 fields is counted. From this count and a knowledge of the microscope field size, the number of organisms per milliliter of milk can be calculated. The direct microscopic count is a rapid method requiring relatively little apparatus or time. It has the disadvantages that both dead and living organisms are counted and that it is relatively insensitive, since only a small amount of milk is examined (0.01 ml), and if the milk is low in bacterial numbers, there are so few organisms per slide that the count is not very accurate. With milks of high bacterial count (over 500,000/ml), the method is just as accurate as viable count methods, and the results are available much sooner. For the average high-quality market milk, the counts are usually so low that the method is probably not worth using. In addition to the quantitative count, it is possible to recognize different types of bacteria by morphological characteristics and hence to obtain at least some idea of the kinds of bacteria contaminating the milk (Figure 15.23).

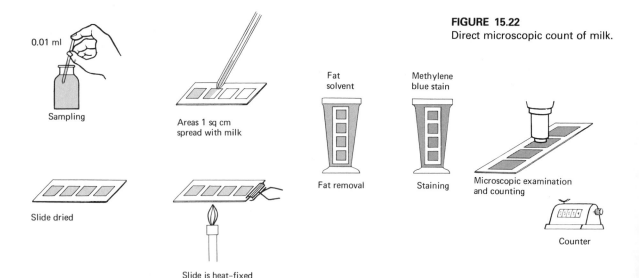

FIGURE 15.22
Direct microscopic count of milk.

0.01 ml

Sampling

Areas 1 sq cm spread with milk

Fat solvent

Fat removal

Methylene blue stain

Staining

Microscopic examination and counting

Slide dried

Slide is heat-fixed

Counter

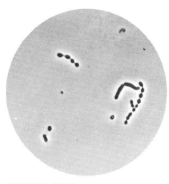

FIGURE 15.23
Typical milk-souring bacteria.

Viable counts are more difficult and time-consuming than direct microscopic counts, but they are more sensitive, indicating the presence of a smaller number of bacteria and, of course, only measuring those organisms that are viable. A viable-count procedure also offers the option of identification of the kinds of organisms present, since colonies can be picked and studied. Further, it is possible to use selective media or conditions so that only certain groups of organisms are counted, such as the coliform bacteria or the thermophilic bacteria. The principles involved in performing viable counts are discussed in Section 4.2.

The *standard plate count* is a carefully standardized procedure designed to permit an estimate of the total number of viable bacteria present. A general-purpose agar culture medium is used, and for government-regulated milk supplies, the medium is specified by the regulating agency.

Since both raw and pasteurized milks will yield plate counts greater than 300/ml, it is essential that dilutions of the sample be made. A standard series of tenfold dilutions is made, and 1-ml samples of each dilution are pipetted into sterile petri plates; melted plate count agar is added and mixed with each diluted sample. The plates are incubated either 72 hours at 30 to 32°C or 48 hours at 37°C, and the number of colonies is counted. For counting, those plates containing between 30 and 300 colonies are selected, since these plates will provide counts with the greatest accuracy (Figure 15.24).

Coliform counts are performed on pasteurized milk samples and other dairy products to detect recontamination after treatment, especially to indicate possible fecal contamination by food handlers. The *Escherichia coli*, or coliform, test for dairy products follows the same principle as that for water, as described in Section 14.3. The number of coliforms is determined by either the MPN technique, using liquid medium in tubes, or the plate-count method, in which colonies in agar are counted. The plate-count method is simpler and more convenient but is less sensitive than the tube method in detecting small numbers of coliforms since sample volumes greater than 1 ml cannot be used. In large culture tubes, volumes as great as 10 ml can be used for the MPN method if the medium is concentrated to correct for this dilution.

FIGURE 15.24
Typical pour plate from a viable count of milk.

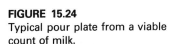

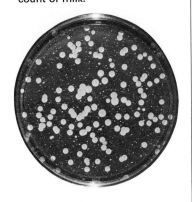

In addition to coliform counts, counts of other organisms can be done, usually for special research purposes. Counts of *Streptococcus agalactiae* are done on milk from cows suspected or known to have mastitis. *Mycobacterium tuberculosis*, the causal agent of tuberculosis, and *Brucella abortus*, the causal agent of brucellosis, are also occasionally assayed for. If difficulties with pasteurization are being experienced, counts of thermoduric and thermophilic bacteria may be done. Although thermoduric bacteria are able to survive temperatures used for pasteurization, they do not necessarily grow at such temperatures, whereas thermophilic bacteria are actually able to grow at pasteurization temperatures and may increase in numbers during the pasteurization process.

It should be emphasized that it is only because of the application of careful and precise microbiological controls on the quality of milk that milk has become a reliable and safe product for human consumption. When communities were generally rural, milk was available locally, and careful quality control was of less importance; but in the present urban world, milk is produced at considerable distances from its place of consumption. It is produced in large volumes, is transported in bulk to central dairies, and may be stored for many days before consumption. The microbiologist plays a vital role in ensuring that the product remains safe and satisfying to consume.

The extremely perishable nature of milk makes it essential to control microbial multiplication from the moment of production. Milk regulations are designed to ensure that the milk delivered to the public is safe to drink. These regulations are therefore public-health regulations and are usually administered by an agency of the government responsible for public-health matters. It is important to emphasize that any regulations established must be so designed that they can be enforced. This means that clearly defined procedures for evaluating the quality of milk must be available. For pasteurized milk, three criteria are used: the standard plate count, the coliform count, and the phosphatase test. Standards, of course, vary with the country, climate, and conditions of manufacture, but standards for pasteurized milk widely accepted in advanced countries are as follows: standard plate count not to exceed 30,000 bacteria/ml, coliform count not to exceed 1/ml, and phosphatase test to be negative. Other regulations may specify the standard and coliform counts of raw milk, the absence of tuberculosis and brucellosis in the dairy herd, and other matters. Milk products are also regulated for chemical quality, such as butter fat and milk solids, but these aspects do not involve the microbiologist.

GOVERNMENT REGULATION OF MILK

Not all microbial action in milk is considered harmful. In the production of yogurt, buttermilk, and other fermented milks, microbial action induces beneficial changes. The microorganisms involved are specific kinds of lactic acid bacteria of the genera *Streptococcus* or *Lactobacillus*, and their function is twofold: (1) to produce lactic acid, which leads to an acidification of the milk; and (2) to produce flavor ingredients that give the final product its distinctive taste. Because of its acidic character, a fermented milk has quite good

FERMENTED MILKS

FOOD MICROBIOLOGY

keeping qualities in comparison to unfermented milk, and historically this was the reason for manufacturing these products. However, at present the keeping qualities are not so important as the flavor that develops, since modern pasteurized fresh milk also has excellent keeping qualities. Buttermilk was originally a by-product of the churning of sour cream into butter, but cultured buttermilk is produced today by the microbial fermentation of skim milk.

Fermented milks are made in most parts of the world, and local products are given special names. *Yogurt* is the most widely consumed form of sour milk in Europe and North America, and its preparation will be discussed below. The lactic acid bacteria grow at fairly high temperatures, producing lactic acid and small amounts of other compounds. *Kefir* is a sour-milk product of southeastern Europe in which lactic acid bacteria combine with a lactose-fermenting yeast to give both a lactic acid and an alcohol fermentation, the final product containing up to 1 percent alcohol. *Koumiss*, a variety of kefir produced in Soviet Central Asia from mare's milk, contains up to 3 percent alcohol, owing to the fact that horse's milk is higher in lactose than is cow's milk, so that more sugar is available for conversion to alcohol. In the Scandinavian countries, a more viscous product is made, called *taette* in Norway or *skyr* in Iceland. In the manufacture of this product, an acid-producing bacterium is combined with a slime-producing bacterium, and it is through the slime produced by the latter bacterium that the viscous consistency of the product arises.

All cultured milks are made in fundamentally the same way. Milk or milk with additives is heated, then cooled, inoculated, and incubated for a certain time at a certain temperature. For the commercial production of yogurt, the milk is heated to 90°C for 15 to 30 minutes, then cooled to 43°C; it is then inoculated with 2 percent culture and held at 43°C until the desired acidity is reached, after which it is cooled to 5°C and held chilled until sold. There are minor differences among particular manufacturers: temperatures, times of incubation, final pH, fat content, and concentrations of additives such as skim milk powder and sucrose.

The organisms involved in production of yogurt are *Streptococcus thermophilus* and *Lactobacillus bulgaricus*; in commercial production, the organisms are generally added in a balanced mixture as a starter culture. *S. thermophilus* grows best at first, removing oxygen and producing the weakly acid conditions that favor the growth of *L. bulgaricus*, which in turn hydrolyses the lactose and the casein to make conditions more acid. Both organisms convert nearly all the milk sugar (lactose) to lactic acid, producing only trace amounts of by-products. These are, however, important by-products, since they give yogurt its characteristic flavor. *S. thermophilus* produces diacetyl, and *L. bulgaricus* produces acetaldehyde. For natural, or plain, yogurt, the incubation takes place in the cartons whereas for fruit yogurts the yogurt is chilled overnight after incubation and mixed with the fruit and syrup the next day.

STARTER CULTURES

In primitive manufacture of fermented milks, the bacteria for the new batch are obtained by adding a small amount of product from the previous batch, but in our modern, more highly controlled industrial processes, carefully selected pure cultures, called starter cultures, are used. By the inoculation of appropriate starter cultures into pasteurized milk, a fermented product can be obtained that is uniform and of high quality. In the manufacture of buttermilk and sour cream, two bacteria are used. One, usually a variety of *Streptococcus lactis* or *S. cremoris*, is responsible for most of the lactic acid produced. The second, usually a variety of *Leuconostoc citrovorum* or *L. dextranicum*, is mainly responsible for the production of flavor and aroma. This second organism is usually a minority member of the starter culture mixture but is nonetheless important. The flavor and aroma are due primarily to the production of diacetyl, and this substance is produced by the bacterium through its action on the citric acid that is always present in small amounts in milk. Starter cultures also play an important role in the manufacture of cheeses, as will be described in Section 15.6.

In many modern dairy operations, starter cultures are not maintained but are purchased in dried or frozen form from commercial firms. In this way, a culture of optimum quality is always available, ensuring uniformity in the product. The milk used for production should be pasteurized and cooled before inoculation. Since it is necessary to add 0.5 to 2.0 percent by volume of starter culture to the product to be fermented, it is usually necessary for the dairy to build up the culture from the initial inoculum, using heated milk as a culture medium. This means that the dairy-plant operator must have some understanding of the manner of handling of cultures and should be familiar with microbial growth curves. Many problems can develop when starter cultures are not properly maintained. The two organisms in the starter can get out of balance, occurring in a proportion different from that in the original starter. Antibiotics carried over into the milk from the cow can inhibit the growth of the bacteria. Viruses (bacteriophages) able to attack the starter bacteria may enter the culture and destroy it or reduce its activity. The quality control of the starter culture is evaluated by determining the rate of development of acid and the degree of acidity obtained, and by checking the odor and flavor of a test fermentation. If difficulty arises, it is best to begin anew with a fresh starter culture.

Frozen starter cultures are finding increasing use in dairy operations. These cultures are prepared commercially in properly balanced proportions and are concentrated and frozen in liquid nitrogen. In this form, they can be shipped from a central factory to virtually any part of the world. Because they are concentrated, they contain sufficient bacteria to inoculate large vats directly; it is not necessary to carry out intermediate culture buildup, thus eliminating one of the major causes of contamination. Frozen starter cultures contain a much higher concentration of viable bacteria than do dried starter cultures, and when the thawed culture is added to the product to be fermented, the process begins immediately. Frozen starter cultures are available for the manufacture of buttermilk, yogurt, sour cream, and a variety of cheeses.

15.6 Cheese

Cheese is a dairy product formed by precipitating the casein of milk in the form of a *curd*. The curd holds most of the fat and other suspended materials of the milk, and the water and dissolved constituents, called the *whey*, are allowed to drain away. The curd is then usually allowed to ripen, and it is during the ripening process that most of the distinctive characteristics of cheese develop. Microorganisms play two roles in cheese manufacture: (1) they are usually responsible for the souring of milk, which causes the production of curd; (2) they play key roles in the ripening process, and hence, microorganisms can be considered responsible for the distinctive characteristics of cheese.

There are literally hundreds of different kinds of cheeses; many are of strictly local production, whereas others are manufactured and marketed over wide areas. Many cheeses can be stored for long periods of time without spoilage and hence can be transported for long distances. It is of course this resistance to spoilage that was originally the main reason for the manufacture of cheese, although in most parts of the world today, cheese is valued because of its flavor rather than because of its storage qualities.

The cheeses that find wide distribution can be grouped as follows:

Unripened, soft cheeses.
    Low fat: Cottage cheese.
    High fat: Cream cheese, Neufchâtel.
Ripened cheeses.
    Soft: Camembert, Brie, Liederkranz.
    Semisoft: Roquefort, Blue, Gorgonzola, Brick, Limburger.
    Hard: Cheddar, Swiss, Gruyère, Provolone.
    Hard grating cheese: Romano, Parmesan.

STEPS IN CHEESE MAKING

The first step is the curdling of the milk. Milk for cheese making is usually whole milk of good quality. It should be tested to be certain that it is not high in bacteria; the methylene blue or resazurin dye-reduction test is suitable for this purpose. A special test to show that the milk will curdle properly is also desirable, since time and money should not be wasted on an unsatisfactory milk. The milk should then be pasteurized if the cheese is not to be aged for a long time, to destroy any pathogenic organisms present. It is usually considered that if the cheese is aged more than 60 days, it need not be made with pasteurized milk, since pathogens will be destroyed during the aging process. A starter culture is then added. The first function of the bacteria is to produce acid, which is necessary for the development of a satisfactory curd. However, the main agency in the development of the curd is not bacteria, but the enzyme rennin, which is added to the milk at the time of the starter culture inoculation. Rennin is an enzyme obtained commercially by making an extract of the lining of cow's stomach, and this enzyme acts on the casein of milk to cause it to coagulate. Curd can be formed through the action of rennin alone or through the action of bacterial acids alone, but the best curds for making most cheeses are achieved through the combined action of rennin and bacteria.

The curd is processed so as to lose water and shrink, and this leads to the production of a curd of desirable consistency (Figure 15.25). The moisture content is determined by the consistency of the curd, which affects how well it retains water, and by the length of time the curd is allowed to drain. The curd is shaped into forms of a desirable size and then salted. Salt not only adds flavor but also promotes the further extraction of water from the curd. Salt is also inhibitory to the growth of many spoilage microorganisms but not to the growth of the ripening organisms. The moisture content of the curd will vary with the cheese variety, the hard cheeses having much less moisture than the soft ones. Unripened cheeses are packaged at this stage, either directly or after mixing with additives, and no more microbial processing is involved.

The liquid remaining after the curd is removed, the whey, may be used to make ricotta or other whey cheeses, which are manufactured by nonmicrobiological processes. The whey is also used as a starting material for several industrial microbial processes, such as in the manufacture of lactic acid and alcohol. The whey may also be discarded, but since it is a potential source of serious water pollution, it must be treated in much the same way as domestic sewage, as described in Chapter 14.

For the manufacture of most cheeses, the next step is ripening, and during this process the cheese undergoes changes in flavor, texture, and consistency (Figure 15.26). Some of the ripening is due to continued action of rennin or of enzymes originally in the milk; however, microorganisms produce the most distinctive changes. The organisms responsible for ripening either are added with the starter

FIGURE 15.25
During cheese-making operations, curd is poured into forms to allow drainage. From N. F. Olson, *Ripened semisoft cheeses*, Chas. Pfizer and Co., Inc., New York, 1969.

FOOD MICROBIOLOGY

FIGURE 15.26

Processes involved in ripening of cheese.

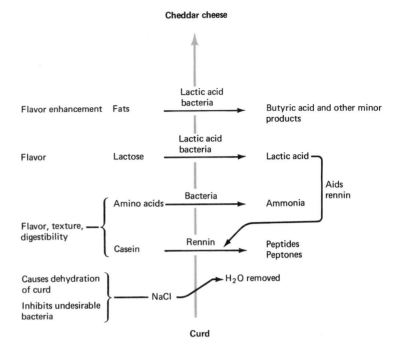

or are spread on the surface of the curd at the initiation of ripening, depending on the type of cheese to be made.

For the hard cheeses, the curd is formed in large blocks carefully wrapped with paraffin or plastic film to prevent surface growth of microorganisms. In general, hard cheeses ripen slowly and are not consumed for months or even years after manufacture. During ripening, excess moisture loss and undesirable surface spoilage are prevented by the airtight coating.

For the soft and semisoft cheeses, ripening actually takes place through the action of microorganisms growing on the surface. Enzymes and flavor components produced by the organisms spread through the cheese and gradually convert the whole block into the desired product. Soft-ripened cheeses are always produced in small blocks so that the microbial products can spread through the cheese rapidly (Figure 15.27a). Organisms involved in the ripening are smeared on the surface, and blocks are not wrapped during ripening but are stored in curing rooms of fairly high humidity to encourage microbial growth. The surface growth, usually called *smear*, is composed of a characteristic mixture of organisms in which yeasts, *Micrococcus* species, and *Brevibacterium linens* are the main components (Figure 15.27b).

In the production of Roquefort, Blue, and Gorgonzola cheeses,

15.6 CHEESE   515

FIGURE 15.27
Curing of soft-ripened cheese.
a Blocks are smeared with
appropriate ripening
microorganisms. *b* A smear of
microorganisms on the outside
of a block of semisoft cheese.
From N. F. Olson, *Ripened
semisoft cheeses,* Chas. Pfizer
and Co., Inc., New York, 1969.

which are semisoft, a fungus, *Penicillium roqueforti,* is the main organism involved. The blocks of curd are inoculated throughout with needles, and the fungal filaments that pervade the cheese give it its characteristic blue color. The flavor is caused mainly by fungal products that accumulate in the cheese.

In Swiss-style cheese, gas-forming organisms develop, and the gas formed causes the production of the characteristic "eyes" of the cheese. The gas is CO_2, which is produced by *Propionibacterium,* an organism that ferments lactic acid with the production of propionic acid, acetic acid, and CO_2. The flavor of Swiss cheese is due in a great degree to the propionic acid produced by this bacterium. This organism is thus responsible for the two main characteristics of Swiss cheese: its flavor and the characteristic holes.

During ripening, microbes do several things. They ferment the sugar and lactic acid remaining in the curd and produce flavor constituents, especially propionic acid and acetic acid. They digest some of the protein of the curd, converting it to amino acids. In soft-ripened cheeses, protein digestion proceeds quite extensively, thus causing the softening of the cheese. The amino acids liberated by the protein digestion contribute to the flavor of the product, but in addition, their microbial decomposition leads to the production of ammonia, hydrogen sulfide, and other constituents that impart distinctive flavors to many cheeses.

In a process as complicated and lengthy as cheese manufacture, it is understandable that defects may develop in the product. Many of these defects are induced by the action of undesirable microorganisms. Of these, the most important are molds, which grow readily on surfaces of cheese exposed to oxygen. Rigid sanitation precautions in the cheese factory and careful wrapping of the product to keep air from it will ensure the absence of mold contamination. Gas formation is in some cases an undesirable change in cheese induced by microorganisms. Sugar-fermenting microbes entrapped in the curd can produce gas, usually CO_2, and cause the formation of holes, cracks, and fissures in the cheese. Microorganisms also produce undesirable flavors and colors in spoiled cheese that will reduce its market value. The cheeses most susceptible to spoilage are the unripened ones, since these are high in moisture content. Such cheeses must be kept under refrigeration until eaten. At the other extreme, the hard-ripened cheeses are of such low moisture content that they can be kept for months or years without serious spoilage.

Another cheese product called *process cheese* differs from the natural cheeses we have just described. One or more kinds of natural cheese are ground or chopped and mixed with other ingredients including whey protein, nonfat milk solids, nondairy foods such as chopped pimento or olive, spices, stabilizers, emulsifiers, and coloring agents, and enough water to produce the desired consistency for cheese spreads. After heating to blend all ingredients and make the mixture homogeneous, the hot fluid cheese is packaged in a variety of forms: 1- and 5-pound blocks, slices, and spreads. Although not sterile, process cheese has a greatly reduced number of microorganisms because of the heating steps in its production, with a resultant increase in shelf life. Microorganisms are not involved in the production of process cheese other than their role in making the original starting material, the natural cheese.

The production of cheeses of consistently high quality is both an art and a science. Because of the crucial importance of microorganisms in cheese manufacture and the large number of ways in which they can get out of control, a knowledge of microbiology is of great importance to the cheese manufacturer and technician.

15.7 Meat microbiology

Meats are among the most nutritious of foods, but they are also among the most susceptible to microbial spoilage. In addition, they are agents of transmission of several serious infectious diseases from animals to humans. Thus, the microbiology of meat and meat products is of considerable practical importance. Most meat is processed and prepared in large central installations called *packing houses*, where slaughtering, cutting, processing, and in some cases packaging of meat is done. Inspection and disease control are under the jurisdiction of qualified veterinarians who have received extensive training in microbiology.

The quality of the meat is closely related to the condition of the live animals at the time of slaughter and the degree of contamination to which the carcass is subjected. The animals to be slaughtered should be carefully inspected while still alive to determine that they are free of infectious diseases. Diseased animals are rejected for slaughter and are either disposed of or moved to an outside location for treatment. Infectious diseases commonly seen include listeriosis (caused by *Listeria monocytogenes*), anthrax (caused by *Bacillus anthracis*), hog cholera (virus-caused), swine erysipelas (caused by *Erysipelothrix rhusiopathiae*), tuberculosis (caused by *Mycobacterium tuberculosis*), and inflammatory diseases such as enteritis, mastitis, pericarditis, and pneumonia (caused primarily by streptococci and staphylococci). Animals infected with these or other pathogens are unfit for slaughter.

Within the slaughterhouse, care should be taken to avoid contamination of the animal with material laden with microorganisms. Soil and dung should be removed from the animal by preliminary washing. Cleanliness in the slaughterhouse is important, and dirt and debris should be cleansed from the floor and walls of the rooms, preferably with a disinfectant such as sodium hypochlorite. The slaughter itself is done by bleeding the animal to death after initially stunning it. Much of the contamination on meat comes from the hide of the animal during the skinning process. Removal of the hide in a manner that restricts the spread of contamination is very important.

After slaughter, every animal should be subjected to postmortem examination (Figure 15.28), with special attention to the lymph nodes, visceral organs, and the exposed parts of the carcass. If disease or other abnormal condition is found during the routine examination, the carcass and its parts are given a more extensive examination. If the abnormal condition is localized in one portion of the carcass, this part may be disposed of and the rest passed on for processing; however, if the whole carcass is to be condemned, it is passed to a holding area for conversion into fertilizer and other nonfood items.

The carcass is normally refrigerated immediately after slaughter in order to prevent spoilage, although because of its bulk, it does not cool instantly. Carcasses are suspended in large cooling rooms held at temperatures of -2 to $-4°C$; care must be taken not to overload the room, as the heat from the warm carcasses will raise the temperature. The carcass is then aged in the cold (always below $3°C$), and the meat gradually becomes more tender, usually reaching an optimum tenderness after 2 weeks of aging. During the aging process, there is always the possibility of bacterial spoilage, especially by psychrophilic bacteria that may contaminate the surface of the

FIGURE 15.28
Sanitary inspection of the carcass in a meat-packing plant. The organs of the animal are carefully examined for evidence of infectious disease.

REFRIGERATION

FIGURE 15.29
Cutting the carcass in a
slaughter house.

MEAT SPOILAGE

meat. Aging is accelerated at higher temperatures, and in one commercial process, the carcass is aged at 15°C for 3 days, using ultraviolet lights to control surface spoilage. Tenderization of meat is even faster at still higher temperatures, and aging at 32 to 43°C has been done experimentally, microbial growth within the meat being prevented either by use of antibiotics or by ionizing radiation. However, because of the uncertainties of the effect of these treatments on the fitness of the meat for human consumption, they have not been considered suitable for routine use. The most common microorganisms of fresh meat are bacteria of the genera *Pseudomonas* and *Achromobacter*, although other organisms are also encountered, including lactic acid bacteria, micrococci, yeasts, and molds. No pathogenic bacteria can grow at the low temperatures at which fresh meats are aged.

After aging, the carcass is cut and marketed (Figure 15.29). Cutting exposes meat to increased contamination by providing greater surface area for microbial growth. Once cut, the meat should be quickly consumed, and it is generally considered that the retail establishment should not display the meat for more than 3 days. In the home, even if the meat is properly refrigerated, it should be consumed within 1 to 4 days of purchase. If the meat must be stored for longer periods, it should be frozen immediately after packaging and kept frozen until needed for cooking. The quality of meat is best retained if it is frozen rapidly and stored carefully wrapped to prevent drying.

The principles of food spoilage that have been discussed earlier in this chapter apply directly to meat. Decomposition of meat protein by bacteria is probably of most significance, and the breakdown of the resulting amino acids leads to the release of bad-smelling substances such as hydrogen sulfide, ammonia, and mercaptans (organic sulfur compounds). This process, called *putrefactive spoilage*, occurs most commonly in the deeper portion of the meat tissues, where anaerobic conditions develop and is greatly hastened by warm temperatures. On the surface of meats, microbial spoilage is usually seen as the development of a *slime*, caused by the massive accumulation of microbial cells. The organisms are mainly *Pseudomonas* species, lactic acid bacteria, micrococci, and yeasts. Surface spoilage proceeds at an appreciable rate even when meats are chilled.

In addition to general spoilage problems, there is the problem of discoloration of meat brought about at least in part by microbial action. The expected color for most meats is red, since this color is associated with freshness. The red color of meat is due to the existence in the meat of a pigment, myoglobin, which in its oxidized form is red and in its reduced form is brownish. Bacteria can cause *greening* of meat by producing hydrogen peroxide, which reacts with the myoglobin to convert it to a greenish pigment. The bacteria responsible for greening are primarily lactic acid bacteria. Another kind of microbial spoilage is the development of *rancid* or tainted flavor owing to the action of microbes on meat fats. Numerous bacteria, molds, and yeasts are lipolytic; that is, they are able to break fats down into fatty acids. The fatty acids so produced may then be oxidized, giving rise to products responsible for the odors and flavors of rancid meat.

ASSESSING THE MICROBIAL CONTENT OF MEAT

The principles involved here are no different from those used for assessing the microbial content of other foods or dairy products. Total microbial counts are made by homogenizing a weighed sample of meat in a sterile diluent, preferably using a high-speed blender. Usually 10 g of sample is homogenized in a sterile blender cup containing 90 ml of sterile diluent. The homogenate is then plated either on a nonselective medium to provide a total count or on selective media to provide counts for specific organisms, such as staphylococci, coliforms, anaerobes, yeasts, or molds.

Surface samples for plating can be taken with sterile swabs or spatulas or by making an impression from the meat surface onto a plate of agar medium or a sterile membrane filter. If the sample is frozen, it should be thawed in its original container in a refrigerator and examined as soon after thawing as possible. Interior samples can be taken after sterilizing the surface of the meat by searing with a flame or painting with iodine or another disinfectant. The meat is then cut into the center with a sterile knife and the sample removed from the desired location. Some suggested standards for allowable microbial counts in meats are included in Table 15.2.

MEAT PRESERVATION AND CURING

Although the primary goal of meat preservation is to retard microbial spoilage, it is also important that meat be handled so that its flavor, odor, texture, appearance, and nutritional value are preserved as much as possible. Since fresh meat is one of the more perishable food items, a major task of the meat industry is to develop and employ methods for adequate meat storage. Two general approaches are possible: preservation and curing. *Preservation* involves such treatments as refrigeration, freezing, canning, drying, addition of chemicals and antibiotics, and irradiation. The *curing* process involves smoking, fermentation, or addition of salts, all of which lead to significant changes in the flavor and texture of the product, as well as to increases in its shelf life.

Refrigeration is the most widely used method of meat preservation, but although the shelf life of the meat is extended, spoilage will eventually occur. Psychrophilic microorganisms, able to grow at refrigeration temperature, are responsible for the spoilage. To keep down the growth of psychrophiles, careful attention to the temperature of the refrigerator is necessary since the product will keep four times as long at 0°C as at 5°C. It is thus desirable to install a thermometer in the meat cooler and to be sure that the temperature is as low as possible.

Meat will *freeze* when its temperature is reduced below about −2°C, but for frozen storage of meat, temperatures of about −20°C are generally used in commercial practice. Even colder temperatures are better, although more expensive to maintain. Freezing has no noticeable effect upon the color, flavor, odor, or juiciness of meat as judged after cooking, although during long-term freezer storage there is a gradual decrease in odor and flavor. However, if meat for freezing is selected and prepared properly and frozen quickly, many kinds can be stored for several months without any significant deterioration in quality. Even psychrophilic microorganisms will not grow in solidly frozen meat, and the main decreases in quality are due to dehydration ("freezer burn") and the development of rancidity due to oxidation. Both of these changes can be eliminated by careful wrapping of the meat before freezing.

Canning of meat involves either moderate heating (about 70°C), which, although insufficient to completely sterilize, is used on cured meats such as hams to extend their shelf life, or more severe heating (121°C), which is applied to uncured meats to produce a sterile product that can be stored indefinitely without refrigeration. Most canned meat products are heated in the can, and because of the bulk of the product, the slow rate at which heat penetrates to the center, and the high pH of meat, fairly long heating times are necessary. Some flavor changes will inevitably take place, but these disadvantages are offset by the greatly lengthened shelf life.

Dehydration of meat is a centuries-old process, still carried out under primitive conditions in many parts of the world using the sun or wood fires as a source of heat. In commercial practice, meat is occasionally dried with hot-air dryers, but recently freeze-drying of meat has become widespread, since the product retains more of the flavor and texture of fresh meat (Figure 15.30). Dehydrated meat can be stored unrefrigerated for long periods of time without spoilage and is ideal when a lightweight product is desired. It is used widely for military rations and on safaris, expeditions, and wilderness trips.

Chemical preservatives have been widely used in the past, but in modern practice addition of chemicals to meats has been greatly curtailed. To be useful, a chemical preservative not only must extend the shelf life of meat but also should not impart an undesirable flavor, odor, or color and should be nontoxic to humans. It is the latter requirement that has curtailed use of chemical preservatives, since the toxicity of many agents upon long-term human consumption has not been established. Agents occasionally added to meats

FIGURE 15.30
The physical state of meat affects its keeping qualities. Ground meat (rear left) spoils faster than whole meat. Dried meat (foreground) keeps indefinitely, even in the absence of refrigeration.

include sodium benzoate, formaldehyde, salicylic acid, sulfite, and boric acid. The antibiotics chlortetracycline and oxytetracycline have been added to raw poultry at levels no greater than 7 ppm, since these antibiotics are destroyed by heat and no significant residue is left after proper cooking. It should be emphasized that no chemical preservative is a substitute for proper sanitation.

Although *ionizing radiation* has been used experimentally to preserve meat, it has not been used commercially. The principles and problems in the use of ionizing radiation to sterilize products are discussed in Section 4.4. The main disadvantages of the use of ionizing radiation are that it is expensive and that undesirable flavors develop.

Primitive *meat curing* involved only the addition of salt (sodium chloride) as a preservative, but in modern curing, sodium nitrate, sodium nitrite, sugar, and sometimes other agents are also added. Nitrite is added to preserve the red color of the meat, and nitrate is added to provide a reserve of nitrite, since nitrate is slowly converted to nitrite in the meat. Sugar is added not only as a flavoring ingredient but also because it promotes reducing conditions in the meat, thus retarding oxidation and color deterioration. Other ingredients sometimes added to the curing mixture are ascorbic acid, which retards oxidation, sodium phosphate, which retards shrinkage in smoked products, and monosodium glutamate, which is a flavor enhancer.

Curing is used most commonly in the production of ham and bacon from pork. Either the curing mixture may be injected into the vascular system of the cut as a brine, or the cut may be dipped in the brine. Alternatively, a dry curing mixture may be rubbed onto the surface of the cut. The treated product is then stored at refrigeration temperature to retard bacterial growth until salt penetration is complete. Preservation from spoilage of cured meat occurs because of the dehydrating effects of the salt on microorganisms present in the meat, as discussed earlier.

Smoking of meat is carried out both to preserve the meat and to impart to it a characteristic flavor. Smoke is generated by controlled combustion of moistened hardwoods (hickory and oak are most commonly used) in a smoke generator. The chemical composition of wood smoke is quite complex, but among the chemicals identified have been fatty acids, phenols, waxes, resins, and aldehydes. Among the latter, formaldehyde is present, and this toxic substance has been identified as the chief bacteriostatic and bactericidal substance in wood smoke. The amount of formaldehyde added to meat during smoking is too small to either impart a bad flavor or cause human toxicity. Curiously, although formaldehyde can be added legally to meat indirectly through smoking, government regulations usually forbid adding it directly from a bottle of the purified chemical.

FOOD MICROBIOLOGY

Although at one time smoking of meat was done mainly to preserve it, today the main function of smoking is to add characteristic flavor.

SAUSAGE

Sausage is prepared from chopped and seasoned meat that is placed in a cylindrical casing derived either from the intestinal tract of an animal (natural casing) or from modified cellulose (artificial casing). These is an enormous variety of sausages, and countries and regions often have specialty sausages of distinctive flavor and character. The meat ingredients in sausage most commonly consist of those parts of the slaughtered animal not readily utilizable in other ways, such as cheek and head meat, trimmings, tripe, and belly, but higher-quality meats may also be used. Sausage may be sold fresh, cooked, or dried. Any meat can be used for making sausage, including pork, beef, lamb, and veal, but the most commonly used meat is pork. The most popular products are fresh pork sausage, which is neither precooked nor smoked, frankfurters and Braunschweiger (liver sausage), which are precooked and smoked, and meat loaves, which are cooked but not smoked.

Some sausages of the dry or semidry varieties owe their distinctive flavor to bacterial fermentation. Fermented sausages include such products as thuringer, cervelat, Lebanon bologna (all semidry or summer sausage), and salami and pepperoni (dry or hard-dry). Although fermented sausages usually are not cooked, if they contain pork they must be given a heat treatment sufficient to destroy the trichina cysts that may be present in pork muscle and can infect humans, causing the disease trichinosis. It is generally considered that heating in a smokehouse at a temperature over 58°C or drying at moderate temperature for at least 25 days is sufficient to destroy the trichinas. If an uncooked sausage type is to be prepared in which neither of these treatments is possible, the pork in the sausage must be derived from swine certified to be free of trichinosis.

Manufacture of fermented sausage is still very much an art, and trade secrets abound. In many sausage types, fermentation is carried out by chance contaminants, although starter cultures are increasingly used in many plants. The microorganisms involved in the sausage fermentation are mainly lactic acid bacteria; their role is to produce lactic acid and flavor ingredients from the sugar that is added to the sausage mix. The advantage of using a starter culture is that it makes possible the manufacture of a product of uniform quality in far less time (only 20 to 40 hours instead of 150 or more hours). *Pediococcus acidilactici* is used as the starter culture for sausage and is supplied to the manufacturer in either lyophilized or frozen form. The ground meats, salts, spices, and some sugar are mixed with the starter culture, and the mixture is stuffed directly into appropriate casings and moved to a warm (27°C) humid area for 12 to 16 hours to permit the organisms to rehydrate and return to the vegetative state. The sausage is then moved to the smokehouse and held at about 40°C with 90 percent humidity until the desired acid production is reached (about 15 to 20 hours, depending on the variety of sausage). A final heating at 55 to 60°C for 4 to 5 hours kills the starter bacteria so that subsequent changes during storage are eliminated. Smoke may be applied during any part of the time in the smokehouse. Sausage fermentation is a much less highly developed process than is the manufacture of fermented dairy products, and it is still mainly carried out by traditional empirical methods, although the increasing use of starter cultures seems likely to make it a much more predictable operation.

15.8 Food sanitation

In restaurants, cafeterias, and other public food services, large numbers of people are served food prepared in central kitchens. Because of the dangers of food-borne poisonings and infections, good sanitation practices are necessary. Effective sanitation begins with cleanliness but extends beyond this to the employment of practices that prevent excessive microbial contamination of foods, utensils, and preparation equipment (pots, kettles, grinders, fryers, and the like). The word *sanitize* is often used in the food industry to describe treatments of equipment and utensils so as to destroy microorganisms. Such treatments rarely sterilize but do reduce the microbial load and virtually always kill all potential pathogens present.

Any establishment offering food to the public should have an effective food-sanitation program. Such a program begins with proper control and training of the food handlers. Food handlers must be free of infectious disease, especially diseases such as typhoid fever, diphtheria, and tuberculosis, the causal agents of which can easily be transferred from the body to food. Discharging wounds or lesions are potentially dangerous since they may be a source of infection of food with food-poisoning staphylococci. Routine medical examinations of food handlers may be of value in detecting infection, but of even greater importance is the education of the food handlers to make them aware of the importance of their health. Infected persons should not handle food, utensils, or equipment, and managers of food establishments should ensure that employees are aware of their responsibilities in this regard.

The best quality food available should be used, and it should be handled and stored in such a manner as to prevent spoilage. Only fresh meats, vegetables, and dairy products should be used, and canned and bottled items should be served soon after opening. Prepared foods present the biggest problem: to be available for quick service, they must be stored in large quantities considerably

Hands washed frequently

Clean tablecloth and napkins used

Coughs and sneezes covered

Tableware handled without touching food surfaces

Butter   Ice

Butter and ice served with utensils

FIGURE 15.31
Proper food-handling practices.

ahead of time, thereby presenting ample opportunity for spoilage. For refrigeration, such prepared foods should be placed in shallow pans, not over 3 inches deep, so that they will cool quickly. Salads and other foods should not be mixed with the hands, but with a large spoon or wooden paddle. Vegetables to be eaten raw should be of high quality and should be thoroughly scrubbed under running water before serving. Steam tables and other devices for keeping food hot before serving should be kept clean and should be drained and cleaned completely once a day. The temperature of the foods on a steam table should always be maintained above 65°C since rapid bacterial growth, especially of thermophiles, can occur at lower temperatures. The water supply used in the restaurant must meet drinking water standards, and if its quality is in doubt, it should be tested bacteriologically. Food displays in the dining room should be protected from flies and contamination by customers. Dining room tables should be free of cracks where food can lodge and permit microbial growth, and tables should be cleaned after each use. The best practice is to provide a clean table cloth for each patron; this is microbiologically the most desirable, as well as esthetically pleasing. Clean dishes and utensils should be stored protected from dust and should not be unduly handled. Employees should handle glasses without touching the rims, should touch only the edges of plates and bowls, should pick up eating utensils (knives, forks, and spoons) only by the handles, should use a scoop to pick up ice, and should use a fork to pick up butter (Figure 15.31). Employees should cover themselves for coughs and sneezes, should always wear head coverings to keep hair out of food, and should wash hands carefully and frequently, and always after using the toilet.

Dishwashing is a critical and major problem, and a well-designed installation is essential. In a small restaurant dishwashing is done manually, but in larger establishments automatic equipment is often used (Figure 15.32). The dishes should be placed in hot water containing soap or detergent at about 50°C until clean to the sight and touch, then rinsed in hot water at 60°C, and finally immersed for at least 2 minutes in water at 75 to 80°C. Since these temperatures will scald the hands if hand washing is used, the dishes should be placed in trays or wire baskets. After washing and rinsing, the dishes should be immersed for at least 2 minutes in a warm chlorine solution having greater than 50 ppm available chlorine. Since chlorine will cause silver to turn black, disinfection of silver utensils should be done by immersing them in a warm solution of an approved quaternary ammonium compound. An alternative to a disinfectant is to boil the utensils for 1 minute. It is preferable that dishes and utensils be air-dried, but if drying cloths are used, they should be clean and used for no other purposes. The same temperatures and procedures should be used in *dishwashing machines* and, indeed, the automatic cycles of such machines are usually so pro-

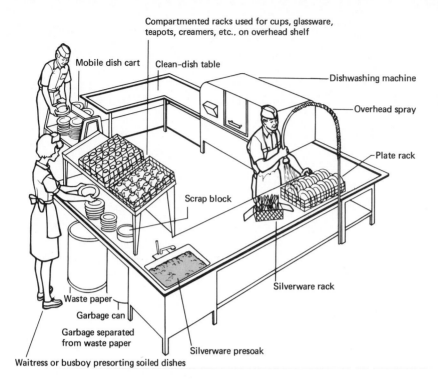

FIGURE 15.32
Recommended procedure for sanitary dishwashing in a restaurant or institutional food-service facility. Adapted from D. R. Turner, *Food science supervisor*, Arco Publishing, Inc., New York, 1969.

grammed. Care should be taken that dishes and utensils are loaded for the machine in such a way that water will reach all parts of them. The machine should be checked from time to time to be certain that it is functioning properly and that the nozzles for water are not clogged.

The effectiveness of dishwashing methods is evaluated bacteriologically by use of a *swab test* (Figure 15.33). A cotton swab is made by twisting nonabsorbent cotton around a wooden applicator stick to form a firm tuft measuring $\frac{3}{16}$ inch in diameter and $\frac{3}{4}$ inch in length. The swab is sterilized by autoclaving it immersed in a tube

FIGURE 15.33
Swab test for counting organisms on utensils.

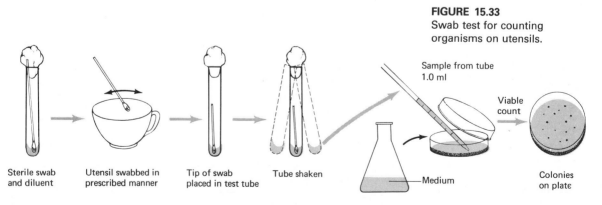

Sterile swab and diluent → Utensil swabbed in prescribed manner → Tip of swab placed in test tube → Tube shaken → Sample from tube 1.0 ml → Medium → Viable count → Colonies on plate

containing 1 ml of diluent (usually sodium phosphate solution) for each utensil to be swabbed. Utensils to be examined are selected at random from clean storage or from the dishwasher. One swab is used for each group of four or more similar utensils. The swab, handled only by the wooden end, is removed from the diluent and squeezed along the inside wall of the tube to remove excess liquid, leaving it moist but not wet; it is then rubbed slowly and firmly three times over the surfaces of the utensils, reversing direction each time. The important surfaces to swab include the upper $\frac{1}{2}$ inch of the inner and outer rims of cups and glasses, the entire inner and outer surfaces of the bowls of spoons and the tines of forks, and the inner surfaces of plates and bowls. The swab is then returned to the diluent and rotated rapidly, the excess water pressed out, and then used for the next utensil. After the four or more utensils have been swabbed, the swab is placed in the diluent and the lower portion broken off and left in the tube. The next steps should be done within 4 hours of swabbing if possible or, if properly refrigerated, within 24 hours. In the laboratory, the swab container is shaken rapidly for at least 10 seconds, then 1 ml of the dilution water is removed and a plate count is performed, using the pour plate method. The plates are incubated for 48 hours at 32 or 35°C, the colonies counted, and the average plate count of organisms per utensil swabbed is calculated. It can generally be assumed that if a count greater than 100 per utensil is obtained, improvement in washing operations should be made.

Waste disposal in eating establishments should be made in cans with tight-fitting lids, and in an area segregated, if possible, from the cooking area. Toilets should be kept in sanitary condition and handwashing facilities should be available; hands should be dried with disposable towels, cloth towel rolls, or electric hand dryers. Adequate dressing facilities should be provided for employees so that uniforms and other coverings worn during food preparation and serving need not be worn on the street. Soiled coats, aprons, and other linens should be kept in covered containers until they are laundered.

In most urban areas, routine inspection of eating establishments is carried out by public-health officials. The importance of frequent, meaningful inspections should be obvious. A standard reporting form is generally used, and any infringements of prescribed procedures reported to the management and to the government agency.

Summary

In this chapter, we have surveyed the many ways in which microorganisms affect the foods we eat. A few microorganisms cause harmful changes in foods, and food spoilage is a serious economic problem. The microbiologist must attempt to identify and erase the source of spoilage and ensure that future spoilage is minimized. Foods may become infected with spoilage organisms at any point along the route from producer to consumer.

Factors affecting microbial growth in foods include the moisture content of the food, the physical structure, the acidity or pH, and the chemical composition or nutritive value of the food.

The presence of pathogenic organisms and their products in foods is an important human concern. Botulism and staphylococcal food poisoning are caused by exotoxins released into foods during the growth of the organisms in the food. Salmonella food infection is caused by ingestion of the organisms with food, followed by growth of the organisms in the host with resultant symptoms of gastroenteritis. Foods can be examined for the presence of microbes by various standard methods, which can determine the total numbers of organisms present or the number of a particular organism of interest.

Food-preservation methods have made it possible to have foods of many types and kinds available to us in all seasons and in regions often far from the source of supply. The most primitive way of preserving foods is by drying. Low-temperature storage is a widely used means of preserving foods. Refrigerator temperatures allow only slow growth of psychrophiles, and freezer temperatures prevent virtually all growth. Fermented and pickled foods are preserved because of their low pH, which prevents growth of spoilage organisms. Foods with high sugar or salt content are preserved because of the dehydrating effect of these additives. Preservation by canning is the most common method and gives products that have a very long shelf life. Canning must be carefully done and controlled; sampling of products for sterility testing and investigation of spoiled products are important aspects of quality control.

Dairy microbiology is involved with the maintenance of high quality in milk and with the production of many milk products. Milk is a very rich medium for the growth of microorganisms. Contaminants in milk include some pathogenic bacteria as well as many nonpathogens, such as the lactic acid bacteria. It is to eliminate the pathogenic organisms present or potentially present that milk is pasteurized. The milk is not sterile after pasteurization, but it is pathogen-free, and its shelf life is greatly lengthened. Milk is tested bacteriologically at many points during processing, and milk standards are set by government regulation.

Much milk is converted into milk products, the most popular of which is cheese. Microbes play a major role in the initial formation of cheese curd and are also responsible for many of the changes that occur during the ripening process, giving the characteristic texture, flavor, and consistency to each type of cheese.

Microorganisms most frequently produce harmful effects in the meat-packing industry. Meats spoil easily unless proper precautions are taken. The slaughtering of animals must be done so as to prevent diseased animals from being used and to avoid contamination of good meat. The carcass is refrigerated immediately and aged at low temperature, after which it is cut and marketed. Microbial spoilage of meat is frequently putrefactive or may result in surface slimes, greening, or rancidity. Meat is most frequently preserved by refrigeration or freezing, but it is also canned, dried, salted, or smoked. Sausages are popular meat products and in some cases are produced by microbial fermentation.

Food sanitation is important in all public food-service areas and especially so in restaurants and cafeterias. Food handlers must be careful not to infect foods. Foods used in the preparation of meals should be of high quality and should be stored so as to prevent spoilage. Equipment and utensils must be handled so as to prevent contamination, and dishwashing should follow prescribed procedures. Microbiological testing of dishes and utensils is a recommended quality control and is generally carried out by public-health agencies.

Study questions

1 Contrast highly perishable, semiperishable, and stable (or nonperishable) foods. List several foods in each category.
2 List five factors that control microbial growth in foods and discuss briefly how each factor affects growth.
3 Discuss the steps you would take to prevent the development of staphylococcal food poisoning; of botulism.
4 Why is *Salmonella* food infection not called a food poisoning?
5 List four ways in which foods can be preserved, and discuss briefly the principle of each procedure.
6 Describe briefly the steps used in the bacteriological examination of a canned food suspected of being spoiled.
7 How is milk pasteurized? What is the main reason that milk is pasteurized?
8 Describe briefly two methods for assessing the bacteriological quality of milk.
9 What is a starter culture? How are starter cultures used in the dairy industry?
10 Discuss the microbiological problems in the slaughter and preservation of fresh meat.
11 Dishwashing is an important aspect of food service. Describe briefly the procedures for effective dishwashing (both manually and automatically). How is the effectiveness of a dishwashing method evaluated bacteriologically?

Suggested readings

Foster, E. M., F. E. Nelson, M. L. Speck, R. N. Doetsch, and J. C. Olson, Jr., *Dairy microbiology*, Prentice-Hall, Inc., Englewood Cliffs, N.J., 1956. Textbook on the roles of microorganisms in the manufacture of dairy products and on the causes and prevention of microbial spoilage of milk and dairy products.

Frazier, W. C., *Food microbiology*, 2nd ed., McGraw-Hill Book Co., New York, 1967. Textbook on food spoilage and preservation and on production of fermented foods using microorganisms.

Jay, J. M., *Modern food microbiology*, Van Nostrand Reinhold Co., New York, 1970. Current brief text on food microbiology.

Pederson, C. S., *Microbiology of food fermentations*, Avi Publishing Co., Inc., Westport, Conn., 1971. Describes the microbial processes involved in the manufacture of fermented foods, such as pickles, sauerkraut, sausage, and cheese.

Thatcher, F. S., and D. S. Clark, (eds.), *Microorganisms in foods: their significance and methods of enumeration*, University of Toronto Press, Toronto, 1968. Procedures for counting different kinds of microorganisms in food ingredients and products.

Our consideration of basic microbiology concludes on a very practical note, with a discussion of the roles of microorganisms in agriculture and industry.

In Chapter 16, we discuss the role of microorganisms in soil formation and fertility, in plant disease, and in animal husbandry. One of the most fascinating agricultural phenomena is the formation of root nodules by bacteria, conferring on the plants the ability to use nitrogen from the atmosphere, thus making them independent of nitrogen fertilizer. Another major area of agricultural microbiology concerns the role of microorganisms in ruminant animals, such as cows, sheep and goats; microbes in the rumen digest the insoluble and

*otherwise indigestible materials of forage crops, permitting rumin-
ants to live on diets composed almost solely of roughage.*

*We consider the large-scale culture of microorganisms for
industrial purposes in Chapter 17. Here we learn about the production of alcoholic beverages by yeast, the commercial production of
antibiotics, and the growing of microorganisms for the production of
human and animal foods. All of these processes require the use of
large-scale industrial fermenters, up to 200,000 liters in size; the
development of processes using such large vessels has been one of the
main challenges and successes in the important field of industrial
microbiology.*

16

**Agricultural
microbiology**

Microorganisms play many important roles in agriculture, both beneficial and harmful. In this chapter, we discuss some of these effects, including the roles of microbes in soil formation and fertility, in compost formation, in beneficial associations with plants, as causal agents of diseases of crop plants, in the control of insect diseases of plants, and as beneficial agents in the digestive processes in cows and other ruminants. A knowledge of the harmful and beneficial effects of microorganisms in agricultural processes can be used to increase crop yields and farm-animal productivity.

**16.1
The soil**

Soil is the basis of agriculture, and microorganisms contribute importantly to both its formation and its fertility. There is an enormous variety of soils, of variable suitability for agriculture. Soils are made up of three components: solid, liquid, and gaseous. The solid materials, both mineral and organic, constitute the soil proper; the liquid and gaseous components, water and air, vary markedly within a single soil. The mineral constituents provide the basic fabric of the soil and vary in size, shape, and chemical composition. The larger particles are sand, the finer particles silt, and the finest clay. The proportion of these different particles greatly affects agricultural utility of the soil. If finer particles predominate, water is

                    AGRICULTURAL MICROBIOLOGY

retained well but the soil tends to become clogged with water, whereas sandy soils are well drained but often become deficient in water. The nonliving organic matter in soil, called *humus*, is important, since it increases the water-holding capacity without causing waterlogging and also improves the soil texture. The living organisms of the soil are numerous and diverse. Bacteria are most numerous, followed by fungi and protozoa, but a large number of other organisms are also present, including algae and invertebrate animals such as earthworms, insects, mites, and millipedes. One of the most important living components is the roots of plants, which ramify extensively through the soil and greatly modify its texture and fertility.

SOIL FORMATION

The formation of soil begins with the breakdown of rock, a process called *weathering* (Figure 16.1). Weathering is a result of three kinds of processes: physical, chemical, and biological. Physical weathering involves the fragmentation of rock owing to freeze and thaw, movement of the earth (as in earthquakes), and other mechanical processes. Chemical weathering involves reactions between substances such as oxygen or water and the minerals of rocks. Biological weathering is due to the action of living organisms, the most important of which are microbes.

The first organisms to become established on a bare rock are photosynthetic: algae, mosses, and lichens (see Section 13.1). The photosynthetic organisms convert carbon dioxide into organic matter, and some of the organic matter formed is excreted and supports the growth of bacteria and fungi. These organisms produce carbon dioxide as a result of respiration, and this CO_2 combines with water to form carbonic acid, a weak acid but one capable over a long

FIGURE 16.1
Steps in the formation of soil.

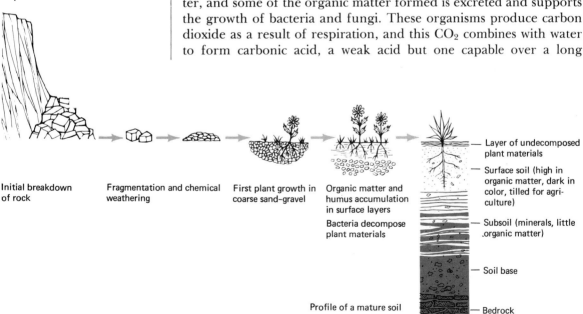

Initial breakdown of rock

Fragmentation and chemical weathering

First plant growth in coarse sand–gravel

Organic matter and humus accumulation in surface layers

Bacteria decompose plant materials

— Layer of undecomposed plant materials

— Surface soil (high in organic matter, dark in color, tilled for agriculture)

— Subsoil (minerals, little organic matter)

— Soil base

Profile of a mature soil

— Bedrock

period of time of dissolving rock. Many of these bacteria and fungi produce organic acids as well as CO_2, and these also contribute to the dissolution of rock. In time a raw soil forms consisting primarily of coarse rock fragments, live and dead organisms, and debris. In this raw soil, small plants can develop, and their roots penetrate into crevices in the rock, causing further breakdown. Organic excretions from the plant roots promote further growth of bacteria and fungi, leading to increased weathering. During the breakdown of rock, minerals are solubilized, further promoting the growth of plants. Organic matter added to soil from plants is broken down by microorganisms and converted into humus. Humus is an important factor in soil fertility since it binds moisture and minerals, thus making plant growth more vigorous, and since it binds soil particles together, thus improving soil texture.

As weathering proceeds, the soil increases in depth, thus permitting the development of larger plants and trees. Soil animals become established and contribute to weathering by keeping the soil mixed and aerated. As water percolates down through the soil, it dissolves some of the minerals and carries them deeper, thus modifying the lower region. Eventually the movement of materials downward results in the formation of layers, leading to what is called a *soil profile* (see Figure 16.1). The time involved in the formation of a soil varies with climate and topography but may be very long, on the order of hundreds of years.

KINDS OF SOILS

Elaborate classification systems for soils have been developed, but we are interested here only in broad groups. The kind of soil found in an area depends on climate, types of plants, type of bedrock, topography of the land, human influences, and other factors. *Podzol soils* are those that form under conifer trees, such as pine, spruce, and fir. The leaves of conifers decompose in the soil in such a way as to produce acids; these acids attack the rock minerals and dissolve important plant nutrients, which are carried down to deeper layers. The surface layers of podzol soils are thus acidic and low in nutrients and are not favorable agricultural soils. They must be fertilized and extensively treated with lime in order to neutralize the acids and make them favorable for crops. *Brown soils* are those that develop under deciduous trees. They are less acid, fewer minerals have been leached out, and they are relatively fertile because of the presence of much organic matter, which holds moisture and minerals. *Grassland soils* are usually higher in organic matter than are brown soils because grasses have extensive root systems, and as the roots die, their organic materials are contributed to the soil. Whereas in brown soils most of the organic matter added to the soil comes from the surface in the form of falling leaves, in grassland soils the organic matter is spread through greater depths. Decomposition of

AGRICULTURAL MICROBIOLOGY

this organic matter leads to the production of an extensive black humus layer, which makes the soil texture quite favorable for agriculture. Most of the extensive grain-producing areas of the world are established on grassland soils. *Tropical soils* differ significantly from soils of temperate climates. If rainfall is high, as in tropical rain forests, extensive leaching of important plant nutrients occurs. When a tropical forest is cleared for agriculture, the soil is initially quite good, as it is high in humus from leaf litter, but in such warm temperatures, the decomposition of humus and remaining leaf litter occurs rapidly, and after several years of farming, the soil is usually depleted of its essential nutrients. Hence, many tropical soils are not good agricultural soils, and cleared areas must be abandoned after a few years. This has led to what is called *shifting agriculture*, which constitutes one reason for the lack of development of stable agricultural communities in many tropical areas.

MICROORGANISMS AND SOIL FERTILITY

Microbes contribute to soil fertility in a number of ways: (1) Through their action in decomposing organic matter they participate in humus formation, and humus contributes significantly to soil texture and the mineral-binding and water-holding capacity of soil. (2) Microbes can cause the release from soil particles of certain minerals that plants need for growth. Such minerals are bound to clay and humus particles; by producing acids, microbes bring about chemical reactions that result in release of these minerals. (3) When plants or animals that live in the soil die, their remains contain significant amounts of mineral nutrients that are bound to organic structures and which, if released, would be utilized as plant nutrient. Microbes decompose the organic materials, releasing these mineral nutrients, a process called *mineralization*. (4) Microorganisms play important roles in the transformation of nitrogen compounds (see Sections 13.5 and 16.4).

16.2 Compost

Compost is a complex mixture of organic materials that is used for fertilization and for improving the texture of soils. It is prepared by piling leaves, straw, and other plant materials in a heap and allowing them to undergo decomposition (Figure 16.2). Within the pile, moist conditions are maintained and microorganisms grow well, decomposing some of the organic matter. Heat, a product of microbial metabolism, is produced, causing the pile to increase in temperature. This process is called *self-heating* and occurs not only in compost but in any situation in which materials rich in organic matter are piled up, such as manure, sawdust, and even coal. During the heating process in compost, thermophilic organisms develop (Figure 16.3). Some of these organisms are able to degrade cellulose and other organic materials in the plant residues. The carbohydrate and

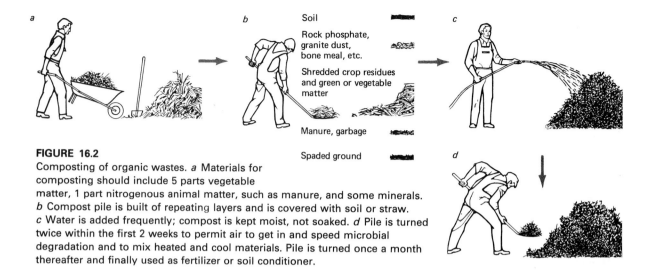

FIGURE 16.2
Composting of organic wastes. *a* Materials for
composting should include 5 parts vegetable
matter, 1 part nitrogenous animal matter, such as manure, and some minerals.
b Compost pile is built of repeating layers and is covered with soil or straw.
c Water is added frequently; compost is kept moist, not soaked. *d* Pile is turned
twice within the first 2 weeks to permit air to get in and speed microbial
degradation and to mix heated and cool materials. Pile is turned once a month
thereafter and finally used as fertilizer or soil conditioner.

Labels in figure: Soil; Rock phosphate, granite dust, bone meal, etc.; Shredded crop residues and green or vegetable matter; Manure, garbage; Spaded ground

fat components of the compost material are rapidly degraded,
whereas the fiber, which contains predominantly woody materials
such as lignin, is less readily decomposed. The final product is
reduced in weight from the original by 25 to 50 percent and is
considerably less bulky. The compact mass of humuslike material
that is produced can be returned to the soil to increase its fertility.

Pesticides are chemicals used to control weeds, insects, fungi, or
other pests. Pesticides are widely used in agriculture, and some of
them are applied year after year to crop land. Although some of
these chemicals decompose rapidly in soil, others are highly persistent; after a number of years of use, residues toxic to animals or
plants may build up in the soil. It has been well established that the
fate of a compound in soil is determined to a great extent by
whether microorganisms able to decompose this compound are
present. Compounds that persist in the soil are resistant to microbial
degradation, whereas those that disappear can be acted upon by
microbes (Figure 16.4).

The decomposition of pesticides by microorganisms occurs
because the microbes can use the compounds as sources of energy,
carbon, or nitrogen. Some organisms are able to completely oxidize
these compounds to CO_2 and water, whereas others are able to
oxidize them only partially. However, some pesticides synthesized by
organic chemists are quite different chemically from the natural compounds in plants and animals. Since these unnatural compounds are
entirely new in nature, it is not very surprising that there may not be
any microorganisms present in the soil that can degrade them.

There are marked differences in the susceptibility of different
pesticides to degradation, and this is reflected in their relative

**16.3
Pesticides**

FIGURE 16.3

Changes that take place in a compost pile during the first 2 weeks. *a* The temperature rapidly increases until it reaches about 70°C, as the result of self-heating. As the organic materials of the compost are consumed, heat production slows, and the temperature gradually decreases to ambient. *b* Thermophilic bacteria, actinomycetes, and fungi, favored by the high temperatures, proliferate rapidly, whereas the mesophilic microbes are destroyed by the heat. As the temperature drops, the mesophilic organisms return and may even reach higher numbers than in the original material. *c* Materials that decompose readily, such as fat and carbohydrate, are rapidly degraded. The ultimate product is enriched in the humuslike characteristic of mature compost.

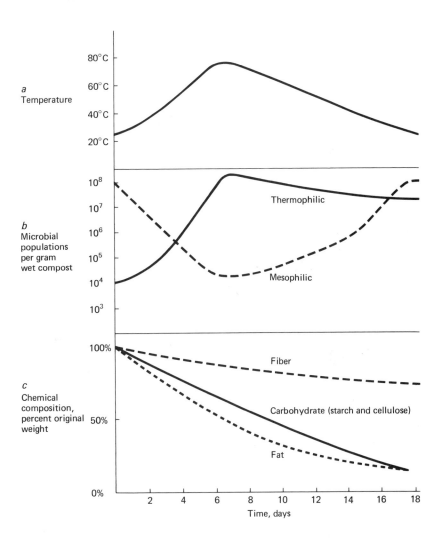

FIGURE 16.4

Pesticide decomposition in soil.

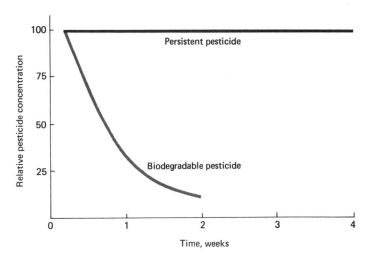

Pesticide	Time for 75 to 100% disappearance	TABLE 16.1
		Persistence of pesticides of various types in soils

Pesticide	Time for 75 to 100% disappearance
Chlorinated insecticides	
DDT	4 years
Aldrin	3 years
Chlordane	5 years
Heptachlor	2 years
Lindane	3 years
Organophosphate insecticides	
Diazinon	12 weeks
Malathion	1 week
Parathion	1 week
Herbicides	
2,4–D	4 weeks
2,4,5–T	20 weeks
Dalapin	8 weeks
Atrazine	40 weeks
Simazine	48 weeks
Propazine	1.5 years

Data from P. C. Kearney, J. R. Plimmer, and C. S. Helling, *Encycl. Chem. Technol.* **18**, 515, 1969.

persistence in soil, as shown in Table 16.1. The decomposition of organic chemicals in soil involves both biological and nonbiological aspects. Some compounds are so unstable that they decompose rapidly in soil without the necessity of microbial action. This is true of the organophosphate insecticides listed in Table 16.1, all of which disappear quite rapidly. At the other extreme, the chlorinated insecticides, such as DDT, are extremely stable in soil and are only slowly decomposed, even in the presence of microbial action; this is reflected in the long persistence of these compounds.

The chemical makeup of a compound can markedly influence its susceptibility to microbial attack. This is seen in the relative persistences of two closely related herbicides: 2,4–D and 2,4,5–T. The second compound differs from the first only by the addition of a single chlorine atom, but as shown in Table 16.1, this markedly influences its persistence. Compound 2,4–D is readily broken down by microorganisms in soil, but the additional chlorine atom inhibits the action of microorganisms on 2,4,5–T.

The breakdown of pesticides in soil through the activity of microorganisms requires that all conditions necessary for good microbial growth in soil be available. Since pesticides are not complete microbial foods, this means that the other nutrients necessary for microbial growth (for example, nitrogen, phosphorus, sulfur, minerals) must be present in sufficient amounts as well as adequate moisture. The presence of other organic compounds in soil often promotes microbial action on pesticides by providing additional sources of food for microbial growth. Finally, it should be emphasized that only a restricted range of microbial species are able to break down a given pesticide. Thus, the proper microbial inoculum must be present. When a pesticide is first used in a particular soil, the proper inoculum probably is absent, so that initial biodegradation is very slow, but gradually, a population adapted to the compound builds up in the soil, so that biodegradation subsequently proceeds much more rapidly.

When using pesticides in agricultural situations, it is essential to know their persistence times, since applications over a period of years can lead to the build-up of perhaps toxic levels of a very persistent pesticide. Thus, it is desirable to assay new compounds for biodegradability before they are made commercially available. Such assays can best be done by adding the compound to a variety of soils and incubating under favorable conditions in the laboratory, then assaying for residual pesticide. Attempts should also be made to develop microbial populations in soils specifically adapted to the compound in order to study the possibility for long-term adaptation. Finally, the persistence of the compound in field trials should be studied before it is used widely in general agricultural practice. If the compound is unusually stable, it may be necessary to forgo its use in agriculture in order to avoid the build-up of toxic levels in soils.

By understanding the principles of biodegradation, the organic chemist may be able to synthesize pesticides that have low persistence rates in soil. We now know that certain types of molecules present more stable structures to microorganisms than others. Chemists can use this information to decide on the appropriate type of compound to synthesize for a specific agricultural task. In this way, the information on microbial degradation of pesticides can be most successfully applied to the problem of pesticide persistence.

16.4 Root-nodule bacteria and nitrogen fixation

In Section 13.5, we discussed the nitrogen cycle and the general importance of microorganisms in nitrogen transformations in nature. One of the most important microbial processes in the nitrogen cycle is nitrogen fixation, the conversion of nitrogen gas, N_2, into combined nitrogen. This process is carried out only by microorganisms, and higher plants are thus dependent on microorganisms for any nitrogen they obtain via nitrogen fixation. There are two main types of nitrogen-fixing microorganisms: free-living and symbiotic. Free-living nitrogen fixers include the blue-green algae and soil bacteria of the genera *Azotobacter* and *Clostridium*. Although these organisms may make significant contributions to the nitrogen budget of soils, by far the most important nitrogen-fixing organisms in soils are those which are symbiotic, living in association with the roots of higher plants.

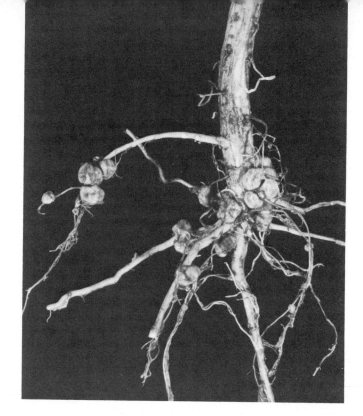

FIGURE 16.5
Soybean root nodules.

The most important symbiotic nitrogen-fixing organisms are members of the genus *Rhizobium*, which live in association with legumes. The main legumes in agriculture are crop plants such as clover, alfalfa, and soybeans, and vegetable crops such as peas and beans. Other legumes of less importance agriculturally include many tropical trees.

ROOT-NODULE BACTERIA OF LEGUMES

FIGURE 16.6
Steps in the formation of a root nodule in a legume infected by *Rhizobium*.

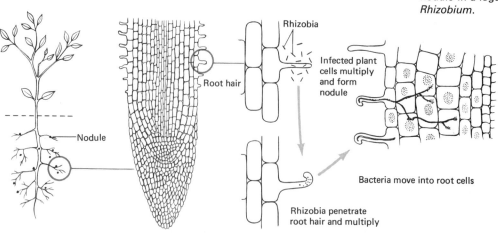

AGRICULTURAL MICROBIOLOGY

Rhizobium cells are Gram-negative motile rods. They are able to infect the roots of legumes and cause the formation along portions of the root system of *root nodules*, which are tumorlike growths in which nitrogen fixation occurs (Figure 16.5). Neither the *Rhizobium* alone nor the legume alone is able to fix nitrogen, yet the interaction between the two organisms leads to the acquisition of this property.

NODULE FORMATION

Nodule formation is a complex process (Figure 16.6). The roots of the legume excrete nutrients that encourage the growth of *Rhizobium* in the zone immediately adjacent to the root. Some of the bacteria then penetrate the root through hairlike projections from the surface of the root (root hairs) and pass down the root hairs to enter root cells. Infection of root cells by bacteria leads to multiple division of the root cells and the formation of the enlarged root-nodule structure. The cells of the root nodules are heavily packed with *Rhizobium*. The *Rhizobium* cells in the nodule are considerably altered in shape from the same bacteria outside the root. Within the root cells, the bacteria are swollen and misshapen and are called *bacteroids* (Figure 16.7).

NODULE FUNCTION

The mature nitrogen-fixing nodule is pink; this color results from the production of a red hemoglobinlike protein (called *leghemoglobin*) within the nodule. Although hemoglobin is very common as a blood protein in the animal world, this is one of the few places in the plant world where it is found. The hemoglobin acts in the nodule to bind O_2 and keep it from interfering with the nitrogen-fixation process. It is well-established that nitrogen fixation is an anaerobic process and thus very sensitive to O_2. Since the soil surrounding the plant root is aerobic, O_2 readily penetrates to the nodule. This O_2 is firmly bound by the hemoglobin, rendering the interior of the nodule anaerobic, so that the nitrogen fixation process occurs at the maximal rate.

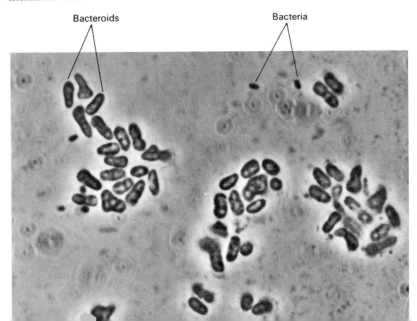

Bacteroids

Bacteria

FIGURE 16.7
Photomicrograph of vegetative bacteria and bacteroids from a crushed root nodule; magnification 2,200×. *Courtesy of F. J. Bergersen.*

Energy for nitrogen fixation in the nodule comes from the plant. Sugars produced in the plant leaves are transported to the nodules and promote nitrogen fixation by serving as nutrients for *Rhizobium* growth. Nitrogen (N_2) passes into the nodule and is converted into amino acids. The amino acids then pass out of the nodule into the roots, stems, and leaves and are used as a nitrogen source for plant growth.

As the root ages, the nodules eventually die and break down, releasing the bacteroids back into the soil. The bacteroids do not multiply there, but there is always a small number of normal *Rhizobium* cells present in the nodule, and these are released also. It is these normal *Rhizobium* cells that multiply in the soil to keep a reservoir of infective cells available. They can serve as the source of infection for a new legume.

There is a surprising specificity for the *Rhizobium*-legume interaction. A single *Rhizobium* strain can infect certain species of legumes but not others; similarly, a given legume can be infected by only one or a very few of the many known strains of *Rhizobium*. A knowledge of which *Rhizobium* strains are infective for a legume is important in agriculture, since it is common practice to inoculate the seed with the appropriate strain before sowing; naturally, the correct strain of bacterium must be used. Cultures suitable for inoculation can be purchased commercially, usually with the seed. Inoculation of seed is especially important if the legume crop has not been previously raised in the area or has not been raised for some time, since a *Rhizobium* suitable for this legume will probably not be already present in the soil. Inoculation is especially important when new lands are planted with legumes for the first time.

In nitrogen-deficient soils, nodulated legumes grow better than nonnodulated plants (Figure 16.8); thus, nodulation is a distinct advantage to the plant. It is also beneficial agriculturally, since a leguminous crop can be grown for a season and then plowed into the soil, thus adding considerable combined nitrogen to the soil. However, it is debatable whether such practice is preferable over using chemically synthesized nitrogen fertilizers. It is usually considered that in regions of high-intensity, high-cost agriculture, such as in North America and Europe, where land costs are high, chemical fertilizer is cheaper, whereas in low-intensity agricultural areas, such as Australia and New Zealand, use of legumes may be better.

In addition to the legume-*Rhizobium* symbiosis, nitrogen-fixing symbioses occur in a variety of nonlegumes, involving bacterial genera other than *Rhizobium*. One example is alder, a common tree that is able to establish itself on poor soils and barren sites. Alder forms

FIGURE 16.8
Nodulated and unnodulated
alfalfa plants growing in
nitrogen-poor soil.

Unnodulated   Nodulated

root nodules in symbiosis with an as yet unknown species of bacterium. Some tropical plants have bacterial nitrogen-fixing symbioses
involving leaf nodules instead of root nodules. In some other cases,
the symbiosis involves a nitrogen-fixing blue-green alga instead of a
bacterium.

16.5
Microbial transformations
of nitrogen fertilizers

Nitrogen fertilizers are widely used in agricultural practice to
increase plant production. Three types of nitrogen fertilizer are
used: nitrate (NO_3^-), ammonia (NH_3), and urea ($H_2N-CO-NH_2$).
All three of these forms are transformed in various ways by microorganisms, and the microbial transformations may be important in
determining the persistence and availability of the fertilizer after its
application to the field.

NITRATE FERTILIZER

Potassium or sodium nitrate is a convenient form of nitrogen fertilizer to add to soils. The nitrate ion is very water soluble, and the
nitrogen is thus rapidly available to plants. In the absence of oxygen,
a variety of bacteria are able to convert nitrate into nitrogen gas, N_2,
thus changing the material into a form unavailable for higher plants.
This process is called *denitrification* and is very undesirable agriculturally. Bacteria that denitrify are carrying out a type of anaerobic
respiration (see Section 6.2), using the nitrate ion instead of O_2 as an

electron acceptor. Thus, for denitrification to occur, soil conditions must be anaerobic, and a source of organic matter must be available for bacterial metabolism.

Since denitrification is undesirable agriculturally, many farming practices are designed to minimize it. Draining of soils to ensure that they do not become waterlogged is of value, since flooded or waterlogged soils quickly become anaerobic and thus favor denitrification. The practice of tilling soil probably inhibits denitrification, since it breaks up large soil clumps within which anaerobic conditions generally develop. However, even in well-drained soils of good texture, some denitrification may occur if large amounts of organic matter are added (as from manure or other organic fertilizer); the organic matter promotes the growth of facultative anaerobic bacteria that use up the oxygen present and create temporary or partial anaerobic conditions. Therefore, a proper carbon-to-nitrogen balance is desirable in agricultural soils.

In large farm practice, the most commonly used nitrogen fertilizer is ammonia, NH_3. Ammonia is preferable to nitrate because on a weight basis it has more nitrogen (the H atoms of ammonia weigh less than the O atoms of nitrate), and it can be more conveniently applied to large fields. Ammonia is normally a gas, but it can be liquefied under pressure, and it is applied to fields in this form. A high-pressure tank is pulled across the field and the ammonia is injected from a nozzle directly into the soil. As soon as the gaseous ammonia contacts the soil particles, it is adsorbed and converted in the soil into the ammonium ion (NH_4^+), which is not gaseous. The water-soluble NH_4^+ can be used directly by higher plants, but most commonly it is converted into nitrate through the action of the nitrifying bacteria, a process called *nitrification*. We discussed nitrification briefly when we considered the overall nitrogen cycle in Section 13.5.

Nitrification is an aerobic process and is thus favored by aeration of the soil. The nitrifying bacteria are widespread in soils and water and are commonly found in fields where ammonia fertilizer is used. Nitrification occurs most readily in soils with neutral to alkaline pH, and is inhibited in acidic soils.

UREA

Urea is an organic form of nitrogen, $H_2N—CO—NH_2$, that is sometimes used as a fertilizer. It is a crystalline solid and is very water-soluble, so that it can be easily applied to soils. It is hydrolyzed in water by the enzyme urease, to yield ammonia and carbon dioxide:

$$H_2N—CO—NH_2 + H_2O \rightarrow 2NH_3 + CO_2$$

AGRICULTURAL MICROBIOLOGY

Urease is a common enzyme in many soil bacteria, and the conversion of urea to ammonia in soil is primarily a bacterial process. The ammonia liberated from urea may then be converted to nitrate by the nitrifying bacteria.

16.6 Plant diseases

Plant diseases caused by microorganisms are of major economic concern to agriculturalists, and a knowledge of methods of control of plant diseases is crucial to ensure high crop yields. Under some conditions, a plant disease can affect the well-being of a whole population or country. The classic case of this was the Irish potato blight, recounted briefly in Section 11.1. As a result of this fungal disease, there was a mass migration of people from Ireland.

Although the principles of host-parasite relations discussed in Chapter 8 are applicable to plant diseases, there are so many differences between plants and animals that a separate discussion of plant diseases seems desirable. Bacteria, fungi, and viruses all cause diseases of plants, although fungi cause the most serious and widespread diseases of crop plants.

NORMAL FLORA OF PLANTS

Plants, like animals, have a normal flora of microorganisms that are not pathogenic. Although microbes can grow on all parts of a plant (flowers, fruits, seeds, leaves, stems, and roots), it is on the roots that the most extensive normal microbial flora develops. The region in the soil immediately around the roots is called the *rhizosphere*, and here large numbers of harmless bacteria and fungi are present, living on nutrients excreted from the roots. Under some conditions, microorganisms are able to penetrate into the living root tissue itself without causing special damage. Trees often have fungi living attached to their roots; this fungus-root association is called a *mycorrhiza*, and mycorrhizae are of considerable benefit to the tree (see Section 13.1).

PLANT ORGANS AND PLANT DISEASES

Plant pathogens may cause disease in plants in a variety of ways: (1) by digesting the contents of host cells; (2) by producing toxins, enzymes, and other substances that kill or damage the function of host cells; (3) by absorbing food materials from the plant and thus depleting the plant's own food supply; (4) by blocking the transportation of food, minerals, and water through the conductive vessels of the plant as a result of microbial growth in the vessels.

Plant diseases are given names based on the organ affected and on the appearance of the diseased organ (Figure 16.9). Thus there are root rots, leaf spots, fruit rots, wilts, cankers, leaf blights,

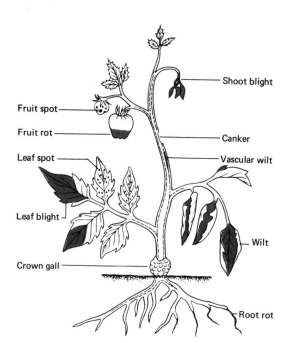

FIGURE 16.9
Typical plant diseases caused by microorganisms.

Shoot blight

Fruit spot

Fruit rot

Canker

Leaf spot

Vascular wilt

Leaf blight

Wilt

Crown gall

Root rot

galls, and mildews. A major group of diseases is the rusts, in which red-colored spores developing on the leaves give the plant a rusty appearance.

Although the microorganism is the primary cause of a plant disease, environmental factors greatly influence its ability to attack the plant. Temperature, moisture, light, soil acidity, and availability of plant nutrients are all factors that influence the severity of disease. Certain diseases may be common in summers that are cool and damp, whereas others may be more common in warm dry summers. Some plant varieties are more susceptible to specific diseases than are others, and disease incidence is thus affected by the variety of plant used.

FACTORS AFFECTING PLANT DISEASE

Control of plant diseases is achieved differently than control of animal and human diseases, since rarely is an individual plant of such value that therapeutic measures are justified. Control of plant diseases is primarily by prevention rather than by treatment.

CONTROL OF PLANT DISEASES

CHEMICAL CONTROL   Chemical methods are employed primarily to protect plants against future disease. Leaf sprays and dusts, seed treatments, soil fumigation, and soil sterilization are some of the treatments used. One of the first chemical treatments used was Bordeaux mixture, a combination of copper sulfate and lime. This was first used as a leaf dust near Bordeaux, France, for controlling powdery mildew of grapes. A number of sulfur preparations are used, including both inorganic and organic forms of sulfur. Inorganic and organic mercury compounds are also highly effective, but because of the high toxicity of mercury to humans and animals, these compounds are used primarily in the treatment of seeds before planting; in that way, none of the mercury reaches the final harvested crop. A number of organic compounds are used, including both synthetic chemicals and antibiotics. Of the antibiotics, the most widely used are streptomycin and tetracycline, which control certain bacterial diseases, and cycloheximide and griseofulvin, which are used to control some fungal diseases (see Section 4.6).

An important factor in the use of chemicals to control plant diseases is the cost of treatment. Usually, when chemicals and antibiotics are used to control human diseases, cost is not the prime factor, but in agriculture, the cost of treatment must be balanced against the anticipated economic gain from its use. Thus, some effective chemicals are not used in practice because they are too expensive to manufacture or apply.

OTHER CONTROL METHODS   Employed in some cases are: (1) mild heat treatments of seeds to eliminate viruses or other pathogens; (2) crop rotation, in which crops not susceptible to the pathogen are planted for a year or two so that the pathogen finds no host to grow on and dies out; (3) quarantine and inspection, to prevent the entry of a pathogen into an area where it has not yet become established; (4) eradication of alternate hosts for a pathogen (these are noncrop plants that serve as reservoirs of infection); (5) appropriate fertilization, tilling, and weed control, which promote vigorous growth of the crop and increase its disease resistance; (6) control of agents such as insects that may disperse the pathogen; (7) use of resistant varieties of the crop.

The field of plant pathology is highly developed, and tremendous advances have been made in the control of plant diseases. These controls have led to marked increases in crop yields and have to a great extent contributed to the agricultural revolution that has taken place in the world in the past 50 years. However, continual vigilance is necessary as new pathogens arise, and research and development aimed at controlling plant diseases must continue. Also, more effective and less toxic chemical agents are needed, so that the use of mercury and copper compounds potentially harmful to humans can be eliminated. In this research effort, the microbiologist and plant pathologist play major roles.

In the manufacture of the fabric linen from flax, the fibers of the flax plant are loosened from the stems by a process known as *retting*, which involves soaking the stems in water. Loosening of the fibers occurs primarily through the action of microorganisms, which produce enzymes that attack the substances holding the plant cells together. Retting can be accomplished by either aerobic or anaerobic microorganisms.

The aerobic process, called *dew retting*, is a very simple procedure, and involves merely laying the flax stems out in a field under fairly humid conditions. Both fungi and aerobic bacteria are involved; one bacterium identified in the process is *Bacillus comesii*. This type of retting is not a well-controlled process, since much depends on atmospheric conditions, and often a fiber of poor quality develops. Because of this, a *vat process* has been developed; the stems are placed in a tank of water that is aerated and inoculated with *B. comesii*. This process is better controlled than the natural one, and a fiber of higher and more uniform quality is obtained.

The *anaerobic retting process* is the traditional European process and is widely used in Belgium, Germany, Holland, Italy, and Ireland. The flax stems are submerged in bundles at the bottom of a slow-flowing river or a pond. Anaerobic conditions quickly develop within the bundles as a result of the action of facultative bacteria, which oxidize the organic matter leaching out of the stems. Anaerobic bacteria such as *Clostridium felsineum* and *C. butyricum*, which are able to attack the cellular cement, then develop and carry out the actual retting. It takes up to 2 weeks for the fibers to become loosened from the stem; they are then peeled away, washed, dried, and spun into thread for weaving. It is important in retting that organisms able to attack the cellulose of the fiber do not develop, since cellulose is the main component of the finished product. If the bundles are soaked too long, cellulose-digesting bacteria may cause deterioration of the fibers, resulting in a low-quality end product. In addition to flax, retting is also used for the preparation of fibers of the hemp plant, used in the manufacture of rope and twine.

Insects cause a wide variety of diseases in plants, and control of insect infestation usually can greatly increase crop yield. Because a number of undesirable effects can arise from the use of chemical pesticides, methods for controlling insects by nonchemical means have been sought. One potentially useful method is the use of microbial insecticides. These are toxins produced by microbial pathogens of insects.

Insects are susceptible to a wide variety of microbial pathogens. In virtually every case, these insect pathogens are quite specific and have no effects on humans or higher animals, which makes them very desirable in comparison to chemical insecticides. One major

TABLE 16.2

Some insect pests that have
been controlled by virus or
bacterial insecticides

Insect	Insecticide
Cotton bollworm	*Bacillus thuringiensis*
	Nuclear polyhedrosis virus
Cabbage looper	*Bacillus thuringiensis*
	Nuclear polyhedrosis virus
Cabbage worm	*Bacillus thuringiensis*
	Granulosis virus
Gypsy moth	Nuclear polyhedrosis virus
	Bacillus thuringiensis
Forest tent caterpillar	*Bacillus thuringiensis*
	Nuclear polyhedrosis virus
Orchard tent caterpillar	*Bacillus thuringiensis*
Codling moth	Granulosis virus
	Bacillus thuringiensis
Spruce budworm	Nuclear polyhedrosis virus
	Bacillus thuringiensis
Japanese beetle	*Bacillus popilliae*

From R. P. Jaques, "Methods and effectiveness of distribution of microbial insecticides," New York Academy of Science *Annals*, **217**, 109–119, 1973.

group of insect pathogens is comprised of certain members of the genus *Bacillus*. During sporulation, these spore-forming bacteria produce a protein that is toxic to specific insects. The most widely used organism is *B. thuringiensis*; this toxin-forming bacillus grows in various moth larvae and causes a fatal disease. Since some moth larvae (for example, cabbage worm, tent caterpillar, gypsy moth) destroy plants, the toxin acts to control these insect infestations.

In actual practice, the toxin-producing organism is cultured in the laboratory and a preparation containing toxin is obtained. This toxin is then formulated into a dust or spray and the product distributed in the same manner as a chemical insecticide. Because of the high specificity of the toxin, there is no danger to the health of human beings or livestock. The material cannot be used in areas where silkworms are raised, however, since these silkworm moth larvae are also quite susceptible. A number of different insect pests can be controlled with the *B. thuringiensis* product (Table 16.2).

Another species of *Bacillus*, *B. popilliae*, has a much more specific action, affecting just the Japanese beetle (see Table 16.2). This agent is thus less widely useful but has proved effective for the single important insect pest that it attacks.

Several viruses have also been found effective in controlled trials. Nuclear polyhedrosis viruses and granulosis viruses affect some of the same insects that are affected by *B. thuringiensis* (see Table 16.2) but have the additional advantage that they actually reproduce in the larvae under field conditions after application, giving some degree of long-term control. However, repeated applications of the viruses are usually necessary in practice because the viruses become inactivated on foliage as a result of exposure to sunlight. Additionally, continued growth of the crop plant leads to the production of new leaves that do not contain the virus.

The microbial agents that have been used so far are essentially insecticides, attacking the insects present at the time of application but having little or no long-term effect. They do not differ in principle from chemical insecticides, except that they are much more specific and have minimal or no harmful effects on the environment. The ideal type of agent for control of insect pests is an agent that multiplies after application and is transmitted from one insect to another. Such an agent would spread through the insect population and would, if sufficiently effective, eventually wipe out the insects in the target area. To date, microbial pathogens with these properties have not been discovered, but the present results are sufficiently encouraging to warrant further research along these lines. Toxin-forming microbes provide an entirely new approach to insect control, and one that may have a great future.

Mycotoxins are chemical toxins produced by fungi; they are discussed briefly in Section 11.1. The best-known mycotoxins are those called *aflatoxins*, produced by the fungus *Aspergillus flavus*. Aflatoxins are found in grains infected by this fungus, and the feeding of such moldy, toxin-containing grain to livestock can result in serious liver disease and death. Aflatoxins were only discovered in the early 1960s, but we already have a good understanding of their nature and significance. The fungus that produces aflatoxins is widespread in the soil and generally causes no harm. Grain becomes infected with the fungus in the field very easily; under improper storage conditions, the fungus can grow in the harvested grain and produce aflatoxins. Grain is usually stored in large storage bins or silos called *grain elevators*. The most important factor determining whether the aflatoxin-producing fungus will grow is the moisture content of the grain. If the moisture content is low and the grain is stored dry, mold growth does not occur. However, if grain is harvested at too high a moisture content, as might occur at the end of an unusually wet season, the fungus may grow in the grain and produce aflatoxins. Susceptible crops include wheat, corn, oats, peanuts, and rice.

Aflatoxins act by affecting the liver. In low doses, they induce the formation of liver cancer (hepatoma), and in higher doses, they cause a general liver toxicity that can lead to death. Animals affected include all cattle, pigs, chickens, turkeys, and horses. Also affected are fish, and extensive damage to rainbow trout in fish hatcheries has resulted from feeding moldy grain. In addition, humans may be affected if they eat foods such as peanut butter or flours prepared from moldy raw materials.

16.9
Mycotoxins and grain storage

The variety of aflatoxins so far identified is large; the major ones include aflatoxin B_1, G_1, M_1, and M_2. All of these can be produced in grain, but of further interest is the fact that aflatoxins M_1 and M_2 can also be produced in the animal body from aflatoxins B_1 and G_1. Tracing the source of aflatoxins in dairy products is complicated by the fact that aflatoxins ingested by the dairy cow can be present in the milk, hence finding their way into milk products (Figure 16.10).

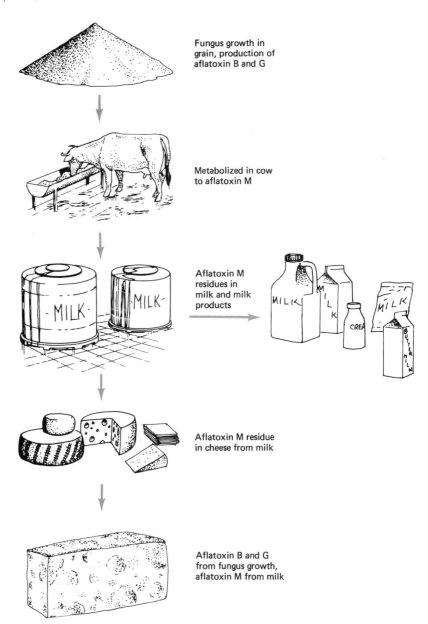

Fungus growth in grain, production of aflatoxin B and G

Metabolized in cow to aflatoxin M

Aflatoxin M residues in milk and milk products

Aflatoxin M residue in cheese from milk

Aflatoxin B and G from fungus growth, aflatoxin M from milk

FIGURE 16.10
Aflatoxin transfer and production in dairy products.

Growth of *A. flavus* is controlled primarily by maintenance of proper storage conditions in grain bins. If the bins are filled with grain at low moisture content (less than 14 percent moisture) and the bins are kept dry, the fungus will not grow. Because of the serious nature of the disease, considerable caution must be exercised to be sure that grain is stored properly. Since grain is often stored in large quantities, mycotoxin development is a worldwide problem and is of considerable economic importance. Fortunately, we have a good knowledge of the proper methods of control of these important infestations.

There are a number of ways in which a knowledge of microbiology can be applied to the care and feeding of livestock. Veterinary medicine is one obvious application, and the principles of host-parasite relations and immunology described in Chapter 8 apply directly to animal diseases. In addition, however, there are several specific aspects of the relationship of microorganisms to livestock, and we shall discuss these here.

16.10 Animal husbandry

Considerable success has been achieved in eradicating from livestock and fowl a number of serious animal diseases that had previously caused significant economic loss. Research continues on many of these diseases because they continue to be widespread in other parts of the world and can possibly be reintroduced into the country, especially in these times of easy transportation. In the United States, research on exotic animal diseases is restricted to a single location, the Plum Island Animal Disease Laboratory, located on an 800-acre island in Long Island Sound. The island location makes it easy to control movement of animals, and exotic diseases can be studied here under strict quarantine. Plum Island is not open to the general public and can be reached only by the laboratory's boats. Test animals coming to Plum Island are first held in quarantine for 2 weeks, then taken to the laboratory through two air locks (Figure 16.11). Once inside the research building animals are never allowed out again.

Diseases under study at Plum Island, and of general concern in livestock health, include some of the most serious diseases that plague livestock worldwide. *Foot-and-mouth disease*, for example, is a highly communicable disease caused by a virus. The virus causes lesions in the mucosa of the mouth, in the skin around the hoofs, and on the teats and udder. The virus spreads throughout the whole body, and all secretions and excretions contain the virus. Foot-and-mouth disease has been introduced into the United States six times in the twentieth century and has been eradicated each time. The last

ANIMAL DISEASES

FIGURE 16.11
Research on exotic animal diseases at Plum Island Laboratory. *a* Test animals coming to Plum Island Laboratory are first held in quarantine for 2 weeks. *b* They are then taken up a ramp into the laboratory. *c* They pass into the laboratory through two air locks. *Courtesy of J. J. Callis, Plum Island Animal Disease Center.*

a

c

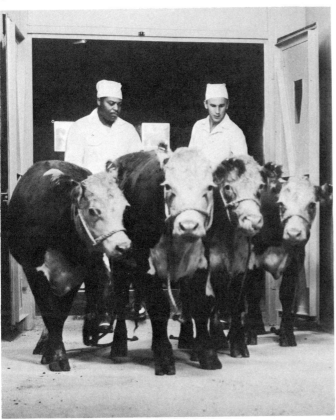

b

outbreak was in California in 1929, when the virus was introduced in ships' garbage. There have been more recent outbreaks in neighboring countries: Canada in 1952 and Mexico in 1953 and 1954. In the United States, a strict quarantine is imposed immediately around any affected area, and all infected animals and exposed susceptible livestock are killed and buried on the premises. After 30 days, a few susceptible animals are introduced onto the farm by the federal government and are observed for disease. If no symptoms develop after another 30 days, permission is granted for introduction of new stock, but the farm is maintained under surveillance for an additional 90-day period. Owners receive payments from the government to compensate for their livestock losses. In countries where the disease is endemic, control is effected by vaccination, but there is no point in vaccinating in the United States, since the disease does not exist here, and the eradication measures when necessary are less costly than universal vaccination.

Other diseases under study at Plum Island include *African swine fever*, a viral disease of pigs; *rinderpest*, a viral disease of ruminants, particularly cattle and buffalo; *contagious bovine pleuro-pneumonia*, a disease of cattle caused by a bacterium of the mycoplasma group; *Teschen disease*, a poliotype disease of swine; and *fowl plague*, a fatal viral disease of chickens and turkeys. In addition to research, the laboratory carries out a diagnostic function and is able to characterize and identify a large number of exotic animal diseases. Because few U.S. veterinarians are familiar with these diseases, it is important to maintain a competent diagnostic service in the country, so that blood and tissue samples may be expertly examined.

Continuous watchfulness is important in maintaining the health of animal populations. In the long run, money spent on diagnosis, quarantine, and eradication is well spent because it prevents the far greater economic losses that would occur as a result of epidemics.

RUMINANTS AND
MICROORGANISMS

Ruminants are animals that possess a special organ, the *rumen*, in which the initial digestion of feeds takes place. Some of the most important domestic animals, the cow, sheep, and goat, are ruminants. In addition, a number of wild animals are ruminants, including the deer, elk, moose, buffalo (bison), antelope, and camel.

Hay and grass, the main feed of ruminants, contain as their main food constituent cellulose, a material that is generally indigestible by higher animals. In humans and other meat-eating animals, cellulose passes through the intestinal tract virtually undigested. In ruminants, on the other hand, cellulose digestion does occur, through the agency of microorganisms living in the rumen.

The rumen is a large saclike organ, about 100 liters (26 gallons) in volume in the cow and 6 liters (1.5 gallons) in volume in the

FIGURE 16.12
Photograph of a cross section through a rumen. The esophagus, where food enters, is at the upper left. From N. V. Benevenga, S. P. Schmidt, and R. C. Laben, *J. Dairy Sci.*, **52**, 1294, 1969.

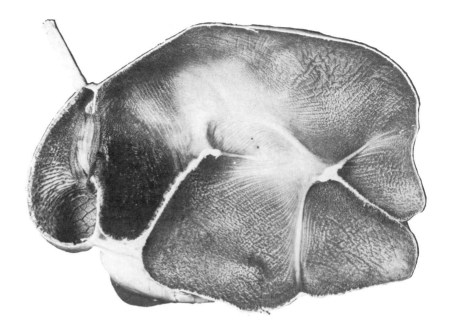

sheep, into which the feed enters first (Figure 16.12). In the rumen, the feed is acted upon by anaerobic cellulose-digesting bacteria. Cellulose is broken down first to glucose, and the glucose is then converted to organic acids, mainly acetic acid. These acids pass out of the rumen into the bloodstream and are then used by the animal as its main source of energy for respiration. These bacteria use as a source of nitrogen the simple inorganic compound ammonia, which is present in the rumen fluid, and convert this into the organic nitrogen of protein. The growth of bacteria can be increased by feeding an additional source of ammonia along with the hay or grass. Bacteria grow to large numbers in the rumen while they are digesting the cellulose and are passed out of the rumen into the stomach and intestine, where they are themselves digested (Figure 16.13). The main source of protein for the animal is thus the rumen bacteria. Through the mediation of the bacteria, the animal is therefore able to subsist on ammonia as the sole source of nitrogen for proteins; all other animals require at least part of their nitrogen in organic form, as preformed amino acids. The rumen bacteria also synthesize all the vitamins the animal needs for growth and survival. Because of the activity of these rumen bacteria, the animal is able to subsist on food materials low in protein, which are quite insufficient for the growth of nonruminant animals.

The microorganisms of the rumen are obligate anaerobic types that are specifically adapted to this habitat and are found nowhere else. Both bacteria and protozoa are found in the rumen. The bacteria are mainly nonsporulating Gram-negative rods and vibrios;

Reactions in the rumen

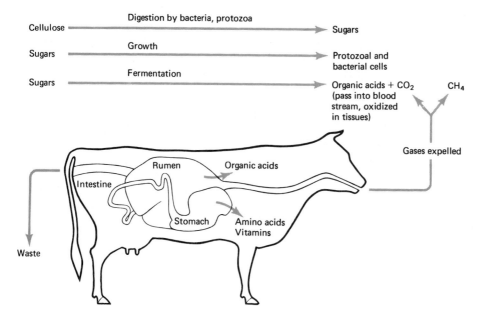

Cellulose $\xrightarrow{\text{Digestion by bacteria, protozoa}}$ Sugars

Sugars $\xrightarrow{\text{Growth}}$ Protozoal and bacterial cells

Sugars $\xrightarrow{\text{Fermentation}}$ Organic acids + CO_2 (pass into blood stream, oxidized in tissues) CH_4

Gases expelled

Rumen   Organic acids

Intestine

Stomach   Amino acids Vitamins

Waste

Reactions in the stomach and intestine

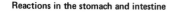

Rumen bacteria $\xrightarrow{\text{Digestion of bacteria by cow}}$ Amino acids, vitamins (assimilated)

FIGURE 16.13
Microbial processes in the ruminant.

the protozoa are ciliates. Although the protozoa are not essential for rumen function, when present they definitely contribute to the process. Their numbers are usually larger in animals fed a good ration than in those fed a poor diet, and their presence may possibly serve as an indication of the well-being of the animal. The protozoa digest cellulose, but they also eat and digest some of the rumen bacteria. Thus the protozoa may serve to keep bacterial numbers in check.

A knowledge of microbial processes in the rumen can be put to practical use in the intelligent management of ruminant animals. As we have seen, the protein value of cattle feeds may be increased by addition of inorganic nitrogen, since the rumen bacteria can convert inorganic nitrogen into protein. The usual nitrogen source fed to ruminants is urea, which is broken down in the rumen to carbon dioxide and ammonia, the ammonia then being converted into bacterial protein. Feeding cereal grains to cattle is a common practice, but this is wasteful and inefficient since grains are already high in protein and contain carbohydrate in the form of starch, which is readily digestible by nonruminants. Thus, there is no advantage in

AGRICULTURAL MICROBIOLOGY

feeding grains to ruminants, and in fact, there is a disadvantage, since ruminants are less efficient in handling these feeds than are nonruminants such as pigs and chickens. In practice, feeding grain to cattle is done primarily to fatten them for slaughter and to ensure that the flavor of meat cuts is superior. If grains are in short supply, they should not be fed to ruminants.

ANTIBIOTICS IN ANIMAL FEEDS

It has been well established that the growth rate of nonruminant livestock, such as pigs and chickens, is increased if they are fed small amounts of antibiotics in their feed. In fact, antibiotic supplement to feeds for these animals has been a widespread practice, since the more rapid growth permits the farmer to take the animals to market earlier. Antibiotics used most frequently as feed supplements are penicillin and tetracycline.

One way in which antibiotics stimulate growth is by reducing or eliminating from the normal flora bacteria that produce small amounts of toxins which decrease growth rate. One organism affected by antibiotic feeds is *Clostridium perfringens*, a bacterium present as a minor member of the normal flora, which produces a series of toxins. With the levels of antibiotic used in feeds, most of the normal flora of the body is unaffected, while *C. perfringens*, which is fairly antibiotic sensitive, is eliminated. With the organism gone, the toxins are also absent and growth is stimulated.

Another mechanism for growth stimulation by antibiotics relates more specifically to pathogenic bacteria. Livestock are frequently infected with pathogens that cause minor, inapparent infections. Such infections may not be sufficiently severe to appear as frank disease but may reduce growth rate. Since livestock are frequently kept in crowded quarters, the chance of transmission of pathogens from one animal to another is high. Antibiotic feed supplements probably keep down the extent and severity of infection by these minor pathogens, thus stimulating growth of the animals. Naturally, antibiotic feed supplements are not used in ruminants, since they would affect the rumen bacteria that are vital to the digestive processes of these animals.

Recently, there has been considerable concern about possible harmful effects of routine antibiotic feed supplements. This has arisen primarily because of the development of antibiotic resistance in bacteria mediated by *resistance transfer factors*. Resistance transfer factors are plasmids that confer antibiotic resistance and are transferrable from one cell to another (see Section 7.7). In the late 1960s, there was an explosive increase in antibiotic-resistance plasmids, and one explanation for their origin was that the use of antibiotic feed supplements had led to the selection of bacteria containing such plasmids. Although there is no direct evidence of the transfer of such plasmids to bacteria infecting humans, such a possibility is theoretically possible. It thus becomes essential to consider whether the obvious benefits of antibiotic feed supplementation may be overshadowed by the possible harmful effects arising from antibiotic-resistance plasmids. One alternative is to use as feed supplements antibiotics which are not used in medicine, thus avoiding the build-up of strains resistant to medically important antibiotics.

Silage is a product made by allowing green hay, grass, cereal crops, or mixtures of these to undergo a bacterial fermentation, thus preserving the feed value of the crop so that it can be fed to animals at seasons of the year when fresh feed is not available. The aim of silage making is to obtain through bacterial action the production of sufficient lactic acid to inhibit the growth and activity of spoilage microorganisms. The production of silage is an old practice and is widely done in temperate climates, especially where an intensive dairy industry has developed. Silage is particularly important in dairy cattle operations, since the animals must remain on good feed year-round. The crudest method of silage production involves packing the crop into a pit dug into the ground; if properly constructed and managed, satisfactory silage can be made in such a pit silo, although the structure is not permanent. The most widely used silo in regions of intensive dairy farming is the tower silo, constructed either of concrete or metal (Figure 16.14). To fill a tower silo, the crop is chopped into small pieces and blown up a pipe to the top. The silo must be packed carefully so that the crop is evenly distributed and no air spaces are present, since the silage fermentation is an anaerobic process; if air is present, the growth of spoilage microorganisms may occur. For making proper silage, the moisture content of the crop is also important. If the crop is too wet the consistency of the silage will be unsatisfactory for feeding, whereas if it is too dry the growth of microorganisms will not be satisfactory and insufficient acid will develop. The top of the silo must of course be covered to protect the product from rain and dirt. In some cases, molasses may be mixed with the crop to provide an additional source of sugar for the acid-producing bacteria in the silage.

The organisms involved in the silage fermentation are mainly lactic acid bacteria of the genera *Streptococcus* and *Lactobacillus*. These bacteria are widespread in the fields and naturally contaminate the crop, although in small numbers. During the early stages of the fermentation, facultative anaerobes develop and consume oxygen, converting the plant mass to anaerobic conditions. Then the lactic acid bacteria take over and produce sufficient acid so that the pH is reduced to about 4. At such pH values, spoilage bacteria cannot develop, and the anaerobic conditions prevent growth of spoilage fungi; thus a stable product is obtained.

FIGURE 16.14
Conventional upright silo.

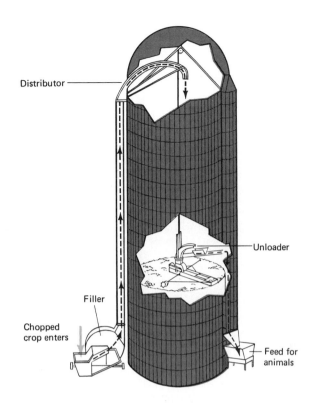

Distributor

Unloader

Filler

Chopped
crop enters

Feed for
animals

Summary

We have seen in this chapter that microorganisms play a variety of beneficial and harmful roles in agriculture. They are essential agents in the processes involved in soil formation and also play major roles in promoting soil fertility. Microorganisms are involved in the breakdown of organic matter and the production of humus, a stable organic fraction that increases the water- and nutrient-holding capacity of the soil. Microbes are also responsible for the decomposition of many pesticide residues in the soil; persistent pesticides are not degraded because no microorganisms able to break them down are present. Microorganisms play major roles in the nitrogen economy of plants, including nitrogen fixation (both free-living and symbiotic), nitrification, and denitrification. Microorganisms cause many serious plant diseases; the fungi are the major plant pathogens. Control of plant diseases by use of chemical agents has greatly increased crop yields. Insect damage to plants can also be controlled by chemicals, but a new and potentially significant means is the use of microbial pathogens of insects. The toxins produced by these pathogens can be formulated into microbial insecticides of high specificity and low human toxicity. Cellulose-digesting microorganisms are key agents in the digestive process in ruminants (cows, sheep, goats) and permit these animals to live on feeds high in cellulose, such as hay and grass, which are not normally digestible by mammals. In addition to digestion of cellulose, the rumen microorganisms produce proteins and vitamins for the ruminant, thus permitting these animals to live on diets deficient in these vital materials. Knowledge of the beneficial and harmful effects of microorganisms has permitted the development of many agricultural practices that have greatly increased crop and animal production.

Study questions

1 Discuss the sequence of steps involved in the transformation of rock to soil. Which organisms are significant at each step?
2 How do podzol soils differ from grassland soils?
3 In what ways do microorganisms contribute to soil fertility?
4 Describe briefly the chemical and microbial changes that take place during the formation of a compost pile. Of what use is compost in agriculture?
5 What is the role of microorganisms in the decomposition of pesticides in soil? Why is it important to know if a pesticide is susceptible to microbial attack?
6 Why is nitrogen fixation so important to soil fertility?
7 What are the root nodules present on leguminous plants? How are they formed? What is their function?
8 In some regions, farmers increase the nitrogen content of their fields by planting legume crops, whereas in other regions, they add combined nitrogen directly in the form of fertilizers. What are the advantages and disadvantages of each method? When might one be preferable to the other?
9 Nitrogen fixation and nitrification tend to offset nitrogen loss due to denitrification. Discuss.
10 List three of the ways in which plant pathogens cause disease in plants.
11 What plant disease controls are most commonly used? What criteria must be met in selection of the control method?
12 Discuss the microbiological aspects of linen manufacture. Why do you think that, in our mechanized age, mechanical processing is not used instead?
13 What properties of microbial insecticides make them desirable for use in agriculture?
14 What are aflatoxins, and why are they a danger to livestock? What control measures are desirable?
15 The protein for a cow's diet is obtained by digestion of the millions of microbial cells that inhabit the rumen, not from hay. True or false? Discuss.
16 What are the effects of antibiotic supplements in animal feeds?
17 Describe the sequence of microbial reactions that occur in the production of silage.

Suggested readings

Agrios, G. N., *Plant pathology*, Academic Press, New York, 1969. Reviews principles of plant pathology and discusses characteristics and causes of major plant diseases.
Alexander, M., *Introduction to soil microbiology*, 2nd ed., John Wiley & Sons, Inc., New York, 1976. The most widely available textbook in soil microbiology.
Callis, J. J., "Plum Island laboratory: Its role in foreign animal disease research," *Agri. Sci. Rev.* **9**, 1–13, 1971. A brief review of the role of the

Plum Island laboratory, together with a discussion of the major exotic diseases of farm animals.

Gray, T. R. G., and S. T. Williams, *Soil microorganisms*, Oliver and Boyd, Ltd., Edinburgh, 1971. Nonchemical discussion of the roles and activities of soil microorganisms.

Hungate, R. E., *The rumen and its microbes*, Academic Press, New York, 1966. An advanced treatise on rumen microbiology.

Jukes, T. H., "Antibiotics in animals feeds and animal production," *BioScience*, **22**, 526–534, 1972. Brief review of the history, rationales, and dangers of antibiotic feed supplementation.

"Microbial insecticides," A symposium, New York Academy of Sciences, *Annals*, Vol. 217, June 22, 1973. A series of articles by the leading researchers, reviewing the fundamental and practical aspects of microbial insecticides.

Mirocha, C. J., and C. M. Christensen, "Fungus metabolites toxic to animals," *Annual Review of Phytopathology*, **12**, 303–330, 1974. Reviews the chemistry, microbiology, and toxicity of mycotoxins.

Poincelot, R. P., *The biochemistry and methodology of composting*, Connecticut Agricultural Experiment Station, New Haven, Conn., Bulletin No. 727, 1972. A technical and practical review of domestic and large-scale composting.

Chapter

17

Industrial microbiology

Industrial microbiology deals with processes involving the large-scale growth of microorganisms for the production of food, feed, vitamins, antibiotics, and industrial chemicals. It is a very diverse field and one of great importance to human economy. The factories and installations for growing microorganisms on a large scale are extensive and expensive, and many new problems arise that are unique to operations on such a scale. Engineers, chemists, and microbiologists work together, the role of the microbiologist being to select and prepare suitable cultures and to monitor the process for contamination and for proper growth and production.

17.1 Industrial fermentation

Industrial products are of three basic types. In the first, the microbial cells themselves are the desired product, such as bakers' or food yeasts. In the second, the desired material is a microbial product, such as an acid (for example, citric acid), an alcohol (for example, ethyl alcohol), an antibiotic, a vitamin, an amino acid, or an enzyme. In the third, called a *bioconversion*, the microorganism converts a chemical compound added to the medium to another compound of economic importance.

In the usual language of the industrial microbiologist, any large-scale microbial process is called a *fermentation*, and the large-scale vessel in which the process is carried out is called a *fermenter*. The size of an industrial fermenter is large, from 5,000 to 50,000 gallons (about 20,000 to 200,000 liters); and a 50,000-gallon fermenter is as tall as a three-story building (Figure 17.1). In such large vessels, the cost of the culture medium is a significant part of the total cost of the operation, and consequently, inexpensive materials are used, such as soybean meal, molasses, cornsteep liquor (a byproduct of the manufacture of corn starch) and distillers' solubles (a by-product of the whiskey-distilling industry). One of the main tasks of the industrial microbiologist is to find an inexpensive culture medium in which the process can be carried out efficiently. In most cases, this requires an understanding of the principles involved in the care and feeding of microorganisms.

STERILIZATION

With the large volumes involved in industrial fermenters, sterilization of the medium is difficult, since the penetration of heat to the

FIGURE 17.1
Large fermenter installation.
Courtesy of Chas. Pfizer and Co., Inc.

center of a large vessel is slow, and consequently, long heating times are necessary. The medium can be sterilized after the fermenting tank has been filled by passing steam under pressure through a metal jacket that surrounds the outside of the vessel. Alternatively, the medium can be sterilized by a process called *continuous sterilization*. In this process, which has some resemblance to the flash pasteurization of milk, the medium is pumped through a heater where it quickly rises to a temperature sufficiently hot so that it is sterilized in a few seconds, and it is then pumped into the empty but previously sterilized fermenter. In some fermentations, sterilization is not necessary, either because the growth of the desired microbe is so rapid that contaminants cannot take over or because the medium or culture conditions are unfavorable for the growth of other microbes. Sterilization is not used, for instance, in the production by yeast of alcohol for distillation.

Great care must be taken to keep the fermenter scrupulously clean. Any culture medium left within the tank or its pipes and plumbing is a potential source of nutrients for the growth of contaminants that may be difficult to eliminate.

Many industrial fermentations are aerobic, and it is important that sufficient air be provided during the process. It is difficult in large fermenters because of the great bulk of the liquid to be aerated. In order to get as much air as possible into the culture, two procedures are used simultaneously: (1) Liquid in the tank is stirred rapidly with a large motor-driven blade. (2) Air is forced into the bottom of the fermenter through small holes so that it rises through the liquid and quickly disperses. The air entering the tank must usually be sterilized, and this can be done by passing it through a filter made of steel wool packed into a long tube. The air is often also passed through a copper tube heated to a high temperature, where incineration of contaminants will occur.

AERATION

Inoculation of a large fermenter is made using a fairly dense, rapidly growing culture, added in a volume equal to 5 to 10 percent of the total volume of fermenter medium. Thus, for a 15,000-gallon fermenter, an inoculum of about 1,500 gallons is required. The inoculum is usually built up in a series of stages, starting with the initial stock culture and increasing step by step in volume. At each stage, it must be certain that the culture is behaving normally and that it is not contaminated. One of the most important roles of the industrial microbiologist is to ensure proper quality of the culture used at all stages of the process.

During fermentation, the process must be carefully monitored and controlled. Heat is produced as a result of microbial metabo-

RUNNING THE FERMENTATION

INDUSTRIAL MICROBIOLOGY

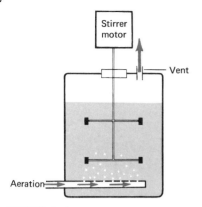

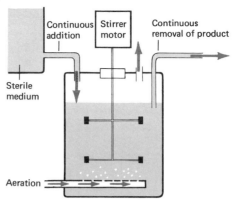

FIGURE 17.2
Comparison of *a* batch fermenter and *b* continuous fermenter.

lism, and to avoid excessive increase in temperature the fermenter must be cooled, usually by circulating cold water through its jacket. Vigorous aeration causes foaming. This foam may be controlled by periodic additions of an antifoam agent such as lard oil or silicone, usually through an automatic foam-control system. Control of pH is often desirable, and this can also be done automatically. Experience will indicate when the fermentation is completed so that the maximum level of the desired product is obtained. At this stage, harvest begins, and the contents of the fermenter are processed in such a way as to remove the desired product.

The fermentation can be carried out as either a batch or a continuous process (Figure 17.2). In a *batch process*, the course of the fermentation is similar to a typical microbial growth cycle. During exponential growth the number of cells increases rapidly, and at nutrient exhaustion the stationary phase is reached. It is quite common for the desired product to appear in the culture only after onset of this stationary phase. In a *continuous process*, medium is pumped into the filled fermenter slowly and continuously, and overflow containing the desired product is removed continuously. The process may be carried out for days or weeks. The advantage of continuous fermentation is that the process can be kept at the stage of maximum production for long periods of time, and harvesting can be done continuously. In a batch process, there is only a small amount of time at which product yield is at maximum, so that the expensive fermenter equipment is tied up for relatively long times in unproductive phases of the cycle. The disadvantages of continuous fermentation are that contamination may ruin the process or that the culture may degenerate during the long course of the process.

Microbial cells are removed from the liquid by filtration or centrifugation. Filtration is usually cheaper and easier, but may not be effective with unicellular microorganisms. With fungi and actinomycetes, which are filamentous, filtration is relatively easy

FIGURE 17.3
Separation of microorganisms
from fermentation broth in a
large rotary vacuum dryer.
*Courtesy of Chas. Pfizer and Co.,
Inc.*

(Figure 17.3), although a filter aid, such as diatomaceous earth, is often used to prevent clogging of the filter. The filter itself is usually a sheet of cotton or synthetic cloth of a tight weave. With unicellular microorganisms, the initial separation of the cells is usually done with a continuous centrifuge. Either the clear filtrate or the cells themselves may be the desired product, and processing details must be designed according to the goal of the fermentation.

A fermentation process that works well in the laboratory or in a small fermentation vessel may work poorly or not at all under similar conditions in large tanks. The reasons for this are often unknown or difficult to determine, and the industrial microbiologist is required to spend considerable research time developing proper conditions for large-scale fermentation. This is called *scale-up* and is of great economic importance. For research on scale-up problems, the industrial company may have a *pilot plant* in which fermenters are used and tested that are intermediate in size between small laboratory equipment and full-size industrial equipment.

The goal of an industrial fermentation is to convert a cheap raw material into a useful end product. The microbiological process must compete with synthetic chemical processes that may be able to accomplish the same task. Where microbial processes excel is in the production of complex organic chemicals such as antibiotics, vitamins, and hormones that are too difficult for the chemist to synthesize inexpensively. The successful management of industrial fermentation requires the combined efforts of chemical engineers and of microbiologists well trained in microbial physiology and genetics.

17.2
Yeasts in industry

Yeasts are the most important and the most extensively used microorganisms in industry (Table 17.1). They are cultured for the cells themselves, for cell components, or for the end products that they produce. Yeast cells are used in the manufacture of bread, and they are also used as a source of food, vitamins, and other growth factors. Large-scale fermentation by yeast is responsible for the production of alcohol for industrial purposes but is better known for its role in the manufacture of alcoholic beverages: beer, wine, and liquors. Production of yeast cells and production of alcohol by yeast are two quite different processes industrially, in that the first process requires the presence of oxygen and hence is an aerobic process, whereas alcoholic fermentation is anaerobic and takes place only in the absence of oxygen. However, the same or closely similar species of yeasts are used in virtually all industrial processes. The yeast *Saccharomyces cerevisiae* was derived from wild yeast used in ancient times for the manufacture of wine and beer, but yeasts currently used have been cultivated in the laboratory for a long time and have been carefully selected for their desirable properties. It is possible to breed yeasts in the laboratory, using genetic hybridization methods, to produce a new strain that contains desirable qualities from two separate parent strains.

BAKERS' YEAST

The baker uses yeast as a leavening agent in the rising of the dough prior to baking. A secondary contribution of yeast to bread making is its flavor. In the leavening process, the yeast is mixed with the moist dough in the presence of a small amount of sugar. The yeast

---

TABLE 17.1
Industrial uses of yeast and yeast products

Production of yeast cells
    Bakers' yeast, for bread making
    Dried food yeast, for food supplements
    Dried feed yeast, for animal feeds
Yeast products
    Yeast extract, for culture media
    Vitamins; B vitamins, Vitamin D, for vitamin pills, etc.
    Enzymes for food industry; invertase, galactosidase
    Biochemicals for research; ATP, NAD, RNA, etc.
Fermentation products from yeast
    Ethanol, for industrial alcohol
    Glycerol
Beverage alcohol
    Beer
    Wine
    Whiskey
    Brandy
    Vodka
    Rum

converts the sugar to alcohol and carbon dioxide, and the carbon dioxide gas expands, causing the dough to rise. When the bread is baked, the heat drives off the carbon dioxide (and incidentally, the alcohol) and holes are left within the bread mass, thus giving bread its characteristic light texture. That yeast contributes more to bread than carbon dioxide is shown by the fact that dough raised with baking powder, a chemical source of carbon dioxide, produces a quite different product than dough raised by yeast. Only the latter bears the name *bread*.

In early times, the bread maker obtained yeast from a nearby brewery, since yeast is a by-product of the brewing of beer. Today, however, bakers use a yeast specially produced for bread making, called *bakers' yeast*. The yeast is cultured in large aerated fermenters in a medium containing molasses as a major ingredient. Molasses, which is a by-product of the refining of sugar from sugar beets or sugar cane, still contains large amounts of sugar that serves as the source of carbon and energy. Molasses also contains minerals, vitamins and amino acids used by the yeast. To make a complete

FIGURE 17.4
Stages in the commercial production of yeast.

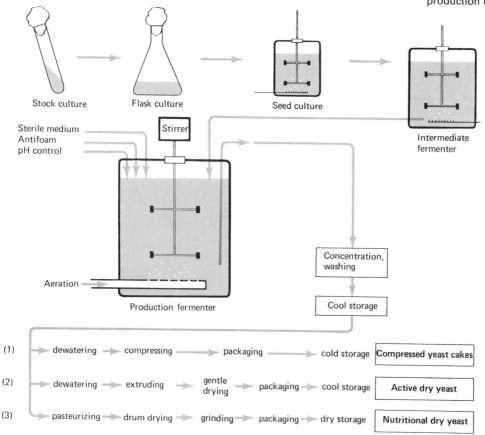

INDUSTRIAL MICROBIOLOGY

FIGURE 17.5
Active dry yeast and cake yeast, two forms in which bakers' yeast is marketed.

medium for yeast growth, phosphoric acid (a phosphorus source) and ammonium sulfate (a source of nitrogen and sulfur) are added.

Fermentation vessels range from 10,000 to 50,000 gallons (40,000 to 200,000 liters). Beginning with the pure stock culture, several intermediate stages are needed to build up the inoculum to a size sufficient to inoculate the final stage (Figure 17.4). Fermenters and accessory equipment are made of stainless steel and are sterilized by high-pressure steam. The actual operation of the fermenter requires special control to obtain maximum conversion of molasses to yeast. It has been found that it is undesirable to add all the molasses to the tank at once, since this results in a sugar excess, and the yeast converts some of this surplus sugar to alcohol rather than turning it into yeast cells. Therefore, only a small amount of the molasses is added initially, and then as the yeast grows and uses this sugar up, more is added. Because yeast grows exponentially, the molasses is also added at an exponential rate. Careful attention must be paid to aeration, since if insufficient air is present, anaerobic conditions develop, and again, alcohol is produced instead of yeast. Production of 1 kg of yeast (dry weight) requires about 9.5 kg of molasses, 2 kg of ammonia, 0.7 kg of ammonium phosphate, and 2.4 kg of ammonium sulfate.

At the end of the growth period, the yeast cells are recovered from the broth by centrifugation. The cells are usually washed by dilution with water and recentrifuged until they are light in color. Bakers' yeast is marketed in two ways, either as compressed cakes or as a dry powder (Figure 17.5). Compressed yeast cakes are made by mixing the centrifuged yeast with emulsifying agents, starch, and other additives that give the yeast a suitable consistency and reasonable shelf life and then forming the product into cubes or blocks of various sizes for domestic or commercial use. A yeast cake will contain about 70 percent moisture and about 2×10^{10} cells per gram. Compressed yeast must be stored in the refrigerator so that its activity is maintained. Yeast marketed in the dry state for baking is usually called active dry yeast. The washed yeast is mixed with additives and dried under vacuum at 25 to 45°C for a 6-hour period, until its moisture is reduced to about 8 percent. It is then packed in airtight containers, such as fiber drums, cartons, or multiwall bags, sometimes under a nitrogen atmosphere to promote long shelf life. Active dry yeast does not exhibit so great a leavening action as compressed fresh yeast but has a much longer shelf life, although it is still best if it is stored at refrigeration temperatures.

Yeast produced as food for humans or as feed for animals can be manufactured in much the same way as described for bakers' yeast, or it may be a by-product of brewing or distilling. The yeast is heat-killed and usually dried. To be acceptable as food, dried yeast must be of proper flavor, color, and nutritional composition and must be free of contamination.

FOOD AND FEED YEAST

Another type of yeast, called *Torula utilis*, has been produced for a feed supplement as a by-product of the manufacture of paper pulp by the sulfite process. In this process, the waste liquid is high in kinds of sugars that are utilized well by *Torula*, although not so well by bakers' yeast. The growth of *Torula* yeast on pulp waste liquid served two purposes: a useful feed product was made, and at the same time, a waste-water treatment was effected, since the sugars in the pulp waste liquid would cause serious water pollution if the liquid were disposed of by dumping it untreated into natural waters. The production of *Torula* by paper-pulp mills was carried out as a continuous process, the waste liquid being pumped continuously into a large fermenter, and the yeast being harvested by removing liquid at a rate that balanced the rate of input. Recently, economic conditions have not been so favorable for the production of *Torula* yeast from pulp waste liquid, and the process is not so extensively operated as it was previously.

There has been considerable research on the production of food and feed yeast from petroleum products. Petroleum products are in huge supply in certain parts of the world, and some yeasts are able to grow on certain petroleum fractions, using them as carbon and energy sources and thus producing a high-protein food. Commercial production of yeast from petroleum is still limited but may develop extensively in the coming years.

The nutritional value of yeast to humans or animals is not so high as that of meat or milk. However, yeast is one of the richest sources of vitamins of the vitamin B group. Although high in protein, yeast protein is deficient in several amino acids essential for humans, most especially the sulfur-containing amino acids. Animals fed on a diet in which all their protein comes from yeast do not grow so well as animals fed on milk or meat protein. Because of these deficiencies, yeast is used primarily as a food supplement for humans, being added to wheat or corn flour to increase the nutritional value of these foods. The advantage of utilizing yeast as human food protein is that yeast makes protein very efficiently from nonprotein nitrogen, whereas beef cattle produce protein much less efficiently, and nonruminants require preformed plant-protein products in their feeds. In parts of the world where plant or animal protein is in short supply, the production of yeast as protein for human food may be of great benefit. Alternatively, yeast may be used as a supplement to animal feeds.

17.3
Alcohol and alcoholic beverages

The use of yeast for the production of alcoholic beverages is an ancient process. Most fruit juices undergo a natural fermentation with the wild yeasts present in the area. From these natural fermentations, yeasts were eventually selected for more controlled production, and today, alcoholic-beverage production is a large industry. The most important alcoholic beverages are wine, produced by the fermentation of fruit juice; beer, produced by the fermentation of malted grains; and distilled beverages, produced by concentrating alcohol from a fermentation by distillation. Production of alcoholic beverages is generally regulated carefully by governments and has become a major source of revenue in many countries.

WINE

Wine is a product of the alcoholic fermentation by yeast of fruit juices or other materials that are high in sugar. Most wine is made from grapes, and unless otherwise specified, the word *wine* refers to the product resulting from the fermentation of grape juice. Wine manufacture occurs in parts of the world where grapes can be most effectively grown. The greatest wine-producing countries, in order of decreasing volume of production, are Italy, France, Spain, Algeria, Argentina, Portugal, the United States, the Soviet Union, Yugoslavia, Romania, Chile, Greece, Hungary, South Africa, West Germany, and Australia. Wine manufacture originated in Egypt and Mesopotamia well before 2000 B.C. and spread through the Mediterranean region, and this is still the largest wine-producing area in the world. Other parts of the world where wine is extensively produced often have a climate similar to that of the Mediterranean, for example, California (United States), Chile, South Africa, and Australia. There is a tremendous variety of wines, and quality and character vary considerably. Dry wines are wines in which the sugars of the juice are practically all feremented, whereas in sweet wines, some of the sugar is left or additional sugar is added after the fermentation. A fortified wine is one to which brandy or other alcoholic spirit is added

after the fermentation; sherry and port are the best-known fortified wines. A sparkling wine is one in which considerable carbon dioxide is present, arising from a final fermentation by the yeast directly in the bottle.

The grapes are crushed by machine, and the juice, called *must*, is squeezed out (Figure 17.6). Depending on the grapes used and on how the must is prepared, either white or red wine may be produced. A white wine is made either from white grapes or from the juice of red grapes from which the skins, containing the red coloring matter, have been removed. In the making of red wine, the skins and seeds are left in during the fermentation. In addition to color differences, red wine has a stronger flavor than does white because of the presence of larger amounts of tannins, which are extracted into the juice from the grape skins during the fermentation.

The yeasts involved in wine fermentation are of two types: the so-called wild yeasts, which are present on the grapes as they are taken from the field and are transferred to the juice, and the true wine yeast, *Saccharomyces ellipsoideus*, which is added to the juice to begin the fermentation. One important distinction between wild yeasts and the true wine yeast is their alcohol tolerance. Most wild yeasts can tolerate only about 4 percent alcohol, and when they produce this much, the fermentation stops. The true wine yeast can tolerate up to 12 to 14 percent alcohol before it stops growing. In unfortified wine, the final alcoholic content reached is determined partly by the alcohol tolerance of the yeast and partly by the amount of sugar present in the juice. The alcohol content of most unfortified wines ranges from 10 to 12 percent. Fortified wines such as sherry have an alcohol content as high as 20 percent, but of course,

INDUSTRIAL MICROBIOLOGY

this is achieved by adding distilled spirits such as brandy. In addition to the lower alcohol content produced, wild yeasts do not produce many of the flavor components considered desirable in the final product, and hence their presence and growth during fermentation is unwanted.

In most wine districts, the true wine yeast is probably present in the environment and becomes inoculated onto the grape along with the wild yeasts. However, to be certain that the true wine yeast predominates during the fermentation, it is the practice in many wine-producing areas to kill the wild yeasts present in the must by adding sulfur dioxide at a level of about 100 ppm. The true wine yeast is resistant to this concentration of sulfur dioxide and can then be added as a starter from a pure culture grown on sterilized or pasteurized grape juice. During the initial stages, air is present in the liquid and rapid aerobic growth of the yeast occurs; then, as the air is used up, anaerobic conditions develop and alcohol production begins. The fermentation may be carried out in vats of various sizes, from 50-gallon casks to 55,000-gallon tanks, made of oak, cement, stone, or glass-lined metal. Temperature control during the fermentation is important, since heat produced during metabolism might raise the temperature above the point where the yeast can still function. Temperatures must be kept below 29°C, and the finest wines are produced at lower temperatures, from 21 to 24°C. Temperature control is best achieved by using jacketed tanks through which cold water is circulated. The fermenter must be so arranged that the large amount of carbon dioxide produced during the fermentation can escape but air cannot enter; this is often accomplished by fitting the tank with a special one-way valve.

After 3 to 5 days of fermentation, sufficient tannin and color have been extracted from the *pomace* (skin, seeds, and pieces of stems), and the wine is drawn off for further fermentation in a new tank, usually for another week or two. The next step is called *racking*; the wine is separated from the sediment (called *lees*), which contains the yeast and organic precipitate, and is then stored at lower temperature for aging, flavor development, and further clarification. The final clarification may be hastened by addition of materials called *fining* agents, such as casein, tannin, or bentonite clay, or the wine may be filtered through diatomateous earth or asbestos. The wine is then bottled and either stored for further aging or sold. Red wine is usually aged for several years (or even much longer) after bottling, but white wine is sold without much aging. During the aging process, complex chemical changes occur, resulting in improvement in flavor and odor, or *bouquet*.

A number of defects arise in wine as a result of microbial action, although the tannins and acids present do tend to retard the growth of many spoilage microorganisms. Pasteur discovered (see Section 1.5) that when wine is heated to 63°C for 30 minutes, all spoilage organisms are destroyed and defects due to microbial growth are eliminated. This discovery was actually the origin of the process of pasteurization, which is of great importance in the wine industry as well as in the dairy industry (see Section 15.5).

The manufacture of alcoholic beverages made from malted grains is called brewing. Typical malt beverages include beer, ale, porter, and stout. *Malt* is prepared from germinated barley seeds, and it contains natural enzymes that digest the starch of grains and convert it into sugar. Since yeasts are unable to digest starch, the malting process is essential for the preparation of a fermentable material from cereal grains. Malted beverages are made in many parts of the world but are most common in areas with cooler climates where cereal grains can be raised well and where wine grapes grow only poorly.

The fermentable liquid from which beer and ale are made is called the *wort* and is prepared by a process called *mashing*. The grain of the mash may consist only of malt, or other grains such as corn, rice, or wheat may be added. The mixture of ingredients in the mash is cooked and allowed to steep in a large mash tub at warm temperatures. There are a number of different methods of mashing, involving heating at different temperatures for various lengths of time; the particular combination of temperature and time used will considerably influence the character of the final product. During the heating period, enzymes from the malt digest the starches and liberate sugars and dextrins, which are fermentable by yeast. Proteins and amino acids are also liberated into the liquid, as are other nutrient ingredients necessary for the growth of yeast.

After cooking, the mash is transferred to a vessel called a *lauter tub*, which serves as a filter, and the liquid wort is separated from the husks and other grain residues. *Hops*, an herb that is derived from the female flowers of the hops plant, is added to the wort at this stage. Hops is a flavoring ingredient, but it also has antiseptic properties, which probably help in controlling contamination in the subsequent fermentation. The wort is then boiled for several hours (Figure 17.7), during which time desired ingredients are extracted from the hops, proteins present in the wort that are undesirable from the point of view of beer stability are coagulated and removed, and the wort is sterilized. Heating is accomplished either by passing steam through a jacketed kettle or by direct heating of the kettle from below by fire. The wort is filtered again and cooled and then transferred to the fermentation vessel.

Brewery yeast strains are of two major types: the top-fermenting and the bottom-fermenting yeasts (Figure 17.8). The main distinction between the two is that top-fermenting yeasts remain uniformly distributed in the fermenting wort and are carried

FIGURE 17.7
Brewing beer. The brew kettle is being filled with wort. *Courtesy of Jos. Schlitz Brewing Co.*

INDUSTRIAL MICROBIOLOGY

FIGURE 17.8
Preparation of yeast culture for inoculation into wort. *Courtesy of Jos. Schlitz Brewing Co.*

to the top by the carbon-dioxide gas generated during the fermentation, whereas bottom yeasts settle to the bottom. Top yeasts are used in the brewing of ales, and bottom yeasts are used to make the more familiar lager beers. The bottom yeasts are usually given the species designation *Saccharomyces carlsbergensis*, and the top yeasts are called *S. cerevisiae.* Fermentation by top yeasts usually occurs at higher temperatures (14 to 23°C) than does that by bottom yeasts (6 to 12°C) and is accomplished in a shorter period of time (5 to 7 days for top fermentation versus 8 to 14 days for bottom fermentation). After completion of lager beer fermentation by bottom yeast (Figure 17.9), the beer is pumped off into large tanks where it is stored at a cold temperature (about −1°C) for several weeks (in German, *lager*

FIGURE 17.9
In these large tanks, the beer fermentation is allowed to proceed for about 2 weeks. *Courtesy of Jos. Schlitz Brewing Co.*

.means "to store"). Lager beer is the most widely manufactured type of beer and is made by large breweries in the United States, Germany, Scandinavia, the Netherlands, and Czechoslovakia. Top-fermented ale is almost exclusively a product of England and certain former British colonies. After its fermentation, the clarified ale is stored at a higher temperature (4 to 8°C), which assists in the development of the characteristic ale flavor.

Prior to filling bottles, cans, or kegs for the trade, beer and ale undergo a final "polishing" that may involve filtration, chill-proofing (the addition of enzymes to digest proteins that might precipitate out when the beer is cooled), carbonation, and pasteurization. The final product must be sparkling clear and of appropriate color if it is to be accepted by the public.

The yeast strain used for the fermentation is usually carefully selected by the brewery, and at least some of the distinctive character of the product of a specific brewery is probably due to the yeast that is used. Some breweries pay close attention to the purity of their yeast strain and begin each fermentation with an inoculum built up to a volume of suitable size from a pure culture. However, in many breweries, the inoculum for one fermentation is derived from yeast saved from the previous fermentation, and a return to a pure culture occurs only if a defect occurs in the product. Bacterial contamination may occur during a fermentation but is not usually a problem since the pH of wort is fairly low (around pH 5), hops are present as an antiseptic, and the wort has been boiled. When bacterial contamination does occur, it is usually due to acid-tolerant bacteria such as *Lactobacillus* and *Acetobacter*. Rarely, wild yeasts may develop during the fermentation and reduce the quality of the product. One role of the microbiologist in the brewery is to control the quality of the yeast used in the fermentation.

A beerlike beverage called *sake* is made in Japan from rice. The agent used for the digestion of the starch is not the enzymes present in malt, but a fungus, *Aspergillus oryzae*. This fungus is grown on steamed rice until the rice is well covered with fungus filaments and the starch-digesting enzyme is present, at which time the product is called *koji*. The koji is then mixed with more steamed rice and water and incubated at low temperature. Several *Saccharomyces* yeasts are present as well as the fungus, and an alcoholic fermentation occurs. The final product is a clear, pale yellow product of high-alcohol content (14 percent or higher), which is sometimes called *Japanese rice wine*, although it is more similar to a beer than to a wine in the manner of manufacture.

DISTILLED ALCOHOLIC BEVERAGES

Distilled alcoholic beverages are made by heating a fermented liquid at such a temperature as to volatilize most of the alcohol, which is then condensed and collected, a process called *distilling*. A product much higher in alcohol content can be obtained than is possible by direct fermentation. Virtually any alcoholic liquid can be distilled, although each yields a characteristic distilled beverage. The distillation of malt fermentations yields *whiskey*, distilled wine yields *brandy*, and distilled molasses yields *rum*. The distillate contains not only alcohol but also other volatile products arising either from the yeast fermentation or from the mash itself. Some of these other products are desirable flavor ingredients, whereas others are undesirable substances called *fusel oils*. To eliminate the latter, the distilled product is almost always aged, usually in wood barrels. During the aging process, fusel oils are removed, and desirable new flavor ingredients develop. The fresh distillate is usually colorless, whereas the aged product is brown or yellow. The character of the final product is determined by the manner and length of aging, and the whole process of manufacturing distilled alcoholic beverages is highly complex. To a great extent, the process is carried out by traditional methods that have been found to yield an adequate product rather than by scientifically proved methods. Whiskey is almost exclusively an Anglo-Saxon (or Gaelic) product. A number of distinct whiskeys occur, usually associated with a country or region. Each of these has a characteristic flavor, owing to the local practices of fermenting, distilling, and aging. Even the word has local spellings, "whisky" being the Scottish, English and Canadian spelling, and "whiskey" the Irish and United States spelling.

It should be noted that it is possible to distill a product in such a way that only alcohol and water are present in the distillate, and all flavor components and fusel oils are absent. This is the way in which *vodka* is manufactured, and to ensure the purity of this product, it is usually filtered through charcoal. Such a product need not be aged to be drunk, but of course, it has only the minimal flavor contributed by the alcohol itself. Vodka can be made from any starchy product, but the most usual starting materials are potatoes or rye malt. A simple solution of alcohol and water is usually called *neutral spirits*, and in addition to being consumed directly, as vodka, it is added to other distilled beverages to dilute them during the blending process, or in the manufacture of gin, cordials, and liqueurs. *Gin* is made by adding a mixture of flavorings to neutral spirits. The principal flavoring ingredient is juniper oil, but there are a variety of others as well. The exact blend of these ingredients is often a guarded trade secret but may include coriander, orange and lemon peel, cassia, angelica, licorice, orris, cardamon, caraway, and cinnamon.

INDUSTRIAL ALCOHOL

A large amount of alcohol is needed for industrial purposes, and this can be prepared by distilling fermented mash. Because such alcohol is not being made for human consumption, many legal restrictions do not apply. Virtually any fermentable carbohydrate source can be used, the main requirement being that it is inexpensive and plentiful. Alcohol produced by yeast fermentation must compete economically with alcohol produced by chemical synthesis. The chemical process uses petroleum products as starting material, and if such products are plentiful, the chemical process is usually preferable to the fermentation process. Because of this, the production of alcohol by fermentation no longer has the prominence in industrial microbiology that it once had, although with the recent sharp increase in the price of petroleum, fermentation alcohol may again become a major industrial product.

17.4 Antibiotic fermentation

We have discussed previously the laboratory study and medical significance of antibiotics. As we noted, antibiotics are chemical substances that are produced by microorganisms and kill or inhibit the growth of other microorganisms. The chemical structures of most medically useful antibiotics are known, and in many cases, chemical synthesis has been achieved. However, only rarely is it possible to synthesize an antibiotic as cheaply and as easily as it is to produce it by microbial fermentation. For this reason, almost all antibiotics are produced commercially by microorganisms, and antibiotic production is one of the most important areas of activity for the industrial microbiologist (see Table 17.2).

Antibiotics differ from other microbial fermentation products such as alcohols, acetone, or lactic acid in that antibiotics are structurally quite complex and are usually produced in much lower concentrations during the fermentation. However, because of the high therapeutic activity and resultant great medical value of antibiotics, they can be economically produced even if yields are relatively low.

Because of the small amounts of antibiotic present in the fermentation liquid, elaborate methods for the extraction and purification of the antibiotic are necessary (Figure 17.10). It is best if the antibiotic is soluble in an organic solvent that is immiscible in water, because then it is possible to extract and concentrate the antibiotic easily (Figure 17.11). If the antibiotic is not solvent soluble, then it must be removed from the fermentation liquid by adsorption or chemical precipitation. In all cases, the goal is to obtain a crystalline product of high purity, although some antibiotics do not crystallize readily and are difficult to purify. A problem of some magnitude is that many cultures produce more than one antibiotic, and it is essential to make a commercial product with only the single desired antibiotic. It has not been uncommon for drug companies to unknowingly produce antibiotics from which not all contaminants have been removed. If the contaminants are toxic or have other

TABLE 17.2
Some antibiotics produced commercially

Antibiotic	Producing microorganism	Some commercial sources
Bacitracin	*Bacillus subtilis*	Commercial Solvents, United States
		Kayakukoseibussitsu, Japan
Chloramphenicol	Chemical synthesis (formerly *Streptomyces venezuelae*)	Parke, Davis, United States
		Sankyo, Japan
Cycloheximide	*Streptomyces griseus*	Upjohn, United States
Cycloserine	*S. orchidaceus*	Commercial Solvents, United States
		Shionogi Seiyaku, Japan
		Sumitomo Kagaku, Japan
Erythromycin	*S. erythreus*	Eli Lilly, United States
		Abbott Laboratories, United States
		Roussel, France
Griseofulvin	*Penicillium griseofulvin*	Glaxo Laboratories, U.K.
		Imperial Chemical Industries, U.K.
Kanamycin	*S. kanamyceticus*	Bristol Laboratories, United States
		Banyu Seiyaku, Japan
		Meiji Seika, Japan
Lincomycin	*S. lincolnensis*	Upjohn, United States
Neomycin	*S. fradiae*	Chas. Pfizer, United States
		E. R. Squibb, United States
		Takeda Yakuhin, Japan
		Upjohn, United States
Nystatin	*S. noursei*	E. R. Squibb, United States
Penicillin	*Penicillium chrysogenum*	Abbott Laboratories, United States
		Banyu Seiyaku, Japan
		Imperial Chemical Industries, U.K.
		Eli Lilly, United States
		Meiji Seika, Japan
		Chas. Pfizer, United States
		E. R. Squibb, United States
		Takeda Yakuhin, Japan
		Bristol Laboratories, United States
		Beecham Laboratories, U.K.
Polymyxin B	*Bacillus polymyxa*	Burroughs-Wellcome, U. K.
		Chas. Pfizer, United States
Streptomycin	*S. griseus*	Eli Lilly, United States
		Merck Sharp & Dohme, United States
		Chas. Pfizer, United States
		E. R. Squibb, United States
		Kyowa Hakko, Japan
		Meiji Seika, Japan
Chlortetracycline (Aureomycin)	*S. aureofaciens*	Lederle Laboratories, United States
Hydroxytetracycline (Terramycin)	*S. rimosus*	Chas. Pfizer, United States
Tetracycline	Chemical derivative of chlortetracycline	Bristol Laboratories, United States
		Lederle Laboratories, United States
		Chas. Pfizer, United States

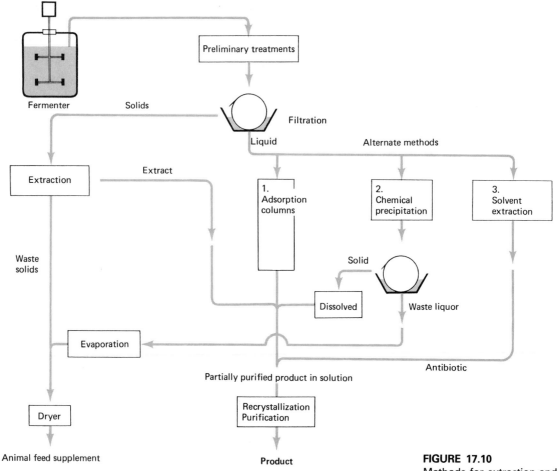

FIGURE 17.10
Methods for extraction and purification of an antibiotic.

undesirable side effects, serious problems can result. Thus, diligent and careful attention to purification methods for antibiotics are essential.

The industrial microbiologist has made a very significant contribution to the antibiotic industry by developing high-yielding processes. For instance, when penicillin was first produced commercially, yields were about $120\,\mu\text{g/ml}$, whereas current yields are well over $8\,\text{mg/ml}$, which is better than a sixtyfold increase. This increase in yield was accomplished by selecting higher-yielding mutant strains of the producing organism, *Penicillium chrysogenum*, by improvements in the fermentation medium, and by careful control of aeration and other aspects of the fermentation process itself.

Antibiotic fermentations are all aerobic and are usually carried out in fermenters of 50,000 to 100,000 liters. Careful attention to sterility is necessary to avoid contamination, and the procedures described earlier in this chapter for sterilizing media and air are used. Another problem that can arise during antibiotic production by bacteria of the genus *Streptomyces* is infection of the culture with virus (bacteriophage). If this happens, lysis of the culture occurs, and antibiotic yields are drastically reduced. Virus infection can be eliminated by selecting strains of the antibiotic-producing organisms that are resistant to the virus. Fortunately, fermentations involving fungi and yeasts do not become infected with virus.

17.5
Vitamins and amino acids
VITAMINS

Vitamins are used as food and feed supplements. Most vitamins are made commercially by chemical synthesis. However, a few are too complicated to synthesize inexpensively, but fortunately they can be made by microbial fermentation. Highly important among these are vitamin B_{12} and riboflavin. Vitamin B_{12} was originally recovered as a by-product of certain antibiotic fermentations, but the industry today uses strains specifically selected for their high yields of the vitamin. One important discovery was that vitamin B_{12} contains cobalt (Co) as an essential part of its structure; yields of the vitamin are greatly increased by addition of cobalt to the culture medium. This is an excellent example of how a discovery in fundamental research (determining the structure of the vitamin) can have important practical consequences. Although the organism most commonly used in its production is *Streptomyces*, vitamin B_{12} can also be prepared commercially in high yields using strains of *Propionibacterium*.

Riboflavin is produced commercially using once of several fungi that produce this vitamin in high yield. *Ashbya gossypii*, a fungus pathogenic for cotton plants, is the fungus most widely used for commercial riboflavin production.

FIGURE 17.11
Installation for the solvent extraction of antibiotics from fermentation broths. *Courtesy of Chas. Pfizer and Co., Inc.*

Although amino acids are obtained in the protein of foods, many plant proteins are deficient in certain amino acids that are essential for human nutrition. If these plant foods are supplemented with the amino acids in which they are deficient, their food value can be considerably increased. Some of these amino acids can be made in high yields by microbial fermentation and are commercially produced. The essential amino acids that have been made by fermentation include lysine, threonine, methionine, and tryptophan. In addition, glutamic acid, a nonessential amino acid, is made in large amounts by microbial synthesis; glutamic acid is added to foods primarily as a flavor-enhancing agent, monosodium glutamate (MSG).

A variety of bacteria can be used to produce amino acids. The important property of high-yielding strains is that they are mutants deficient in the mechanisms that regulate the rate of amino acid synthesis. In normal microorganisms, amino acid synthesis is carefully regulated so that amino acids are synthesized at a rate just sufficient to balance the rate of protein synthesis. Thus, free amino acids do not accumulate, since they are immediately converted into protein. The mechanisms that regulate amino acid synthesis are genetically controlled, and it is possible to isolate mutants in which a regulatory mechanism is deranged. In such a mutant, overproduction of amino acid can occur, and the amino acid spills out of the organisms into the medium. In other fermentations, where a substance is produced in unusually high yields (for instance, vitamins or antibiotics), similar derangement of regulatory mechanisms may also occur.

17.6
Microbial bioconversion

One of the most far-reaching discoveries in industrial microbiology is the discovery that microorganisms can be used to carry out specific chemical reactions that are beyond the capabilities of organic chemistry. The use of microorganisms for this purpose is called *bioconversion* and involves the growth of the organism in large vats, followed by the addition at an appropriate time of the chemical to be converted. Following a further incubation period during which the chemical is acted upon by the organism, the fermentation broth is extracted and the desired product purified. Although in principle bioconversion may be used for a wide variety of processes, it has been used practically only in the production of certain steroid hormones.

Steroids are important hormones in humans and are produced by the body to regulate various metabolic processes. Some steroids are also used as drugs in human medicine. One group, the adrenal cortical steroids, have been found to reduce inflammation and hence are effective in controlling the symptoms of arthritis and allergy. Another group, the estrogens and androgenic steroids, are involved in human fertility, and some of these can be used in the control of fertility. Steroids can be obtained by complete chemical synthesis, but this is a complicated and expensive process. Certain key steps in chemical synthesis can be carried out more efficiently by microorganisms, and commercial production of steroids usually has at least one microbial step.

A large number of fungi and bacteria have been discovered that can perform steroid transformations. Organisms used commercially include the fungi *Rhizopus nigrificans* and *Curvularia lunata*, and the bacteria *Streptomyces roseochromogenes* and *Corynebacterium simplex*.

17.7
Enzyme production by microorganisms

Enzymes are found in all organisms, and each organism produces a large variety, most of which are produced only in small amounts for use in cellular processes. However, certain enzymes are produced in much larger amounts by some organisms, and instead of being held within the cell, they are excreted into the medium. Extracellular enzymes are usually involved in digesting insoluble nutrient materials such as cellulose, protein, and starch, the products of digestion then being able to pass into the cell, where they can be used as

TABLE 17.3
Microbial enzymes and their application

Enzyme	Source	Application	Industry
Amylase (starch-digesting)	Fungi	Bread	Baking
	Bacteria	Starch coatings	Paper
	Fungi	Syrup and glucose manufacture	Food
	Bacteria	Cold-swelling laundry starch	Starch
	Fungi	Digestive aid	Pharmaceutical
	Bacteria	Removal of coatings (desizing)	Textile
Protease (protein-digesting)	Fungi	Bread	Baking
	Bacteria	Spot removal	Dry cleaning
	Bacteria	Meat tenderizing	Meat
	Bacteria	Wound cleansing	Medicine
	Bacteria	Desizing	Textile
	Bacteria	Household detergent	Laundry
Invertase (sucrose-digesting)	Yeast	Soft-center candies	Candy
Glucose oxidase	Fungi	Glucose removal, oxygen removal	Food
		Test paper for diabetes	Pharmaceutical
Pectinase	Fungi	Pressing, clarification	Wine, fruit juice

nutrients for growth. Some of these extracellular enzymes are used in the food, dairy, pharmaceutical, and textile industries and are produced in large amounts by microbial synthesis (Table 17.3).

Enzymes are produced commercially from both fungi and bacteria. The process is usually aerobic, and culture media similar to those used in antibiotic fermentations are employed. The enzyme itself is generally formed in only small amounts during the active growth phase but accumulates in large amounts during the stationary phase of growth. As we have seen (Section 6.5), induced enzymes are produced only when the substrate they attack is present in the medium, and thus either the substrate must be present as one of the main medium ingredients or it must be added as a supplement.

17.8 Vinegar

Vinegar is the product resulting from the conversion of ethyl alcohol to acetic acid by bacteria of the genus *Acetobacter*. Vinegar can be produced from any alcoholic substance, although the usual starting materials are wine or alcoholic apple juice (cider). Vinegar can also be produced from a mixture of pure alcohol in water, in which case it is called *distilled* vinegar, the term "distilled" referring to the alcohol from which the product is made rather than the vinegar itself. The character of the vinegar will depend greatly on the starting material used. Vinegar is used as a flavoring ingredient in salads and other foods, and because of its acidity, it is also used in the preservation of foods, a process called *pickling* (see Section 15.4). Meats and vegetables properly pickled in vinegar can be stored unrefrigerated for years.

ACETIC ACID BACTERIA

The acetobacters are an interesting group of bacteria. Although aerobic, they differ from most other aerobes in that they do not oxidize their energy sources completely to CO_2 and water. Thus, when provided with ethyl alcohol, they oxidize it only to acetic acid, and this product accumulates in the medium. Acetobacters are fairly acid tolerant and are not killed by the acidity that they produce. There is a high oxygen demand during growth, and the main problem in the production of vinegar is to ensure sufficient contact between the bacteria and air.

VINEGAR PRODUCTION METHODS

There are three different processes for the production of vinegar. The *open-vat*, or *Orleans, method* was first method used and still finds some use in France where it was developed. The wine is placed in shallow vats with considerable exposure to the air, and *Acetobacter* develops as a slimy pellicle on the top of the liquid. This process is not very efficient, since the only place that the bacteria come in contact with the air is at the surface. In the *trickle method*, the contact between the bacteria and air is increased by trickling the alcoholic liquid over beechwood twigs or wood shavings that are packed loosely in a vat or column while a stream of air enters at the bottom and passes upwards. The bacteria grow upon the surface of the wood so that they are maximally exposed both to air and liquid. The vat is called a *vinegar generator* (Figure 17.12), and the whole process is operated in a continuous fashion. The life of the wood shavings in a vinegar generator is long, from 5 to 30 years, depending on the kind of alcoholic liquid used in the process. The third process is the *bubble method*. This is basically a submerged fermentation process, such as already described for antibiotic production. Efficient aeration is even more important with vinegar than with antibiotics, and special highly efficient aeration systems have been devised. The process is operated in a continuous fashion: alcoholic liquid is added at a rate just sufficient to balance removal of vinegar, the whole process being operated to ensure that all the alcohol is converted to acetic acid. The efficiency of the process is high, and 90 to 98 percent of the alcohol is converted to acid. One disadvantage of the bubble method is that the product must be filtered to remove the bacteria, whereas in the open vat and trickle methods the product is virtually clear of bacteria since the cells are bound in the pellicle in the former and remain adherent to the wood chips in the latter.

Although acetic acid can be easily made chemically from alcohol, vinegar itself is a distinctive product, the flavor being due in part to other substances present in the starting material. For this reason, the fermentation process has not been supplanted by a chemical process.

FIGURE 17.12
Diagram of one kind of vinegar generator. The alcoholic juice is allowed to trickle through the wood shavings, and air is passed up through the shavings from the bottom. Acetic acid bacteria develop on the wood shavings and convert alcohol to acetic acid. The acetic solution accumulates in the collecting chamber and is removed periodically. The process can be run semicontinuously. *Courtesy of Food Engineering.*

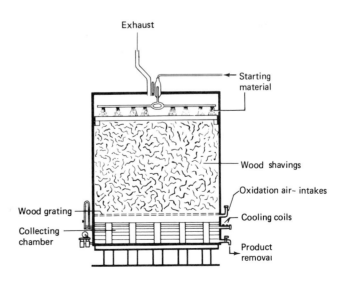

A variety of chemicals are produced by microorganisms in sufficient yields so that they can be manufactured commercially by fermentation. *Citric acid*, used widely in foods and beverages, *itaconic acid*, used in the manufacture of acrylic resins, and *gluconic acid*, used in the form of calcium gluconate to treat calcium deficiencies in humans and industrially as a washing and softening agent, are produced by fungi. *Sorbose*, which is produced by allowing *Acetobacter* to oxidize sorbitol, is used in the manufacture of *ascorbic acid* (vitamin C). *Gibberellin*, a plant growth hormone used to stimulate growth of plants, is produced by a fungus. *Dihydroxyacetone*, produced by allowing *Acetobacter* to oxidize glycerol, is used as a suntanning agent. *Dextran*, a gum used as a blood-plasma extender and as a biochemical reagent, and *lactic acid*, used in the food industry to acidify foods and beverages, are produced by lactic acid bacteria. *Acetone* and *butanol* were formerly produced in large amounts by fermentation of *Clostridium acetobutylicum* but are now produced mainly from petroleum by strictly chemical synthesis.

Of the foregoing, citric acid is perhaps the most interesting product to consider here since it was one of the earliest successful aerobic fermentation products. Citric acid was formerly made commercially in Italy and Sicily by chemical purification from citrus fruits, and for many years, Italy held a world monopoly on citric acid, which resulted in relatively high prices. This monopoly was broken when the microbiological process using the fungus *Aspergillus niger* was developed, and the price of citric acid fell drastically. Today, virtually all citric acid is produced by fermentation. The process is carried out in large aerated fermenters, using a molasses–ammonium salt medium. One of the key requirements for high citric acid yields is that the medium must be low in iron; therefore, most of the iron is removed from the medium before it is used. There has been considerable fundamental research on the citric-acid fermentation process, and some of this work has led to great improvements in the efficiency of the industrial process.

17.9
Citric acid and other fermentation products

Microorganisms can be grown to produce food for human beings, and we have discussed the production of food and feed yeast. In recent years, there has been considerable interest in the expanded production of microorganisms as food, especially in parts of the world where conventional sources of food are in short supply. Perhaps the most important potential use of microorganisms is not as a complete diet for humans but as a protein supplement. It is usually protein that is in shortest supply, and it is in the production of protein that microorganisms are perhaps the most successful. In many cases, microbial cells contain greater than 50 percent protein, and in at least some species, this is complete protein, containing all

17.10
Food from microorganisms

of the amino acids essential to humans. The protein produced by microbes as food has been called *single-cell protein* to distinguish it from the protein produced by multicellular animals and plants.

The only organism presently used as a source of single-cell protein is yeast, as already mentioned, but algae, bacteria, and fungi have also been considered. The following are desirable properties that an organism should possess to be most useful as a source of single-cell protein: (1) rapid growth; (2) simple and inexpensive medium; (3) efficient utilization of energy source; (4) simple fermentation system; (5) simple processing and separation of cells; (6) nonpathogenic organism; (7) harmless when eaten; (8) good flavor; (9) high digestibility; (10) high nutrient content.

ALGAE

Algae show promise as sources of single-cell protein because in the presence of light they can grow on a completely inorganic medium and yield a product high in protein. For their production, a large outdoor tank is used, completely open to light and the air, into which a nutrient medium is pumped (Figure 17.13). One of the best nutrient sources for algae is domestic sewage, and of course if sewage is used it becomes more or less purified as the algae grow. In fact, a sewage lagoon, as described in Section 14.14, is essentially a large algal culture system. Of course bacteria also grow in the sewage, but this turns out to be useful, since they decompose the organic matter of sewage to carbon dioxide, which is then used by the algae in photosynthesis.

The harvesting of algae from the system makes use of a large centrifuge, and the pellet from this centrifugation is then dried. Unfortunately, the dried product cannot be consumed directly by humans, since the algal cells are fairly indigestible because of the presence of a thick cellulose cell wall. They are more digestible to animals such as cattle and sheep, which perform a special cellulose fermentation in the rumen (see Section 16.10).

FIGURE 17.13
A large pilot plant used for production of single-cell protein from algae. *Courtesy of W. J. Oswald.*

Bacteria have many potential advantages as a source of single-cell protein. They grow rapidly, can give high yields, are high in protein, and are nutritionally versatile. Many bacteria use petroleum products as sources of energy and carbon, and because large supplies of petroleum are found in many parts of the world, bacterial single-cell protein made from this source is potentially quite valuable. Another virtue of the bacterial product is that bacteria are fairly readily digested by humans, in contrast to yeast and algae, which are rather poorly digestible.

The bacterium that has been most widely studied as a source of single-cell protein is *Pseudomonas aeruginosa*, a common soil organism. However, some strains of this organism are pathogenic for humans, so that extensive testing will be necessary before it can be utilized in large-scale commercial operations.

Several kinds of fungi are sources of human food, of which the most important are the mushrooms. Mushrooms have been used as food by humans for many years. Both wild mushrooms and those grown commercially in special mushroom beds are used, although only the latter are produced and eaten extensively. The manner of formation of the mushroom fruiting body is discussed in Section 2.7.

The mushroom commercially available in most parts of the world is *Agaricus bisporus*, and it is generally cultivated in mushroom farms. The organism is grown in special beds, usually in buildings where temperature and humidity are carefully controlled. Since light is not necessary, mushrooms may even be grown in basements of homes. Another favored spot for mushroom culture is a cave. Beds are prepared by mixing soil with a material very rich in organic matter, such as horse manure, and these beds are then inoculated with mushroom "spawn." The spawn is actually a pure culture of the mushroom fungus that has been grown in large bottles on an organic-rich medium. In the bed, the mycelium grows and spreads through the substrate, and after several weeks it is ready for the next step, the induction of mushroom formation. This is accomplished by adding to the surface of the bed a layer of soil called "casing soil." The appearance of mushrooms on the surface of the bed is called a "flush" (see Figure 17.14), and when flushing occurs the mushrooms must be collected immediately while still fresh. After collection they are packaged and kept cool until brought to market. Several flushes will take place on a single bed, and after the last flush the bed must be cleaned out and the process begun again.

There have been some attempts to raise the mushroom fungus in aerated fermentation vats. In this case, the characteristic mushroom structure does not develop, but the mushroom flavor does, and the goal would be to use such material as a flavor ingredient in soups and sauces. However, no commercial process for deep-vat

FIGURE 17.14
A commercial mushroom bed.
Courtesy of Lee C. Schisler.

mushroom culture is yet available. Filamentous fungi other than mushrooms have also been tested for commercial deep-vat culture but have not been found satisfactory.

Although mushrooms make flavorful food, their digestibility and nutritional value are not very high. They are low in protein content and deficient in certain essential amino acids; they are also not exceptionally rich in vitamins. The mushrooms and the filamentous fungi are definitely inferior to yeast as food sources, although they are certainly welcome as flavoring ingredients.

Summary

We have seen in this chapter that microorganisms play many significant roles in industrial activities. The economic benefit from the proper control of microorganisms for industrial purposes is great, and future important developments can also be anticipated. A large-scale microbial process is usually called a fermentation, and the vessel in which it is carried out is a fermenter. The size of an industrial fermenter is large, and many complications arise in its operation that are not found in laboratory culture of microorganisms, such as cost of medium ingredients, difficulties of sterilization, avoidance of contamination, and harvesting of product. Yeasts are the most important and most extensively used microorganisms in industry.

Yeast cells are produced in large amounts for the baking and animal-feed markets and for the production of such alcoholic beverages as beer, wine, and distilled liquors. The second largest industrial use of microorganisms is in the production of antibiotics; the most common organisms involved here are in the bacterial genera *Bacillus* and *Streptomyces* and in the fungal genus *Penicillium*. Other products manufactured industrially with the aid of microorganisms include vitamins, amino acids, steroid hormones, enzymes, and acids. Some microorganisms other than yeasts that have been used as sources of food include algae, certain bacteria, and mushroom fungi.

Study questions

1 What is a fermenter? How is it used in industrial microbiology? Describe briefly how it is constructed.
2 Discuss the problems involved in sterilizing industrial fermenters.
3 Contrast the ways in which yeast is used in the baking and in the brewing industries.
4 Describe four different products manufactured through the use of yeast. Discuss briefly the process involved in the manufacture of each product.
5 From what you know about the two substances, which do you think would have historically been the first product to have been discovered: wine or beer?
6 How do wild yeasts and wine yeasts differ? How are wild yeasts eliminated from the wine-making process?
7 What is the role of malt in the brewing process? of hops?
8 Contrast top and bottom fermentation as they apply to beer. In which process is lager beer made?
9 How are microorganisms used in the manufacture of antibiotics? What problems arise in the manufacture of antibiotics that do not arise during the manufacture by fermentation of alcohol?
10 How are microorganisms used in the manufacture of vitamins? amino acids? enzymes?
11 What is meant by microbial bioconversion? Describe one product that is made by this procedure.
12 Vinegar production is a highly aerobic process. Why? Describe several methods used to ensure that O_2 is available during the vinegar-making process.
13 What is meant by single-cell protein? How might it be made using algae? bacteria?

Suggested readings

Advances in applied microbiology, Academic Press, New York. This book appears annually and contains reviews on various topics of interest to industrial microbiology.

Casida, L. E., Jr., *Industrial microbiology*, John Wiley & Sons, Inc. New York, 1968. This brief textbook gives some of the principles of industrial microbiology.

Developments in industrial microbiology, American Institute of Biological Sciences, Washington, D.C. This publication appears annually and contains the proceedings of the general meetings of the Society for Industrial Microbiology. Short technical papers on various subjects of interest to industrial microbiology.

INDUSTRIAL MICROBIOLOGY

Peppler, H. J., *Microbial technology*, Van Nostrand Reinhold Publishing Corp., New York, 1967. Excellent reviews of industrial fermentation processes, with individual chapters on different microbial products.

Prescott, S. C., and C. G. Dunn, *Industrial microbiology*, 3rd ed., McGraw-Hill Book Co., New York, 1959. Although dated, still the only available large textbook on industrial fermentations. A good starting place in any literature review, as it covers the historical aspects very well.

Progress in industrial microbiology, Churchill Livingstone, Edinburgh. This book appears annually and contains reviews on various topics of interest to industrial microbiology.

Rose, A. H., and J. S. Harrison (eds.), *The yeasts*, Vol. 3: *Yeast technology*, Academic Press, London, 1970. Chapters by experts on various aspects of the use of yeasts in industry. Good chapters on wine and beer yeasts.

Smith, J. E., and D. R. Berry (eds.), *The filamentous fungi*, Vol. 1: *Industrial mycology*, John Wiley & Sons, Inc., New York, 1975. This book contains chapters by experts on various aspects of the use of filamentous fungi in industry. Discusses production of enzymes, organic acids, single-cell protein, mushrooms, and the use of fungi in bioconversion.

APPENDIX

Kingdom: Procaryotae
　Division I: The Cyanobacteria
　Division II: The Bacteria

　Part 1: Phototrophic bacteria
　　Order 1: Rhodospirillales
　　　Family I: Rhodospirillaceae
　　　　Genus I: *Rhodospirillum*
　　　　Genus II: *Rhodopseudomonas*
　　　　Genus III: *Rhodomicrobium*
　　　Family II: Chromatiaceae
　　　　Genus I: *Chromatium*
　　　　Genus II: *Thiocystis*
　　　　Genus III: *Thiosarcina*
　　　　Genus IV: *Thiospirillum*
　　　　Genus V: *Thiocapsa*
　　　　Genus VI: *Lamprocystis*
　　　　Genus VII: *Thiodictyon*
　　　　Genus VIII: *Thiopedia*
　　　　Genus IX: *Amoebobacter*
　　　　Genus X: *Ectothiorhodospira*
　　　Family III: Chlorobiaceae
　　　　Genus I: *Chlorobium*
　　　　Genus II: *Prosthecochloris*
　　　　Genus III: *Chloropseudomonas*

　　　　Genus IV: *Pelodictyon*
　　　　Genus V: *Clathrochloris*

　Part 2: Gliding bacteria
　　Order I: Myxobacterales
　　　Family I: Myxococcaceae
　　　　Genus I: *Myxococcus*
　　　Family II: Archangiaceae
　　　　Genus I: *Archangium*
　　　Family III: Cystobacteraceae
　　　　Genus I: *Cystobacter*
　　　　Genus II: *Melittangium*
　　　　Genus III: *Stigmatella*
　　　Family IV: Polyangiaceae
　　　　Genus I: *Polyangium*
　　　　Genus II: *Nannocystis*
　　　　Genus III: *Chondromyces*
　　Order II: Cytophagales
　　　Family I: Cytophagaceae
　　　　Genus I: *Cytophaga*
　　　　Genus II: *Flexibacter*
　　　　Genus III: *Herpetosiphon*
　　　　Genus IV: *Flexithrix*
　　　　Genus V: *Saprospira*
　　　　Genus VI: *Sporocytophaga*

Abstracted by permission from *Bergey's Manual of Determinative Bacteriology*, 8th ed. R. E. Buchanan and N. E. Gibbons, eds. © 1974 by The Williams & Wilkins Company, Baltimore.

Family II: Beggiatoaceae
  Genus I: *Beggiatoa*
  Genus II: *Vitreoscilla*
  Genus III: *Thioploca*
Family III: Simonsiellaceae
  Genus I: *Simonsiella*
  Genus II: *Alysiella*
Family IV: Leucothrichaceae
  Genus I: *Leucothrix*
  Genus II: *Thiothrix*
Families and genera of uncertain affiliation
  Genus: *Toxothrix*
  Family: Achromatiaceae
    Genus: *Achromatium*
  Family: Pelonemataceae
    Genus: *Pelonema*
    Genus: *Achroonema*
    Genus: *Peloploca*
    Genus: *Desmanthos*

Part 3: Sheathed bacteria
  Genus: *Sphaerotilus*
  Genus: *Leptothrix*
  Genus: *Streptothrix*
  Genus: *Lieskeella*
  Genus: *Phragmidiothrix*
  Genus: *Crenothrix*
  Genus: *Clonothrix*

Part 4: Budding and/or appendaged bacteria
  Genus: *Hyphomicrobium*
  Genus: *Hyphomonas*
  Genus: *Pedomicrobium*
  Genus: *Caulobacter*
  Genus: *Asticcacaulis*
  Genus: *Ancalomicrobium*
  Genus: *Prosthecomicrobium*
  Genus: *Thiodendron*
  Genus: *Pasteuria*
  Genus: *Blastobacter*
  Genus: *Seliberia*
  Genus: *Gallionella*
  Genus: *Nevskia*
  Genus: *Planctomyces*
  Genus: *Metallogenium*
  Genus: *Caulococcus*
  Genus: *Kusnezovia*

Part 5: Spirochetes
  Order I: Spirochaetales
  Family I: Spirochaetaceae
    Genus I: *Spirochaeta*
    Genus II: *Cristispira*
    Genus III: *Treponema*
    Genus IV: *Borrelia*
    Genus V: *Leptospira*

Part 6: Spiral and curved bacteria
  Family I: Spirillaceae
    Genus I: *Spirillum*
    Genus II: *Campylobacter*
  Genera of uncertain affiliation
    Genus: *Bdellovibrio*
    Genus: *Microcyclus*
    Genus: *Pelosigma*
    Genus: *Brachyarcus*

Part 7: Gram-negative, aerobic rods and cocci
  Family I: Pseudomonadaceae
    Genus I: *Pseudomonas*
    Genus II: *Xanthomonas*
    Genus III: *Zoogloea*
    Genus IV: *Gluconobacter*
  Family II: Azotobacteraceae
    Genus I: *Azotobacter*
    Genus II: *Azomonas*
    Genus III: *Beijerinckia*
    Genus IV: *Derxia*
  Family III: Rhizobiaceae
    Genus I: *Rhizobium*
    Genus II: *Agrobacterium*
  Family IV: Methylomonadaceae
    Genus I: *Methylomonas*
    Genus II: *Methylococcus*
  Family V: Halobacteriaceae
    Genus I: *Halobacterium*
    Genus II: *Halococcus*
  Genera of uncertain affiliation
    Genus: *Alcaligenes*
    Genus: *Acetobacter*
    Genus: *Brucella*
    Genus: *Bordetella*
    Genus: *Francisella*
    Genus: *Thermus*

Part 8: Gram-negative, facultatively anaerobic rods
  Family I: Enterobacteriaceae
    Genus I: *Escherichia*
    Genus II: *Edwardsiella*
    Genus III: *Citrobacter*
    Genus IV: *Salmonella*
    Genus V: *Shigella*
    Genus VI: *Klebsiella*
    Genus VII: *Enterobacter*
    Genus VIII: *Hafnia*
    Genus IX: *Serratia*
    Genus X: *Proteus*
    Genus XI: *Yersinia*
    Genus XII: *Erwinia*
  Family II: Vibrionaceae
    Genus I: *Vibrio*
    Genus II: *Aeromonas*

Genus III: *Plesiomonas*
Genus IV: *Photobacterium*
Genus V: *Lucibacterium*
Genera of uncertain affiliation
  Genus: *Zymononas*
  Genus: *Chromobacterium*
  Genus: *Flavobacterium*
  Genus: *Haemophilus* (*H. vaginalis*)
  Genus: *Pasteurella*
  Genus: *Actinobacillus*
  Genus: *Cardiobacterium*
  Genus: *Streptobacillus*
  Genus: *Calymmatobacterium*
Parasites of *Paramecium*

Part 9: Gram-negative, anaerobic bacteria
  Family I: Bacteroidaceae
    Genus I: *Bacteroides*
    Genus II: *Fusobacterium*
    Genus III: *Leptotrichia*
  Genera of uncertain affiliation
    Genus: *Desulfovibrio*
    Genus: *Butyrivibrio*
    Genus: *Succinovibrio*
    Genus: *Succinomonas*
    Genus: *Lachnospira*
    Genus: *Selenomonas*

Part 10: Gram-negative cocci and coccobacilli (aerobes)
  Family I: Neisseriaceae
    Genus I: *Neisseria*
    Genus II: *Branhamella*
    Genus III: *Moraxella*
    Genus IV: *Acinetobacter*
  Genera of uncertain affiliation
    Genus: *Paracoccus*
    Genus: *Lampropedia*

Part 11: Gram-negative anaerobic cocci
  Family I: Veillonellaceae
    Genus I: *Veillonella*
    Genus II: *Acidaminococcus*
    Genus III: *Megasphaera*

Part 12: Gram-negative, chemolithotrophic bacteria
  a: Organisms oxidizing ammonia or nitrite
    Family I: Nitrobacteraceae
      Genus I: *Nitrobacter*
      Genus II: *Nitrospina*
      Genus III: *Nitrococcus*
      Genus IV: *Nitrosomonas*
      Genus V: *Nitrosospira*
      Genus VI: *Nitrosococcus*

Key to the 19 parts

I. Phototrophic **Part 1**

II. Chemotrophic

    A. Chemolithotrophic

        1. Derive energy from the oxidation of nitrogen, sulfur, or iron compounds; do not produce methane from carbon dioxide

            a. Cells glide **Part 2**
          aa. Cells do not glide
              b. Cells ensheathed **Part 3**
             bb. Cells not ensheathed **Part 12**

        2. Do not oxidize nitrogen, sulfur, or iron compounds; produce methane from carbon dioxide **Part 13**

    B. Chemoorganotrophic

        1. Cells glide **Part 2**

        2. Cells do not glide (exceptions in Part 19)

            a. Cells filamentous and ensheathed **Part 3**
          aa. Cells not filamentous and ensheathed
              b. Products of binary fission not equivalent (have appendages other than flagella and pili or reproduce by budding) **Part 4**
             bb. Not as above
                c. Cells not rigidly bound
                    d. Cells spiral-shaped, have cell wall **Part 5**
                  dd. Cells not spiral-shaped, no cell wall **Part 19**
              cc. Cells rigidly bound
                d. Gram negative
                    e. Obligate intracellular parasites **Part 18**
                  ee. Not as above
                    f. Curved rods **Parts 6 & 9**
                  ff. Not curved rods
                      g. Rods
                        h. Aerobic **Part 7**
                        hh. Facultatively anaerobic **Part 8**
                        hhh. Anaerobic **Part 9**
                      gg. Cocci or coccobacilli
                        h. Aerobic **Parts 7 & 10**
                        hh. Anaerobic **Part 11**
                dd. Gram positive
                e. Cocci
                    f. Endospores produced **Part 15**
                    ff. Endospores not produced **Part 14**
                ee. Rods or filaments
                    f. Endospores produced **Part 15**
                    ff. Endospores not produced
                      g. Straight rods **Parts 16 & 17**
                      gg. Irregular rods (coryneform) or tend to form filaments or filamentous **Part 17**

Glossary

ABO SYSTEM   The most important classification for designating blood groups.

ACTIVATED-SLUDGE PROCESS   An oxidative treatment of sewage; waste is mixed and aerated in tanks, and microbes degrade the organic materials present.

ADJUVANT   A substance that improves the effectiveness of an antigen without being an antigen itself.

AEROBE   An organism that can use O_2 as an electron acceptor in metabolism.

AEROSOL   A suspension of particles in air.

AFLATOXIN   A mycotoxin produced by *Aspergillus flavus*.

AGGLUTINATION   The reaction between antibody and cell-bound antigen that results in clumping of the cells.

AKINETE   A resting spore of a blue-green alga.

ALGAE   Eucaryotic photosynthetic organisms, including unicellular and multicellular forms. All contain chlorophyll.

ALGAL BLOOM   Extremely heavy algal growth in a body of water; algae float to the surface and the water resembles pea soup.

ALLERGY   An unusual sensitivity to a substance normally harmless to others. The host forms antibodies against an allergen and reacts whenever exposed to the allergen.

ANABOLISM   The biochemical processes involved in the synthesis of cell constituents from simpler molecules; usually requires energy.

ANAEROBE   An organism that grows or metabolizes without oxygen (air). *Obligate anaerobes* live only in the absence of oxygen; *facultative anaerobes* grow either with or without oxygen.

ANAEROBIC RESPIRATION   The oxidation of a substrate using an electron acceptor other than oxygen. Alternate electron acceptors include nitrate, sulfate, CO_2, and ferrous iron.

ANAPHYLACTIC SHOCK   An extreme antigen-antibody reaction in allergy that produces acute asthma and sometimes results in death; of most concern in drug allergies.

ANTIBIOTIC   A chemical agent produced by one organism that is harmful to other organisms.

ANTIBODY   A specific protein (*immunoglobulin*) found mainly in blood serum, formed against a specific antigen and reactive with it.

ANTIGEN   A substance that induces specific antibody formation; usually a protein or polysaccharide.

ANTISEPTIC   An agent that kills or inhibits growth, for use on skin or mucous membranes but not to be used internally.

ANTITOXIN   An antibody active against a toxin.

ASEPTIC   Uncontaminated.

ASEPTIC TECHNIQUE   The procedures used in handling cultures, media, and equipment so that only the desired organisms (if any) are present, with no contaminants.

ATTENUATION OF VIRULENCE   Loss of a pathogen's ability to cause disease, generally due to overgrowth of non-virulent strains, in laboratory culture. Attenuated strains are often used as vaccines.

AUTOCLAVE   An apparatus for sterilizing by heat under steam pressure.

AUTOTROPH   An organism able to utilize CO_2 as sole source of carbon.

BACTEREMIA   The presence of bacteria in the blood.

BACTERIA   Small procaryotic organisms, which may have a spherical, rod, or spiral shape.

BACTERICIDAL   Capable of killing bacteria.

BACTERIOPHAGE   A virus that infects bacteria.

BACTERIOSTATIC   Capable of inhibiting bacterial growth without killing.

BIOCHEMICAL OXYGEN DEMAND (B.O.D.)   The amount of oxygen consumed by bacteria in a quantity of water, determined by the amount of oxidizable organic matter in the water.

BIOCONVERSION   The microbial conversion of a chemical to one of economic importance.

BIODEGRADABLE   Capable of being broken down by living organisms; usually used in reference to manmade organic compounds such as pesticides.

BIOGEOCHEMISTRY   The study of the combined effects of biological and geological activities on chemical changes observed in nature.

BIOSPHERE   The total assemblage of all living organisms on earth.

BIOSYNTHESIS   The synthetic reactions of cell metabolism; see *Anabolism.*

BLUE-GREEN ALGAE   Photosynthetic procaryotes that contain chlorophyll and phycocyanin pigments and often move by gliding; cyanobacteria.

BROWNIAN MOTION   Vibratory, haphazard motion observed in microscope preparations (as distinguished from true motility).

BUDDING   The process of cell division in which the mother cell retains its identity, and the daughter cell forms by growth of a new cell upon one part of the mother.

CALORIE   A unit of heat or energy; that amount of heat required to raise the temperature of one gram of water by 1°C.

CAPSULE   Gummy material secreted in a compact layer outside the cell wall.

CARRIER   An individual who harbors a virulent pathogen but does not exhibit symptoms of disease; may be a source of infection.

CATABOLISM   The biochemical processes involved in the breakdown of organic compounds, usually leading to the production of energy.

CATALYST   A substance that promotes a chemical reaction without itself being changed in the end.

CELL   An individual biological unit, capable of independent function and able to divide to form two new identical cells.

CELL MEMBRANE   A thin envelope surrounding the cell through which food materials pass in and waste materials and other metabolic products pass out.

CHEMICAL OXYGEN DEMAND (C.O.D.)   The amount of oxygen consumed by a chemical oxidizing agent in a quantity of water, determined by the total amount of oxidizable organic material in the water.

CHEMOAUTOTROPH   An organism that utilizes CO_2 as its sole carbon source and an inorganic compound as its energy source.

CHEMOPROPHYLAXIS   The use of a drug or antibiotic to prevent future infections in people who are unusually susceptible.

CHEMOTHERAPY Treatment of an infectious disease by drugs that act against the pathogen but do not harm the host.

CHLORINE RESIDUAL A specified chlorine concentration that remains in a water supply to help protect against contamination.

CHLOROPLAST The green, chlorophyll-containing organelle in eucaryotes; site of photosynthesis.

CHROMOSOME The structure that contains the DNA in eucaryotes, usually complexed with histones.

CILIUM A short, filamentous structure that beats with many others to make a cell move; found only in certain protozoa.

COAGULASE A substance produced by *Staphylococcus aureus* that causes plasma to clot.

COCCUS A spherical bacterium.

COLIFORM TEST Prescribed methods for analysis of water for the presence of coliform bacteria (gram negative, lactose-fermenting bacteria such as *Escherichia coli*). May use either the *MPN* or the *MF procedure*.

COLONY A population of cells arising from a single cell, growing on solid medium.

COMPLEMENT A complex of proteins in the blood serum that acts in concert with specific antibody in certain kinds of antigen-antibody reactions.

COMPLEMENT FIXATION The utilization of complement during an antigen-antibody reaction. Removal of complement can be measured and is a sensitive indicator of occurrence of an antigen-antibody reaction.

COMPOST A mixture of organic materials used as a soil conditioner and fertilizer; prepared by microbial degradation of plant materials.

CONIDIA Asexual reproductive structures of fungi, formed at the tip of conidiophores; spread by air currents.

CONJUGATION The transfer of genetic information from one cell to another by cell-to-cell contact.

CONTAGIOUS (INFECTIOUS) DISEASE An infection transmitted through the air, by water, food, objects, direct contact, or by insects or animals.

CULTURE A particular strain or kind of organism growing in a laboratory medium.

CYANOBACTERIA See *Blue-green algae.*

CYTOPLASM Cellular contents, excluding the nucleus, inside the plasma membrane.

DENITRIFICATION The conversion of nitrate into nitrogen gases under anaerobic conditions, which results in loss of nitrogen from ecosystems.

DENTAL CARIES Tooth decay.

DENTAL PLAQUE A coating on teeth that consists of organic materials and bacteria.

DERMATOPHYTES Fungi that cause superficial mycoses.

DISINFECTANT An agent that kills microorganisms; may be harmful to human tissue, and thus should be used on inanimate objects only.

DISPERSAL The spread of organisms through the environment by air, water, food, animals, or human contact.

ECOSYSTEM The total community of organisms living in a particular habitat, together with their physical and chemical environment.

ELECTRON ACCEPTOR A substance that accepts electrons during an oxidation-reduction reaction.

ELECTRON DONOR A substance that gives up electrons during an oxidation-reduction reaction; an energy source.

ELECTRON-TRANSPORT (OXIDATIVE) PHOSPHORYLATION The synthesis of high-energy phosphate bonds within the electron-transport particle.

ENDEMIC A disease that is constantly present in a population. Compare *Epidemic.*

ENDOSPORE A structure produced inside some bacteria, highly resistant to heat and chemicals; can germinate to form a new cell.

ENDOTHERMIC REACTION A chemical reaction that requires input of energy to proceed.

ENDOTOXIN A cellular component of certain pathogens released upon cell death, which then attacks and damages tissue cells. Compare *Exotoxin.*

ENTERIC Intestinal.

ENTEROTOXIN A toxin affecting the intestine.

ENZYME A protein functioning as the catalyst of living organisms, which promotes specific reactions or groups of reactions.

EPIDEMIC A disease occurring in a greater-than-usual number of individuals in a community at the same time. Compare *Endemic.*

EPISOME A procaryotic nonchromosomal genetic element (plasmid) that may under certain conditions become integrated into the chromosome.

EUCARYOTIC CELL The cell type found among all "higher" organisms (algae, fungi, protozoa, plants, animals); contains mitochondria, membrane-bounded nucleus, chromosomes.

EUTROPHICATION Nutrient enrichment of natural waters, usually from manmade sources, which frequently leads to excessive algal growth.

EXPONENTIAL GROWTH The phase of most rapid growth, in which the progressive doubling of cell number results in a continually increasing rate of growth.

EXOTHERMIC REACTION A chemical reaction that proceeds with the liberation of heat.

EXOTOXIN An extracellular product of some pathogens; spreads to distant tissues where it causes cellular damage. Compare *Endotoxin.*

FACULTATIVE A qualifying adjective indicating that an organism is able to grow either in the presence or absence of an environmental factor; e.g. "facultative anaerobe," "facultative psychrophile."

FEEDBACK INHIBITION The inhibition by an end product of the biosynthetic pathway involved in its synthesis.

FERMENTATION The oxidation of organic compounds occurring in the absence of any external electron acceptor.

FERMENTATION (INDUSTRIAL) A large-scale microbial process.

FERMENTER A large vessel (5,000 to 50,000 gallons) in which industrial fermentations are carried out.

FLAGELLUM A long whiplike organ of motility. The flagella of procaryotes and eucaryotes differ in structure.

FLUORESCENT ANTIBODY An antibody labelled with fluorescent dye; useful in staining procedures to identify bacteria or viruses that have particular antigens.

FOOD INFECTION Illness arising from ingestion of food containing pathogenic organisms such as some salmonellas.

FOOD POISONING Illness arising from ingestion of food containing toxins from the growth of organisms such as some staphylococci or clostridia.

FREEZE DRYING See *Lyophilization.*

FUNGI Eucaryotes that often show mycelial, spreading growth; included are the molds, yeasts, and mushrooms.

GAMMA GLOBULIN A protein fraction of blood serum that contains the antibodies.

GANGRENE The occurrence of dead tissue resulting from bacterial invasion and interference with nutrition of tissue.

GASTROENTERITIS A food infection causing inflammation of the lining membrane of the stomach and intestines and resulting in vomiting and diarrhea.

GENE A unit of heredity; a segment of DNA specifying a particular protein or polypeptide chain.

GENERATION TIME The time required for formation of two cells from one, or for the microbial population to double.

GENETIC MARKER Any mutant gene useful in genetic analysis.

GENOTYPE The genetic complement of an organism.

GENUS A group of related species.

GERMICIDE An agent that kills microorganisms. See also *Antiseptic* and *Disinfectant.*

GLYCOLYSIS The stepwise breakdown of glucose by fermentative reactions.

GRANULOCYTE One type of phagocyte, numerous during acute phase of infection.

GROWTH An increase in the number of cells or in microbial mass.

GROWTH FACTOR An organic compound that is required in very small amounts as a nutrient.

GROWTH RATE The amount of increase in cell number or mass per unit time.

HEMOLYSIS Dissolution of red blood cells, caused by specific bacterial toxins (*hemolysins*).

HETEROCYST A specialized cell in filamentous blue-green algae frequently associated with nitrogen fixation.

HETEROTROPH An organism that uses an organic compound as carbon source.

HOST An organism that supports the growth of a parasite.

HUMUS The nonliving, brown, partially decomposed organic matter in soil; increases water-holding capacity and improves soil texture.

HYPERIMMUNE Having a very high antibody titer.

HYPERSENSITIVITY Unusual and extreme allergic reaction to some antigen to which the host was previously exposed.

HYPHA A filament that is the usual form of fungal cellular structure.

IN VITRO In glass, in culture.

IN VIVO In the body, in a living organism.

IMMUNITY Defense system against disease, including innate host defense mechanisms and acquired defense, such as antibodies.

IMMUNIZATION The induction of specific immunity by injecting antigen or antibodies.

IMMUNOGLOBULIN Proteins in the blood serum that are antibodies.

INCINERATION Killing of organisms by burning.

INDICATOR ORGANISM An organism associated with the intestinal tract whose presence in water indicates fecal contamination.

INDUCTION The process by which an enzyme is synthesized in response to the presence of an external substance, the inducer.

INFECTION Growth of a microorganism in a host, often but not always causing harm (disease) in the host.

INFLAMMATION A response of host tissues to injury, microbial infection, or presence of foreign matter, characterized by swelling, redness, pain, and accumulation of phagocytes at the site.

INOCULUM Material used to initiate a microbial culture.

INTERFERON A naturally occurring, nonspecific antiviral agent, produced as a result of virus infection or the presence of some nucleic acids, that interferes with virus multiplication.

INVASIVENESS The degree to which an organism is able to spread through the body from a focus of infection.

KOCH'S POSTULATES Experimental criteria, devised by Robert Koch, to demonstrate that a specific disease is caused by a specific organism.

LAG PHASE The earliest phase of growth, during which the cell number stays constant while cells adjust to the new medium.

LATENT VIRUS A virus present in a cell but not causing any detectable effect.

LEUKOCIDIN A microbial substance able to destroy phagocytes.

LEUKOCYTE A white blood cell, usually a phagocyte.

LICHEN A regular association of an alga with a fungus, usually leading to the formation of a plantlike structure.

LITHOTROPH An organism that can obtain its energy from oxidation of inorganic compounds.

LYMPH A fluid similar to blood but lacking the red-cell component; flows within the lymphatic system.

LYMPHOCYTE A leukocyte, often involved in antibody formation.

LYOPHILIZATION The preservation of a food or culture of organisms by drying under vacuum while the product is frozen.

LYSIS Rupture of a cell, resulting in loss of cell contents.

LYSOGENIC Describing a bacterium that contains a temperate bacteriophage integrated into the cell DNA.

LYSOZYME An enzyme in body fluids, active in host defense; able to kill invading organisms by digesting cell wall.

MACROMOLECULES Very large polymers in cells, including proteins, polysaccharides, and nucleic acids.

MACROPHAGE A type of phagocyte; includes *wandering cells* in the blood and lymph, and *histiocytes* in the reticuloendothelial system.

MEMBRANE FILTER (MF) METHOD A method for counting organisms, especially in dilute samples; the liquid sample is passed through a membrane filter, which is placed on agar medium to permit the organisms to grow and be counted.

MENINGITIS Infection accompanied by inflammation of the meninges, the membranes surrounding the brain and spinal cord.

MESOPHILE An organism that grows best at temperatures between room temperature (18–25°C) and body temperature (37°C).

MESSENGER RNA (mRNA) An RNA molecule containing a base sequence complementary to DNA; directs the synthesis of protein.

METABOLISM All biochemical reactions in a cell, both anabolic and catabolic, by which an organism converts nutrients into cell material and energy.

METACHROMATIC GRANULE See *Volutin.*

MICROBIAL LEACHING The extraction of minerals from low-grade ores by action of bacteria, especially *Thiobacillus ferrooxidans.*

MICROMETER One-millionth of a meter, or 10^{-6} m (abbreviated μm), the unit used for measuring microbes. Formerly called micron.

MINERALIZATION The conversion of organically bound elements into inorganic forms, which are then available for metabolism.

MINIMUM INHIBITORY CONCENTRATION (MIC) The lowest concentration of an antibiotic that inhibits growth of a test organism in the tube dilution test.

MITOCHONDRION Intracellular organelle in eucaryotic cells; the site of respiratory activity and energy production.

MONOMER A small unit that is repeated many times in formation of a polymer.

MORBIDITY The incidence of disease in a population, including fatal and nonfatal diseases.

MORTALITY The incidence of diseases causing death in a population.

MOST PROBABLE NUMBER (MPN) METHOD A method for counting organisms, using liquid media; statistical evaluation of growth allows estimates of cell number.

MUTAGEN Any agent that induces mutation.

MUTANT A strain genetically different from its parent because of mutation.

MUTATION A sudden inheritable change in the phenotype of an organism.

MYCELIUM The spreading, often highly branched mass of hyphal filaments typical of vegetative fungal growth.

MYCORRHIZA A symbiotic association between plant roots and fungi.

MYCOSIS Fungal disease of humans; may be widely disseminated in the host (*systemic*) or restricted to the skin, hair, or nails (*superficial*).

MYCOTOXIN Toxin produced by fungus; formed when grains or some other products become moldy.

NEUTRALIZATION Reaction between antibody and soluble antigen, such as a toxin. Viruses are also neutralized by antibodies.

NITRIFICATION Conversion of ammonia to nitrite and the further conversion of nitrite to nitrate.

NITROGEN FIXATION The conversion of nitrogen gas into combined form; it may be done by either *free-living* or *symbiotic* organisms.

NUCLEIC ACID A polymer of nucleotides, either deoxyribonucleic acid (DNA) or ribonucleic acid (RNA).

NUCLEUS Membrane-enclosed structure containing the genetic material (DNA) organized in chromosomes.

OPSONIZATION Combination of antibody with a cell in the presence of complement, increasing susceptibility of the cell to phagocytosis.

OPTICAL DENSITY (O.D.) A means of expressing numerically the turbidity of a suspension, such as a bacterial culture. (As cell number increases, so does O.D.).

OXIDATION The process by which a compound gives up electrons, acting as an electron donor, and becomes oxidized.

OXIDATION-REDUCTION (REDOX) REACTION A coupled pair of reactions, in which one compound becomes oxidized while another becomes reduced and takes up electrons released in the oxidation reaction.

PANDEMIC A worldwide epidemic.

PARASITE An organism that lives in and damages another organism (its host); all of its nutrients come from the host.

PASTEURIZATION A mild heat treatment of liquids to kill all pathogens and reduce the total bacterial count.

PATHOGEN A parasite that causes disease.

PATHOGENICITY Ability of an organism to cause disease.

PERITRICHOUS FLAGELLATION Having flagella attached to many places on the cell surface.

PESTICIDE A chemical used to control weeds, insects, fungi, or other pests.

pH An expression used to indicate the degree of acidity or alkalinity of a solution: pH 7 is neutral; acids have pH less than 7; bases have pH more than 7.

PHAGE See *Bacteriophage*.

PHAGOCYTE A body cell able to ingest and digest foreign particles.

PHAGOCYTOSIS The ingestion of particles or parasites by cells (phagocytes) of the animal body.

PHOSPHORYLATION The addition of phosphate to a molecule, often to activate it prior to oxidation.

PHOTOAUTOTROPH An organism that uses CO_2 as sole carbon source and light as its energy source.

PHOTOPHOSPHORYLATION The synthesis of high-energy phosphate bonds as ATP, using light energy.

PHOTOSYNTHESIS The conversion of light energy into chemical energy that can then be used in the formation of cellular constituents from CO_2.

PHOTOTROPH An organism that obtains its energy from light.

PLASMA The blood fluid that remains after cellular components are removed; contains antibodies.

PLASMA CELL A lymphocyte that produces immunoglobulins.

PLASMID A procaryotic extrachromosomal genetic element not essential for growth. Compare *Episome*.

PLAQUE A localized area of virus lysis on a lawn of cells. See also *Dental plaque*.

PLEOMORPHIC Able to undergo change in shape; existing in more than one form.

PNEUMONIA An infection that results in inflammation of the lungs.

PODZOL A soil type formed under conifer trees; acidic and low in nutrients.

POLAR FLAGELLATION The condition of having flagella attached at one or both ends of the cell.

POLY-β-HYDROXYBUTYRIC ACID (PHB) A polymer of β-hydroxybutyric acid; an energy-storage compound in procaryotes.

POLYMER A large molecule made up of repeating small units (monomers).

POLYRIBOSOME An aggregate of ribosomes, held in linear sequence by messenger RNA.

PRECIPITATION An antibody-antigen reaction that results in formation of an aggregate or precipitate.

PROCARYOTIC CELL Cell type found in all bacteria and blue-green algae, characterized by a nuclear region not bounded by a membrane.

PROPHYLAXIS Treatment, usually immunologic, designed to protect an individual from a future attack by a pathogen.

PROTOPLASM The complete cellular contents: plasma membrane, cytoplasm, and nucleus; usually considered to be the living portion of the cell, thus excluding those layers peripheral to the plasma membrane.

PROTOZOA Eucaryotic single-celled animals, including amoebas, flagellates, ciliates, and sporozoans.

PSYCHROPHILE Any organism that grows best at cold temperatures.

PURE CULTURE An organism growing in the absence of all other organisms.

PYOGENIC Pus-forming; causing abscesses.

PYROGENIC Fever-inducing.

QUARANTINE Restriction of movement of individuals with very infectious diseases to prevent spread of disease to others in a population.

RECOMBINATION The process in which genetic elements from two parents are brought together in one unit.

REDUCTION The process by which a compound accepts electrons to become reduced.

REPRESSION The process by which the synthesis of an enzyme is inhibited by the presence of an external substance, the repressor protein.

RESPIRATION Oxidation of organic compounds in which molecular oxygen serves as the electron acceptor.

RETICULOENDOTHELIAL SYSTEM The system of fixed phagocytes (histiocytes) able to remove particles or microorganisms from the blood stream.

RETTING A process for production of linen fibers from flax; flax stems are soaked and bacteria decompose the substances holding the linen fibers together.

RHIZOSPHERE The region around the plant root immediately adjacent to the root surface where microbial activity is unusually high.

RIBOSOME A cytoplasmic particle composed of RNA and protein; part of a cell's protein-synthesizing machinery.

RUMEN A special organ in ruminant animals in which initial digestion occurs by anaerobic fermentation.

SEPTICEMIA Invasion of the bloodstream by microorganisms; bacteremia.

SEROLOGY The study of antigen-antibody reactions in vitro.

SERUM The fluid portion of blood that remains after the blood cells and materials responsible for clotting are removed; contains antibodies.

SINGLE CELL PROTEIN The protein produced by microbes for use as food.

SLUDGE DIGESTION An anaerobic system for treatment of solid wastes from sewage; the material is degraded by fermentation reactions.

SPECIES A collection of closely related strains; one member of a genus.

SPORE A resistant structure formed by a cell, including endospores, conidia, cysts, akinetes.

STARTER CULTURE Selected pure culture, available commercially, for use as direct inoculum in the production of fermented milk and meat products.

STATIONARY PHASE Growth-cycle phase when rapid cell division has ceased; cells divide slowly or not at all.

STERILE Free of living organisms.

STERILIZATION Treatment resulting in death or removal of all living organisms in a material.

STOCK CULTURE A culture carried in a culture collection for future study or reference.

SUBSTRATE The compound undergoing reaction with an enzyme.

SUBSTRATE-LEVEL PHOSPHORYLATION The synthesis of high-energy phosphate bonds through reaction of inorganic phosphate with an activated (usually) organic substrate.

SYMBIOSIS Association of two organisms in which the relationship is mutually beneficial.

SYSTEMIC Not localized; an infection disseminated widely throughout the body is said to be systemic.

TEMPERATE VIRUS A virus that, upon infection of a host, does not necessarily cause lysis but may become integrated into the host genetic material. See also *Lysogenic*.

THERMAL DEATH TIME The heating time required (at a given temperature) to kill *all* the organisms present.

THERMODURIC Able to survive specific high temperatures but not grow.

THERMOPHILE Any organism that grows best at warm temperatures.

TOXIGENICITY The degree to which an organism is able to elicit toxic symptoms.

TOXIN A microbial substance able to induce host damage.

TOXOID A toxin modified so that it is no longer toxic but is still able to induce antibody formation.

TRANSCRIPTION The synthesis of a messenger RNA molecule complementary to one of the two double-stranded DNA molecules.

TRANSDUCTION The transfer of genetic information via a virus particle.

TRANSFER RNA (tRNA) A type of RNA involved in the translation process of protein synthesis; each amino acid is combined with one or more specific transfer RNA molecules.

TRANSFORMATION The transfer of genetic information via free DNA.

TRANSLATION The process during protein synthesis in which the genetic code in messenger RNA is translated into the polypeptide sequence in protein.

TRICARBOXYLIC ACID (TCA) CYCLE (CITRIC ACID CYCLE) The series of steps by which pyruvate is oxidized completely to CO_2, also forming $NADH_2$ which allows ATP production.

TRICKLING FILTER An oxidative process for the treatment of sewage; liquid sprayed onto a rock bed is oxidized by microbes growing on the rock surface.

VACCINATION Treatment to render an individual resistant or immune to a particular infectious disease.

VACCINE Material used to induce active immunity; a harmless preparation containing antigens from the pathogen, often as a suspension of killed or attenuated microorganisms.

VECTOR An agent, usually insect or other animal, able to carry pathogens from one host to another.

VIABLE COUNT The enumeration of living organisms in a given sample, usually by plating in agar.

VIRULENCE The degree to which a given pathogen can cause disease; it is a function of both *invasiveness* and *toxigenicity*.

VIRUS A noncellular, parasitic, infectious agent able to reproduce only in living host cells; too small to be seen in the light microscope.

VITAMINS Compounds required in small amounts for growth; some microorganisms cannot synthesize all those required and must have them added to the medium.

VOLUTIN A reserve of inorganic phosphate stored within the cell and stainable by basic dyes; also called metachromic granules.

ZOONOSES Diseases transmitted from animals to humans.

Index